COMMUNICATION AND CONTROL IN SOCIETY

COMMUNICATION AND CONTROL IN SOCIETY

Edited by
KLAUS KRIPPENDORFF
Associate Professor of Communications
University of Pennsylvania, Philadelphia

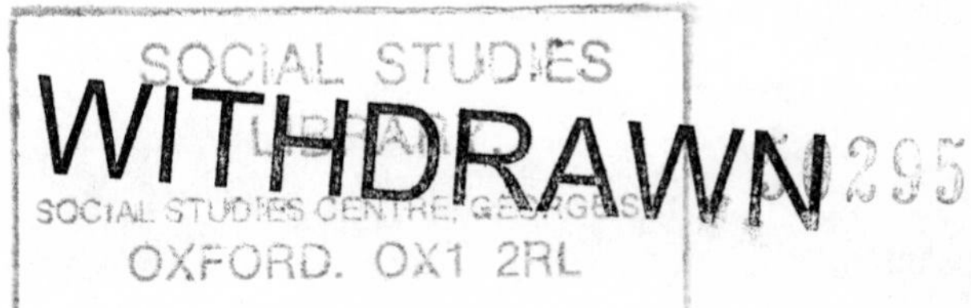

GORDON AND BREACH SCIENCE PUBLISHERS
New York London Paris

Gordon and Breach Science Publishers, Inc.
One Park Avenue
New York NY10016

Gordon and Breach Science Publishers Ltd
42 William IV Street
London WC2N 4DF

Gordon & Breach
7–9 rue Emile Dubois
Paris 75014

Library of Congress Cataloging in Publication Data

Main entry under title:

Communication and control in society.

Based on a conference held at the University of Pennsylvania, and sponsored by the American Society for Cybernetics and three schools of the University of Pennsylvania.

1. Cybernetics – Congresses. 2. Social systems – Congresses. I. Krippendorff, Klaus. II. American Society for Cybernetics. III. Pennsylvania. University.
Q300.C64 001.53 77–18586
ISBN 0 677 05440 8

Library of Congress catalog card number 77–18586. ISBN 0 677 05440 8. Printed in Great Britain.

FOREWORD

This volume grew out of a conference held at the University of Pennsylvania, Philadelphia. It was sponsored by the American Society for Cybernetics and three Schools of this University, the Annenberg School of Communications, the College of Engineering and Applied Science, and the Wharton School.

The American Society for Cybernetics initiated the idea of bringing social scientists and engineers together under the common theme of communication and control with emphasis on social phenomena. Such meetings are not new. They go back at least to the meetings sponsored by the Macy Foundation in the early 1950's, where, under the leadership of Norbert Wiener and Warren S. McCulloch, cybernetics was found to transcend the boundaries of the natural sciences, the life sciences and the social sciences.

Despite this transdisciplinary history and its declared concern with *all* communication and control processes in man, in machines, and in societies alike, cybernetics has flourished more in engineering than in the social sciences. I do not believe this to be inherent in cybernetics. Rather, this seems to stem from at least two extraneous factors: one is a certain property of the physical world and the other a prevailing cultural bias. Regarding the first, it is well recognized that natural and man-made physical systems are structurally more stable, more transparent, and more readily subject to willful manipulation (including decomposition into parts and isolation from disturbances) than are social systems. It follows that natural science data are more readily amenable to formal mathematical analysis and synthesis than are data in the social sciences. Secondly, western modes of thinking have given preference to solving social problems by technological means, a fact which explains why much of our energy, education, and funding is directed toward engineering rather than toward social research. As a consequence, cybernetics has played a key role in the development of computer, communication and information sciences, and systems design, while making only conceptual contributions in the social sciences. It is worth noticing, however, that other scientific approaches have fared far worse. In fact, simple natural science methods in the social sciences, – stimulus-response behaviorism in psychology, for example – have by and large proven unproductive. This experience does not apply to cybernetics.

Today, the cultural bias favoring technological solutions is undergoing critical examinations. Questions are being raised as to whether technology is perhaps a main source of the social problems it is designed to solve. We now experience these problems on all levels, including those of the family, the city, the economy, the ecology and the global community. Questions are also being raised as to whether the rationality of man might not be disfunctional in the long run. Rationality, according to which we perceive and decide among alternative courses of action in the context of envisioned purposes, inevitably leads to planning and towards fixing ever longer-ranging futures. In the rich ecological environment with which we interact, this may prove counter-adaptive and ultimately self-destructive. Related to this concern is the question of whether the human mind is capable of the kind of cybernetic thinking that would be required for man to live in the complex socio-technical world he has helped, for the most part unintentionally, to create.

One must point out that these very questions and concerns arise primarily from cybernetic ideas. For example the mechanism underlying the acceleration of technological development is now identified as a deviation-amplifying (positive) feedback mechanism, whose principal properties are well understood but whose details still need further inquiry. Quite a number of social problems are linked to such self-stimulating, circular flows of information: arms races, population growth, fads and fashion, urban crime, etc. Only the advent of cybernetics has made it possible to explain such problems as emerging from circular causes. The question of whether man's rationality (when applied iteratively over a large range of human activities) has counter-adaptive consequences pertains to theories of adaptation, the law of requisite variety, information theory and, perhaps, to game theory, all of which have evolved in cybernetic pursuits. The limits to an understanding of the complex socio-technical fabric of society are in part a question of epistemology and in part a question of organization theory. Both of these areas have received a major impetus from cybernetics, mediated through its contributions to artificial intelligence, information processing and theories of complex organization.

It is apparent from these developments that cybernetics, having matured in engineering, is now directing its attention to the more difficult problems within society. Although the appropriate theoretical path is poorly marked and the goal of empirical research is not entirely transparent, the conference on *communication and control in social processes* attracted a remarkable variety of excellent people from anthropology, biophysics, communication, computer and information science, control engineering, decision sciences, economics, education, individual and social psychology, international relations, journalism, literature, management science, philosophy and political science. They presented papers to each other, held seminar-like discussions and met in spontaneously arranged workshops to discover considerable cross-disciplinary commonality.

Naturally such "social entropy" contains a lot of noise, which accounts in part for the excitement experienced by these participants but which may make it difficult to trace the thread that runs through their contributions. To reduce this noise, these proceedings are organized into several topical sections, each of which is provided with a separate introduction. The topics concern philosophical aspects, data analysis, incentives and informational controls in organizations, control theory, large social systems, small human systems, and knowledge structures in society. A statement containing policy recommendations developed during a final seminar-type meeting concludes the volume.

In retrospect, one hopes that the enthusiasm created by this conference is contageous and will lead to further inquiry. However, more specific, definite, and practical results than this volume can offer. What this conference did provide was a platform for discussion of new and powerful conceptual frameworks for understanding social processes – whether or not one calls them cybernetic. It is for us all, especially the reader of this volume, to develop these ideas into a truly transdisciplinary approach capable of solving some of the more pressing social problems of our time.

I owe special thanks to the members of the Steering Committee, who worked relentlessly on the preparation of the conference: Paul R. Kleindorfer and Frederick Betz of the Wharton School, Fred Haber and Noah S. Prywes of the College of Engineering and Applied Science, and Henry Teune of the Faculty of Arts and Sciences, all at the University of Pennsylvania. I am also grateful to Thomas H. Wickenden II of the Annenberg School of Communications, who ungrudgingly gave his time in supervising the organizational aspects of the conference and to Terry Mocker who typed parts of the manuscript and the correspondence.

Klaus Krippendorff
The Annenberg School of Communications
University of Pennsylvania

TABLE OF CONTENTS

PHILOSOPHICAL ASPECTS OF CYBERNETICS

CYBERNETIC DATA ANALYSIS

INCENTIVES AND INFORMATIONAL CONTROL

TOWARDS SOCIAL CONTROL THEORY

KNOWLEDGE STRUCTURES IN SOCIETY

IMPLICATIONS FOR POLICY

PHILOSOPHICAL ASPECTS OF CYBERNETICS

INTRODUCTION

Beyond its often highly formal and technical contributions to knowledge, cybernetics has always had an affinity to philosophy. This stems in part from the quite radical reconceptualization of reality proposed by the science of communication and control which poses epistemological questions, questions of the codetermination of mind and the manmade world, questions of control and of genesis, and many more. The religious implications of such questions were recognized already in the writings of Norbert Wiener and his contemporaries.

The three papers in this section are not the only ones with philosophical content, but they have more of this than the others. They stem from quite different events during the conference, they are authored by quite different individuals, and they take up somewhat different issues. Heinz von Foerster's paper was delivered as a presidential address and still carries with it the flavor of a vivid after-dinner presentation. Anthony Wilden's paper grew out of a seminar-type workshop he organized during the conference, and C. West Churchman's paper was formally read to the conference as a whole. Von Foerster has contributed to cybernetics virtually from its inception just as has Churchman did in the area of operations research, where he is now a proponent of the systems approach. Wilden, on the other hand, belongs to a new generation of thinkers and is trying to develop an ecological, philosophical and radical perspective for cybernetics and systems science.

Heinz von Foerster's paper may be seen as raising issues of a cybernetic semantics or of a cybernetic epistemology: the role of an observer relative to the observed, the role of the language used by an observer to describe the observed, the role of the observer of an observer, and so forth. Some of the issues and paradoxes a philosopher will encounter in such situations are old indeed. However, von Foerster proposes new and typically cybernetic solutions to these old problems derived from a theory of infinite recursion and a calculus of self-reference both of which were developed only recently. On these grounds, von Foerster suggests that social cybernetics be a second-order cybernetics, a cybernetics of cybernetics, or a cybernetics of observing systems rather than of observed systems.

In his paper, Anthony Wilden undertakes the task of examining what he calls the deep structure of cybernetics (including systems theory and contemporary approaches to ecology) in relation to society. By deep structure is meant the web of underlying, primarily epistemological, assumptions that are taken for granted when applying cybernetics to solve practical problems, when proposing theories of selective portions of the real world, and when employing a cybernetical perspective for understanding the predicament of man. Wilden sees this deep structure as embedded in certain fundamental ideas and ideologies permeating society and as supporting certain institutional forms. Central to his analysis is the role of the creator, the notion of design and of control from the outside (whether the controller is God, the ruling class or man exploiting his environment), which he sees as a fundamental obstacle to the proper understanding of systemic behavior in nature and in society. Naturally, such issues impinge upon religion and lead Wilden to compare western concepts of autonomy, free will, and socio-economical controls with Eastern, particularly Chinese, concepts of man as part of an ecological system of which God is not separate but a part. This analysis not only criticizes a philosophical misuse of cybernetics but also sheds new light on the potential power to overcome some basically epistemological but, in the end, socio-economical obstacles to systems thinking.

C. West Churchman wrestles with much the same problem except, perhaps, that he seeks psychological rather than social explanations. Churchman starts by examining what cybernetics is; and he identifies it, much as Wilden does, with a science of model building. He then asks what a cybernetician does, and, climbing up the ladder of logical types, why he does what he does. In answer to the latter question Wilden would point to ideological frames of reference within society, leaving open questions of their origin or *raison d'être.* In contrast to this approach, Churchman chooses to examine the work of several thinkers: E. A. Singer, Jung, the writers of the Bhagavad Gita, Paul, and Locke among others. He comes to the conclusion that limitations to infinite regression in reasoning and justification often lie in tautologies and in basic human characteristics such as the search for agreement and the natural tendency to be political, religious, and aesthetical. This approach also leaves open the question of what molds these characteristics and, thus, points back to von Foerster's paper for an answer.

All three papers are short and suggestive of further search and inquiry. The first two give adequate bibliographies to quench the curiosity they may have aroused in the reader.

CYBERNETICS OF CYBERNETICS

HEINZ VON FOERSTER
University of Illinois, Urbana

Ladies and gentlemen – As you may remember, I opened my remarks at earlier conferences of our Society with theorems which, owing to the generosity of Stafford Beer, have been called "Heinz von Foerster's Theorems Number One and Number Two". This all is now history [1; 10]. However, building on a tradition of two instances, you may rightly expect me to open my remarks today again with a theorem. Indeed I shall do so but it will not bear my name. It can be traced back to Humberto Maturana [7], the Chilean neurophysiologist, who a few years ago, fascinated us with his presentation on "autopoiesis", the organization of living things.

Here is Maturana's proposition, which I shall now baptize "Humberto Maturana's Theorem Number One":

"*Anything said is said* by *an observer.*"

Should you at first glance be unable to sense the profundity that hides behind the simplicity of this proposition let me remind you of West Churchman's admonition of this afternoon: "You will be surprised how much can be said by a tautology". This, of course, he said in utter defiance of the logician's claim that a tautology says nothing.

I would like to add to Maturana's Theorem a corollary which, in all modesty, I shall call "Heinz von Foerster's Corollary Number One":

"*Anything said is said* to *an observer.*"

With these two propositions a nontrivial connection between three concepts has been established. First, that of an *observer* who is characterized by being able to make descriptions. This is because of Theorem 1. Of course, what an observer says is a description. The second concept is that of *language*. Theorem 1 and Corollary 1 connect two observers through language. But, in turn, by this connection we have established the third concept I wish to consider this evening, namely that of *society*: the two observers constitute the elementary nucleus for a society. Let me repeat the three concepts that are in a triadic fashion connected to each other. They are: first, the observers; second, the language they use; and third, the society they form by the use of their language. This interrelationship can be compared, perhaps, with the interrelationship between the chicken, and the egg, and the rooster. You cannot say who was first and you cannot say who was last. You need all three in order to have all three. In order to appreciate what I am going to say it might be advantageous to keep this closed triadic relation in mind.

I have no doubts that you share with me the conviction that the central problems of today are societal. On the other hand, the gigantic problem-solving conceptual apparatus that evolved in our Western culture is counterproductive not only for solving but essentially for perceiving social problems. One root for our cognitive blind spot that disables us to perceive social problems is the traditional explanatory paradigm which rests on two operation: One is *causation*, the other one *deduction*. It is interesting to note that something that cannot be explained – that is, for which we cannot show a cause or for which we do not have a reason – we do not wish to see. In other words, something that cannot be explained cannot be seen. This is driven home again and again by Don Juan, a Yaqui Indian, Carlos Casteneda's mentor [2; 3; 4; 5].

It is quite clear that in his teaching efforts Don Juan wants to make a cognitive blind spot in Castaneda's vision to be filled with new perceptions; he wants to make him "see". This is doubly difficult, because of Castaneda's dismissal of experiences as "illusions" for which he has no explanations on the one hand, and because of a peculiar property of the logical structure of the phenomenon "blind spot" on the other hand; and this is that we do not perceive our blind spot by, for instance, seeing a black spot close to the center of our visual field: we do not see that we have a blind spot. In other words, we do not see that we do not see. This I will call a second order deficiency, and the only way to overcome such deficiencies is with therapies of second order.

The popularity of Carlos Castaneda's books suggest to me that his points are being understood: new paradigms emerge. I'm using the term "paradigm" in the sense of Thomas Kuhn [6] who wants to indicate with this term a culture specific, or language specific, stereotype or model for linking descriptions semantically. As you may remember, Thomas Kuhn argues that there is a major change in paradigms when the one in vogue begins to fail, shows inconsistencies or contradictions. I however argue that I can name at least two instances in which not the emergent defectiveness of the dominant paradigm but its very flawlessness is the cause for its rejection. One of these instances was Copernicus' novel vision of a heliocentric planetary system which he perceived at a time when the Ptolemaeic geocentric system was at its height as to accuracy of its predictions. The other instance, I submit, is being brought about today by some of us who cannot – by their life – pursue any longer the flawless, but sterile path that explores the properties seen to reside within objects, and turn around to explore their very properties seen now to reside within the observer of these objects. Consider, for instance, "obscenity". There is at aperiodic intervals a ritual performed by the supreme judges of this land in which they attempt to establish once and for all a list of all the properties that define an obscene object or act. Since obscenity is not a property residing within things (for if we show Mr X a painting and he calls it obscene, we know a lot about Mr X but very little about the painting), when our lawmakers will finally come up with their imaginary list we shall know a lot about them but their laws will be dangerous nonsense.

With this I come now to the other root for our cognitive blind spot and this is a peculiar delusion within our Western tradition, namely, "objectivity":

"The properties of the observer shall not enter the description of his observations."

But I ask, how would it be possible to make a description in the first place if not the observer were to have properties that allows for a description to be made? Hence, I submit in all modesty, the claim for objectivity is nonsense! One might be tempted to negate "objectivity" and stipulate now "subjectivity". But, ladies and gentlemen, please remember that if a nonsensical proposition is negated, the result is again a nonsensical proposition. However, the nonsensicality of these propositions either in the affirmative or in their negation cannot be seen in the conceptual framework in which these propositions have been uttered. If this is the state of affairs, what can be done? We have to ask a new question:

"What are the properties of an observer?"

Let me at once draw your attention to the peculiar logic underlying this question. For whatever properties we may come up with it is we, you and I, who have to make this observation, that is, we have to observe our own observing, and ultimately account for our own accounting. Is this not opening the door for the logical mischief of propositions that refer to themselves ("I am a liar") that have been so successfully excluded by Russell's Theory of Types not to bother us ever again? Yes and No!

It is most gratifying for me to report to you that the essential conceptual pillars for a theory of the observer have been worked out. The one is a calculus of infinite recursions [11]; the other one is a calculus of self-reference [9]. With these calculi we are now able to enter rigorously a conceptual framework which deals with *observing* and not only with the observed.

Earlier I proposed that a therapy of the second order has to be invented in order to deal with dysfunctions of the second order. I submit that the cybernetics of observed systems we may consider to be first-order cybernetics; while second-order cybernetics is the cybernetics of observing systems. This is in agreement with another formulation that has been given by Gordon Pask [8]. He, too, distinguishes two orders of analysis. The one in which the observer enters the system by stipulating the *system's* purpose. We may call this a "first-order stipulation". In a "second-order stipulation" the observer enters the system by stipulating *his own* purpose.

From this it appears to be clear that social cybernetics must be a second-order cybernetics – a *cybernetics of cybernetics* – in order that the observer who enters the system shall be allowed to stipulate his own purpose: he is autonomous. If we fail to do so somebody else will determine a purpose for us. Moreover, if we fail to do so, we shall provide the excuses for those who want to transfer the responsibility for their own actions to somebody else: "I am not responsible for my actions; I just obey orders." Finally, if we fail to recognize autonomy of each, we may turn into a society that attempts to honor commitments and forgets about its responsibilities.

I am most grateful to the organizers and the speakers of this conference who permitted me to see cybernetics in the context of social responsibility. I move to give them a strong hand. Thank you very much.

REFERENCES

[1] Beer, S., *Platform for Change:* 327, New York: Wiley, 1975.

[2] Castaneda, C., *The Teachings of Don Juan: A Yaqui Way of Knowledge,* New York: Ballantine, 1969.

[3] Castaneda, C., *A Separate Reality,* New York: Simon and Schuster, 1971.

[4] Castaneda, C., *Journey to Ixtlan,* New York: Simon and Schuster, 1972.

[5] Casteneda, C., *Tales of Power,* New York: Simon and Schuster, 1974.

[6] Kuhn, T., *The Structure of Scientific Revolution,* Chicago: University of Chicago Press, 1962.

[7] Maturana, H., "Neurophysiology of cognition", in Garvin, P. (Ed.), *Cognition, A Multiple View:* 3–23, New York: Spartan Books, 1970.

[8] Pask, G., "The meaning of cybernetics in the behavioral sciences (the cybernetics of behavior and cognition: extending the meaning of 'goal')" in Rose, J. (Ed.), *Progress in Cybernetics,* Vol. 1: 15–44, New York: Gordon and Breach, 1969.

[9] Varela, F., "A calculus for self-reference", *International Journal of General Systems,* **2**, No. 1: 1–25, 1975.

[10] Von Foerster, H., "Responsibility of competence", *Jounral of Cybernetics,* **2**, No. 2: 1–6, 1972.

[11] Weston, P. E. and von Foerster, H., "Artificial intelligence and machines that understand", in Eyring, H., Christensen, C. H., and Johnston, H. S. (Eds.) *Annual Review of Physical Chemistry,* **24**: 358–378, Palo Alto: Annual Review Inc., 1973.

CHANGING FRAMES OF ORDER: CYBERNETICS AND THE *MACHINA MUNDI*[1†]

ANTHONY WILDEN
Simon Fraser University, Burnaby, B.C., Canada

---

1 THE PROBLEMATIC

> . . . cruel Works
> Of many Wheels I view, wheel without wheel, with cogs tyrannic
> Moving by compulsion each other, not as those in Eden, which
> Wheel within wheel, in freedom revolve in harmony and peace.
>
> WILLIAM BLAKE: *Jerusalem* (1820)

Cybernetics, allied with systems theory and with contemporary approaches in ecology, is not simply a science or a methodology, but more significantly, a point of view. If it seems astonishing that the epistemological position we have called "cybernetic" for some thirty years is still unfamiliar to many people, we must remember that, for anyone raised in what we may term the Newtonian–Cartesian *epistemology* of science and the one-dimensional, atomistic *ideology* of our society, the nonmechanistic cybernetic perspective may at first sound contradictory or unscientific, or even teleological and anthropomorphic.

Cybernetics is explicitly a science of models. But apart from any concerns about the complex relationship between our social ideology and the contemporary discourse of science, the fact that cybernetic models have mostly had their source in the study of *engineered* systems continues to create problems about the applicability of cybernetic (and systems) theory to the nonengineered reality which surrounds us.

Ashby quite rightly pointed out in 1956 that his closed-system cybernetics did not owe its basic postulates and theorems to any other discipline [2]. Unfortunately, this *ab initio* construction has not always been true of the *application* of cybernetic principles to real, nondesigned systems. Ashby was careful to avoid such misapplications, but as I have argued elsewhere, he quite unintentionally[2] exemplified this problematic in several instances [43: 138–140, 371–374].

† Numbered notes are collected at the end of each chapter under the heading "Endnotes"

More recently, Jean Piaget has made some somewhat dubious associations between cybernetics, cognitive development, social structure and his own "genetic" or "constructivist" structuralism [27].

We have also seen anti-Freudian psychotherapists attempting to use a systems–cybernetic perspective within what appears to be, in the last analysis, the neo-Freudian epistemological and ideological framework of the "autonomous ego" [42]. Sometimes we find the traditional Cartesian dichotomy between the subject and the object hidden behind the expressions "system" and "environment". In economics, we may find in the literature examples of cybernetically inclined researchers apparently rewriting Adam Smith's *Wealth of Nations* (1776) without realizing it – without realizing that their conception of economic "self-regulation" is no less mysterious and providential than the moralistic model of the "hidden hand" underlying Smith's analysis of the surface structure of liberal capitalism [cf. 49: 216–230, 308–343, 532–534; 46]. Philosophers, too, can fall into a similar trap, unconsciously using cybernetic terms to reformulate Hegel's rationalist and idealist dialectics of the "cunning of reason". Ethologists and ecologists sometimes run into trouble with apparent isomorphies between goalseeking behavior in different systems, isomorphies which turn out to be mere superficial similarities. This is especially the case when they attempt to cross the clearly defined boundary between natural and social ecosystems [18: 3–41; 43: 233–273]. Influenced by systems analysis and systems engineering, some ecologists can be heard implicitly imputing a design to nature and to society, and consequently speaking in unrecognized teleological terms, rather than in consciously teleonomic ones. These problems call to mind the Providence that served as a guiding and controlling principle for the classical economists, as well as its historical corollary in biology, the beneficient Mother Nature who served with such regularity as an explanatory principle in the work of Charles Darwin [15].

Moreover, before Maruyama [21], amongst others, provided positive feedback (deviation amplification) with a proper place in cybernetic theory (1963), the almost exclusive concern of cyberneticians for mechanisms and processes implementing or displaying primary negative feedback (deviation suppression) made cybernetics more a science of control as such, than a perspective that could also accommodate the various kinds of amplification and accumulation we find in the real world (e.g. the accumulation of productive capacity in our economic system over time [46]).

The crux of this problematic can be easily stated (which is not to imply that it can be easily resolved). Like other general theories, cybernetics, systems theory, and ecological explanation can usefully and legitimately be applied to complexes of system–environment relations differing widely in their *order* or kind of complexity and in the *scope* or extent of their organization. Any of these ecosystems will almost surely itself involve levels and kinds of complexity [20].

The point is that if the theoretical approach applied to any ecosystem does not itself manifest a logical typing[3] of orders of variety which closely approximates the logical typing of the variety of the ecosystem under study, then the approach will be forced willy-nilly into an epistemological or ideological reductionism.

This is a reduction of complexity (and a consequent loss of information) quite different from the consciously methodological reductions necessary in scientific investigation. In essence it involves crossing boundaries between orders of complexity in the ecosystem without realizing it. Lest the word "reductionism" conjure up misleading images of Procrustes' bed, it should be emphasized that this kind of reduction of the "data" to fit the approach is not quantitative, but qualitative; and that it does not simply involve the adjustments in a single dimension which occupied Procrustes, but rather an unrecognized switching *between* dimensions. The consequence is a "flattening out" of the various orders of complexity in the ecosystem, orders which in living and social systems are always arranged in hierarchies or heterarchies.

As an illustration of this one-dimensionalization, we may take Edmund Burke's notorious dictum: "The laws of commerce are the laws of Nature, and therefore the laws of God." Similarly, in the 1970 edition of his *Economics*, Paul Samuelson does much more than simply mix his metaphors:

> Takeovers, like bankruptcy, represent one of Nature's methods of eliminating deadwood in the struggle for survival. A more efficiently responsive corporate society can result. But, without public surveillance and control, the opposite can also emerge. The Darwinian jungle is not guaranteed to produce a happy ending [34: 505].

My point, however, is not to fill this paper with a catalog of examples. I would rather approach the relationship between cybernetics and society by means of a historical parable. This, I hope, will indicate more by the multiplicity of evocation than by the simplicity of criticism what I see as the problematic of cybernetics today.

That this should be a historical approach is essential to the argument, because it is not enough that systems–cybernetic models should be, as it were, "non-disciplinary" in Ashby's well-taken sense.

The systems–cybernetic perspective must be both interdisciplinary and transdisciplinary, and both semiotic and dialectical. On the one hand, in order to both uncover and bridge the gaps between the disciplines, it must draw on the social sciences, the experimental sciences, and the arts. Specifically it must be philosophical, historical, communicational, ecological, economic, and anthropological. On the other hand, in the hope of introducing new order into the currently disordered state of the discourse of science, the perspective must develop a semiotic vocabulary abstract enough – of a high enough logical type – to allow translation into the specifics of any given area of study. The vocabulary may well be borrowed from any disciplinary dialect, but it must in its use transcend the limitations of a jargon.

Lastly, it must also attempt to transcend its academic origins by having some manifest instrumentality in what Marx called "the language of real life"; the day-to-day communication (*Verkehr*) of the socioeconomic system. It must in other words be ultimately translatable into the language of that complex of relations to which no single term does adequate justice: the production, reproduction, and exchange relationships of the classes, the races, and the sexes within-society-in-history-in-organic-and-inorganic-nature (socioeconomic ecosystems).

2 MODELS AND METAPHORS

Concepts that are basically shorthand for process elude verbal definition. . . . It is only that which has no history which can be defined.

NIETZSCHE

Whether a society speaks science or myth or both, its single concern about the realm of ideas or ideology can be only that its dominant epistemology and ideology must perform a long-range adaptive function. This adaptive function must of course match the long-range survival values embodied in the material reality of human interaction which permitted the system of ideas to arise in the first place. Whatever other functions they serve, science and myth must ultimately be viewed as instrumental aspects of the systemic organization which they serve (but do not directly control). Rappaport [31] has shown conclusively in his now classic cybernetic–ecological study of the Tsembaga in New Guinea that this kind of matching of long-range survival functions in myth, ritual, and reality has indeed existed.

Other ecologically oriented anthropologists have provided evidence to suggest that if there is one feature which distinguishes capitalist and state capitalist societies from all others, it is that our kind of socioeconomic organization, with its accompanying ideology, is the exception rather than the rule [12; 13; 29; 33; 39]. On the one hand, what appears to be exceptional about us – although I cannot argue the case in detail here – is that our socioeconomic organization appears to have passed from adaptivity to counteradaptivity within a single century (Section 8 in this paper). On the other, what makes us peculiar is that the ideology which has been dominant in our society since the capitalist revolution has not only never matched its socioeconomic reality, but it is now also in the process of trying to catch up with the present and the future by valorizing various aspects of the unrecoverable past [47].

This valorization extends all the way from the current and profitable epidemic of nostalgia – "The great thing about the good old days was that we didn't feel nostalgic" – through the resurgence of forms of utopian and agrarian socialism and their primitivist religious equivalents, and eventually to demands for that economic condition most dreaded by the "progressive" establishment of the nineteenth century: the steady state economy [7].

This said, we may be permitted to go back to reexamine a venerable and essentially cybernetic construction of social and natural reality without being nostalgic about it: the theological–alchemical model of the cosmos in the middle ages.

In a celebrated passage from *On Learned Ignorance* (1440), Nicholas of Cusa restated a medieval definition of God's locus in the cosmos that had been an intellectual commonplace since at least the time of Alan of Lisle (d. 1202) and Jean de Meun's contribution to the *Roman de la rose* (c. 1277):

> In consequence, there will be a *machina mundi* whose center, so to speak, is everywhere and whose circumference is nowhere, for God is its circumference and its center, and He is everywhere and nowhere.[4]

Cardinal Nicholas was a mystic, a mathematician, a philosopher, a theologian, and perhaps the first experimental biologist. (He studied the respiration of plants.) He, with his colleagues, probably understood better than many of us today in just what sense the scientific discourse of any age or culture is, in the last analysis, only a moderately consistent code of metaphors, amenable to different translations in different times and places.

We have no difficulty in giving Cusa's description of God's relationship to the structure (*machina*) of the cosmos the serious examination it deserves. It is not of course a statement about God, but rather a statement about the dominant structure of medieval socioeconomic organization. It invites several interrelated levels of analysis, all of which, it seems to me, both require and illuminate the systems–cybernetic perspective.

Like his forbearers and contemporaries, Nicholas of Cusa understood his society as an ordered subsystem operating within an organically organized whole governed by a *hierarchy of constraints.*[5] The ultimate constraint on all communication in the system (production, reproduction, exchange, maintenance, interaction) is embodied in a mysterious principle called God. No subsystem in the whole can "go outside" or transcend this constraint of constraints without becoming effectively extinct.

Taking the metaphor seriously, *as a metaphor*, we note at once that "God" simply symbolizes the ultimate constraint on all past, present, and future behavior on this planet, the constraint we now call the principle of entropy. As Wiener once remarked, entropy has always had a somewhat magical and mystical aura to it. But the parallel is even more marked. For Nicholas, God is the locus where there is "the highest concordance". In it the locus of the coincidence of antitheses (*coincidentia contrariorum, oppositorum*), the concordance of differences (*De concordantia catholica,* 1433). This might well be a statement about a system at maximum entropy (complete randomness), for at entropy all differences are equal. A system of completely random diversity cannot be distinguished from a completely determined single-state system (unity).

Even without this particular parallel, the concept of a hierarchically constrained universe is an elementary ecosystemic principle, and one that most readily distinguishes the rules governing the behavior of living and social systems from the linear or efficient causality which is assumed to operate in the isolated, mechanical, equilibrium systems of a Newtonian universe [cf. 2: 127; 3: 405–415; 43; 47].

3 IDEOLOGY AND ORGANIZATION

Every sign . . . is a construct between socially organized persons in the process of their interaction. Therefore, the forms of signs are conditioned above all by the social organization of the participants involved and also by the immediate conditions of their interaction. When these forms change, so does sign . . . [Therefore:] (1) Ideology may not be divorced from the material reality of sign (i.e. by locating it in the "consciousness" . . .); (2) the sign may not be divorced from the concrete forms of social intercourse . . .; (3) Communication and the forms of communication may not be divorced from the material basis.

V. N. VOLOSHINOV: *Marxism and the Philosophy of Language* (1929)

The ideological and epistemological construct medieval and early renaissance society called "God" performed an essential and recognized socioeconomic function for the survival of the system. It performed a function which the constraint on growth we call the order–disorder relations of (planetary) entropy has not so far been permitted to do in our form of economic organization [43; 47]. Like the "heavenly influences" in the pantheon of the alchemists, the astrologers, and the natural magicians, the "God" of the theologian and the mystic is simply a set of ideological *representations*[6] of the real and material conditions which govern the *mode of production* dominant in medieval times and the *reproduction* of medieval society itself.[7]

As in any social system of any kind, these ideological constructs *re*-present to consciousness (and to the unconscious), at the level of ideas and images, the *semiotic* organization of the medieval economic structure by the *information* flowing through the system at all levels. The various levels and types of information (semes or signs) which both (positively) control and (negatively) constrain the behavior of every individual subsystem in the whole are encoded in various sign systems; in money, in capital, in commodities, in artefacts, in the patterning of the natural landscape by human activities, and in the structure of social relations.

Neither God nor the dominant ideology he serves are the actual principles of socioeconomic organization in the system (any more than any ideological construct ever is in any known social system). They do, however, function as a set of *metastatements* about the actual organizing principles at work at the level of the *deep structure* of the system. As such, these metastatements are signs about the behavior of other signs. The ideological signs are available to consciousness in verbal and nonverbal forms: in speech, in concepts, in images, in art, in writing, and in other aspects of patterned diversity. Their relationship to the deep structure of the system can be described as that of a set of *messages* to a *code*.

A code may be defined as a set of rules governing the permissible construction of messages in the system. A code mediates the relationships between the goal-seeking sender–receivers that employ it. Thus a code or set of codes is the basis of the creative principle that makes messages and relationships *possible* in the first place, at the same time as the constraints embodied in a code make an even greater variety of qualitatively different messages and relationships *impossible* in the system as it stands [48].

Such messages will remain impossible – and indeed unimaginable for the sender–receivers mediated by a given code – unless some particular combination of processes and events sets off a reordering or restructuring of the code. In this eventuality, existing or novel variety in the environment of the sender–receivers may consequently pass from the status of *noise* (disorder unacceptable or invisible to the system) to that of *information* (acceptable or usable order), and become incorporated in the coding arrangements of a qualitatively distinct system. In adaptive cybernetic systems, internally or externally generated noise does not therefore lead necessarily to breakdown. If a system can restructure its codes at relevant levels so as to generate new order out of disorder [3; 41], it will undergo an evolution or a revolution, as the case may be. In a properly dialectical sense, the system will have passed through an *Aufhebung* by means of which the old order forms the basis of and at the same time gives way to the new; and the whole, as a whole, survives.

4 STEADY STATE

The Creator made man of all things, as a sort of driver and pilot, to drive and steer the things on earth, and charged him with the care of animals and plants, like a governor or steward subordinate to the chief and great King.
PHILO JUDAEUS (30BCE–50CE)

There was a god, either maker or governor or both, of all this whole engine of the world.
THOMAS MORE: *Heresyes* (1529)

We used to be reminded by Robert Browning that when “God’s in His heaven, all’s right with the world”. But this cliché would hardly have been understandable at all for many people in the medieval context. Whatever we may now find objectionable in the feudal arrangements of the medieval *communitas*, perhaps the most significant aspect of its organic relationship to its interventionist deity was that, if God was anywhere, he was most certainly not simply in heaven, looking down on the world – as he was to be for post-seventeenth-century society. On the contrary, as the quotation from Nicholas of Cusa makes clear, even at the moment when the medieval world order was about to be transcended by the "commercial revolution" associated with the "age of discovery", God was still perceived by the theological establishment as *both* in heaven *and* in the world. There was no barrier or gulf separating humanity from God; rather there was an intimately linked "great chain of being" ascending through all the species from the lowest matter to the highest Form, to the absolute first and last represented by God as the *primum mobile* [19].

In the words of the mathematician, philosopher, and economist Nicole Oresme (c. 1325–1382), in his commentary (1377) on Aristotle's *De Caelo*:

> . . . God does not need the heavens or any other place for Himself because He is everywhere – both outside and inside the heavens . . . The heavens are moved by immaterial and incorporeal or spiritual things called intelligences or separate substances . . . Spiritual things are indivisible and neither occupy nor fill any place . . . Each intelligence is whole and wholly in every part – however small – of the heaven that it moves, just as the human soul is whole and wholly in every part of the human body – except that the soul is in the body by information (*informacion*), and an intelligence is in its heaven by a mutual fitting together (*apropriacion*) [26: Folios 69d–70a].

What we immediately see in these images of a perfectly ordered world organism is another statement about the dominant pattern of production and social relations in medieval times. All relationships are conceived to be truly interrelational; the dominating pattern is that of a hierarchical unity of differences and a reciprocity of oppositions. Sympathy matches antipathy; difference is mediated by similitude; patterns are repeated through analogy and emulation; and the whole is bound together in the reciprocal coming and fitting together of mutual convenience. *Similitudo, analogia, aemulatio, convenientia*: the dominant theme is that every apparently separate entity shares its being in some way with all the others; that everything knows its place (in every sense); and that the cosmos is ordered as it is because all beings literally *co-operate* together in the whole ([9: 17–44; 30; on the logical typing of cooperation and competition in social systems; see [48]).

The actual socioeconomic reality is less pleasantly harmonious, of course. It is true that in medieval society cooperation predominates over competition in production, just as the exchange of use values dominates the production and exchange of exchange values, but this cooperation is hierarchically enforced by the medieval class structure. Nevertheless, what does emerge from comparing the ideology with the reality is that medieval society approximated a steady-state economy. The feudal mode of production was as exploitative as any other class-centered system. But it was not predominantly based on the maximization of gain through the institutionalization of competitive *uncertainty*,[5] with the resulting necessity of continuous linear expansion, as is our own. On the contrary, except for the still subordinate activities of the merchant class, the feudal economy was based on the maximization of *certainty* through cyclic repetition. As represented as late as 1416 in the iconic images of the *Très riches heures du duc de Berry*, for example, the medieval economy was in essence a homeostatic system whose fluctuations were keyed more directly to the succession of the seasons than to its own generation of various types of disorder.

The rising merchant class was already shaking this system at its roots in the fifteenth and sixteenth centuries. By the seventeenth century, the still existing feudal mode of production was clearly occupying a subordinate status in the system. Its ideological representatives no longer form an integral part of the religious, political, and scientific establishments, except insofar as they play out their role as "conservatives", inveighing against technological innovation (Montaigne), for instance, or resisting the expansion of credit (the repeated condemnations of usury).

In science, a century after the birth of Galileo, there is still a sufficient confusion about the pace and the actual characteristics of ongoing socioeconomic changes for Giambattista della Porta's *Magia naturalis*, first published in 1558, to be translated into English (1658) [30]. The steady-state images remain – "The Sun is the Governor of Time and the Rule of Life", says Porta – but they have been overtaken by events. Western Europe is on its way towards breaking down or bypassing the traditional constraints on production; it is about to invent the uniquely modern ideology of "progress" – progress unlimited in time or space – to justify the economic necessities of expansion at the basis of the new order [28; 45; 46].

5 THE LOCUS OF CONTROL

Systems [of explanation] in many respects resemble machines. A machine is a little system created to perform, as well as to connect together, in reality, those different movements and effects which the artist has occasion for. A system is an imaginary machine invented to connect together in the fancy those different movements and effects which are in reality performed. ADAM SMITH: *Essays on Philosophical Subjects* (1795)

Let us return once more to Nicholas of Cusa's image of a God who is both the center and the boundary of the system he has created. Precisely because he is mystically inclined, Cusa is looking far more deeply into the biosocial ecosystem than those amongst his predecessors and contemporaries who were content to conceive of God, government, and the emerging nation state on the model of the patriarchal family – as represented by the Church hierarchy and the Western monarchies, for example (see [35], for an analysis of the metaphorical remnants of this position).

If God is everywhere in the system, then God, the ultimate constraint, is not a controlling agency external to the system, as is a household thermostat, for example. In other words, the primary locus of constraint and control in this medieval system is exactly where it actually is in all non-engineered living systems: *it is the structural relations of the system itself.* Like Cusa's God, structural relations are everywhere, but since (like memory) they cannot be located or localized, we may also say that they are nowhere at the same time. Constraint and control, as in natural ecosystems and in most social ecosystems, lie in the hierarchical and heterarchical networks of the system itself, *both* at the level of the individual subsystem *and* at the level of the whole.

This informational conception of the *immanence* of the locus of constraint and control in the relations between the "partials" of an ecosystem is a conception we in the West have had to rediscover in this century, once the Newtonian energy-entity equilibrium models of social and biological reality were found wanting. Not surprisingly, we find practically no trace of any similarly atomistic epistemology in Chinese science and ideology. In 1956, Needham noted the similarity between the geometric metaphor of the circle or sphere with its center everywhere and its circumference nowhere (taken from Pascal's *Pensées*) and the Chinese aphorism "Wu chi erh thai chi". We do not know precisely what this proposition meant, but it may be approximately translated as: "That which has no pole. And yet itself the supreme pole!" [23]. Others have compared the saying to the "Ungrund doch Urgrund" of the mystic Jakob Böhme (1575–1624).

But Needham points out, correctly I think, that the Chinese proposition should be taken as a metaphor of the reality of Chinese society and its basic epistemological premisses, rather than as a mere poetical, mystical, or theistic statement. The Chinese organization of socioeconomic reality has been for some thousands of years quite overtly systemic, ecological, and cybernetic.[8] The dominant epistemology in China never gave up the principle that the "order of things" was embodied and imbedded in the structure of natural and social reality. Neither nature nor society were considered to be ultimately governed by any potentially external principle equivalent to a divine ruler or lawgiver. As in Western alchemy and natural magic, what ultimately ruled the cosmos was *Li*: Form or Pattern (information). Consequently, the Chinese never produced the essentially juridical notions of "natural law" and the "laws of nature", implying as they did a Creator or Lawgiver, and, as a result, an anthropomorphic *designer* [23: 288–291, 344–345, 460–465].

The Western association of creator, law, design, and external control (whether the controller is assumed to be God, the ruling class, or the government) has in fact been the greatest of ideological and epistemological obstacles to the proper understanding of systemic behavior, whether in nature or in society. This is not the place to detail the socioeconomic reasons for this difference between East and West. Nor can I analyze here the paradoxical but explicable fact that an anti-ecosystemic epistemology, ideology, and pattern of economic behavior has been *necessary* for the short-range survival of capitalism for the past three centuries [43; 47; 48]. But we have every reason to suspect that the only explanation of why we are now discovering these ecosystemic conceptions is that our economic system is working its way towards a crisis of such grave proportions that it both generates and needs them. Indeed, in its own behavior, we may assume that it is already ahead of us in this respect, and that in developing these ideas, we are in effect adapting to its adaptations.

6 DETERMINISM AND FREE WILL

[The factory system] involves the idea of a vast automaton, composed of various mechanical and intellectual organs, acting in uninterrupted concert for the production of a common object, all of them being subordinated to a self-regulated moving force. . . . Three distinct powers concur to their vitality [manufactures] – labour, science, capital; the first destined to move, the second to direct, the third to sustain. When the whole are in harmony, they form a body qualified to discharge its manifold functions by an intrinsic self-governing agency, like those of organic life.

ANDREW URE: *The Philosophy of Manufactures* (1835)[9]

Taken as such, the underlying organic model used by Ure in his description of the early modern factory seems if anything more holistic and less mechanistic than Nicholas of Cusa's *machina mundi*. In reality, of course, the opposite is the case. Whereas *machina* could once stand as an equivalent of "structure" or "frame", and "engine" still retain its semantic connections with the Latin *genius* (spirit), the word "organ" now stands for a machine or for the alienated human being attending it. The natural and social ecosystems, as well as their inhabitants, could now indeed be profitably represented in science and ideology as Newtonian machines, driven by the linear and efficient causality of a one-dimensional universe.

It comes as no surprise that the nineteenth century invented the word that most aptly labelled the machine perspective, the word "determinism" (as distinct from fatalism). After the Great Depression of the 1840s, "determinism" became almost the equivalent of a political slogan in the ideological and epistemological conflicts stemming from the competitive struggle for economic and political power in nineteenth-century society. Insofar as it represented science (physics, biology, evolutionary theory), determinism occupied one pole of the opposition between science and religion. Insofar as it represented technological "progress" achieved by understanding the "laws of nature", determinism lay on the side of the manufacturer against the landowner. And insofar as determinism, along with mechanism, was the watchword of the "progressives" in nineteenth-century science, it did battle against the mystical, animistic, ethical, and religious attitudes summed up in the expressions "teleology", "vitalism", and "free will".

From the systems-cybernetic perspective, however, the debate between the determinists and the vitalists in biology, and between determinism and free will in ethics, provides us with little by way of an explanation of behavior in living and social systems. In the socioeconomic and historical context that makes individuals possible – the context which both provides them with humanity and socializes them to be what they are – the activities of the individual person are *neither* determined in the classic sense, *nor* the products of free will. The supposed polarity between these terms or conditions is in fact an ideological illusion, and the debate itself can be seen as representing one of the basic – and paradoxical – relations inherent in social democracy.

If we are led to feel that we must choose to say that in society we are *either* free *or* not free, then it makes no difference which pole of the paradox we choose. Alone, each alternative is contradicted by experience. Similarly with the either/or of determinism. As in all such questions, the very act of choosing one alternative will require us to turn back to choose the other, and so on *ad infinitum*. The question of free will versus determinism is a double bind, a paradoxical injunction [3: 201–227, 309–337; 42; 45; 48].

The real question is not whether we are free or not free, but *what* we are free to do – and the one great freedom ideologically guaranteed us by the commoditization of labor under capitalism is that we should be free to sell our labor potential (our creativity) at the best price, and with "equal opportunity" (to compete).

The middle ages had similarly to deal with the relationship between freedom and determinism, except that this was not the Newtonian determinism of *efficient* causes, but rather the Aristotelian determinism of *final* causes; the teleological determinism represented by God in his omniscience. In the medieval context, however this relationship does not have the characteristics of a double bind. On the contrary, the theological-alchemical model of the cosmos successfully neutralized any potential paradoxical relation between human will and God's will by means of the doctrines of "faith" and "grace", which allowed the Christian to chose *both* sides of the question at the same time. Symbolized in the Trinity, with the "only begotten son" as the man who was God, the very idea that there could be two sides on such a question was itself absurd (e.g. the "Credo *quia* absurdam" attributed to Tertullian).

If we now translate the problematic back into the cybernetic interpretation of the medieval cosmos with which we began, however, we can see why a both-and choice, *not dependent on faith*, was indeed possible (as also in China). There is involved here only an apparent opposition between two poles, rather than a real one. The relevant terms are not in fact "free will" and "determinism", but "goalseeking" and "constraint", between which there is no contradiction. Once translated in this way, the whole question disappears, to be replaced by a definition of an ecosystem in cybernetic terms. For an ecosystem is simply a "phase space" of hierarchically ordered constraints within which individual goalseeking systems are free to live and move and have their being. Individual goalseekers may follow any number of trajectories within this ordered space as long as no trajectory breaks the boundaries defined by a given level or set of constraints. (These constraints include those embodied in each subsystem's relationships with other goalseeking subsystems.) To go outside the boundary is to invite death or extinction – with this exception, nevertheless, that if in the process of breaking through any particular boundary, a restructuring or transcendence of its associated constraints takes place, then some aspect of the system will have undergone the morphogenesis of evolution or revolution, and it will be possible for a new system to emerge from the old [43: 353–377].

7 THE SPLITTING OF THE ECOSYSTEM

I am come in very truth leading to you Nature with all her children to bind her to your service and make her your slave. FRANCIS BACON: *The Masculine Birth of Time, Or the Great Instauration of the Dominion of Man over the Universe* (1603)

The commands through which we exercise our control over our environment are a kind of information we impart to it. . . . In contol and communication we are always fighting nature's tendency to degrade the organized and to destroy the meaningful: the tendency . . . for entropy to increase. NORBERT WIENER: *The Human Use of Human Beings* (1954)[10]

This restructuring of constraints was precisely what was accomplished in the three or four centuries of change (c. 1500–1800) which eventually produced that particular morphogenesis in the deep structure of Western society we now label the capitalist revolution.

The essence of this change can be captured in the metaphor of "infinite progress" (infinite growth, the infinite production and accumulation of exchange values). This restructuring required that the emerging system transcend the constraints on growth represented metaphorically in the medieval steady-state economy by the figure of God. And we do indeed find the sign of this emergence quite starkly represented in the idea that the God who had once been immanent in the person and in the social universe (as in nature) had now decided to withdraw his presence from the world. The God who had always been, in one commonplace image, the stage manager of the *theatrum mundi*, now became a mere spectator.

This novel status, that of the "hidden God" [14], is prefigured by Nicholas of Cusa in the way he employs an image from Isaiah to refer to God, but without any hint of alarm, as a *deus absconditus.* In contrast, by the time of Pascal and his fellow Jansenists living out the rapid scientific, technological, and social changes of the seventeenth century, the "Vere tu es deus absconditus"[11] had taken on a new and much more frightening meaning. The idea that God had abandoned all relationship to humanity and left it to its own devices – the implication that the coherence of the old ecosystem had been split asunder – this had become part of a growing tragic vision about the future of humanity.

Today, living as we do in an economic system which has discovered how to exploit all of its environments – geographic, human, natural, and temporal – and which continues to do so without limit, we necessarily remain much closer to Pascal's tragic vision of the splitting of the ecosystem and to Bacon's triumphal declaration of the war on nature than to the Chinese and the medieval conception of the unity between and within society and the natural world [17; 43].

Pascal lived in a century which, unlike the nineteenth century, had little idea of where it was going. It took the political and economic crises of the twentieth century to allow intellectuals to rediscover the Pascalian feeling of "abandonment", the expression which became the watchword of existentialists like Heidegger and Sartre in the thirties and the fifties. The deepening crises of the sixties and the seventies have only reinforced this feeling that somehow our society has lost its way. Looking back, we see that Pascal's "spectator God" was as good as dead. Looking forward, we may well suspect that the modern substitute for "god in his heaven" – I mean the utopia promised to all by progress through economic growth – is equally moribund.

If it is, we have no idea what to replace it with. We are not faced here with some sort of "eternal return" in history – a cyclic process which would allow us to assume that we face the same old problems and can therefore resolve them by applying the same old solutions. On the contrary, whatever problems the young capitalist revolution created for itself in the seventeenth century, it still had adequate flexibility – and adequate ecotime and ecospace – to go beyond them. Now that we live in a system which has apparently discovered the ultimate in exploitation, the exploitation of its own future and its own generations to come, we may well wonder whether the kind of systemic restructuring it is apparently heading for is even *imaginable* from our position within its ongoing processes, much less "engineerable".

8 FUNCTIONAL CONTROLS AND STRUCTURAL CONSTRAINTS

This question of future change returns us to the problematic with which we began: the applicability of various approaches calling themselves "cybernetic" to the cybernetics of real life in living and social systems. Manifest as it is in the work of many reformers and futurologists, amongst others, we seem still to be saddled with one version or another of the traditional "technocratic" perspective on socioeconomic change. This viewpoint carries with it explicit or implicit assumptions about the "designed" origins of social systems (respectably ensconced, for example, in the eighteenth-century version of the "social contract", derived from Locke), as well as correlative projections about the effectiveness of consciously engineered solutions to socioeconomic crises (as in the now-forgotten flood of utopian, anarchist, and agrarian socialist writings of the 1820's and 1840's, for instance). As I have tried to point out, these assumptions and projections, in the modern period, are intimately connected with the metamorphosis of the metaphor of the Divine Creator and Lawgiver into that of the Divine Artificer, and thence – by the mediation of the man-made machine system – into that of the "Humanistic Engineer". (The other side of this coin, and an enduring present conflict, might be called the metamorphosis of the Hidden God into the Laissez-Faire Economist.)

Whatever the apparent accomplishments of the recent past (and our present economic system is a mere three centuries old), it seems unlikely that the *long-range* adaptivity of the global economy – its long-range adaptive stability in relation to all of its various environments – can be maintained very much longer by the surface-structure adaptations and "error-corrections" we are familiar with. The most obvious example of this type of adaptation – immediate "corrections" for immediate problems, with little concern for the whole structure of the system in its environments, and even less for the historical processes of which it is the result – is the relatively recent practice of conscious "tuning" of the economy by governments and their agents. As even Keynes himself can be seen to have recognized, however, this surface-structure "tuning" may in fact amount to no more than a repeated application of unproven patent medicines to what are in actuality emerging *symptoms* of structural contradictions and paradoxes in the system as a whole. The remedies are temporary and "physiological", rather than "morphological" [43: 334–335].

Moreover, other economists are now beginning to suspect that the repeated application of short-range remedies to long-range problems does not simply result in a failure to resolve many significant problems. The overall result is more invidious. Misguided attempts at solutions may actually aggravate the overall situation of impending crisis: short-range adaptations may in fact multiply over time – and as a result of their own effects – into states of long-range maladaptation or counteradaptivity (cf. *The Bank Credit Analyst*, October, 1974). Significantly enough, also, even the more grandiose schemes concerned to avoid what are perceived to be approaching socioeconomic and/or ecological disasters usually share the now traditional "technocratic" attitude. Instead of examining the question of the *restructuring* of the system, these approaches tend to concern themselves only with the surface-structure question of the *redistribution* of the present inputs and outputs of the system, leaving its fundamental organization, and of course its ideological and economic values, essentially untouched – as if they were *beyond question* [36].[12]

The well-known Meadows' study, *The Limits to Growth* [22], for example, employs an overtly systemic and cybernetic approach to the problem of the various environmental limits on economic growth on this planet. On close examination, however – and whatever may be the other faults of the study (some of which are quite significant) – we find that it does not escape the trap of inadvertently treating the "system" as if it were just another "environment" to be controlled. In other words, implicit in the study is the unfounded assumption that the "controls" are somehow external to the "system" (e.g. the "political-legislative" process is somehow separate from the basic economic process, and not subordinate to it). In this sense, *The Limits to Growth* is a striking example of what we may now label the "knob-twiddling" approach to the political economy of socioeconomic change. Without meaning to, it unconsciously perpetuates what are in essence *ideological* illusions about the locus of control in complex adaptive systems. Correlatively, it reinforces equivalent illusions about the real role of ideas, policies, and "planning" in human systems. This type of approach effectively limits itself to a search for the socioeconomic equivalents of the "thermostats" in the system (the black boxes which are assumed to control it). The implication is that, once these "controllers" are found – or, in other versions of the same approach, once the ones society believes in have been properly influenced by scientific reasoning – then their sensitivity, their design(s), and/or their "value-settings" can be modified so that the output of the system will be more appropriately distributed, and the crisis will pass.

What is left out of the analysis of the system in such studies, and what is missing from the various policies they put forward, is the starkly simple fact that the envisaged or actual "controllers" have a merely *functional* relationship to the system as a whole. As in any known society, these functional controls are subordinate to already-given *structural constraints*. The "controllers" operate at the level of the *messages* which constitute the surface structure of the system, whereas the structural constraints operate at the level of its deep structure, the level of its constraining *codes*. We do not know, in any long-term sense, just what contribution functional adjustments in the surface structure of societies do in fact make to deep-structure (morphological) change. Little current historical knowledge is of help here, partly because the two levels of change go hand in hand (if not in synchrony), and partly because most of the history we were socialized in shares the same surface-structure approach.

All that can be said with any certainty at this point, is that functional adjustments belong to the domain of the symptoms of crises, not to the domain of the real problems which generate them. And this, in the last analysis, is the crux of the entire issue, for – if we may borrow an image from epidemiology – it is a standard axiom in the treatment of illness, that the only time it is legitimate to treat the symptoms, rather than the disease, is when the disease is beyond a cure.

ENDNOTES

[1] Some of the detailed argument underlying the perspective of this paper was worked out during research on the National Science Foundation project: "The Design and Management of

Environmental Systems", directed by W. E. Cooper and H. E. Koenig in the Department of Electrical Engineering and Systems Science at Michigan State University [cf. 4; 5]. Their help is gratefully acknowledged, as is financial assistance for continued research from the President's Research Fund at Simon Fraser University.

[2] E.g. [2: 271–272; 69]. In the last analysis, however, it is not really of great significance what anyone *intends* or *means*; in the end it is what they *say* and *do* that counts. Any "calculus of intention" or motivation is in fact a psychological construct with no exit, for, from this perspective, all motivations are equal. Such a position is of little help in a scientific and critical understanding of human reality; its total circularity and self-closure makes effective judgments impossible. It is in any case a usually unrecognized component of the ideology of "freedom" in our society (cf. Section 6) – and, as we know, the road to the reinforcement of the status quo is paved with good intentions. In dealing with writing, then, we do not deal with the author, but rather with the *text*. Otherwise we fall into what literary criticism has long called the "intentional fallacy".

[3] See Bateson's use and interpretation of this concept [3: 279–308], taken from Russell; or the discussion between Alice and the White Knight about the song he is going to sing, in Carroll's *Through the Looking-Glass* (Chapter 8). The discussion concerns the *name* the song is *called*, the *name* of the song, *what* the song is called, and what the song *is*. The essence of the problem of logical typing is that whereas the relationship between items of the same logical type is perfectly straightforward, the relationship between logical types themselves is not, and is usually the ground of paradox [cf. 43: 12–18, 110–124, 185–188, 414; 48].

[4] This notion is restated in his self-consciously mystical *The Vision of God, Or The Icon* (1453). It is a modification of Alan of Lille's seventh "theological rule": "God is an intelligible sphere (*sphaera intellegibilis*), whose center is everywhere (*ubique*), and whose circumference is nowhere (*nusquam*)", derived from the pseudo-Hermetic *Book of the Twenty-Four Philosophers*, and known to Alexander of Hales, Vincent de Beauvais, Bonaventura, and Thomas Aquinas, amongst others [6: 353]. It was an aphorism favored by the anti-Cartesian Pascal (1623–1662), split as he was between the "old faith" and the "new science", and has been analyzed in terms of the philosophy of science by Koyré [16: 5–27].

[5] Hierarchies of complexity, hierarchies of rules and metarules, hierarchies of constraints, and so forth are not to be confused *per se* with socioeconomic hierarchies involving people. In historical terms, socioeconomic hierarchies rise and fall, or undergo metamorphosis over time. They are consequently to be regarded as special cases of hierarchy: as *heterarchies* (i.e. as systemic networks in which the dominant locuses of constraint and control immanent in the system may change place and function – and their relative logical typing – in the overall structure through time: [43: 248]). See also endnote 7.

[6] As Althusser usefully summarized it [1: 231, 233]: "An ideology is a system (with its own logic and rigor) of representations (images, ideas, or concepts as the case may be) with a historical existence and a role in the heart of a given society. . . . As a system of representations, ideology is distinct from science in that the practical-social function is more important in it than its theoretical or knowledge function. . . . An ideology is profoundly *unconscious*. . . . It is above all as *structures* that these representations are imposed on the immense majority of people, without passing through their consciousness. They are cultural objects which are perceived/accepted/submitted to, and they act functionally on people by a process which escapes them" (translation modified).

[7] Most complex socioeconomic systems display (coexisting) dominant and subordinate modes of production. Their relationship (their relative logical typing) becomes more evident in times of crisis or change. A mode of production consists of the *means of production* (the

natural ecosystem, the human population, a given technology) and the *social relations of production* (i.e. the relations stemming from the way the components of the means of production are organized). The dominant mode of production in the middle ages hinges on the lord-serf relationship in agriculture, itself dominated by the exchange of *use values* rather than by *exchange value* as such. Subordinate to it one finds wage labor based on exchange value in the crafts and in the mines, for example. Capital exists also, but it has yet to become a commodity. All the components of the capitalist revolution exist, but only where the commoditization of capital, land, and labor is nearly universal and fully dominant (c. 1800) can we legitimately describe the economy as capitalist.

[8] If Needham is correct in his interpretation of the "south-pointing carriage", the Chinese also invented the first closed-loop negative feedback device [24: 286–303], antedating the level regulator for water clocks invented by Ktesibios in the third century BCE.

[9] The supreme advantage and indeed the principle of the factory system, explains Ure, is that it allows children, trained to superintend a single "self-regulating mechanism", to replace craftsmen. This "union of capital and science" will do away with the problem that "by the infirmity of human nature . . . the more skillful the workman, the more self-willed and intractable he is apt to become, and, of course, the less fit a component of a mechanical system, in which, by occasional irregularities, he may do great damage to the whole" [38: 20]. In 1831, Ure patented a "self-acting heat governor", a thermostatic device based on the thermocouple.

[10] See Farrington [8: 62]; and Leiss [17] on Bacon's role as an ideological champion of the use of the new science of the seventeenth century in the (equally new) domination of nature. On the relationship between sexist images of the alienation of nature, the mind/body split, and other exploitative or oppressive dichotomies expressed in Western ideology, see [43: 217–225, 131]. C. S. Lewis once remarked that what we call "Man's power over Nature" usually turns out to be "a power exercised by some men over other men with Nature as its instrument" (quoted in Leiss, [17: 195]). Note also in Wiener's remarks the one-dimensionalization of the logical typing of complexity in nature. In not respecting the boundary between organic and inorganic systems (DNA), Wiener fails to recognize that organic systems are not only the source and the ground of all meaningful organization, but that they are also responsible for its maintenance and reproduction. The entropy they create is continuously neutralized by the input of energy into the closed system of the biosphere by the sun. The cosmic or solar entropy Wiener refers to is irrelevant in human concerns, when compared with the increase in planetary entropy being produced by the global economic system (not by "nature") [46; 47].

[11] "Truly Thou art a hidden God" (Pascal's *Pensées*, Nos 366, 591–599; Isaiah, 45: 15).

[12] Apart from the fact that their cybernetic model is concerned primarily with economic *exchange*, rather than also with the fundamental processes – the production of commodities and the reproduction of the commodity system – which constrain all exchanges, Shaw and Sposato use a telling analogy from Wiener's writings: his comparison of the economic process to a game of Monopoly. Monopoly is a zero-sum game of real-estate speculation, not a representation of the *non-zero-sum* activities of real monopolies in the economic environment of production and consumption. (In the short-run, at least, corporations are well aware of the fact that in any significant zero-sum competition with their multiple environments – a situation in which the *global* socioeconomic system may well find itself – the "winning" system is necessarily doomed to the equivalent of extinction, after the event.) However familiar to a particular class of people it may be – the class of home-buyers and mortgage holders – real estate speculation is a minor and subordinate component of our mode of

production.

The comparison, moreover, is not an innocent one, any more than Monopoly is an innocent game. The "dog-eat-dog" values of Monopoly are undoubtedly useful in the reproduction of the dominant ideology of "free and equal competition" – along with the "survival of the fittest" and the "dream of the chance of 'success' " – in our society. But as a statement about economics, the analogy with Monopoly is worse than misleading. Precisely because we consider it "only a game", we usually fail to see that Monopoly supports the ideology of competition by basing itself on a logical and ecological absurdity. It is assumed that the winning player, having consumed all the resources of all the opponents, can actually survive the end of the game. In fact this is impossible. Unlike a true predator, which never competes with its prey, the Monopoly winner is "unfit" and must consequently die – because in the context of the resources provided by the game, the winner has consumed them all, leaving no environment at all (no other players) to feed on.

REFERENCES

[1] Althusser, L., *For Marx*, New York: Vintage, 1969.

[2] Ashby, W. R., *An Introduction to Cybernetics*, London: Chapman & Hall, 1956.

[3] Bateson, G., *Steps To An Ecology of Mind*, Los Angeles: Chandler, New York: Ballantine, 1972.

[4] Connor, L. J., Holtman, J. B., Hughes, H. and Tummala, R., "Beef feedlot design and management", Agricultural Economics Staff Paper, East Lansing: Michigan State University, Mimeo, 1973.

[5] Cooper, W. E., Edens, T. C., Koenig, H. E. and Wilden, A., "Toward environmental compatibility", East Lansing: Michigan State University, Mimeo, 1973.

[6] Curtius, E. R., *European Literature and the Latin Middle Ages*, New York: Harper Torchbooks, 1953.

[7] Daly, H. E. (Ed.), *Toward a Steady-State Economy*, San Francisco: Freeman, 1973.

[8] Farrington, B., *The Philosophy of Francis Bacon*, Chicago: Phoenix, 1964.

[9] Foucault, M., *The Order of Things*, New York: Pantheon, 1970.

[10] Gieidion, S., *Mechanization Takes Command*, New York: Norton, 1969.

[11] Godelier, M., "Structure and contradiction in *Capital*", *in* Blackburn, R. (Ed.), *Ideology in Social Science*, London: Fontana, 1972.

[12] Godelier, M., "Anthropology and biology", *International Social Science Journal,* **26**, No. 4: 611–635, 1974.

[13] Godelier, M., "Modes of production, kinship, and demographic structures", *ASA Studies*, London: Malaby Press, 1975.

[14] Goldman, L., *Le Dieu caché*, Paris: Gallimard, 1955.

[15] Hyman, S. E., *The Tangled Bank*, New York: Grosset and Dunlap, 1966.

[16] Koyré, A., *From the Closed World to the Infinite Universe*, New York: Harper Torchbooks, 1958.

[17] Leiss, W., *The Domination of Nature*, New York: Braziller, 1972.

[18] Lévi-Strauss, C., *The Elementary Structures of Kinship*, Needham, R. (Tr.), Boston: Beacon Press, 1971.

[19] Lovejoy, A. O., *The Great Chain of Being*, New York: Harper Torchbooks, 1960.

[20] Marney, M. C. and Smith, N. M., "The domain of adaptive systems", *General Systems Yearbook*, 9: 107–131, 1964.

[21] Maruyama, M., "The second cybernetics: deviation-amplifying mutual causal

processes", *in* Buckley, W. (Ed.), *Modern Systems Research for the Behavioral Scientist:* 304–313, Chicago: Aldine, 1968.

[22] Meadows, D. H., Meadows, D. L., Randers, J. and Behrens, W. W., *The Limits to Growth*, London: Earth Island Press, 1972.

[23] Needham, J., *Science and Civilisation in China*, Vol. 2, Cambridge: Cambridge, University Press, 1956.

[24] Needham, J., *Science and Civilisation in China*, Vol. 4, Part 2, Cambridge: Cambridge University Press, 1965.

[25] Nicholas of Cusa, *Philosophische Schriften: I,* Petzelt, A. (Ed.), Stuttgart: Kohlhammer, 1949.

[26] Oresme, N., *Le Livre du ciel et du monde*, Menut, A. D., and Denomy, A. J. (Eds.) Madison: University of Wisconsin Press, 1968.

[27] Piaget, J., *Structuralism*, New York: Basic Books, 1970.

[28] Pollard, S., *The Idea of Progress*, Harmondsworth: Penguin Books, 1971.

[29] Polyani, K., Arensberg, C. M. and Pearson, H. W. (Eds.), *Trade and Market in the Early Empires*, Chicago: Gateway Press, 1957.

[30] Porta, G. D., *Natural Magick*, New York: Basic Books, 1957.

[31] Rappaport, R. A., *Pigs for the Ancestors*, New Haven: Yale University Press, 1968.

[32] Rappaport, R. A., "The sacred in human evolution", *Annual Review of Ecology and Systematics, 2:* 23–44, 1971.

[33] Sahlins, M., *Stone Age Economics*, Chicago: Aldine-Atherton, 1972.

[34] Samuelson, P. A., *Economics*, 8th ed., New York: McGraw-Hill, 1970.

[35] Schon, D. A., *Invention and the Evolution of Ideas*, London: Social Science Paperbacks, 1967.

[36] Shaw, A. and Sposato, D., "Implications of an alternative world exchange system", *Transactions of the New York Academy of Sciences,* **35**: 557–572, 1973.

[37] *The Compact Edition of the Oxford English Dictionary (1884–1928)*, Oxford: Clarendon Press, 1971.

[38] Ure, A., *The Philosophy of Manufactures*, with an Appendix to the 1860 ed. by Simmonds, P. L., (Ed.), London, 1861.

[39] Vayda, A. P. (Ed.), *Environment and Cultural Behavior*, Garden City, N.Y.: Natural History Press, 1969.

[40] Voloshinov, V. N., *Marxism and the Philosophy of Language*, Matejka, L. and Titunik, I. R., (Eds. and Trs.), New York: Seminar Press, 1973.

[41] Von Foerster, H., "On self-organizing systems and their environments", *in* Yovits, M. C., Jacobi, G. T. and Goldstein, G. D. (Eds.), *Self-Organizing Systems*, Washington: Spartan Books, 1959.

[42] Watzlawick, P., Beavin, J. and Jackson, D. D., *The Pragmatics of Human Communication*, New York: Norton, 1967.

[43] Wilden, A., *System and Structure: Essays in Communication and Exchange*, London: Tavistock, New York: Barnes and Noble, 1972. Second edition, revised, 1978.

[44] Wilden, A., Review of Leiss, 1972, *Psychology Today* (October), 1972.

[45] Wilden, A., Review of Bateson, 1972, *Psychology Today* (November): 138–140, 1973.

[46] Wilden, A., "Piaget and the structure as law and order", *in* Riegel, K., and Rosenwald, G. C. (Eds.), *Structure and Transformation:* 83–117, New York: Wiley Interscience, 1975.

[47] Wilden, A., "Ecology, ideology, and political economy", Burnaby, B. C.: Simon Fraser University, Mimeo, 1975.

[48] Wilden, A. and Wilson, T., "The double bind: logic, magic, and economics", *in*

Sluzki, C. and Ransom, D. C. (Eds.), *The Double Bind: The Foundation of the Communicational Approach to the Family*, New York: Grune and Stratton, 1976: 263–86.

[49] Wills, G., *Nixon Agonistes: The Crisis of the Self-Made Man*, New York: Signet, 1969.

THE DOG THAT BELONGED TO HIMSELF, OR ON THE NATURE OF THE MIND THAT BELIEVES THE REAL WORLD, IT IS ESSENTIALLY CYBERNETIC

C. WEST CHURCHMAN
University of California, Berkeley

1 PROLOGUE

At a recent conference in Paris, we were discussing the question of whether world models make sense, i.e. are useful ways of planning mankind's future. Jay Forrester commented that it was not a question of whether we develop models or not, because we all use models of some type in making decisions; thus models are essential, and the only question is what kind of model we should use. Martin Shubik later intervened by asking Forrester a well-known and old question: Tell me, "Where is fancy bred? Or in the heart or in the head?" Forrester elected not to reply, perhaps because for him there was only one reply.

2 ON "CYBERNETICS"

Philosophers often like to take interesting concepts of the disciplines and generalize on their meaning; thereby they think they can begin to grasp the essence of the concept. Thus "ecology" is a perfectly respectable concept of biology, but we philosophers want to say that the concept stands for a lot more than a technical description of a species, its habitat and its environment; ecology, to us, connotes the general idea of how humans ought to interact with their environment, which "environment" includes future generations. Ecology, then, is essentially a branch of ethics.

Similarly, I know that cybernetics has a perfectly respectable meaning amongst its practitioners, and connotes a model of some aspect of reality which employs the mathematics of informational feedback. But the philosopher wants to generalize on this technical concept, so as to say that cybernetics is the process of learning (acquiring knowledge) while acting. Cybernetics, thus viewed, is a branch of epistemology: How is it possible that knowledge is created by action?

3 THE CYBERNETICIAN

You can now appreciate the trickster nature of the philosopher, who delights in paradox. For now the question is: Who gains the knowledge by action? If you go to the texts, you are apt to conclude that the answer is: the controller, who is a "box" receiving inputs emanating from the effects of previous decisions and who makes new decisions partially on the basis of these inputs. But strangely enough, this picture leaves out a very important "box", namely, the cybernetician himself, who surely is the master controller since he designs the cybernetic controller. What inputs does he, the cybernetician receive, and how does he use these to modify his own design? Or, what is the nature of the cybernetician's mind?

The question is paradoxical, because to respond to it we seem to need another mind, who now diagrams a reality consisting of the cybernetic model, the cybernetician and his inputs and outputs; and then another mind, and so on. But paradox need not be addressed directly. Instead, we can ask much plainer questions, in the form of a PhD qualifying examination; e.g.

a) Explain why Stafford Beer's mind led him to believe that society is like Ashby's model of a brain, and why did Ashby's mind lead him to create the brain model?

b) Why does Jay Forrester's mind believe that we all use models in decision making?

c) What is there about James Miller's mind that caused him to conclude that there are seven levels of living systems, starting with the cell and ending in the supra-nation?

d) Russ Ackoff's psyche which produced a teleological model of human decision making which "sweeps in" the whole of mankind.

4 AN EMBARRASSMENT

The last qualifying exam question is a bit embarrassing to me, because my own psyche was allied with Russ's in producing such a model. I have some ideas, therefore, as to how one might go about responding to question 4. The hero or culprit was E. A. Singer, Jr., who got his inspiration from a tautology, a very common beginning point for thinkers. In speculating on what each of us humans must want in our lives, Singer writes:

> Of one who has been granted a wish to become a more powerful chess player and thereafter his wish to become a more powerful flute player, what would one say in the end; has he been increased in power, or multiplied in powers? With no more than this to guide us the latter no doubt is all we should feel justified in affirming; but if we were asked, not which of the two we had prayed for but for which of the two we should pray, could we hesitate? Not if there is any soundness to the "ancient wisdom of childhood", learned from a thousand fairy godmothers who have left no godchild untested on this very point. Their lesson is always the same: "With only one wish to be had, choose rather the power to get whatever you may come to want than the pleasure of having any dearest thing in the world." Our modern at any rate takes this ancient wisdom to have touched the bottom of things; he takes the deepest wish in any man, the common wish of all man, to be no other than the wish for more power – the wish to grow more powerful [1: 145–147].

There is a tautology lurking behind these remarks, because to define "desire" one must specify the conditions under which an investigator would be willing to say that an individual desires some outcome. The reply must be "whenever he is aware of the consequences of his choices and whenever he chooses that action, that leads to this outcome and not to others". In other words, the reason we all would wish that all our wishes come true, follows from the definition of "wish". Nevertheless, Singer does develop an astonishingly complete theory of an "enabling ethics" out of his tautology. The event is reminiscent of Kant's efforts in the *Foundations of the Metaphysics of Morals*, where he evolves a very beautiful concept of the kingdom of ends ("never treat another as means only, but as an end withal") out of a rather bland tautology (the will can have no phenomenal cause).

5 JUNG

It was Jung's *Psychological Types* that made me suspect that Singer's model of humanity might be questioned. Briefly, in the *Types*, Jung postulates four basic functions of the mind: thinking, intuition, sensation and feeling. He also develops the theory that when thinking is the "dominant" function, then feeling is apt to be unconscious and "underdeveloped". The thinking type, who can't escape the impact of feeling, tries to categorize feelings, shape them into moral or ethical prescriptions, give them operational definitions. At long last, I realized that all this attempt to classify and understand feelings was inappropriate – from a feeling point of view – and hence got nowhere at all.

Of course, one could continue the PhD qualifying examination on the Jungians:

e) What caused Jung to write the *Types* and later to go beyond "types" to "archetypes"?

f) Why does James Hillman believe he can legitimately explain all the intellectual enterprise of the disciplines in terms of archetypes (his latest effort is the metaphor of economics)? I can hear the cultural anthropologist whispering his response to these tantalizing questions; so

g) Why does an anthropologist believe he can explain Jung's work in terms of Jung's cultural background?

6 YOU ARE NOT YOUR MIND

But why stop at Jung and his Western ideas of the human conscious and unconscious? Go back to the Sanskrit texts, to learn that you are not your body (and hence *not* a cybernetic machine), nor are you your mind (and hence *not* a purposive entity). "The seers say truly that he is wise who acts without lust or scheming for the fruit of his acts; – turning his face from the fruit, he needs nothing; the Atman is enough; he acts and is beyond action." Thus the *Bhagavad-Gita* (Bk. 14). In the end, teleology is dead. But then,

h) explain why the writers of the *Upanishads* and the *Bhagavad-Gita* were persuaded that there is a reality beyond teleology?

7 PAUL

I cannot stop my incessant questioning without adding one more example. It is St. Paul in the first century, trying to develop a cybernetic model for the little Christian communities. What could be more up-to-date than "for as in one body we have many members, and all members have not the same function; so we, being many, are yet one body in Christ, and everyone members one of another". (Romans, 12.) And in Corinthians 13 he concludes that all "models" are imperfect (we're still repeating his warning today): (paraphrased) "for our models are imperfect and our forecasts are imperfect". But he adds a note of optimism, that says that our models are indeed approximations, and as they get better, their imperfection will pass away. However, the necessary condition for human progress along these lines is not a tightening of our models, or the collection of better data, but rather the basic attitude of all humans, which is faith, hope and love. So

i) Explain why Paul felt that – –

8 THE INQUIRING SYSTEM

In the Design of Inquiring Systems I struggled a bit with these questions, by trying first to describe five types of inquiring system, and then to indicate why none of these really captures the essence of what goes on in a modeling mind. Perhaps, for present purposes, the most important of these types is the Lockean inquirer, which relies on a strong and ineffable force called agreement. We intellectuals seem to be able to build these Lockean communities with great success. In many ways they are unassailable, because they create their own paradigms of model building and data collection, which cannot be attacked from outside. In my experience, the responses to all the questions raised above must include the phrase "because he (she) belonged to a Lockean community".

9 POLITICS

But now I believe we can begin adding to the list of what can be said about the cybernetician-model-building. First, there is politics, but not in the dreadful and overdone sense in which today's politicians exhibit it, where ugly power dominates. Rather, I'm using it in Aristotle's sense, as the urge to speak (or write, or act, or influence) publicly. In this sense politics is pervasive, in every street encounter, family dinner table, bar, assembly, whatever. It will dominate this conference, as each delegate tries to get the floor, some of them over and over, and on and on. To speak – or influence – publicly is a basic feeling of all humanity, and is especially felt by our cybernetician. So add to the list "because he (she) is political".

10 MORALITY

But in addition there is another feeling, of the difference between good and bad, and between right and wrong. All of us have it, all the time. The feeling is inadequately or feebly represented by moral codes. It is the feeling of injustice, or justice, of moral breakdown or elevation. At the Paris meeting referred to earlier, one of the speakers tried to tell us that the major problem of the world today is the need for more energy. A young man from an African country suggested that this was a problem for the "over developed" countries, but for the world as a whole the major problem was oppression. So, add "because he has a moral sense".

11 RELIGION

This may seem an odd one to add to our list, but the oddity disappears once we think of religion as one individual's relationship to the grand, and in the case of our cyberneticians, the grand design of all human destiny. This is a religious idea, the idea of one individual's relating himself to the whole history of mankind. To be sure, he may do this by creating an idea of a God, but he need not do so. I only have to ask you to realize the immensity of the image of one model builder who is struggling to deal with all the major problems of mankind, to convince you that his image has all the religious strength of a Teresa in ecstasy. So, "because he is religious".

12 AESTHETICS

The less I say about this last, the better, because talk about the aesthetic quality of our lives runs the immediate danger of being unaesthetic. But I am sure that aesthetics applies to most model builders because it represents their own unique way of imaging the real world. But what this way is, is quite obscure. Things might be improved if we could get them to tell us a myth along with every model, but I doubt if they would or could. So I could add to the list, but I won't. I'll just add that I'm sitting in my study in the evening of a beautiful September day, surrounded by redwood trees; that might help explain why I think this last is so important.

13 SO WHAT?

So, for the sake of those of us who are driven by curiosity, could the model-builders tell us something about themselves – how they got to where they are? It would be lovely if they would.

REFERENCE

[1] Singer, E. A., Jr., *On the Contented Life*, New York: Henry Holt, 1936.

CYBERNETIC DATA ANALYSIS

INTRODUCTION

There are fundamental differences between the kind of knowledge that enables someone to create and set into motion something new from given and well understood parts and the kind of knowledge that enables someone unobtrusively to observe and to predict the behavior of a portion of his environment. The paradigm of inquiry in engineering favors the development of the former kind of knowledge while the paradigm of inquiry in the social sciences emphasizes the latter. The contrast between "synthesis" and "analysis" does not do justice to this difference although it lies at the root of it. For whatever reason, early cybernetics captured the imagination of engineers more so than that of social scientists and thereby helped to shape the monstrous marvel of technology. But in the course of this association cybernetics failed to develop, even for its most basic concepts, techniques of measurement, data collection and analysis which are central to understanding the functioning of our own societies. The papers in this section aim to overcome some of these deficiencies and to make cybernetic thinking more palatable to social scientists.

The concept of feedback, with which the first paper in this section is concerned, may serve as a case in point. This concept is fundamental to cybernetics and has provided the basis of many technological developments like computers, like communication networks and the plethora of control devices for the automation of production, of defense, and of various business activities. The concept of feedback first made inroads into the social sciences after the publication of works by Wiener [4; 5] and Ashby [1; 2], but it remained a verbal concept. Analytical techniques for identifying feedback in data, not to mention relevant data themselves, were simply not available. It is fair to say that, as a result, this concept has not gained the status in social theory it deserves.

It is only now that social scientists are discovering a "harder" concept of feedback through an analytical technique called path analysis. The aim of this technique is to ascertain and to quantify from observational data the strength of causal connections among variables. Half a century ago, S. Wright, the initiator of this technique, was mainly using it to discern relatively simple and uni-causal connections. This conceptualization has been quite sufficient for analyzing data which incorporate independent (or experimenter controlled) and dependent (or resultant) variables, a distinction that is still maintained by much of experimental physiology and experimental individual and social psychology. However, the extension of this technique and its application to large networks in biology and econometrics has naturally led to the discovery of larger cycles of causal connection and to the concept of mutual causation, which is nothing other than what cyberneticians call feedback. The first paper of this section, by Malcolm E. Turner, develops a theoretical foundation for this most recent development in path analysis and thereby gives the concept of mutual causation or feedback a solid analytical base. Applications may be seen in an earlier paper by the same author [3]. Path Analysis thus provides an instrument to explore possible mechanisms of control underlying the behavior of natural systems.

The next two papers are concerned with analyses of social structures. Again. for the engineer structure may be nothing more than a wiring diagram. In society the analogues of wires are communication channels, patterns of influence, constraints mediated by rules adopted, implicitly or explicitly, by small groups or incorporated formally into large institutions. Although a material base is undeniable, social structures are less visible – often symbolic – but nevertheless effective. The ability to analyze these analogues of wiring diagrams is a prerequisite to understanding larger organizations in terms of the interaction among their parts. The two papers in question are both concerned with the kind of structures that link human individuals. Although there are other ways of defining social structure, the ones analyzed here are easily conceptualized and consistent with current social science theory. However, as both papers indicate, the analysis of even these obvious kinds of social structure is not a simple matter and calls for considerable analytical innovation.

William D. Richards and Georg Lindsey have been working for a number of years to develop methods for analyzing communication networks that span a large number of people, usually working in natural settings such as business enterprises, military organizations and the like. They realize, all analyses of communication networks are subject to restrictions which stem from the fact that the number of possible links grows exponentially with the number of individuals in the net and quickly approaches computational limits. The techniques outlined in their paper are applicable to social networks involving up to 4,000 persons which is a surprisingly large number. They aim at identifying in the data cliques, bottlenecks, hierarchies, etc., and they provide the user with a variety of local and global statistical indices of such networks without making any normative assumptions.

Kenneth D. Mackensie also gives an overview of recent developments in the analysis of social structures. However, his data concerns smaller numbers of individuals in committees, problem solving and decision making groups, etc. Because the number of possible links among these individuals is significantly less than in Richards' data, Mackensie's techniques allow further inquiries into the content of those structures, changes in group processes, adoption of new structures, development of social roles, and so forth. His paper discusses some of the many explanatory functions served by structural concepts in a variety of social theories, and he contrasts his approach with those that make structure a prime mover, giving it a status independent of the group members in interaction.

The fourth paper in this section applies cybernetics to the very process of generating data: measurement. Customarily, data are regarded as "given off" by a process, perhaps with the help of an observer. Once data are collected, they are then analyzed as if they were unshakable facts. This is the position taken in the natural sciences and it is also evident in the three preceding papers in this section. Richard O. Mason presents several illustrative examples of data collection to show that data are not free of assumptions, values and beliefs underlying the process of collection and cannot always be taken at face value. For example, there is the problem of developing a measure as initially called for by a given theory or problem, but which, in the course of its operationalization, requires stepwise adjustments of both the theory and the process to be assessed. There is the problem of requiring consensus about a measure among interested parties who are affected in opposite ways by the proposed measure. Also, there is the problem of measuring an entity that is under the instrumental control of the parties being assessed. These situations and presumably several others involve circular reasoning. Traditional unidirectional modes of reasoning, deduction and induction in particular, cannot cope adequately with such situations and lead to accusations of biased data. To resolve some of these measurement problems, Mason develops a dialectical scheme which would account for the operationalization of measures that synthesize conflicting alternatives. This mode of reasoning is proposed as the cybernetic companion of deduction and induction in the measurement process.

The last paper in this section responds to the observation that the link between data collection and decision making processes in society has become increasingly confounded by technological and institutional developments. The demand for more complex models is not the only reason for this confounded relation. New and sophisticated computational technology is increasingly deployed and causes problems in understanding analytical results. More specialists and different kind of specialists – developers of data analysis methodology, data analysts, interpreters of analyses and decision makers – mediate the information that is gathered and processed and thereby impose their own institutional values and systematic biases. In his paper, Roy E. Welsch considers this process as a communication channel. He proposes several analytical concepts and applies them to a large number of recent developments in data analysis techniques pertaining to the four groups of individuals described above. His analysis makes transparent deficiencies in recent developments and clarifies current problems in data handling techniques.

REFERENCES

[1] Ashby, W. R., *An Introduction to Cybernetics*, London: Chapman & Hall, 1958.
[2] Ashby, W. R., *Design for a Brain*, 2nd ed., New York: Wiley, 1960.

[3] Turner, M. E. and Stevens, C. D., "The regression analysis of causal paths", *in* Blalock, H. M., Jr. (Ed.), *Causal Models in the Social Sciences:* 75–100, New York: Aldine, 1971.
[4] Wiener, N., *Cybernetics*, Cambridge, Mass.: M.I.T. Press, 1948.
[5] Wiener, N., *The Human Use of Human Beings*, New York: Houghton Mifflin, 1950.

THE STATISTICAL ANALYSIS OF MUTUAL CAUSATION

MALCOLM E. TURNER, JR.
University of Alabama in Birmingham

INTRODUCTION

Mutual causation concerns interactive influence or feedback between variables in a natural system. The favorite mathematical tool for exploring such relations has been systems of ordinary differential equations, especially in such fields as engineering, chemistry and physiology. However, for large systems with limited information about exact mechanisms of action, the method of differential equations offers insurmountable difficulties. For this reason, a simpler tool, called "path analysis", or the "method of dynamic equations" has been developed in biology and economics to supply an exploratory device for examining mutual causation. As this conference attests, the device of path analysis is being used increasingly for the purpose of exploring social processes.

The present paper presents (1) a straightforward description of the popular technique of linear path analysis, and (2) an extension to one of many possible schemes of nonlinear model, the so-called "multiple-process" law [8].

Applications to social processes will be left to others. For examples, see the volumes by Blalock [1] and Jencks [4].

LINEAR PATH REGRESSION

1 *The Origins of Path Regression Analysis*

Sewall Wright [13] conceived of the notion and term of path regression analysis. The notion has been much extended and developed in the econometric literature. See especially Hood and Koopmans [3] and Wold [12]. Wright himself has come to favor the use of standardized variates, the analysis of which he refers to as "path analysis". Summaries of Wright's views may be found in his review papers, Wright [14; 15]. Tukey [6] has argued against the use of standardized variates even when these have a rational basis, i.e. when the "independent" variables have normal distributions. When the independent variables are "fixed", i.e. have no probability distribution, then standardization as Wright advocates is meaningless.

For elementary presentations of the regression method, see Kempthorne [5] and Turner and Stevens [10].

In this section will be presented a recapitulation of the main results of linear path regression analysis. An extension to the non-linear case using the "single process law" will be described in the next section.

2 *The Structural Equations*

Let us suppose that we have p "secondary" variables which are causally related to one another and to a set of q "primary" or "controllable" variables. We imagine that the q primary variables (denoted $\xi_a, \xi_b, \ldots, \xi_q$) have values which can be chosen at will by the experimenter. The p secondary variables (denoted by $\eta_1, \eta_2, \ldots, \eta_p$) are then supposed to be completely determined. We further assume that all rates of change are constant. Thus, the matrix of partial derivatives η_{ij} is H where

$$H' = \left[\begin{array}{cccc} \eta_{1a} & \eta_{2a} & \cdots & \eta_{pa} \\ \eta_{1b} & \eta_{2b} & \cdots & \eta_{pb} \\ \cdot & \cdot & & \cdot \\ \cdot & \cdot & & \cdot \\ \cdot & \cdot & & \cdot \\ \eta_{1q} & \eta_{2q} & \cdots & \eta_{pq} \\ \hline \eta_{11} & \eta_{21} & \cdots & \eta_{p1} \\ \eta_{12} & \eta_{22} & \cdots & \eta_{p2} \\ \cdot & \cdot & & \cdot \\ \cdot & \cdot & & \cdot \\ \cdot & \cdot & & \cdot \\ \eta_{1p} & \eta_{2p} & \cdots & \eta_{pp} \end{array}\right] = \left[\begin{array}{cccc} \alpha_{1a} & \alpha_{2a} & \cdots & \alpha_{pa} \\ \alpha_{1b} & \alpha_{2b} & \cdots & \alpha_{pb} \\ \cdot & \cdot & & \cdot \\ \cdot & \cdot & & \cdot \\ \cdot & \cdot & & \cdot \\ \alpha_{1q} & \alpha_{2q} & \cdots & \alpha_{pq} \\ \hline 1 & \alpha_{21} & \cdots & \alpha_{p1} \\ \alpha_{12} & 1 & \cdots & \alpha_{p2} \\ \cdot & \cdot & & \cdot \\ \cdot & \cdot & & \cdot \\ \cdot & \cdot & & \cdot \\ \alpha_{1p} & \alpha_{2p} & \cdots & 1 \end{array}\right]$$

and where the "path regression coefficients", α_{ij}, are constant. Solution of (1) yields the p "structural equations"

$$\begin{aligned} \eta_1 &= \alpha_1 + \alpha_{1a}\xi_a + \alpha_{1b}\xi_b + \ldots + \alpha_{1q}\xi_q + \alpha_{12}\eta_2 + \alpha_{13}\eta_3 + \ldots + \alpha_{1p}\eta_p \\ \eta_2 &= \alpha_2 + \alpha_{2a}\xi_a + \alpha_{2b}\xi_b + \ldots + \alpha_{2q}\xi_q + \alpha_{21}\eta_1 + \alpha_{23}\eta_3 + \ldots + \alpha_{2p}\eta_p \\ &\cdot \\ &\cdot \\ &\cdot \\ \eta_p &= \alpha_p + \alpha_{pa}\xi_a + \alpha_{pb}\xi_b + \ldots + \alpha_{pq}\xi_q + \alpha_{p1}\eta_1 + \alpha_{p2}\eta_2 + \alpha_{p3}\eta_3 + \ldots \end{aligned} \qquad (2)$$

where $\alpha_1, \alpha_2, \ldots, \alpha_p$ are constants of integration. These equations may be written in matrix notation as

$$\left[\begin{array}{ccccc|ccccc} \alpha_1 & \alpha_{1a} & \alpha_{1b} & \cdots & \alpha_{1q} & 0 & \alpha_{12} & \alpha_{13} & \cdots & \alpha_{1p} \\ \alpha_2 & \alpha_{2a} & \alpha_{2b} & \cdots & \alpha_{2q} & \alpha_{21} & 0 & \alpha_{23} & \cdots & \alpha_{2p} \\ \cdot & & & & & \cdot & & & & \\ \cdot & & & & & \cdot & & & & \\ \cdot & & & & & \cdot & & & & \\ \alpha_p & \alpha_{pa} & \alpha_{pb} & \cdots & \alpha_{pq} & \alpha_{p1} & \alpha_{p2} & \alpha_{p3} & \cdots & 0 \end{array}\right] \left[\begin{array}{c} 1 \\ \xi_a \\ \xi_b \\ \cdot \\ \cdot \\ \cdot \\ \xi_q \\ \hline \eta_1 \\ \eta_2 \\ \cdot \\ \cdot \\ \cdot \\ \eta_p \end{array}\right] = \left[\begin{array}{c} \eta_1 \\ \eta_2 \\ \cdot \\ \cdot \\ \cdot \\ \eta_p \end{array}\right]$$

or

$$(A_1, A_2)\begin{pmatrix} \xi \\ \eta \end{pmatrix} = \eta. \tag{3}$$

Therefore, the "structure" of the causal system is completely described by Eq. (3).

3 . *The Reduced Structural Equations*

From Eq. (3) we get by multiplication

$$A_1\xi + A_2\eta = \eta \tag{4}$$

and

$$\eta - A_2\eta = A_1\xi$$

$$(I - A_2)\eta = A_1\xi$$

If $(I - A_2)$ is nonsingular we get the "reduced structural equations"

$$\eta = (I - A_2)^{-1} A_1\xi. \tag{5}$$

Thus, the secondary variables η are expressed as linear functions of the primary variables ξ and it is seen that the necessary and sufficient condition for this to be possible is that $(I - A_2)$ be nonsingular.

4 *The Equations of Identification*

Let B be a $p \times (q + 1)$ matrix of regression coefficients as follows:

$$B = \begin{bmatrix} \beta_1 & \beta_{1a} & \beta_{1b} & \cdots & \beta_{1q} \\ \beta_2 & \beta_{2a} & \beta_{2b} & \cdots & \beta_{2q} \\ \cdot & \cdot & \cdot & & \cdot \\ \cdot & \cdot & \cdot & & \cdot \\ \cdot & \cdot & \cdot & & \cdot \\ \beta_p & \beta_{pa} & \beta_{pb} & \cdots & \beta_{pq} \end{bmatrix}$$

Now we will state the "identification equations" as follows:

$$B = (I - A_2)^{-1} A_1. \tag{6}$$

The reduced structural equations (5) may then be written

$$\eta = B\xi \tag{7}$$

the structural equations for an ordinary multiple regression system.

We note that it may or may not be possible to solve (6) for (A_1, A_2) uniquely or at all. If it is not possible to find a solution at all, the system is said to be "under-identified". If a unique solution exists the system is said to be "just-identified", and if more than a single solution exists, the system is said to be "over-identified".

If we expand the right-hand side of (6) we will find, in general, that some of the β's are zero as a consequence of the causal restrictions on the α's (in particular, some of the causal pathways will be absent and the corresponding α's will be zero). Let us agree to omit all those ξ's from the rows of (7) having zero β's as coefficients. Then, we would replace (7) by

$$\begin{aligned} \eta_1 &= \beta_1' \xi_1 \\ \eta_2 &= \beta_2' \xi_2 \\ &\cdot \\ &\cdot \\ &\cdot \\ \eta_p &= \beta_p' \xi_p \end{aligned} \tag{8}$$

where each β_i and ξ_i is of order less than or equal to $q + 1$.

5 *Estimation in the Case of Under- and Just-identified System*

The under- and just-identified cases are of especial interest because in these cases regression type estimators are available. We consider two possibilities:

a) All β's are nonnull and

$$\mathbf{y}_j = \eta_j + \epsilon_j$$

$$\mathbf{x}_j = \xi_j$$

$j = 1, 2, \ldots, n$.
Then by substitution (7) becomes

$$\mathbf{y}_j = B\mathbf{x}_j + \epsilon_j \tag{9}$$

or

$$\mathbf{y}_j' = \mathbf{x}_j'B' + \epsilon_j'. \tag{10}$$

For n observations we have

$$\begin{bmatrix} \mathbf{y}_1' \\ \mathbf{y}_2' \\ \cdot \\ \cdot \\ \cdot \\ \mathbf{y}_n' \end{bmatrix} = \begin{bmatrix} \mathbf{x}_1' \\ \mathbf{x}_2' \\ \cdot \\ \cdot \\ \cdot \\ \mathbf{x}_n' \end{bmatrix} B' + \begin{bmatrix} \epsilon_1' \\ \epsilon_2' \\ \cdot \\ \cdot \\ \cdot \\ \epsilon_n' \end{bmatrix}$$

or

$$Y = XB' + E. \tag{11}$$

Now, if the $n \times p$ elements of the "error" matrix are uncorrelated and have constant variance σ^2 then the minimum variance unbiased estimator is given by the well-known expression

$$\hat{B}' = (X'X)^{-1} X'Y. \tag{12}$$

If in addition E is distributed as a multivariate normal distribution, then, of course, $\hat{B}'$ is the sufficient estimator for B'. In the case of just-identification B' may be replaced by $\hat{B}'$ in (6) and estimates of (A_1, A_2) solved for. In the case of under-identification, it will not be possible to solve for all elements of (A_1, A_2) explicitly unless a sufficient number of *a priori* constraints can be assumed.

b) More often than not restrictions on A_1 and A_2 will force some β's to be zero and then we must omit the corresponding x's from the estimation equations. In this case a separate regression analysis must be performed for each y variable.

Hence, we get the individual estimators

$$\begin{aligned}
\hat{\beta}_1 &= (X_1'X_1)^{-1}X_1'y_1 \\
\hat{\beta}_2 &= (X_2'X_2)^{-1}X_2'y_2 \\
&\cdot \\
&\cdot \\
&\cdot \\
\hat{\beta}_p &= (X_p'X_p)^{-1}X_p'y_p.
\end{aligned} \tag{13}$$

The remarks following (12) concerning distribution and identification apply similarly to (13).

For treatment of the over-identified case see Hood and Koopmans [3]. Refer also to Turner and Stevens [10].

6 *Classification of Path Schemata*

There are several types of causal relationships which are conveniently distinguished. First of all there is the situation in which a single secondary variable (or response variable) is determined by one or more primary variables. This is the case of "ordinary" regression, the single structural equation corresponding to that of the usual multiple regression model. A second case is the situation in which two or more secondary variables are jointly determined by some or all of a common set of primary variables. We term this case the case of "joint" regression. In a third situation there is a chain of cause and effect leading from one or more primary variables to a secondary variable and then on to still another "secondary" variable. This case we term the case of "chain" regression. A final type involves cycles of causation or "feedback" from one response variable to another and back again after possibly passing through one or more other secondary variables. This is the case of "cyclic" regression. All of these cases individually or in combination have been synoptically treated in the previous sections of this chapter. Here we wish only to note that the several cases are distinguished by having different kinds of $(I - A_2)$ matrices. For this reason we term $(I - A_2)$ the "classification" matrix. The type of regression is identified as in Table 1.

TABLE 1

Classification of path schemata

Regression type is	$(I - A_2)$ is
Ordinary	1
Joint	I
Chain	Triangular
Cyclic (feedback)	Nontriangular

It is evident that any scheme not involving feedback will possess a triangular classification matrix. In this case the identification Eq. (6) are of a particularly simple variety and algorithms have been devised by Wright and his followers for writing down these equations from inspection of a diagram representing the flow of cause and effect. A statement of these algorithms for the linear path *regression* analysis is given by Turner and Stevens [10].

7 *Some Examples of Particular Path Schemata*

We will consider several very simple cases of linear causal networks, nonlinear analogues of which we will consider in the next section.

a) The case of two primary variables and one secondary variable. We represent the flow of cause and effect by a "path diagram" as follows

$$\xi_a \xrightarrow{\alpha_{1a}} \eta_1 \xleftarrow{\alpha_{1b}} \xi_b$$

The structural equation is

$$\eta_1 = \alpha_1 + \alpha_{1a}\xi_a + \alpha_{1b}\xi_b \tag{14}$$

or

$$(\alpha_1, \alpha_{1a}, \alpha_{1b} \mid 0) \begin{pmatrix} 1 \\ \xi_a \\ \xi_b \\ \hline \eta_1 \end{pmatrix} = (\eta_1)$$

and

$$(\beta_1, \beta_{1a}, \beta_{1b}) = (1)\,(\alpha_1, \alpha_{1a}, \alpha_{1b}) \tag{15}$$

from (6). Thus, we get the obvious result that if one obtains the usual multiple regression estimates $\hat{\beta}_1, \hat{\beta}_{1a}, \hat{\beta}_{1b}$ by regressing y_1, on x_a and x_b then we would estimate the path coefficients by equating according to (15):

$$\hat{\alpha}_1 = \hat{\beta}_1$$
$$\hat{\alpha}_{1a} = \hat{\beta}_{1a}$$
$$\hat{\alpha}_{1b} = \hat{\beta}_{1b}.$$

b) A case in which both joint and chain regression occurs. Consider the path diagram

$$\begin{array}{ccc} \xi_a & \xrightarrow{\alpha_{1a}} & \eta_1 \\ & \overset{\alpha_{1b}}{\nearrow} & \downarrow \alpha_{21} \\ \xi_b & \xrightarrow[\alpha_{2b}]{} & \eta_2 \end{array}$$

Proceeding as before we obtain

$$I - A_2 = \begin{pmatrix} 1 & 0 \\ -\alpha_{21} & 1 \end{pmatrix}$$

and according to (6) we get identification equations

$$\begin{pmatrix} \beta_1 & \beta_{1a} & \beta_{1b} \\ \beta_2 & \beta_{2a} & \beta_{2b} \end{pmatrix} = \begin{pmatrix} 1 & 0 \\ -\alpha_{21} & 1 \end{pmatrix}^{-1} \begin{pmatrix} \alpha_1 & \alpha_{1a} & \alpha_{1b} \\ \alpha_2 & 0 & \alpha_{2b} \end{pmatrix}$$

$$= \begin{pmatrix} 1 & 0 \\ \alpha_{21} & 1 \end{pmatrix} \begin{pmatrix} \alpha_1 & \alpha_{1a} & \alpha_{1b} \\ \alpha_2 & 0 & \alpha_{2b} \end{pmatrix}$$

$$= \begin{pmatrix} \alpha_1 & \alpha_{1a} & \alpha_{1b} \\ \alpha_2 + \alpha_{21}\alpha_1 & \alpha_{21}\alpha_{1a} & \alpha_{2b} + \alpha_{21}\alpha_{1b} \end{pmatrix} \tag{16}$$

Now (16) is solved for the path coefficients and the β's are replaced by regression estimates to give estimates of the path coefficients.

c) An example of feedback. The path diagram of a simple case of cyclic regression is given below.

$$\begin{array}{ccc} \xi_a & \xrightarrow{\alpha_{1a}} & \eta_1 \\ & \alpha_b \uparrow & \downarrow \alpha_{21} \\ \xi_b & \xrightarrow{\alpha_{2b}} & \eta_2 \end{array}$$

We have according to (6)

$$\begin{pmatrix} \beta_1 & \beta_{1a} & \beta_{1b} \\ \beta_2 & \beta_{2a} & \beta_{2b} \end{pmatrix} = \begin{pmatrix} 1 & -\alpha_{12} \\ -\alpha_{21} & 1 \end{pmatrix}^{-1} \begin{pmatrix} \alpha_1 & \alpha_{1a} & 0 \\ \alpha_2 & 0 & \alpha_{2b} \end{pmatrix}$$

$$= \frac{1}{1 - \alpha_{12}\alpha_{21}} \begin{pmatrix} 1 & \alpha_{12} \\ \alpha_{21} & 1 \end{pmatrix} \begin{pmatrix} \alpha_1 & \alpha_{1a} & 0 \\ \alpha_2 & 0 & \alpha_{2b} \end{pmatrix}$$

$$= \frac{1}{1 - \alpha_{12}\alpha_{21}} \begin{pmatrix} \alpha_1 + \alpha_{12}\alpha_2 & \alpha_{1a} & \alpha_{12}\alpha_{2b} \\ \alpha_2 + \alpha_{21}\alpha_1 & \alpha_{21}\alpha_{1a} & \alpha_{2b} \end{pmatrix}$$

Regression estimates of all path coefficients are then easily found.

NONLINEAR PATH REGRESSION

1 *The Structural Equations for Pathways Following the Single Process Law*

The ideas of path analysis presented in the last chapter can be extended to the case of nonlinear causal relationships between variables. In this chapter we will study the situation in which the relationships can each be represented by the single process law [8]. The general approach is valid for other relationships as well.

As in the linear case we describe the structure of a hypothetical network of causal pathways by a system of partial differential equations. In the linear case each of the partial derivatives are constant – the constants being termed "path coefficients". In the nonlinear case the analogues of the path coefficients (i.e. the partial derivatives) are not constant.

Let the matrix of partial derivatives be given, as before, by

$$H' = \left[\begin{array}{cccc} \eta_{1a} & \eta_{2a} & \cdots & \eta_{pa} \\ \eta_{1b} & \eta_{2b} & \cdots & \eta_{pb} \\ \vdots & \vdots & & \vdots \\ \eta_{1q} & \eta_{2q} & \cdots & \eta_{pq} \\ \hline \eta_{11} & \eta_{21} & \cdots & \eta_{p1} \\ \eta_{12} & \eta_{22} & \cdots & \eta_{p2} \\ \vdots & \vdots & & \vdots \\ \eta_{1p} & \eta_{2p} & \cdots & \eta_{pp} \end{array}\right] = \left[\begin{array}{c} H'_\xi \\ H'_\eta \end{array}\right]$$

and a matrix of rate coefficients be

$$\Delta' = \left[\begin{array}{cccc} \delta_{1a} & \delta_{2a} & \cdots & \delta_{pa} \\ \delta_{1b} & \delta_{2b} & \cdots & \delta_{pb} \\ \cdot & \cdot & & \cdot \\ \cdot & \cdot & & \cdot \\ \cdot & \cdot & & \cdot \\ \delta_{1q} & \delta_{2q} & \cdots & q_{pq} \\ \hdashline 1 & \delta_{21} & \cdots & \delta_{p1} \\ \delta_{12} & 1 & \cdots & \delta_{p2} \\ \cdot & \cdot & & \cdot \\ \cdot & \cdot & & \cdot \\ \cdot & \cdot & & \cdot \\ \delta_{1p} & \delta_{2p} & \cdots & 1 \end{array}\right] = \begin{bmatrix} \Delta'_\xi \\ \Delta'_\eta \end{bmatrix}$$

and diagonal matrices be

$$D_\xi = \begin{bmatrix} \xi_a & 0 & \cdots & 0 \\ 0 & \xi_b & \cdots & 0 \\ \cdot & \cdot & & \cdot \\ \cdot & \cdot & & \cdot \\ \cdot & \cdot & & \cdot \\ 0 & 0 & \cdots & \xi_q \end{bmatrix}, D_\mu = \begin{bmatrix} \mu_a & 0 & \cdots & 0 \\ 0 & \mu_b & \cdots & 0 \\ \cdot & \cdot & & \cdot \\ \cdot & \cdot & & \cdot \\ \cdot & \cdot & & \cdot \\ 0 & 0 & \cdots & \mu_q \end{bmatrix}$$

$$D_\eta = \begin{bmatrix} \eta_1 & 0 & \cdots & 0 \\ 0 & \eta_2 & \cdots & 0 \\ \cdot & \cdot & & \cdot \\ \cdot & \cdot & & \cdot \\ \cdot & \cdot & & \cdot \\ 0 & 0 & & \eta_p \end{bmatrix}, D_\alpha = \begin{bmatrix} \alpha_1 & 0 & \cdots & 0 \\ 0 & \alpha_2 & \cdots & 0 \\ \cdot & \cdot & & \cdot \\ \cdot & \cdot & & \cdot \\ \cdot & \cdot & & \cdot \\ 0 & 0 & \cdots & \alpha_p \end{bmatrix}$$

Let us now consider the multivariate analogue of the single process law [8] given by

$$H' = \begin{bmatrix} H'_\xi \\ H'_\eta \end{bmatrix} = \begin{bmatrix} (D_\xi - D_\mu)^{-1}\,\Delta'_\xi\,(D_\eta - D_\alpha) \\ (D'_\eta - D_\alpha)^{-1}\,\Delta'_\eta\,(D_\eta - D_\alpha) \end{bmatrix}. \tag{1}$$

The analogy in form to the univariate case [8] is evident.

A solution of (1) is

$$\begin{aligned}
\eta_1 &= \alpha_1 + \beta_1(\xi_a - \mu_a)^{\delta_{1a}}(\xi_b - \mu_b)^{\delta_{1b}} \ldots (\xi_q - \mu_q)^{\delta_{1q}}(1)(\eta_2 - \alpha_2)^{\delta_{12}} \ldots (\eta_p - \alpha_p)^{\delta_{1p}} \\
\eta_2 &= \alpha_2 + \beta_2(\xi_a - \mu_a)^{\delta_{2a}}(\xi_b - \mu_b)^{\delta_{2b}} \ldots (\xi_q - \mu_q)^{\delta_{2q}}(\eta_1 - \alpha_1)^{\delta_{21}}(1) \cdots (\eta_p - \alpha_p)^{\delta_{2p}} \\
&\cdot \\
&\cdot \qquad\qquad\qquad\qquad\qquad\qquad\qquad\qquad\qquad\qquad (2) \\
&\cdot \\
\eta_p &= \alpha_p + \beta_p(\xi_a - \mu_a)^{\delta_{pa}}(\xi_b - \mu_b)^{\delta_{pb}} \ldots (\xi_q - \mu_q)^{\delta_{pq}}(\eta_1 - \alpha_1)^{\delta_{p1}}(\eta_2 - \alpha_2)^{\delta_{p2}} \ldots (1).
\end{aligned}$$

Thus, Eq. (2) are the "structural equations" for the causal network arising from allowing each pathway to individually follow the single process law [8].

2 *Identification*

If one eliminates the η's from the right-hand side of (2) in the spirit of the preceding section to form "reduced equations", the problem of identification arises once more. Generally, one does not place restrictions on the scale factors α or μ when describing the causal network. When a particular pathway is nonexistent, this fact is represented by setting the appropriate exponent δ equal to zero. Hence, we will not be surprised to discover that the problem of identification primarily concerns these exponents, although the constants of integration (β's), signifying initial conditions of the system, also may be involved.

Let us transform (2) by taking logarithms. We set

$$\begin{array}{lll}
\eta_1' = \log(\eta_1 - \alpha_1) & \xi_a' = \log(\xi_a - \mu_a) & A_1 = \log \beta_1 \\
\eta_2' = \log(\eta_2 - \alpha_2) & \xi_b' = \log(\xi_b - \mu_b) & A_2 = \log \beta_2 \\
\cdot & \cdot & \cdot \\
\cdot & \cdot & \cdot \\
\cdot & \cdot & \cdot \\
\eta_p' = \log(\eta_p - \alpha_p) & \xi_q' = \log(\xi_a - \mu_a) & A_p = \log \beta_p
\end{array}$$

and $A_{ij} = \delta_{ij}$. Then (2) becomes

$$\begin{aligned}
\eta_1' &= A_1 + A_{1a}\xi_a' + A_{1b}\xi_b' + \ldots + A_{1q}\xi_q' + A_{12}\eta_2' + \ldots + A_{1p}\eta_p' \\
\eta_2' &= A_2 + A_{2a}\xi_a' + A_{2b}\xi_b' + \ldots + A_{2q}\xi_q' + A_{21}\eta_1' + \ldots + A_{2p}\eta_p' \\
&\cdot \\
&\cdot \qquad\qquad\qquad\qquad\qquad\qquad\qquad\qquad (3) \\
&\cdot \\
\eta_p' &= A_p + A_{pa}\xi_a' + A_{pb}\xi_b' + \ldots + A_{pq}\xi_q' + A_{p1}\eta_1' + A_{p2}\eta_2' + \ldots
\end{aligned}$$

Now the transformed structural Eq. (3) are identical with the corresponding linear structural equations.

Since this is true, we can apply all of the results of the earlier chapter regarding identification and reduction directly to the present problem.

3 *Some Examples*

In this section we will consider the same path diagrams as considered in the linear case.

a) The "multiple regression" analogue with diagram as follows

$$\xi_a \xrightarrow{\eta_{1a}} \eta_1 \xleftarrow{\eta_{1b}} \xi_b$$

The structural equation is then

$$\eta_1 = \alpha_1 + \beta_1(\xi_a - \mu_a)^{\delta_{1a}}(\xi_b - \mu_b)^{\delta_{1b}}, \tag{4}$$

b) The second example with diagram

$$\xi_a \xrightarrow{\eta_{1a}} \eta_1, \quad \xi_b \xrightarrow{\eta_{1b}} \eta_1, \quad \eta_1 \xrightarrow{\eta_{21}} \eta_2, \quad \xi_b \xrightarrow{\eta_{2b}} \eta_2$$

and structural equations

$$\begin{aligned} \eta_1 &= \alpha_1 + \beta_1(\xi_a - \mu_a)^{\delta_{1a}}(\xi_b - \mu_b)^{\delta_{1b}} \\ \eta_2 &= \alpha_2 + \beta_2(\xi_b - \mu_b)^{\delta_{2b}}(\eta_1 - \alpha_1)^{\delta_{21}} \end{aligned} \tag{5}$$

yields by substitution or by the methods of the last chapter the reduced structural equations

$$\begin{aligned} \eta_1 &= \alpha_1 + \beta_1(\xi_a - \mu_a)^{\delta_{1a}}(\xi_b - \mu_b)^{\delta_{1b}} \\ \eta_2 &= \alpha_2 + \beta_2\beta_1{}^{\delta_{21}}(\xi_a - \mu_a)^{\delta_{21}\delta_{1a}}(\xi_b - \mu_b)^{\delta_{2b}+\delta_{21}\delta_{1b}}. \end{aligned} \tag{6}$$

c) The feedback example

$$\xi_a \xrightarrow{\eta_{1a}} \eta_1, \quad \xi_b \xrightarrow{\eta_{2b}} \eta_2, \quad \eta_1 \xrightarrow{\eta_{21}} \eta_2, \quad \eta_2 \xrightarrow{\eta_{12}} \eta_1$$

produces structural equations

$$\begin{aligned} \eta_1 &= \alpha_1 + \beta_1(\xi_a - \mu_a)^{\delta_{1a}}(\eta_2 - \alpha_2)^{\delta_{12}} \\ \eta_2 &= \alpha_2 + \beta_2(\xi_b - \mu_b)^{\delta_{2b}}(\eta_1 - \alpha_1)^{\delta_{21}} \end{aligned} \tag{7}$$

and reduced structural equations

$$\eta_1 = \alpha_1 + \beta_1{}^{1/(1-\delta_{12}\delta_{21})}\beta_2{}^{\delta_{12}/(1-\delta_{12}\delta_{21})}(\xi_a - \mu_a)^{\delta_{1a}/(1-\delta_{12}\delta_{21})} \cdot (\xi_b - \mu_b)^{\delta_{12}\delta_{2b}/(1-\delta_{12}\delta_{21})}$$
$$\eta_2 = \alpha_2 + \beta_1{}^{\delta_{21}/(1-\delta_{12}\delta_{21})}\beta_2{}^{1/(1-\delta_{12}\delta_{21})}(\xi_a - \mu_a)^{\delta_{21}\delta_{1a}/(1-\delta_{12}\delta_{21})} \cdot (\xi_b - \mu_b)^{\delta_{2b}/(1-\delta_{12}\delta_{21})}. \qquad (8)$$

Comparison of the results for these three examples with the corresponding linear results will serve to illustrate the analogies. The important point to note is that, under suitable conditions, reduced structural equations can be found, even in this nonlinear case, such that each secondary variable can be expressed as an explicit function of only the primary variables.

4 *The Gaussian Iterant for Maximum Likelihood Estimators*

In considering estimation of parameters for the linear case, we restricted our attention to systems in which reduced equations could be found and in which a state of under- or just-identifications existed. We similarly confine our consideration to these situations for the nonlinear case.

After reduction we have p independent nonlinear regression equations, each of the form

$$\hat{y} = B_o + B_1(x_a - M_a)^{D_a}(x_b - M_b)^{D_b} \ldots (x_q - M_q)^{D_q}. \qquad (9)$$

As usual, we linearize by expanding in a Taylor's series and neglect terms beyond those containing the first derivatives. First, we must obtain trial estimates M_{ao}, $M_{bo}, \ldots M_{qo}, D_{ao}, D_{bo}, \ldots, D_{qo}$. We then compute

$$X_{10} = (x_a - M_{ao})^{D_{ao}}(x_b - M_{bo})^{D_{bo}} \ldots (x_q - M_{qo})^{D_{qo}}$$
$$X_{20} = X_{10}/(x_a - M_{ao}) \qquad X_{q+2,o} = X_{10}\log(x_a - M_{ao})$$
$$X_{30} = X_{10}/(x_b - M_{bo}) \qquad M_{q+3,o} = X_{10}\log(x_b - M_{bo}) \qquad (10)$$
$$\vdots \qquad\qquad \vdots$$
$$X_{q+1,o} = X_{10}/(x_q - M_{qo}) \qquad X_{2q+1,o} = X_{10}\log(x_q - M_{qo})$$

for all observational vectors.

In addition, we define

$$
\begin{aligned}
B_2 &= -B_{10}D_{ao}(M_a - M_{ao}) & B_{q+2} &= B_{10}(D_a - D_{ao}) \\
B_3 &= -B_{10}D_{bo}(M_b - M_{bo}) & B_{q+3} &= B_{10}(D_b - D_{bo}) \\
&\;\;\vdots & &\;\;\vdots \\
B_{q+1} &= -B_{10}D_{qo}(M_q - M_{qo}) & B_{2q+1} &= B_{10}(D_q - D_{qo}).
\end{aligned}
\tag{11}
$$

Now we have by use of Taylor's series

$$
\begin{aligned}
\hat{y} \doteq \hat{y}_o + \hat{y}_{Boo}(B_o - B_{oo}) + \hat{y}_{B1o}(B_1 - B_{10}) &+ \hat{y}_{Mao}(M_a - M_{ao}) \\
&+ \hat{y}_{Mbo}(M_b - M_{bo}) \\
&\;\;\vdots \\
&+ \hat{y}_{Mqo}(M_q - M_{qo}) \\
&+ \hat{y}_{Dao}(D_a - D_{ao}) \\
&+ \hat{y}_{Dbo}(D_b - D_{bo}) \\
&\;\;\vdots \\
&+ \hat{y}_{Dqo}(D_q - D_{qo})
\end{aligned}
\tag{12}
$$

and by substitution we obtain the linear equation

$$
\hat{y} = B_o + B_1X_{10} + B_2X_{20} + \ldots + B_{2q+1}X_{2q+1} \tag{13}
$$

as an approximate substitute for (9). By the usual methods of linear regression we now obtain the estimates $B_o, B_1, \ldots, B_{2q+1}$ and from (11) we get the improved estimates $M_a, \ldots, D_q$. These can now be used as our "trial" estimates and the entire process repeated until convergence.

If any subset of parameters is known *a priori* then, as in the simple regression case, the appropriate X's are simply omitted and the analysis carried out without them. For example, if it is known that all processes have vertical asymptotes zero then we omit X_{20} through $X_{q+1,o}$.

When there is more than a single response variable the application of (13) separately to each provides multiple estimates for the vertical asymptotes M_a,

$M_b, \ldots, M_q$. In order to provide unique, efficient estimates of these values we fit, instead of (13), the family of hyperplanes

$$\hat{y}_i = B_{0i} + B_{1i}X_{10i} + B_2X_{20i} \ldots + B_{q+1}X_{q+1,0i} + B_{q+2,i}X_{q+2,0i} + \ldots + B_{2q+1,i}X_{2q+1,0i} \tag{14}$$

where $i = 1, 2, \ldots, p$ and $B_2, B_3, \ldots, B_{q+1}$ are common for all i. Linear regression theory provides the minimum variance unbiased estimators for fitting (14).

Asymptotic tests and confidence limits are found as before by applying linear results to the final iteration.

REFERENCES

[1] Blalock, H. M., Jr. (Ed.), *Causal Models in the Social Sciences*. Chicago, Aldine–Atherton, 1971.

[2] Deming, W. E., *The Statistical Adjustment of Data*, New York: Wiley, 1943.

[3] Hood, W. C. and Koopmans, T. C. (Eds.), *Studies in Econometric Method*, Cowles Commission Monograph No. 14. New York: Wiley, 1953.

[4] Jencks, C., *Inequality: A Reassessment of the Effect of Family and Schooling in America*. New York: Basic Books, 1972.

[5] Kempthorne, O., *An Introduction to Genetics Statistics*, New York: Wiley, 1957.

[6] Tukey, J. W., "Causation, regression, and path analysis", *in* Kempthorne, O., *et al.*, (Eds.), *Statistics and Mathematics in Biology:* 35–66, Ames: The Iowa State College Press, 1954.

[7] Turner, M. E., and Stevens, C. D., "The regression analysis of causal paths", *Biometrics*, **15**: 236–258, 1959.

[8] Turner, M. E., Monroe, R. J., and Lucas, H. L., Jr., "Generalized asymptotic regression and non-linear path analysis", *Biometrics*, **17**: 120–143, 1961.

[9] Turner, M. E., Monroe, R. J., and Homer, L. D., "Generalized kinetic regression analysis: hypergeometric kinetics", *Biometrics*, **19**: 406–428, 1963.

[10] Turner, M. E., and Stevens, C. D., "The regression analysis of causal paths", *in* [1: 75–100], Reprint of 1959 Biometrics paper with a previously unpublished "Appendix on Exact Confidence Regions".

[11] Williams, E. J., *Regression Analysis*, New York: Wiley, 1959.

[12] Wold, H., "Causal inference from observational data: a review of ends and means", *J. Roy. Stat. Soc.*, Ser. A. (General), **119**, Pt. I: 28–61, 1956.

[13] Wright, S., "Correlation and causation", *J. Agri. Res.*, **20**: 557–585, 1921.

[14] Wright, S., "The method of path coefficients", *Ann. Math. Stat.*, **5**: 161–215, 1934.

[15] Wright, S., "The interpretation of multivariate systems", *in* Kempthorne, O., *et al.*, (Eds.), *Statistics and Mathematics in Biology:* 11–33, Ames: The Iowa State College Press, 1954.

SOCIAL NETWORK ANALYSIS: AN OVERVIEW OF RECENT DEVELOPMENTS

WILLIAM RICHARDS
Simon Fraser University, Burnaby, Canada

and

GEORG LINDSEY
Stanford University, Stanford California

Communication networks have been described by Pool [28; 3] as the "thread" that holds social systems together. An analysis of these networks can, therefore, provide a characterization of the system's structure. If techniques can be developed which allow descriptions of social systems based upon their communication patterns – patterns which are emergent, *a posteriori* system properties rather than imposed, *a priori* expectations – great improvements in methods of modeling large-scale systems may become possible. This paper describes a method for the analysis of communication networks (herein called NEGOPY – the *H* technique) which may address this consideration. Our discussion is presented as a general overview; more complete theoretical discussions can be found in Richards [30; 33; 34]; a very applied, practical discussion is found in Monge and Lindsey [25].

CONCEPTUAL FRAMEWORK

Several inherent problems exist in the analysis of communication nets. First, the size and complexity of the analytic problem pose a very real barrier to research. With 100 persons, for instance, each of the 100 could talk to 99 others. Thus, 9,900 possible connections exist. With a 5,000-person net nearly 25,000,000 possible links exist; if a full matrix were used to represent these contact patterns, the processing capacity of most present day computers would be exceeded. This difficulty has been overcome by an alternate conceptualization of the problem, which has allowed the development of a computer program (described later) which can handle over 4,000 persons.

A closely related second problem involves the different research strategies which have been used to handle this complexity. While many different methods of describing (or modeling) communication networks exist, there are few standards, or guidelines, for choosing the better or more appropriate of several methods. For example, Mears [22] has delineated one method of modeling communication structure in large organizations. He proposes but does not support the generalization that since most work is done in small five or six person groups, a large organization can be conceptualized as merely a collection of these smaller groups. To improve the communication and thus improve efficiency we merely examine and modify communication patterns within these small units. Notions such as the "wheel", "comcon", etc., are useful in such modifications. While this method does provide a simplification of sorts, it does so at the expense of throwing away a great deal of information, i.e. communication links to members of other groups. If this method could be legitimately applied, then generalizations from laboratory studies of communication nets could be utilized to improve communication flow in small groups.

Mears' treatment is one example of the many studies of this type which are based upon a paradigm roughly analogous to the mechanistic or reductionistic model of science. It assumes that understanding is possible by taking the process apart, looking at the separate parts, and putting it back together again. The necessity of looking only at the parts stems from the fact that the complexity of the whole, functioning system is far too great for existing analytic methods. Division into parts is relatively arbitrary, and all the information due to the interaction of the parts is lost. The *H* method, on the other hand, searches for parts (groups) which result from the application of a set of straightforward, explicit criteria to the particular system being considered. It does this by an examination of the total set of interactions among the elements as they function in the whole, operating system; an examination which is conducted independently of any prior expectations concerning the structure of the system. While there is as yet no accepted "standard" for social network analysis, we may suggest a set of criteria that appear to be useful in real-life situations and sensible in terms of the logical basis upon which they rest.

We suggest first that any such criteria must be applied to an *a posteriori* description of the system, i.e. the system *as it is*, rather than an *a priori* specification, i.e. the system as someone thinks it *should be*. Secondly, if they are to be "standards", these criteria must be explicit and complete. Perhaps one reason network analysis has remained more at the level of art than science is that previous conceptualizations have been ambiguous, thus requiring subjective decisions to be made during any application. Thirdly, the criteria should be formulated specifically to deal with the problems faced in the study of large, complex systems; forced adaptations of other less suitable methods of analysis will not suffice.

In delineating such criteria a standard strategy is to examine existing literature.

Massive amounts of empirical data have been gathered on communication networks. Two considerations, however, preclude the use of most of this information. The first is that most empirical investigations considered small groups of three, four, or five people. Not only is there no general agreement whether generalization is possible across these three group sizes [10], but even if there were, it is doubtful that these findings could be extended to systems having several hundred members. Five-person groups are simply too small to allow the kinds of things commonly observed in larger systems, e.g. hierarchical organization, to occur.

Secondly, according to Collins and Raven [10], an unfortunate state of affairs is quite prevalent throughout the entire network literature. They say, "It is almost impossible to make a simple generalization about any variable without finding at least one study to contradict the generalization. [10: 146]". We contend that one factor contributing to this equivocal state of affairs is an improper conceptualization of nonlinear dynamic processes as linear, static cause-and-effect relationships. A shift in analytic perspective may possibly rectify this situation.

In addition to the literature mentioned above, which results mainly from experimental investigations of communication networks [3; 4], another area of network research is provided in field/survey studies. The sociogram, developed by Moreno [26], has evolved into a number of techniques for the description of system structure. The major intent of a sociogram is to identify cliques or clusters of people who communicate primarily with each other. Closely related are Flament's [13] "kernels". Methods for locating the various parts or groups within a communication network may utilize graphic methods [26], matrix algebra [17; 12; 9; 21; 37], or formal graph theory [14; 13; 11].

A more general area of literature which does provide some insight into methods for describing large complex systems is systems theory. The description of systems theory given by Buckley [8] leads us to believe that this field may provide some guidelines in the area of articulating a network. According to Buckley, systems theory contends with:

> wholes and how to deal with them as such; the general analysis of organization – the complex and dynamic relations of parts, especially when the parts are themselves complex and changing and the relationships are nonrigid, symbolically mediated, often circular, and with many degrees of freedom; problems of intimate interchange with an environment, of goal seeking, or continued elaboration and creation of structure, or more or less adaptive evolution; the mechanics of "control" of self-regulation of self-direction [8; 2].

The notions of "wholes", "parts", and "structure", are, then, considered of primary importance by Buckley. Von Bertalanffy [7] defines general systems theory as "a science of 'wholeness' [7; 37]" which deals with "organized wholes". Similarly, Rapaport [29] cites as one element of four constituents of a system definition, "A structure, i.e. recognizable relationships among the elements which are not reducible to mere accidental aggregation of elements [23; 21]".

Systems theory has not presented a "new" concept in insisting that analysis proceed from emergent system properties; rather, systems theory has revived and revitalized an important concept which became apparent around the turn of the century. For example, discussions of the necessity of a "holistic" approach can be found in biology [7; 5; 6; 36], evolution theory [35], psychology [37; 18], personality theory [1], etc. The basic problem is well articulated by Smuts [35]:

> This system process cannot be fully defined unless the structure of the system is known; that is, until its fundamental component parts have been identified. However, these parts are neither unchanging or infinitesimal nor do they interact only in pairs. The unitary analysis of a complex system involves the identification within the whole, not of constant entities but of units of formative process, and even in the ultimate analysis these units have a finite extent both in space and time [35; 50].

However, a specification of exactly how one proceeds to find these "units of formative process" has not been adequately established. Indeed, Krippendorff [19] mentions this very problem of how the "parts" of complex systems are to be identified as one of the major issues facing systems theory.

The development of the technique may then be seen as a complimentary adjunct to systems theory. Systems theory provides some abstract notions of how complex organizations should be handled; the analysis provides one very specific method of handling a complex organization which takes into account some of these notions.

NETWORK ANALYSIS: NEGOPY – THE *H* TECHNIQUE

The critical distinguishing feature of the analysis is the method by which communication groups are identified. In this method no decision is made as to what constitutes a communication group (or clique) until the entire pattern of interrelations between individuals has been considered. Thus, if persons in the network left or were replaced, or if measures were taken at different points in time, different communication groupings would likely emerge.

Due to the fact that division into parts could only take place after descriptive data were obtained, and due to the fact that these groupings or structures would change as the system changed, this method of analysis is considered to more adequately reflect emergent properties of a system than techniques which merely impose a structure before analysis begins. We have seen that systems theory embodies a set of general guidelines for describing emergent properties of systems in discussions of "wholes" or "holism". In order to have emergent properties, a system must be characterized by some form of organization or order. The conceptualization of order as negative entropy suggested the name NEGOPY. The technique which employs the emergent principles has been called the "*H*" technique from this notion of "holistic" [20]. In the emergent or *H* technique, division into parts has been described as proceeding *a posteriori*. In other words, an *a priori* decision of how to divide the system into parts is inappropriate. First, all relationships in the organization must be considered; division may then proceed along lines which are appropriate to that organization. To find the communication groups or cliques within the network, a consideration must be made of all the persons interacting in order to describe (not prescribe) the structure which is present.

In this context, the analytic techniques presented here are well-suited for their task. The measurement process used is one that focuses on the relationships between individual members of the system. The data obtained describe the entire set of relationships among the members, in the context of the intact, functioning system. The analytic methods used were designed specifically for this kind of data, preserving intact units at multiple levels of analysis.

An exploration of the conceptual basis of the systems approach resulted in a confirmation of several ideas which appeared much earlier in the sociometric literature [36; 16; 26; 21]. For example, the model outlined here is roughly hierarchical, with the system as a whole being composed of groups or cliques, which are made up of sets of individuals working together. Individual people in the system can fill any of several roles in terms of the way they contribute to the overall functioning of the system. They can be isolates, for example, or participants of various types. Participants are either group members or linkers, i.e. liaisons or bridges [16; 37].

The underlying concept here is one of order or structure, in terms of a differentiation of the whole, into parts having specialized functions [27]. As mentioned earlier, this approach is not new. There have, however, been recent advances in an understanding of the nature of structure [2; 31], the kinds of things leading to the development of structure [24], and the ways in which structure can be studied [32].

Once the relevant systems concepts were clarified, their implications could be examined. These implications were found to be far-reaching indeed – demanding a radical shift in analytic techniques. This is so because the structural problem is basically a topological one, where the information describing the system in terms of components and subcomponents is clearly nominal data. This suggested that an analysis method based on a topological model would be better-suited, both conceptually and operationally, than traditional methods based upon distance paradigms (for example, multidimensional scaling techniques like factor analysis), which assume more than nominal data and produce other than topological representations of the system.

The first stage in NEGOPY, the computerized version of the *H* technique [25; 33], is a topological process, using many concepts drawn from classical sociogram analysis [16; 37], graph theory [12; 13; 14], matrix theory [9; 12; 21; 37], and set theory. These concepts are drawn together into a heuristic pattern-recognition algorithm, which produces a primarily topological solution [33].

After the structure of the system has been "mapped out", other, more conventional, statistical methods may be used to describe properties of various aspects of the system. We thus have a conventional statistical analysis imposed on a topology.

TOPOLOGICAL STRUCTURAL ANALYSIS

Our present analytic capabilities center around a cluster of specially designed computerized methods. The main computer program is NEGOPY, a network analysis program capable of efficiently analyzing data descriptive of systems having up to 4,096 members [33]. Since the program was based on an algorithm designed specifically for topological structural analysis of large complex systems, it produces results which are readily used by investigators of large systems, rather than results which must be forced into a topological format by complicated interpretative methods. The efficiency of this program is due to the fitting of the algorithm with the data analyzed, the analytic model, and the goals of the analysis. For this reason, NEGOPY is at least ten times as efficient for this type of analysis as most multidimensional scaling routines. In multidimensional methods, a Euclidian distance paradigm is utilized, and results which are very difficult to interpret are produced.[1]

The goals of the program are two-fold: (1) to produce a topological description of the network under investigation, i.e. a list of the groups in the system and a description of the roles of all the individual members in the system, and (2) to calculate a number of statistics descriptive of several parts of the system at various levels of analysis.

An explicit set of goals was needed in order to develop a computerized method of analysis. This explicitness was especially important for the structural aspects of the problem. The result of the reconceptualization is the following set of definitions and criteria:

I) *Nodes* may be of two types – *participants* and *nonparticipants*. Nonparticipants are either not connected to the rest of the network or are only minimally connected. They include:

A) Isolate type one. These nodes have no links of any kind.

B) Isolate type two. These nodes have one link.

C) Isolated dyad. These nodes have a single link between themselves.

D) Tree node. These nodes have a single link to a participant, and have some number of other isolates attached to them.

II) Participants are nodes that have two or more links to other participant nodes. They make up the bulk of the network in most cases, and allow for the development of structure. They include:

A) Group Member. These nodes have most of their interaction with other members of the same group.

B) Liaison type one. These nodes have most of their interaction with members of groups in general, but not with members of any one group.

C) Liaison type two. These nodes have less than half of their interaction with members of groups. These nodes usually have most of their links with other liaisons, and may thus be considered to be "indirect" liaisons.

III) To be called a group, a set of nodes must satisfy these five criteria:

A) There must be at least three members.

B) Each must satisfy the criterion for group membership.

C) There must be some path, lying entirely within the group, from each member to each other member. (This is called the connectiveness criterion.)

D) There may be no single node (or arbitrarily small set of nodes) which, when removed from the group, causes the rest of the group to fail to meet any of the above criteria. (This is called the critical node criterion.)

E) There must be no single link (or subset of links) which, if cut, causes the group to fail to meet any of the above criteria. (This is called the critical link criterion.)

The classification of the members of the system in terms of these specifications is accomplished by a two-stage process. First, an approximate solution is obtained by applying a pattern-recognition algorithm to the results of an iterative operation which treats each relationship (link) between a pair of nodes as a sort of vector. This representation is consistent with the topological model being used, since the vectors have two aspects: direction and magnitude. The "direction" of each vector is operationalized as a nominal variable indicating to whom the link goes, while the "magnitude" is operationalized as the strength of the relationship, i.e. the extent to which the behavior of the involved nodes is constrained or influenced because of the relationship. The result of this process is a tentative description of the system's structure. Because this method is an approximate heuristic method, rather than an exact mathematical method, the solution is only an approximation.

An exact solution is obtained by applying the various criteria described earlier to the tentative solution obtained in the first stage. This allows adjustment to an exact solution to be made. Again, several heuristic devices are utilized to maximize the efficiency of the algorithm.

STATISTICAL ANALYSIS

Once the structure of the system has been determined, the calculation of any desired statistics is straightforward. In network analysis, as in any other area, there is an unlimited number of statistics that could be computed for any given network, depending on the viewpoint of the observer of the system (the analyst) and his objectives. If progress is to be made in the understanding of networks and how they work, however, it is essential that the statistics used in one study be comparable to those used in others. For this reason a set of three types of *descriptive statistics* is suggested in [32] and briefly described here.

First is a set of *parametrics*, which are themselves not of direct interest, but which are used as "scale factors", allowing all networks to be described on the same scales in such a way that the values obtained will be absolutely comparable, regardless of the size (N = number of nodes) or linkage (L = number of links) of the system. The parametrics include relevant values for both size and linkage at each of three levels of analysis: the whole system, the group, and the individual node.

Second is a set of *completeness metrics*, all of which express some observed value in terms of a proportion of the maximum that value could take. Here the appropriate parametrics are used to standardize the calculation by defining the metric in the form $M = f(x, g)$; so that g is a parametric in the appropriate N and L; f is the equation for the particular metric, defined in terms of the parametric g; x is the set of relevant conditions specific to this particular situation; and M is the final value for the metric. An example of this form of a metric equation, together with a graphic representation of the results, is shown in Figure 1.

Included in the set of completeness metrics are: *connectiveness*, or density, the extent to which the members of a particular unit are linked to the other members of the same unit; *connectedness*, the extent to which this element is linked to other member elements of the same unit; *integrativeness*, the extent to which the units linked to this unit are linked to each other; and certain structural indicators, which refer to the extent to which constraint or differentiation is observed in various subsets of the system.

The third set of metrics all refer to the extent to which units vary in the degree to which they show some property. They are thus called *dispersion* metrics. There are two types of metrics in this class – the difference being found in the way the desired values are calculated. Those of the first type are all expressed as variances, calculated as mean squared deviations.[2] Those of the second type [cf. 24] are entropy- or uncertainty-measuring metrics, and are calculated as logarithmic information theoretic indices of distributional redundancy,[3] i.e. as indicators of the extent to which an event is predictable, given a description of either all occurrences of events or a set of past occurrences of events. The information theoretic measures are included in the set of dispersion

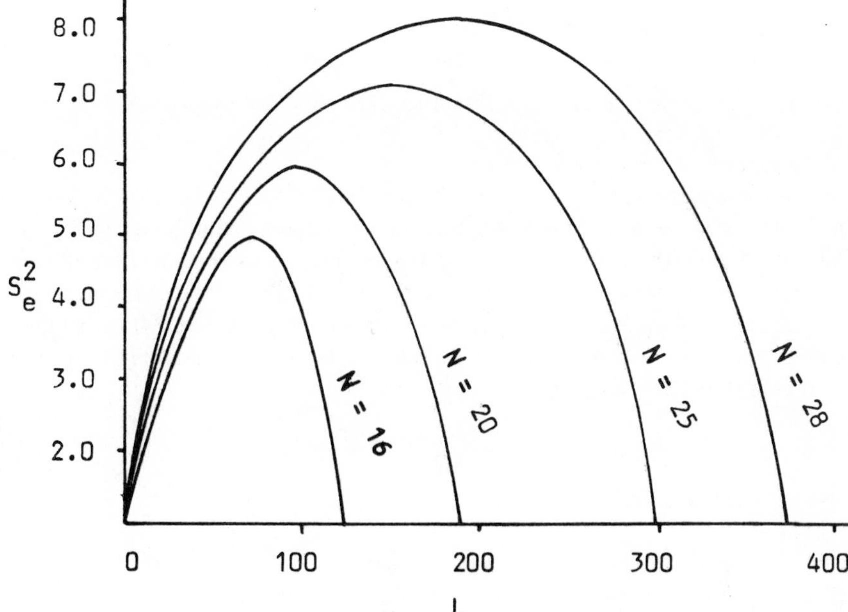

FIGURE 1a Plot of expected variance against number of links for Ns (number of nodes) of 16, 20, 25, and 28. Note that each network requires a new graph.

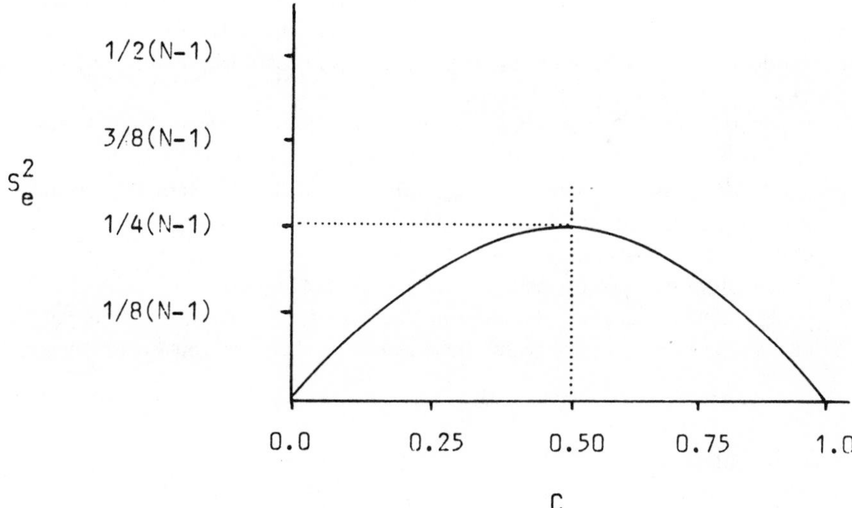

FIGURE 1b Generalized plot of expected variance against N and C (system connectiveness with respect to nodes). Note that the maximum value for S_e^2 of $\frac{1}{4}(N-1)$ is at the point where $L = \frac{1}{4}(N(N-1))$. At this point, the observed number of links will be one-half of the maximum possible, and C will be 0.5. Note also that *all* networks, regardless of size (N) and linkage (C), are described by this single graph. Thus, absolute comparisons between networks are possible with this form of description.

metrics because they refer to the extent to which relative frequencies of occurrence vary from event to event within the set of all possible events.

The set of dispersion metrics includes: the variance in the number of links each node has; the variance in the entries of a given row or column of the distance matrix for a subset of the network; the variance in row or column means for any distance matrix; the variance in the relative frequencies or strengths of the links to a given node; and so on. Also included are information theoretic measures of the extent to which the source or receiver of a given message is predictable; the extent to which the interactions among a set of nodes are dominated by a subset of these nodes [24], and so on.

INFERENTIAL STATISTICS

The metrics described above are all descriptive statistics, i.e. they are used to describe a system under investigation. In addition to the simple descriptive statistics is a set of statistics used for testing hypotheses of various types. These inferential statistics all make use of a model system of some type; for example, the network predicted by a random (unconstrained) model, or the network predicted by using another observed network as a model. Inferences are made by comparing some aspect of an observed network to the same aspect of a predicted network and testing the difference for significance. If the difference is significant, the model used to generate the predicted values is rejected as providing an explanation of the observed network.

Typically, these tests use either the t-test, for working with summary values, or the F-test, for working with variances. Comparisons that can easily be made by matching an observed network to a random one include tests of the variance in the l_i's (l_i is the number of links with node i) and the amount of constraint or structuring in the observed network.

Comparisons can also be made between a subset of an observed network (treating the subset as a sample) and the whole network (treating it as the population) on any dimension for which there is a value for each individual member.

CONCLUSION

We have described a method for modeling social systems which we feel tends to capture more emergent systems properties than prior conceptualizations. A needed next step in the development of this research program is to relate the endogenous variables described in this paper to exogenous factors. Thus the empirical utility of the H-technique must be demonstrated. Our preliminary applications, such as analyses of large organizations like banks and hospitals, have produced insightful and useful data concerning the functioning of these organizations. At an empirical, real-world level, then, utility seems promising.

The potential uses of network analysis are enormous. For example, in "satellite communication" network strategies may be useful in determining optimal locations for ground stations, i.e. perhaps ground stations should be placed within cliques, in order to minimize the cost of the more expensive terrestial links. Network analysis strategies may further refine notions of knowledge structures in society, and may eventually lead to more efficient human resource information retrieval. More scientific and precise descriptions of "invisible colleges" and related invisible institutions may be described and discussed. Thus with the refinement of these techniques we may be on the verge of an important scientific advance, i.e. new insights into the way organizations work may be possible.

Describing correspondences he has received, Senator Mondale notes the response from a prominent social scientist:

> The behavioral sciences, in my judgment, are in no real position at this point to give any hard data on social problems or conditions. There are many promises and pretensions; however, when it comes to delivery, what is usually forthcoming are more requests for further research . . . [15; 114–115].

It is our belief that this impotence has resulted partially from a misconception of social systems, and it is our contention that the techniques described herein may vastly improve methods for describing and analyzing such systems.

ENDNOTES

[1] For a more detailed comparison of analytic techniques see Richards [34].

[2] That is, as $S_x{}^2 = \dfrac{\Sigma(X_i - \bar{X})^2}{N}$.

[3] That is, as $H = -\Sigma p_i \log_2 p_i$ for absolute uncertainty, or as

$$H_{\text{rel}} = \frac{-\Sigma p_i \log_2 p_i}{\log_2 N}$$

for relative uncertainty.

REFERENCES

[1] Angyal, A., *Foundations for a Science of Personality*, New York: Commonwealth Fund, 1941.

[2] Barnett, G., "The nature of random structures", unpublished manuscript, Michigan State University, 1973.

[3] Bavelas, A., "Communication patterns in task-oriented groups", *Acoustical Society of America Journal*, **22:** 725–730, 1950.

[4] Bavelas, A., "A mathematical model for group structures", *Applied Anthropology*, **7:** 16–30, 1948.

[5] Bertalanffy, L., "Der Organismus als physikalisches System betrachtet", *Die Naturwissenschaften*, **28:** 521–531, 1940. Cited by L. Bertalanffy, *General Systems Theory*. New York: Braziller, 1968.

[6] Bertalanffy, L., "The theory of open systems in physics and biology", *Science*, **3:** 23–29, 1950. *In* Emery, F. E. (Ed.), *Systems Thinking*. Harmondworth, England: Penguin Books, 1969.

[7] Bertalanffy, L., *General Systems Theory*, New York: Braziller, 1968.

[8] Buckley, W., *Sociology and Modern Systems Theory*, Englewood Cliffs, N.J.: Prentice-Hall, 1967.

[9] Chabot, J., "A simplified example of the use of matrix multiplication for the analysis of sociometric data", *Sociometry,* **13:** 131–140, 1950.

[10] Collins, B. E., and Raven, B., "Group structure: attraction, coalitions, communication and power", *in* Lindzey, G. and Aronson, E. (Eds.), *The Handbook of Social Psychology*: 102–204, Reading, Mass: Addison-Wesley, 1969.

[11] Feather, N. T., "A structural balance model of communication effects", *Psychology Review*, **71:** 291–313, 1964.

[12] Festinger, L., "The analysis of sociograms using matrix algebra", *Human Relations*, **2:** 153–158, 1949.

[13] Flament, C., *Applications of graph theory to group structure*, Englewood Cliffs, New Jersey: Prentice-Hall, 1963.

[14] Harary, F., Norman, R. Z. and Cartwright, D., *Structural Models: An Introduction to the Theory of Directed Graphs.* New York: Wiley, 1965.

[15] Harris, F., *Social Science and National Policy*, Chicago: Aldine, 1970.

[16] Jacobsen, E. W., and Seashore, S. E. "Communication practices in complex organizations", *Journal of Social Issues*, **7:** 28–40, 1951.

[17] Katz, L., "On the matrix analysis of sociometric data", *Sociometry*, **10:** 233–241, 1947.

[18] Koehler, W., *Gestalt Psychology*, New York: Liveright, 1947.

[19] Krippendorff, K., "The 'Systems' approach to communication", *in* Ruben, B. D. and Kim, J. Y. (Eds.), *General Systems Theory and Human Communication*: 138–163. Rochelle Park N.J.: Hayden Book Co., 1975.

[20] Lindsey, G. N., *M. A. Thesis*, San Jose State University, 1974.

[21] Luce, R. D., "Connectivity and generalized cliques in sociometric group structure", *Psychometrika*, **15:** 169–190, 1950.

[22] Mears, P., "Structuring communication in groups", *Journal of Communication*, **24**: 71–79, 1974.

[23] Monge, P., "The study of human communication from three system paradigms", *Doctoral dissertation*, Michigan State University, 1972.

[24] Monge, P. R., "The evolution of structure", unpublished manuscript, Michigan State University, 1971.

[25] Monge, P. and Lindsey, G., "Communication patterns in large organizations", unpublished manuscript, San Jose State University, 1973.

[26] Moreno, J., *Who Shall Survive?*, Washington, D.C.: Nervous and Mental Disease Monograph, No. 58, 1934.

[27] Piaget, J., *The Construction of Reality in the Child*, New York: Basic Books, 1959.

[28] Pool, I., "Communication systems", *in* Pool, I. and Schramm, W. (Eds.), *Handbook*

of communication: 3–26. Chicago: Rand McNally, 1973.
[29] Rapaport, A., "Modern systems theory – an outlook for coping with change", *General Systems*, **15**: 15–26, 1970.
[30] Richards, W., "Communication networks", a paper prepared for the International Communication Associations's Communication Audit Project: December 1976.
[31] Richards, W., "Network analysis in large complex organizations: the nature of structure", paper presented to the International Communication Association, New Orleans, April, 1974.
[32] Richards, W., "Network analysis in large complex organizations: metrics", paper presented to the International Communication Association, New Orleans, April, 1974.
[33] Richards, W., *A Manual for Network Analysis*, a report of the Institute for Communication Research, Stanford University, June, 1975. Revised, Department of Communication, Simon Fraser University, Burnaby, B.C., 1978.
[34] Richards, W., *Communication Networks and Network Analysis*, Burnaby, B. C.: Department of Communication, Simon Fraser University, mimeo, 1977.
[35] Richards, W., Farace, R., and Danowski, J., "Negopy – program description", unpublished manuscript, Michigan State University, 1973.
[36] Smuts, J., *Holism and Evolution*, New York: Macmillan, 1926.
[37] Weiss, R. S., *Process of Organization*, Ann Arbor: University of Michigan, Institute for Social Research, 1956.
[38] Wetheimer, M. *In* Ellis, W. (Ed.), *A Source Book of Gestalt Psychology*, New York: Harcourt-Brace, 1938.
[39] Whyte, L., *The Unitary Principle in Physics and Biology*, New York: Holt, 1949.

WHERE IS MR STRUCTURE?

KENNETH D. MACKENZIE
University of Kansas, Lawrence and Organizational Systems, Inc.

INTRODUCTION

Basically, a system can be loosely described in terms of its entities or parts, the relationships between these entities, and the processes engaged in by the system. There are vast numbers of different systems and accordingly, large numbers of different types of entities, relationships, and processes. Structure is defined in terms of a set of entities and the relationships between these entities. We speak of structures and we study structures for many reasons. We believe that structure can summarize many salient aspects of a system, as in the rank structure of an organization. We feel that the structure is an aggregate characteristic of a system and that understanding structure can provide clues about the effects and determinants of aggregation. We also harbor the assumption that in some cases the aggregate exhibits behaviors or processes that are in some sense different from what one would expect from the disjointed analysis of the entities. Structures serve as a theory by which we organize our analyses of a system. The study of structure has also become important because of the many studies on structure. Structures are a perceived reality and we seek to understand how they form and why they change.

Structures can have substructures and any given system can have many structures. Within any given system, the set of entities for the structure for one purpose may be different from the set of entities defined for another purpose. Most processes can be described in terms of stages and relationships between stages. These are also structures. Furthermore, as we shall show, the structure of a group process has entities that are themselves structures defined in terms of specific types of interactions among the group members.

It seems clear to me that the concept of structure is important and that it is worthwhile to study structures by inventing theories and subjecting these to experimental test. While my concern is for human groups and studying a theory of structural change for groups, these studies are conceptually just a special case of a much larger class of structures. Despite these limitations on the scope of this work, I have reached a number of methodological conclusions that have broader theoretical implications. Let us examine a few of these methodological issues before discussing some ideas in a theory of group structures.

1) There is a useful set of relationships between theory, problem, purpose, measures, models, conditions under which a study is made, and the results and implications of the study. There is, in other words, a need to structure the analysis of structures in specific terms. Most analyses of structures are unstructured and naive. It is rare for an author to state his theory, define his problem, state his purposes, define his measures, describe his model, specify his conditions, label his results, and draw theoretical implications to his results. In most studies there is no close relationship between the concept of structure introduced in one part of an article and the measures used in the study. There is a strong tendency to proceed with measurement before defining one's terms. It is difficult to see how results can be cumulated about structure when each analysis is unstructured within itself and in relation to the many other studies about structure.

2) There is a basic interchangeability between the entities of a structure and the relationships between these entities. Even in the "simple" case of a steel bridge where there is a physical reality to a structure, there are problems. For example, consider a bridge as an abstraction consisting of girders and rivets. Which are the entities and which are the relationships? In one conception, the rivets connect the girders. In this case the girders are the entities and the rivets are the relationships. In another concept, the girders connect the rivets in which case the rivets are the entities and the girders the relationships. Yet in another view the relationships are mechanical forces and both the rivets and the girders are the entities.

It is not clear from any general definition of structure which is the set of entities and what are the types of relationships that serve to connect them. Going beyond the definition requires a concatenated sequence of conventions, methodological and theoretical commitments and specific decisions that are dependent upon the purposes, theories, and special conditions for making the study.

Aside from bridges, most social structures do not have "objective" physical reality. The set of entities, for example, persons in a group, may be "real" but usually the relationships are less tangible. The measured and perceived relationships exist for an interval of time because of the categories and observational lens that we employ to examine them. In a human group, we may speak of authority and influence as the relationships but how can we touch and see them except as the abstraction that they represent? When we speak of a hierarchy or perhaps a centralized system, it is very difficult to sense the physical reality of such structures. The conceptual abstraction called structure can capture properties of interaction sequences and can be employed to encapsulate some properties of a system. However, such structures are ephemeral and may not even exist at a given instant in time.

STRUCTURE AS AN EXPLANATORY VARIABLE

Any given system can have many structures and each structure its own substructures. And, within a given system, the structure for one purpose may be very different from those defined for another. In fact, the identity of the entities for one structure may itself be composed of structures of different types of entities. We can devise different measures of structure and we can build different theories about structures. However, there are always the very difficult conceptual problems of relating the abstract concepts of structure to another set of abstractions called behavior. Often the only bridging principal is the mind of the analyst.

We are often asked to believe that, somehow, an entity called structure can determine behavior. I have searched in vain to meet this Mr Structure. I am told about him repeatedly. I am impressed by his ubiquitousness. I can infer some of his traits. He is described using any number of traits. For example, Mr Structure is strong–weak, open–closed, rigid–loose, friendly–unfriendly, democratic–authoritarian, human–inhuman, tall–flat, centralized–decentralized, large–small, lean–fat, directive–laissez faire, costly–cheap, and efficient–inefficient. I have searched for but never met this metaphysical being who can decide and cause so many things.

Now, it is a simple matter to demonstrate that a concept of structure can be used in order to summarize certain aspects of behavior. It is quite another task to show exactly how Mr Structure can cause behavior. If I may be permitted an analogy with mechanics, I cannot understand how if I "move" Mr Structure, a certain specific response or sequence of responses will occur. After over a decade of search to meet Mr Structure, I have not been able to find him. Perhaps Mr Structure is like the Loch Ness monster in that failure to find him does not establish his nonexistence. However, I have come to the conclusion that we don't really need him for scientific purposes. Moreover, I believe that the repeated conjuring up of his presence stems from either laziness or a fundamentally incorrect conception about living systems.

My quarrel with Mr Structure as an explanatory variable is deeper than whether or not a specific representation of him is or is not a good predictor of some behavior. I shall get around to this type of question later. Here, I am more interested in the efficacy of the type of reasoning which, for a complex system of behaving entities, posits that one can extent stimulus-response analysis to these classes of problems without substantial revision. We all understand the idea that in some cases there is a set of phenomena called stimuli that cause a set of responses. That is, there is a function of some form that maps elements from the set of stimuli to the set of responses. When it is also believed that the stimuli cause the responses, we have a stimulus-response explanation. Further, if over time the basic function relating the stimuli to the responses does not change, we have what can be labeled a *passive process*. This is opposed to what can be labeled an *active process*, wherein the function and stimuli, and the responses can change. It is possible for passive processes to be dynamic as in the case of a path of a torpedo. The same laws are governing the trajectory along its path. However, if the torpedo can alter its path according to contingent signals, the trajectory becomes a more active process. Human groups are very active.

One can certainly see cases where the assumption of simple passive process stimulus-response models are useful. For example, the study of short-term duration responses to simple stimuli (such as a rap of a hammer just below one's knee causing an involuntary movement of one's leg) can be approximated by a passive stimulus-response model. The study of closed systems may provide other cases. The study of physical systems may also involve mainly passive processes. However, human groups who interact over a longer time period and who have to cope with many exogeneous and endogeneous contingencies are highly active. It seems reasonable to argue that we should use active process models to study active process behaviors. As I see it, the main problem with the concept of a Mr Structure is that explanations based upon him are inherently process passive. It is also reasonable to suggest that it is not sensible to postulate passive process explanations for active processes.

THE EXISTENCE OF MULTIPLE STRUCTURES

Persons in a group employ a variety of media in order to interact on a larger variety of topics. They will selectively employ different media and channels to interact with different people on different issues. For each issue and time interval, the pattern of the interaction of a group can be defined. This interaction pattern for the group is called a *group structure*. There are many issues and topics and the set of participants interacting about each can vary with the topic. Hence, any group can have many different group structures during any given time interval and different sets of structures for different time intervals. Furthermore, these group structures can and do change over time, even for a specific topic.

Even in a little organization, existing for a few short hours within the confines of a laboratory setting, there are a multitude of different structures. They also change these structures in amazingly precise ways. Larger scale organizations such as markets also display these same properties of multiple structures and change. Thus, it does not make a great deal of sense to speak of group structure as if it were an entity, Mr Structure, that is a cause of behavior. It is more accurate and useful to see the different structures and to recognize their tendency to change. These structures describe behavior rather than causing them and the more fruitful theoretical questions involve changes in these structures. While behavior can affect behavior, if one focuses upon how behavior affects behavior, there is no logical, empirical and theoretical reason, outside of custom and the restrictions imposed by systems of thinking, to invoke Mr Structure as the prime "mover and shaker".

To this writer, Mr Structure's status as an entity capable of causing behavior has about the same status as phlogiston, the ether, and the funiculus. Mr Structure is an imaginary theoretical construct existing in many minds. That is the answer to the question, "Where is Mr Structure?" It is the belief in Mr Structure and the resulting organizational deformations such as creating an authority system before specifying the purpose and functions of the system that provide a stimulus to many organizational problems. Despite this, relying on Mr Structure to explain behavior takes our attention away from the scientific problems about group structure. Furthermore, widespread misconceptions about structure can cause more problems. We should try to get to the source of the phenomenon.

However, because so many social scientists have for so long relied on Mr Structure to "explain" troublesome data it is necessary to offer alternative theoretical formulations for the phenomenon in whose name he has been evoked so many times. This is, of course, an extremely difficult task which has been the main goal of my research since 1963. I have just completed a two volume work on this topic [2; 3] and I am currently working on a third volume. I suspect that there will also be more than three in the future. I should like to share some of these ideas in the remaining portion of this paper. M. Lippitt and I discuss an application of these ideas in developing "A Theory of Committee Formation" [1, in this volume]. This companion paper illustrates one of the many future applications that will be in the third volume. It also describes some of the problems that arise because of Mr Structure type of thinking. The pair of papers, taken together, is an introduction to some of the ideas that have been developed as a replacement for Mr Structure.

A CONCEPT OF STRUCTURE

We begin by considering two entities, x_1 and x_2 who interact on an abstraction called a channel. There are two parts of the channel between x_1 and x_2. There is the half-channel over which x_1 directs his interactions to x_2 and there is the half-channel over which x_2 directs his interactions to x_1. A channel can be conceived of as a medium by which two persons interact. There can be different channels for different types of interaction. To start with, let us assume that we are speaking of only one medium and one specific topic. We shall relax these restrictions later.

Let $X_n = (x_1, x_2, \ldots, x_i, \ldots, x_n)$ be the list of n participants in a group who are interacting. For each of the $n(n-1)/2$ possible pairs of channels there are two half-channels. Let ij designate the half-channel from x_i to x_j and ji designate the half-channel from x_j to x_i. We can define a relationship between any pair in X_n on the half-channels. Thus, the relationship between x_i and x_j is designated by r_{ij} for ij and r_{ji} for ji. We can describe all $n(n-1)$ relationships in a square array called a group structure $S = (X_n; R)$ where R is an nxn matrix of the r_{ij}'s. We write R as

$$R = \text{“Sender”} \begin{array}{c} \\ x_1 \\ x_2 \\ \cdot \\ \cdot \\ \cdot \\ x_i \\ \cdot \\ \cdot \\ \cdot \\ x_n \end{array} \overset{\text{“Receiver”}}{\begin{array}{c} x_1 \quad x_2 \ldots x_j \ldots x_n \\ \left[\begin{array}{cccc} r_{11} & r_{12} \ldots & r_{1j} \ldots & r_{1n} \\ r_{21} & r_{22} \ldots & r_{2j} \ldots & r_{2n} \\ \cdot & & & \\ \cdot & & & \\ \cdot & & & \\ r_{i1} & r_{i2} \ldots & r_{ij} \ldots & r_{in} \\ \cdot & & & \\ \cdot & & & \\ \cdot & & & \\ r_{n1} & r_{n2} \ldots & r_{nj} \ldots & r_{nn} \end{array}\right] \end{array}}$$

There are many possible lists X_n and many possible tables of relationships, R. In general the task of choosing X_n and R depend upon the theory, purpose, codes, measures, models, the time interval, and the resources of the analyst. One example is the five person group who exchange data in an experiment. Here X_n becomes the list of the five persons and each r_{ij} could be the number of data messages sent from x_i to x_j. There are many other specifications possible. However, I don't want to dwell on these problems here.

STRUCTURE AS A NEED SATISFYING INTERACTION PATTERN

There is a curious feature of $S = (X_n; R)$ which is the key to understanding group structure. I call this *absolute control*. Examine the first row of R. This row consists of the half-channels originating from x_1. Person x_1 can always open the half-channel to any other in X_n if he wants to badly enough. Person x_1 can always send to any other. He may not be able to get x_2 to listen and he may suffer great hardships in opening a half-channel but he can open it and he can use it. The receiver may choose not to respond and he may raise the costs to x_1 for exercising his right. But if x_1 wants to use any of his half-channels he can.[1] Similarly, x_2 can do the same for his half-channels (represented by the second row in R). In fact, each $x_i \in X_n$ *can* use any of his half-channels. We say that each $x_i \in X_n$ is an absolute controller of his half-channels.

If we allow a group to interact over a series of related problems, they often develop stable interaction patterns. An interaction pattern is called stable if the entries in R do not change over a specified series of problems. This stability represents a *consensus* on how to interact. It can be broken if any member desires to change it. Each of the n absolute controllers has to agree to open and close those half-channels under his control. To achieve a stable structure, there are $n(n-1)$ decisions that have to be consistent. Each structure represents $n(n-1)$ decisions. The stable ones are the most interesting because they raise questions about how and under what conditions, the agreements are struck among the n absolute controllers. The key phenomenon in a theory of group structures is change or no change in R. The active processes by which R is changed obey remarkable orderliness and can be modeled, controlled, and analyzed with accuracy, at least for the studies that I have conducted. Given the absolute control of each member, the consensus must satisfy member needs if it is to be stable. For if it does not satisfy needs, it can be changed. This is why it is useful to consider a structure as a need satisfying interaction pattern.

There can also be more than one structure operating in a group during any given time period. These different structures can have different characteristics. For example, it is not uncommon for a group to be decentralized for one structure and fully centralized on another. It is also possible to have operating structures, say for socializing and play, that involve a subset of the X_n that are working on completing a task.

STRUCTURES AND PROCESSES

The concept that a group can have a number of structures operating within a given time interval raises a number of questions. Do the processes for structural change apply to all of the structures? Best evidence supports the affirmative. Do all structures change at the same rate? The evidence here supports the negative. Among the set of possible questions is whether or not there are some systematic relationships among these structures and task processes. There are but they are not the passive stimulus-response relationships incorporated in Mr Structure type explanations. The relationships between structures and task processes are of particular relevance to this paper and the one by Lippitt and Mackenzie. Let us consider them in more detail.

Suppose that we are interested in the structures of a problem solving group. We can describe the problem solving task in terms of sequences of transitions between states called *group milestones*. Group milestones represent stages of the progress of a group from the initial presentation of a problem to the completion of an acceptable end product. The nature of the problem constrains the definition of the milestones and the specific description of the milestones varies with the type of group problem being solved. These milestones are like stepping stones across a creek where one travels from one side to the other by reaching the stones in some order. In our case of groups, each milestone corresponds to a phase or stage of the problem solving processes. In order to move from one milestone to reach another, the group must behave. When the group members behave they interact. For each group milestone there is a pattern of interaction that is relevant to the reaching of the specific milestone. Consequently, there is a relevant *group milestone structure* for each group milestone. In order to make a judgment for *reaching* a milestone we state a criterion to the group milestone structure. If this criterion is met, we state that the milestone has been reached. We can note the time and message number for each reaching.

Let $M = (M_1, M_2, \ldots, M_m, \ldots, M_F)$ be the list of group milestones for a specific problem where M_1 is the first milestone at the beginning of the problem solving process and M_F is the final milestone reached at the end of the problem solving process. The list of milestones, M, can be ordered so that the subscript becomes larger as the task process gets closer to the final milestone, M_F. A *task process structure*, S_M, can be defined where $S_M = (M; T)$ where T is the table of possible transitions from one milestone to another. An entry t_{ij} in T is unity if it is possible to proceed from milestone M_i to another M_j, and zero otherwise. One interesting feature of the task process structure, S_M, is that the entities in S_M are group milestone structures. Thus, S_M is a structure of structures.

STRUCTURES, PROCESSES AND ROLES

The actual task processes of a group can be described using the group milestone structures and the sequence in which they are reached. We can differentiate the behaviors of the group in terms of whether they are (a) timely, (b) untimely, and (c) redundant. A behavior is *timely* if it is not redundant and not untimely. A behavior is *untimely* if it occurs after the relevant group milestone has been reached. A behavior is *redundant* if it is not untimely and other than the first occurrence of a timely behavior. A timely group milestone structure's matrix only has 0 or 1 entries. The second, third, etc., occurrences of a behavior can be placed in a redundant group milestone structure. Similarly one can make a structure for the untimely behaviors for each group milestone.

By suitable rewriting of the group milestone structures one can define timely, untimely, and redundant group role matrices. A *group milestone role matrix* has its rows defined by the participants and the columns by the half-channels. The various group milestone role matrices can be adjoined to form a *group role matrix.* The group role matrix can be subdivided into its timely, redundant, and untimely parts.

One can make analyses of each role matrix in order to compact it, reorder it, and to do special calculations such as assigning responsibility. The group role matrix can also be used to determine the level of each person and relationship such as immediate *superior-subordinate*, "uncle" relationships between adjacent levels that do not follow the immediate superior-subordinate relationships, and within level relationships, called *cousin* relationships. Out of this, one can calculate measures for the degree of hierarchy and efficiency that have proved invaluable in analyzing group processes. Recent work, not in the two cited volumes, has extended these ideas to studying group disturbances caused by member turnover, and "meta" processes of change in the task processes.

The multivariate, dynamic concept of group structures has led to the uncovering of striking relationships between structures, roles, and problem solving processes that have proven useful in laboratory studies. These have led to experiments and these to improved theory. One interesting result is the close (correlations average about .95 across 13 studies) relationship between our process based concept of hierarchy and the degree of efficiency. Furthermore, the degree of hierarchy which gives so much specific detail on the task processes, is less than or equal to the efficiency. This in turn suggests a new array of approaches to the study of system design. For one thing, the relationships between hierarchy and efficiency and between hierarchy and process and structure suggests an optimization principle: maximization of the degree of hierarchy. I am actively engaged in applying these ideas to the design of organizations.

STRUCTURAL CHANGE

The structures and group role matrices can be employed to provide a baseline from which to assess change, study its determinants, and to test various possible change processes. By definition, a timely group role matrix, $R_T{}^*$, has zero or one entries. An entry, $r_{ia}{}^*$, is one if person x_i performs activity a and is zero otherwise. The matrix, $R_T{}^*$, can be defined for any given time interval and compared to the matrix in another time interval. Some of the entries remain the same and others change from zero to one and some change from one to zero. The nature of a change in task-oriented group behavior can be quickly spotted simply by comparing a pair of timely role matrices. One can also examine changes in redundant and untimely group role matrices.

Having pinpointed the changes, it is possible to trace them back to changes in each group milestone structure. Then one can posit mechanisms of change processes and examine several possible determinants for the rate and timing of specific changes that are of theoretical or empirical interest. The theoretical development of some ideas for this type of problem occupies four chapters of [2] and the testing of these ideas four chapters of [3]. While it is clearly infeasible to go into the details and specific arguments for these developments in a brief introductory paper, it is still possible to sketch a few of the ideas. These include behavioral constitutions, adoption-diffusion models and mapping functions.

Behavioral Constitutions

Behavioral constitutions are essentially algebraic influence information processing devices invented to describe the manner in which interacting groups process influence attempts. Several were inferred by laborious analyses of the actual sequence of messages exchanged by subjects in laboratory experiments. The author would see a message sequence and then the result in terms of a change or no change in the entries of the timely group role matrix. He would then examine another sequence and another change. Slowly, over a period of several years patterns began to emerge and these were generalized into a rule system that the subjects seemed to follow when they processed influence attempts. The resulting system is called a behavioral constitution. These behavioral constitutions were then tested by examining data previously collected and by conducting new studies. The result has forced the author to consider radically new (at least to him) concepts of how problems get solved. The outputs from a behavioral constitutional analysis of data also led to models and new ideas for change processes. For example, data from the constitutions are used as inputs for models predicting the timing and rates of change in group structures. It also became apparent that previous emphasis on teleological or hedonic models is not really necessary to explain the data and, in fact, seriously hamper the analysis by imposing unrealistic data requirements.

There are probably many different behavioral constitutions. I report two in [2] and one in [3]. I shall briefly describe a behavioral constitution for changes in the *timely group role matrix*, $R_T{}^*$. The entries in $R_T{}^*$ that are being discussed by the group are the *issues*. Those members whose behavior can be affected by the changes are the *controllers*. The list of controllers is the *control set*. Only members of the control set are *eligible* to vote and only eligible votes are processed. There are three types of votes: explicit votes, implicit votes, and preemptions. *Explicit* votes are influence attempts that are explicitly stated. *Implicit* votes are less direct and are often made by performing or not performing an activity. *Preemptions* are influence attempts which, by themselves, can determine the state of a set of entries in $R_T{}^*$.[2] It is useful to classify votes according to whether they are made by a person or received from another. Each entry in an issue of $R_T{}^*$ can be classified according to whether the state is known or in doubt. If a state is known we say that it has been *elected*. If a state is in doubt, we say that it is in *recall*. A state is elected "open" if the corresponding entries are 1 and "closed" if zero. Recalls can be recalls to open or recalls to close. The state of any issue is assumed to begin in an elected state. Any vote that causes this elected state to be in doubt is a recall. The effect of a recall is to set off an election to remove the doubt. There can be long sequences of elections for any given issue. There are *voting rules* for determining the state of an issue. Voting rules and definitions are governed by a set of axioms. The axioms can be employed to derive a set of outcomes for any sequence of votes. Persons may attempt to alter the issues by *amendment* in order to change the control set.

Ordinary voting systems used at meetings are seen to be special cases wherein at some point in time formal votes are recorded and compared with a previously agreed decision rule such as majority. Such procedures are allowed by the behavioral constitution. However, a behavioral constitution would also analyze the influence attempts before, during, and after the meeting. It also permits many different issues to be considered simultaneously.

Data from experiments do not infirm the behavioral constitution and employing one allows quite precise analysis of the behavior. The record of votes can also serve to analyze other processes. For example, if we know the outcome of a voting sequence, we can compare each person's votes with the outcome. Votes that are consistent with the final outcome are called *favorable*. Votes that are inconsistent with the final outcome are called *unfavorable*. The relative intensity of favorable and unfavorable votes can be used to study the rate of adoption of a structure, the relative influence of each member and perhaps leadership processes. Lippitt and Mackenzie will employ an extension of a behavioral constitution called a treaty, in order to frame our analysis of the active process of committee formation.

Rate of Adoption of Structure

Experimental data demonstrate that some groups adopt their "stable" structures more quickly than others. Some take a couple of minutes and some take hours. This variation in adoption rates can be explained by combining an analysis of change with relative intensity of favorable and unfavorable votes. The result of an analysis of change of group structures is a diffusion model whose main parameter is the slope on semi logarithmic paper. The relative intensity of favorable and unfavorable votes are considered the main determinants of the variation in these slopes. This model does a reasonably good job in explaining these data. Mackenzie and Barron [4] use a similar model to analyze the acceptance of optimal operations research techniques in business game teams.

The analysis of change comes from a mathematical framework which has several distinct subproblems. The first subproblem is to decide what is actually being changed. In the case of structural change, we use the states of open or closed half-channels. For example, a 5-person all-channel has 20 open channels. If this group becomes a wheel then there are only 8 open half-channels. The change in this case is the closing of 12 half-channels. The second decision is the set function to apply to the changes. One can use Boolean or non-Boolean set functions. In the case of structural adoption, we use a 0 or 1 set function. In other cases we may decide to impose other set functions. We may also weigh some behaviors more than others. The result of the first two decisions is a number. Each sequence of behaviors can generate a series of such numbers. Our task now is to invent a process model whereby these numbers can change. The result is a process model. For the case of structural change, we derived a diffusion model. This process model will have certain parameters that need to be interpreted with respect to the phenomena being studied. In our case, the main parameter is the slope of the diffusion curve. The next decision is to observe variations in the parameter and to postulate a set of determinants for these changes. Then we compare the results from the models with actual data. In many cases, there will be discrepancies that are sufficient to force a reexamination of the whole process of formulation. In other cases the model can result in a "good" fit. If this fit is too good, one should reexamine the data and models. This can often be done by conducting a new study. This procedure is explained in Chapter 8 of [2] and illustrated in Chapters 12 and 15 of [3].

Predicting the Occurrence of Structural Change

Models for the rate of adoption of a group structure do not specify exactly when a structure is adopted and what structure becomes adopted. The rate of adoption tells us how fast the group adopted its structure in relation to when the change process began. The change processes do not all start at the same time and the rate of adoption apparently depends upon the relative intensity of favorable and unfavorable votes.

However, one can observe that a group structure became centralized on problem 5, that it remained centralized until problem 8, and then became noncentralized on the next set of problems. Clearly, we need to be able to explain the "twists and turns" of group structure as the groups evolve and face new contingencies. This explanation of when a structure is going to change (or not change) must be made in terms of the existing conditions in a group.

There are many groups facing quite different conditions. This diversity makes it very difficult to construct models that fit all possible cases. However, it is possible to construct models that can be used in any given case. As in the case of the analysis of change, the most important problem is to develop ways of thinking about problems. As I understand these problems, we need to discover methods of analyzing active processes and to move away from the more traditional stimulus-response formulations. The method of positing a set of "independent" variables that cause "dependent" variable changes is simply inadequate for tracking the many transitions and new contingencies that arise during the interval between the "stimulus" and the "response". I see a long sequence of stimuli and a whole class of contingent responses when I observe group behavior. While in some cases variable x affects variable y, in a group process there is usually more than one x and y and the variables that are relevant depend upon the sequences of behavior. I am arguing that we need newer concepts of active processes rather than patching up older ones with multivariate analyses. The multivariate analyses just make the stimulus and the response more complex but retain the basic passive process stance. They do not examine the problem that once a process gets going it may invoke its own variables and that the relative importance of each is contingent upon the past and anticipated behavior sequences.

Consider a group of five persons who interact in a communications network environment to solve a preset series of problems. The task of the group can be described by the sequence of milestones. They have another task of deciding how to organize themselves. Assume that they, as a group, want to solve these problems as fast as possible. Assume further that the milestones can be grouped into two phases: data and answer formation. We observe that all such groups either centralize or not. Our problem as analyst is to be able to predict these outcomes on problem $T + 1$ in terms of a set of other variables that we know from problem T. That is, there is a $y(T + 1) = \langle y_1(T + 1), y_2 \rangle$ where

$$y_1(T+1) = \begin{cases} 0 \text{ if the structure is predicted to be non-centralized on problem } T+1 \\ 1 \text{ if the structure is predicted to be centralized on problem } T+1 \end{cases}$$

$$y_2 = \begin{cases} 0 \text{ if the data phase structure is being predicted} \\ 1 \text{ if the answer phase structure is being predicted} \end{cases}$$

We set a set of variables $X(T)$ and a model f such that

$$y(T + 1) = f(X(T)).$$

The vector $X(T)$ has eight components: (i) $x_1(T)$, the phase on problem T, (ii) $x_2(T)$, the current structure, (iii) $x_3(T)$, the voting history, (iv) $x_4(T)$, the current state of the election outcome, (v) $x_5(T)$, the feasibility of the recall state, (vi) $x_6(T)$, the elected state of the structure, (vii) $x_7(T + 1)$, the predicted capacity for change on the next problem based upon performance on previous problems, and (viii) $x_1(T + 1)$, the phase for which the prediction is to be made on the next problem. Each of these variables has specific definitions and conditions for which it takes on different values. For example, $x_7(T + 1)$ depends upon a whole series of calculations on member capacity to process information (the problem of maximum span of control).

The logic of the variables and specific situation suggests that there is a set of contingent relationships among these variables. These contingencies allow one to construct a network whose vertices are the variables in $X(T)$. The network contains 131 possible paths which is much smaller than the 2^8 possibilities (each $x_i(T)$ is defined to be binary). The function, f, becomes the network. The function, f, is a type that I call a *mapping function*.

A mapping function is a generalization of a stimulus-response function in that (1) the relationships among the "independent" variables are explicitly stated and made a part of the function, (2) the stimulus becomes the path through the network, and (3) different variables involve different analyses. The mapping function, f, just described has been used to follow the actual changes in group structures as a group worked on a series of eight problems of one type and a series of four of another. It has also been used to cause structural change.

Other mapping functions for other situations have been constructed (for example, the formation of interpersonal hostility) and I doubt that this particular mapping function is general. For example, Lippitt and Mackenzie [1, in this volume] present a mapping function to describe the choice of committee formation. Nevertheless, I think that this type of formulation is a harbinger of future models of social processes.

SUMMARY

Task processes, behavioral constitutions, change processes, mapping functions, and other ideas have been invented to explain changes in group processes and group structures. These are a long way from Mr Structure types of analyses. They are more precise, they predict better, they raise more questions, they generate new theory, they can be experimentally tested, and they open up to more careful analyses some of the age-old problems of disorganized and organized behavior. This new paradigm focuses on the processes of behavior which determine effects rather than trying to explain effects without knowledge of the intervening processes. I believe that, in spite of the numerous shortcomings of this theory of group structures, it provides a sturdy framework for building better theories about social systems.

An important development in a theory of group structures is the concept of a timely group role matrix, $R_T{}^*$. This concept incorporates both the many structures and task processes. The entries in $R_T{}^*$ are derived from actual behavior rather than assumed behavior. The entries in $R_T{}^*$ describe whether or not a person is performing or taking responsibility for performing each act. $R_T{}^*$ also shows, at a glance, task process relationships among the group members. Experiments have demonstrated how $R_T{}^*$ develops and the remarkable relationships between hierarchy and efficiency for groups that evolve their structures and task processes simultaneously.

Unfortunately, many organizations such as universities and government bureaus often work out an authority system that is inconsistent and mostly independent from the task processes. Many large universities, for example, have layers of administrators whose connection with the task processes of teaching and research is often quite tenuous. Many times the response to a new problem is to set up a "structure" of offices before there is much known about the task processes. While the Mr Structure concept of structure implied in such cases is deficient and the results are often comical, wasteful, and even tragic, the phenomenon is observed often and needs to be studied. I believe that in many cases it is the control system that causes the majority of the control problems. It is quite possible that by removing some controllers and some control systems, the organization would be in better control in that the authority-power relationships would be consistent with the task process relationships.

In the paper with Lippitt, we argue that inconsistency between the task process system and the authority system creates a succession of a type of problem called an *authority-task problem* (ATP). We use the concept of an ATP to study the processes of committee formation. We believe that the increase in the number of committees to solve ATPs is a direct result of inadequate concepts of structure wherein the authority role system is inconsistent with the task role system. The taxpayer part of me laments this trend but the research part of me finds it fascinating.[3] I believe that ATPs are often traceable to a Mr Structure concept of structure. Even though Mr Structure may only exist in a few minds, he apparently exists in enough minds to be a problem. I suppose that just as Sherlock Holmes had his Moriarity, I am stuck with Mr Structure. Moriarity was magnificently clever and evil. Mr Structure is a "pragmatist" and a muddler. Compared to Moriarity, Mr Structure is rather nondescript. I envy Sherlock Holmes his grander nemesis.

ENDNOTES

[1] In a laboratory communication network experiment the experimenter imposes physical barriers such as metal partitions. Ordinarily, the subjects will use the half-channels made available by the experimenter. But, in some cases, I have observed subjects climbing over, under, and around the partitions. Even if I further restrict interactions to written messages, subjects can whistle, yell, stomp their feet, shoot rubber bands, send gliders, etc.

[2] Authority can be described in terms of the right to make certain preemptions. Power can be described in terms of the ability to make a preemption "stick". Authority and power may be only loosely related, especially when the authority is not consistent with the task processes.

[3] A partisan in a modern university who is interested in group behavior can find as great a variety of rare and exotic groups as a zoologist can find specimens in a zoo. Moreover, our creatures of interest mutate faster, often have unlimited appetites for resources, pollute more, and are often more dangerous. I find myself partially immobilized by simultaneously feeling outraged and morbidly curious.

REFERENCES

[1] Lippitt, M. and Mackenzie, K. D., "A theory of committee formation", in this volume.

[2] Mackenzie, K. D., *A Theory of Group Structures, Volume I: Basic Theory*, New York, N.Y.: Gordon and Breach, Science Publishers, 1976.
[3] Mackenzie, K. D., *A Theory of Group Structures, Volume II: Empirical Tests*, New York, N.Y.: Gordon and Breach, Science Publishers, 1976.
[4] Mackenzie, K. D. and Barron, F. H., "Analysis of a decision making investigation", *Management Science, 17*, 4: B–226 – B–241, 1970.

THE ROLE OF DIALOGUE IN THE MEASUREMENT PROCESS

RICHARD O. MASON

University of California at Los Angeles, and National Science Foundation, Washington DC

THE PROBLEM

There is a fundamental problem in "data analysis" that is essentially inherent in the term itself. That is, there is a tendency to take data as *given* and then proceed to analyze it. The question explored in this paper concerns just how "given" any set of data are and what might we do to help guarantee the truthfulness of data.

The discussion will be in part philosophical but it has some very strong practical implications. Data is collected in order to engage in a reasoning process. The purpose is to search for items of information that can be used to yield new items of information. These new items, in turn, will be used to make predictions and decisions and ultimately to serve as the basis for taking action. Action taking, of course, is a very practical activity. So, it will behoove us to look a little more closely at the concept of "data", and especially measured data since there seems to be something about quantification which blinds us to questions of underlying assumptions. A brief illustration will help.

According to the Bureau of Census there were 25,522,000 people living in poverty in March 1971. This datum derives from a series of observations in accordance with some very specific definitions and rules. For example, a person is either unrelated or in a family, families being defined as "a group of two or more persons related by blood, marriage or adoption and residing together". Any person or family earning less than $1,576 per annum (the threshold level for a 65-year-old female living alone on a farm) is automatically considered to be impoverished and there are additional threshold scales above this depending on the number of people in the family, their age, whether they are farm or nonfarm and whether the head of the household is male or female. However, any person or family earning more than $6,486 per annum (a nonfarm family of 7 or more with a male head of household) is not considered to be in poverty.

Now, many of us in the inflation ridden economy of today would question that these procedures capture the real poverty level today. We might argue, for example, that various cost of living factors are not included in the measure.

Beginning such a line of questioning would involve us in a dialogue or debate about the measurement process. We will not engage in a full-blown debate about measures of poverty nor criticize the existing poverty measures here. Rather, the measures will be used as an illustration of the role of dialogue in the measurement process and to serve as a backdrop from which to make some general observations. The discussion will follow from principles of measurement.

STEPS IN MEASUREMENT

There are at least three steps in any measurement process: .

a) The conception of an underlying dimension. (The dimension is relevant to some current or future problematic situation.)

b) The definition of units and scales of the dimension.

c) Execution of operations to determine the numerosity of units in a given object or collection of objects.

All three steps are difficult and subject to differing points of view. However, the first step, the conception of an underlying dimension, is the most critical and, hence, the source and object of most debate. The conception of a dimension (a construct or a measure space) requires the judgmental act of *distinction*. Among the vast and chaotic set of possibilities the measure must be able to identify those things to be included within the measure and those things which should not. That is, one must be able to set boundaries around the phenomena to be measured. Exactly how these conceptual boundaries or dimensions are set is not an idle matter. It impinges critically on the vital interests of those concerned with the measure's use, be it for scientific, policy and decision making, or for general public information purposes. Consequently, in a very real way debate, active or silent, is an integral component in any measurement process.

AN ILLUSTRATION

For example, in the creation of poverty measures we might imagine that the following hypothetical debate occurred:

QUESTION (Q): What is poverty?

MEASURER (M): Poverty is a deficiency in the necessary or desirable qualities of life in our culture.

Q: How do we measure it?

M: Well, in a capitalist exchange economy one's annual income in dollars is the best indicator of one's ability to have necessities and luxuries. Clearly, anyone who earns less than $1,576 a year simply could not purchase enough calories of food nor clothing and shelter to live comfortably.

Q: Perhaps, but ours is a culture in which the family is a central institution. People live together. Doesn't that affect their comparative poverty level? You know the old saw "Two can live cheaper than one".

M: Yes, we can incorporate a graduated scale which adjusts for the fact that two people living together do not need to earn twice as much as they would singly in order to achieve the same standard of living. Indeed the scale accommodates family sizes from 1 to 7 members.

Q: OK. But then some families are headed by a man, others by a woman. Does that make a difference?

M: Yes, a female headed household generally requires less income to live. There's an adjustment for that also.

Q: How about age?

M: Oh yes, elderly people receive aid, care and other benefits in our society. Consequently, a provision is made to reduce the poverty threshold level slightly if the head of house is 65 years or older.

Q: One last question. Many people are able to secure resources and benefits directly without going through the exchange economy and therefore their real standard of living is not reflected in their income figures. I'm thinking particularly of farmers who grow their own food. How do you account for this?

M: Of course it is very difficult to capture all of these possibilities and we really do not attempt to do so. However, we separate farm from non-farm families and the upper limit of the official poverty standard for farm families is somewhat lower. This accounts for the fact that they require less dollar income to live.

We could extend this debate considerably. Many readers, I'm sure, might add questions and responses that are perhaps more provocative than those already posed. However, by now the main point of the illustration should be clear. The process of setting dimensions and units for measurement is fundamentally judgmental. It is based on the values, beliefs and concerns of those involved in defining the measure. As William James [2] put it, measurement is a "teleological instrument". "This whole function of conceiving, of fixing, and molding facts to meanings, has no significance apart from the fact that the conceiver is a creature with partial purposes and private ends". Measures vary with the end the measures have in view. And the fact that some one goes to the effort to measure suggests that the phenomenon is important to him and that some use will be made of the measure.

Since the process is judgmental no truly analytical ways exist for completing the process. Yet measurements are the basis for many important decisions, decisions which affect many different lives in many different ways. Poverty figures, for example, are used to guide welfare and other social policy decision making. So, the question becomes: How can we aid the judgmental processes which underlie measurement? Much has been done on the mathematics of measurement, what can be done to improve the judgments on dimensions and units?

My response is that we must learn to design and structure the debate underlying a measure so that the richest set of possibilities is generated and best synthesis is achieved. An approach for designing such a debate is developed later in the paper. First, however, let us review a few short cases to clarify some of the issues involved.

Case 1

About two years ago a research team began a survey of the "state of the art" and current research in progress in the field of computer aided design and manufacturing (CAD/CAM). In order to complete their task they created a hierarchical taxonomy which subdivided the manufacturing enterprise into a series of elemental goal or process oriented tasks. Each item of current research was then classified as to its contribution to each task. An elaborate cross-indexing scheme was employed. One of the results of the effort was to show that rather heavy research efforts were underway to apply CAD/CAM to some selected manufacturing tasks and there were virtually none in other areas.

The team was aware of the difficulties involved in subsuming the research under its categories and undertook the process with great care. However, they were astonished at the line of criticism that was directed at their final report. Most people criticized the "differentia" employed in the classification scheme. The team believed its taxonomy was rooted in the logic of the manufacturing process and not subject to emotional debate. Not so, said the critics. Some researchers whose work was classified argued that their research was *different* from other's work subsumed under the same category, regardless of what classification scheme was used. Some researchers suggested a different, and for them more realistic logical model of manufacturing, which would result in somewhat different categories. Since most researchers like to be at the leading edge of their field a few criticized the taxonomy because it made those researchers "look good" who were working or had announced plans to work in categories with minimal or no activity. Those who were working in the "same old" areas did not look so good.

The team had not adequately confronted these issues earlier and the credibility of its results suffered accordingly.

Case 2

Several years ago a colleague and I were working with a large timber, lumber, pulp and paper company. The company was decentralized into a lumber division and a timber division both of which had strong profit incentive programs. Top management made a good part of its salary in incentive bonuses.

The management science group at corporate headquarters had just completed a large-scale linear programming model of the entire corporate production system starting with the source timber lands and ending in the final product markets such as lumber, furniture, paper, books, packaging, etc. In order to run the model they had to estimate its coefficients including one crucial cost coefficient that occurs at the point in the production process where logs are peeled. The peeled logs go to make lumber (lumber division) and the chips and bark are used to make paper (paper division). Joint costs had to be assigned to both peeled logs and chips. The measurement problem was, "By what principle do we assign costs to lumber and to chips?"

Historically, chips had been substantially a free good. Prior to the development of their use to make paper they were simply burned as surplus. But once the measurement question was posed it could not be ignored. Initially, the management science team set a price based on the production capacity used to process the chips. As it turned out this was a modest cost figure. However, the paper division's management complain vigorously because this cost figure adversely affected their profit ratings. The paper division management's position was that they had created utility where no utility existed before by using these chips. Consequently, they should not now have to pay a premium for the "privilege" of providing this contribution. The MS group countered that they could sell all the chips they could get to the Japanese at a rate over 3 times that which they suggested charging the paper division. The division replied that it was company policy not to sell to the Japanese or to favor the company's competitors. Also, they claimed that since the Japanese government in effect subsidized much of the paper industry, it wasn't proper to charge the division the market price set by the Japanese. The MS group retaliated that company policy *does*, however, permit going out of the lumber business altogether. In this case the logs could be totally converted to chips and the division would have to pay the full cost, a cost which proved to be higher than the original estimate but less than the Japanese offer.

About this time the debate became so heated that corporate management, recognizing that if it were carried further it would destroy the company, scuttled the model. The implications of the alternative measures were simply too threatening to the organization, and the company management had no way short of a revolution for dealing with them.

Case 3

As part of a research program to evaluate the use of NASA's Earth Resource Technological Satellite, I surveyed some applications of satellite and aircraft remote sensing in the field of agriculture. One project involving forecasting the grape crop was especially interesting in terms of the debate it created.

Crop forecasts are a very important aspect of agriculture information systems. They are used by bankers, farm implement companies, canners, government officials, etc., to plan their activities. The forecasts are refined as the cycle from growing through to consumption progresses. It is well known that all of these forecasts and updating refinements are difficult to make and subject to considerable error. However, in the California grape market there was an additional complication contributing to the error. Grape growers have two basic options. They can elect to sell grapes on the open market or they can set them in the sun to dry and thereby produce raisins. Consequently, not only must natural events be forecasted, but also the grower's decisions must be forecasted in order to estimate the supply of grapes that will reach the market.

To solve this problem some researchers working with the Department of Agriculture came up with a rather creative solution. They commissioned airplanes to fly over the fields and to photograph them at periodic intervals. By means of an analyses of the aerial photography the researchers could determine how many raisin drying pans the growers were using and when they put them out. Since the pans were of generally standard sizes it was possible to estimate the amount of grapes dried per pan, multiply by the total number of pans and obtain an estimate of the raisin output. The total grape forecast minus the raisin estimate yields the new grape market supply forecast.

It would seem that a more accurate figure of this nature would have greater utility and that the researchers efforts would be applauded. Such was not the case. Grape growers, wholesalers and wineries alike complained when the first experiment was run.

The grape growers resented the invasion of privacy. They knew and trusted the field methods used to make forecasts previously. These served as the basis for bank loans, government activity and other business decisions and consequently were important to them. This new information potentially upset an information balance in a critical element of their business and they were very uncomfortable with it.

The wineries were even more incensed. They mounted heavy lobbying efforts with the legislature to have the program stopped. The wineries were basically monopsonists in their supply markets. The lack of accurate information improved their power to set prices. The new measurements threatened to change the way market prices were determined. Consequently, they felt they would lose out, or at least not gain, in the process. The measurement program became so politically hot that it was eventually dropped.

These three cases illustrate several of the levels at which debate can occur in the measurement process. Debate can occur when the basic distinctions are made which define the dimensions that will be measured (Case 1 and the Poverty illustration). Debate can occur when the alternative units and concepts used to measure within the dimensions are determined and decided upon (Case 2). And, finally, debate can occur when the measurement data itself is published and employed in the decision process (Case 2 and Case 3). Thus, debate can occur at all stages in the measurement process. This points to the need for a procedure for handling this debate and making it more efficient. This requirement, as we shall see, calls for a new kind of reasoning.

DEDUCTION, INDUCTION AND DIALECTIC

In the history of reasoning two paradigms have dominated. One is deduction; the other induction. Deductive arguments generally proceed from major premises and minor premises to conclusions. The archetype for a deductive argument is:

	Form	*Example*
	A (assumption)	All men are mortal
and	D (data)	Socrates is a man
Therefore	C (conclusion)	Socrates is a mortal.

In the pure form of the argument the data are taken as *given*. The conclusion is true if and only if the assumptions and data are true. Induction, on the other hand, is a process by which data about some members of a class is used to arrive at conclusions or explanations concerning more or all members of a class. Accordingly, the following archetype typifies induction:

	D_1
and	D_2
	.
	.
	.
and	D_n
Therefore	C

Inductive arguments, of course, are very important in practical affairs. Most of our current knowledge (and all of our forecast knowledge) is essentially based on induction. Induction, however, is not a complete reasoning form. As with deduction, in an inductive argument the data are also taken as *given*.

Both induction and deduction are very useful. They are frequently employed together in most complex scientific and practical situations. Indeed, they often mutually support each other. Collectively, however, they are incomplete as forms of reasoning. Both induction and deduction seek valid or probable conclusions *given* the data. They do not provide a guarantee for the data which they employ. In order to complete the reasoning process, an additional mode is required. For historical reasons I will call this additional mode of reasoning dialectic.

A SCHEME FOR DIALECTICS

In what follows, an outline for another form of reasoning will be sketched out. The method can serve as a checking process on data and as a means of structuring debate in the measurement process. The method is different from and a compansion to deduction and induction.

The approach is dialectical. An argument can be said to be dialectical whenever a situation is examined systematically and logically from two or more points of view. Dialectical arguments seek an answer to the question, "What does it mean to say that . . .?" A dialectic argument applied to data might proceed as follows:

1) Select the data item D_1 (e.g. the number of people in poverty, the level of CAD/CAM research pertaining to a manufacturing task, the cost of wood chips, or the number of grapes being converted to raisins). D_1 is considered to be a proposition.

2) D_1, in turn, is implied by an underlying set of assumptions or a system S_1. S_1 includes the formal definition of terms and units used in the measurement process and the rules used to employ them. S_1 should be specified as completely as possible.

3) The act of distinction which resulted in the formation of S_1 essentially selected a specific set of definitions and rejected or ignored all others. A relation of opposition or otherness was formed in the process. Since selections of this nature are based on purposes, values and beliefs, the third step is to identify the basic value and belief system W_1 which underlies S_1. In social systems the cultural, social and psychological history surrounding the measurement will serve as clues to W_1. W_1 represents the worldview, or root metaphor, behind the system and its data.

4) The fourth step is to create an opposing point of view. This can be done in one (or any combinations) of three ways:

a) Select an alternative data item D_2 of the same form as D_1 (e.g. another estimate of the number of people in poverty). Repeat steps 2 and 3 to discover S_2 and W_2.

b) Select an alternative system S_2. This can often be done by negating or redefining some of the premises in S_1. The relevant D_2 and W_2 are then found.

c) Select an alternative worldview W_2 based on another set of metaphysical values and beliefs and derive its S_2 and D_2.

Criteria for choosing W_2, S_2 and D_2 are that they must be plausible, credible and in social–political as well as logical opposition. That is, they should have some force behind them in the real world.

5) Place (W_1, S_1, D_1) and (W_2, S_2 and D_2) in opposition to one another and systematically explore their respective implications and differences. This can take the form of a programmed debate in which each element of each system is interpreted and the strongest possible case is made for including it instead of its counterpart in the system. The strongest case is also made for the counterpart.

6) The sixth step is based on a theory of systems. S_1 and S_2 as systems in opposition are at the same level in the hierarchy of world systems. This relationship suggests that there must exist a supra system S^1 that implies both S_1 and S_2. It will be based on its own, expanded worldview W^1. Step six is to discover or create on S^1 and W^1. This will be the result of synthetic thinking and requires imagination and insight. The implications of S^1 are then derived.

7) The seventh step involves judgment. The measurer must now choose the "best" system S^* based on the insight and discoveries obtained in the dialectic process. This is, of course, the most hazardous step in the process. However, there is a substantial history of opinion which holds that good judgments are made in the context of opposition. In designing the dialectic we have tried to insure that the best context of opposition possible would be developed.

8) Finally, S^* is used to obtain a new measure D^*. D^* may now be employed in deductive and inductive arguments.

SUMMARY

In this paper some of the problems of using and understanding data have to be discussed. All measurements, it was asserted, have a system of purposes, values and beliefs underlying them and therefore are fundamentally based on judgment.

Some illustrations and cases were given. We found that deduction and induction were inadequate to the task of guaranteeing data. Both methods take data as given. A dialectical mode of reasoning, based on the thesis, antithesis, synthesis triad of Hegel, was developed and proposed as a companion mode of reasoning to be used with deduction and induction.

Dialectic is a structured way of improving dialogue in the measurement process. Churchman [1: 149–205] has developed an underlying rationale for this approach. Mason [3] has interpreted it and applied it in a decision-making context.

REFERENCES

[1] Churchman, C. W., *The Design of Inquiring Systems*, New York: Basic Books, 1971.

[2] James, William, *The Principles of Psychology*, Vol. 53, *Great Books of the Western World*, Chicago: Enzyclopaedia Britanica, Inc.: 1952.

[3] Mason, R. O., "A dialectical approach to strategic planning", *Management Science* 15, No. 8: B–403 – B–414, 1969.

DATA ANALYSIS, COMMUNICATION AND CONTROL

ROY E. WELSCH
Massachusetts Institute of Technology, Cambridge

1 INTRODUCTION

Many times in the past few years, I have felt that I existed mainly to fill out questionnaires for researchers trying to discover something about me or my role in various social processes. These days I ignore all but those with well-designed questions, assuming that the rest will send a follow-up questionnaire, if they are at all serious. This almost never happens. Often I wonder just how my answers are used to understand a particular social process and whether my life was changed in a negative way because of inadequate data analysis.

Probably twice a week, I am asked (or told) to act on the basis of some data that has been collected and perhaps (not very often) analyzed. Most of this I ignore, too. The statistics produced by the Federal Government are harder to ignore. If the inflation rate is ten per cent and my raise is five per cent, I am likely to act. A single number (unfortunately with no measure of precision attached), which we will presume for now is the result of careful data analysis, not only communicates something about the social processes around me, but almost compels me to act.

Data is one raw material we use to investigate social processes and programs. Some data is a thoughtless (and often essentially useless) by-product of the social process. For example, observational studies and non-randomized field trials are often used to evaluate social programs, but past experience [6] indicates that we would be better off with more carefully controlled randomized field trials. Many medical researchers have already learned the lesson of randomization, perhaps because they realize that clinical trials may cost lives and cannot be wasted. In any case, this raw material (naturally we would like high-grade ore) needs processing, in many different ways, before we can begin to extract indications and evidence from it and before we can integrate it into the decision-making processes of society.

2 COMMUNICATION AND CONTROL

Data analysis is what we do with data to help extract and communicate information about the process that generated the data. Data analysis then is a communication link in most social processes. How data analysis is performed and presented helps to determine how (and how much) information flows in society, and this information helps determine the behavior of social processes and our ability to control them. Data analysts cannot afford to lose sight of this fact: it might be argued that undue emphasis on the rigid models of mathematical statistics did cause some of us to lose sight of this broader role of data analysis.

For purposes of this paper I have divided those involved in the world of data analysis into four groups: developers of data analysis methodology, data analysts, interpreters of analyses, and decision-makers. The lines are often blurred and some of us can count ourselves in all four groups. The following sections of this paper will discuss aspects of data analysis related to all four groups. It has been my experience that if the members of the first group fail to consider the needs of the other three, much of their research fails to have a significant impact on communication in society and the control of social processes.

3 PROBLEMS

There are problems with data analysis as it is practiced and used today. At each level – decision-maker, interpreter, analyst, and developer – we see unfilled needs. In this section some of these problems are presented. I do not mean to imply that there is nothing right about data analysis as it is practiced today. My main point is that there is definitely room for improvement.

Communication

All of us can rightfully wonder if statisticians know very much about communication, especially when we examine how the results of data analysis are used in society as a whole. Consider the summary numbers printed in precise columns on a page of computer output from a typical statistical routine. Do multiple correlation coefficients and χ^2 statistics really tell the story? Perhaps they do to a few people, but we cannot afford to think of just this small group alone. What good is high quality data analysis if it fails to get the message in the data across to society and decision-makers?

Complex Models

The increasing emphasis on systems, feedback, and control, has meant that the models we build to describe social processes have become increasingly complex. They are large, nonlinear, and interdependent. Standard calibration techniques like single stage least-squares often fail to provide reasonable results in either the analysis of model structure or prediction. Must we be forced to use linear models because we cannot calibrate nonlinear models? We also need more help in deciding which variables to include and how they should be reexpressed to simplify our analysis.

Large Data Bases

While we still need more and better data bases in some areas, a large number of cases exist where the size of the data base we now have is almost overwhelming. We have trouble getting a feel for the structure of these sets of data. Often we do not have a good prior feeling about the models needed to describe the process that generated these data. We would like to have the data help us derive reasonable models. Do we have to worry about data-dredging or can we take a more relaxed attitude and explore the data in a sensible and sensitive way? Are large data bases likely to be well behaved, or do we need to consider more deeply the effects of blunders and mistakes?

One-Pass Processing

All too often we see data that has been "analyzed" by one pass through a statistical package. When reading we often pause to reread a paragraph or sentence in order to summarize it in our own words. We may go over it several times before we feel we have understood its significance. Is there any real reason why we should not look at our data several different times in several different ways? What do we do if we get different results on some of these journeys through the data?

Rigid Assumptions

Many of the statistical methods that we use today have been derived under rigid sets of assumptions and then shown to be optimal when these conditions hold. With more complex models and data bases it has become increasingly difficult to justify some of these assumptions. Data analysis cannot afford to be locked in by the need for computational ease or mathematical simplicity. Can we tolerate methods that are violently affected by small changes in these assumptions? For example, linear programming techniques provide solutions to certain kinds of constrained optimization problems. If some of the coefficients (derived from data) used in the model are very sensitive to one data point, then what good does it do to apply an elaborate optimization technique which can be very sensitive to a change in a single coefficient?

Resistance

If we forego mathematical simplicity and computational ease, what should the axioms of data analysis be? One emerging principle is that of resistance [16]: a technique is resistant if the summary conclusions it produces change little in the face of gentle perturbations in much of the data or violent perturbations in a little of the data. For example, the median is resistant while the mean is not. We may have to give up something to achieve this for a broad spectrum of data analysis situations. What is a reasonable price to pay for the insurance and stability obtained by using resistant methods?

Validity

Many of us have discovered the hard way that a model can be "optimally calibrated" on a given data base and then fail rather spectacularly when used for prediction or checked on a different data base. There are many reasons for this, but we need to have ways to assess the validity of our data analysis. Are we to take seriously the "inflation rate statistics" published monthly if there is no indication of their precision? At least most political polls now contain some indication of precision. Data analysis suffers when measures of validity are lacking or are removed by those who interpret and publish the results of data analysis.

Prior Information

Data analysis cannot operate in a vacuum away from those who are familiar with the process that generated the data. Often those close to the data will be far more helpful in suggesting models and constraints than any exploratory look at the data by a data analyst. Sometimes we may want to give formal weight to these prior ideas in the data analysis process. While Bayesian statisticians have formulated many ways to do this, far too little of this research has been implemented in ways that are useful to the average analyst who is working with data and with those who have prior information. How can we use prior information to help validate the results of data analysis?

Access to Methods

One of my great frustrations is to tell colleagues about a new technique that would really help with the analysis of their data, only to see later that they did not try it because of the difficulties involved in getting the method to the data. (Getting the data to the method is another, usually equally frustrating, possibility.) We need somehow to bring analyst, data, method, and models together via a communication system that will allow creative data analysis to take place.

Responsive Data Analysis Research

A significant portion of those involved in the development of new data analysis methods should be responsive to the needs of data analysts and decision-makers who are working with new kinds of data and models. Part of the problem is communication and feedback among these groups. In some cases developers are also analysts and analysts are also decision-makers. This is often extremely useful and needs to be encouraged. But we need to make better use of the communication channels already open to us (conferences, on-line consulting) to foster the responsive development of data analysis methodology.

4 PROGRESS

In recent years progress has been made on many of the problems outlined in the last section. The computer and the communication and control systems associated with computers have made much of this progress possible. Equally important is the fact that the development of data analysis techniques has become somewhat of a distinct discipline linked to, but separate from, mathematics and mathematical statistics. New data analysis methods get tested by computer simulation and field experience and not just by adherence to rigid mathematical models. The primary purpose of data analysis research is to develop techniques to analyze data rather than to apply mathematics to the problems that arise in data analysis.

The following summary of progress reflects my own views and experience. For convenience I have broken this section up into the same headings as the last one. It will be readily apparent that many of the new techniques and ideas cut across these boundaries.

Communication

The most important progress here has been in graphics. Perhaps I am biased because I am in a school of management, where it seems that the singularly most effective way for a management scientist to communicate with a manager or potential manager is with a picture, graph or display. The development of graphics has been helped immensely by the availability of low-cost graphics terminals for interactive computer systems. Since more and more data analysis is being carried out on interactive time-sharing systems, it is most unpleasant to have to wait for a plot from a passive device or take numbers from the terminal and make a plot by hand.

I have participated in the development of a low-cost interactive graphics system at the National Bureau of Economic Research Computer Research Center. The system, called CLOUDS, was originally designed to plot and manipulate p-dimensional point clouds and, in particular to fully exploit the proven utility of the basic scatter plot. Groups of commands are designed to scale, project, rotate, mask, overlay, zoom, connect points with lines, provide labels and text, and work with individual points.

The last group of commands has proved to be especially useful because with a scatter plot on the screen, the cursors can be used to add, move, identify, label, and delete points, while a keyboard command allows the user to find where a particular point has been plotted. It is also possible to mark different groups of

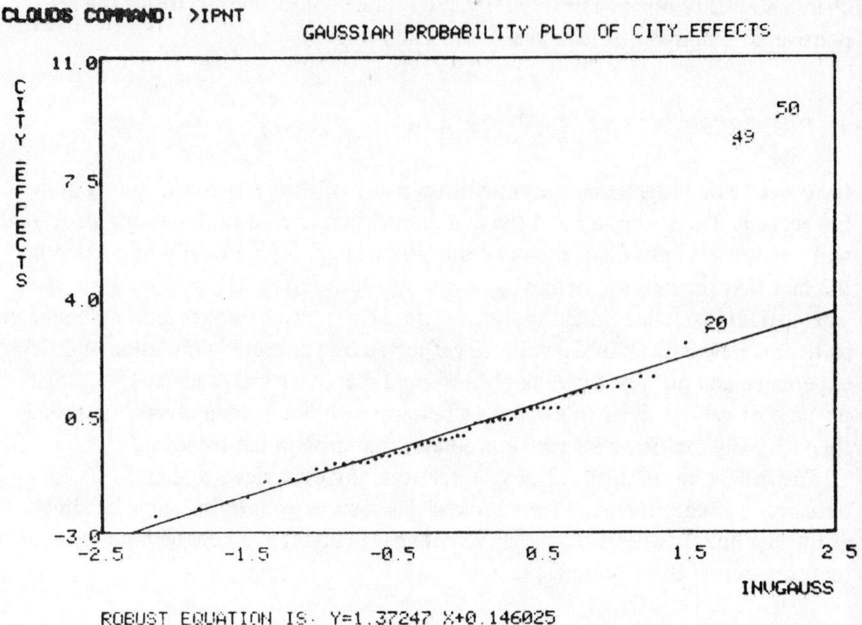

FIGURE 1 Identifying points with the CLOUDS command IPNT.

points (say by years, scale, or cluster) with different symbols to facilitate observation and identification. An example identifying outliers and finding points is provided in Figures 1 and 2 while Figure 3 shows how marked points can provide additional information on a scatter plot. For a detailed description of this system see [15; 19].

CLOUDS is often used in its own right, but the commands have been used even more as building blocks to implement the plots developed by Tukey [17] and others, many of which bear little resemblance to point clouds.

Today, most people who use data analysis in their work are either not acquainted with the possibilities of graphic data analysis or believe the cost is too high. As a consequence, potential ideas for new pictures are suppressed or never formulated. We have found that once a user sees a demonstration of interactive

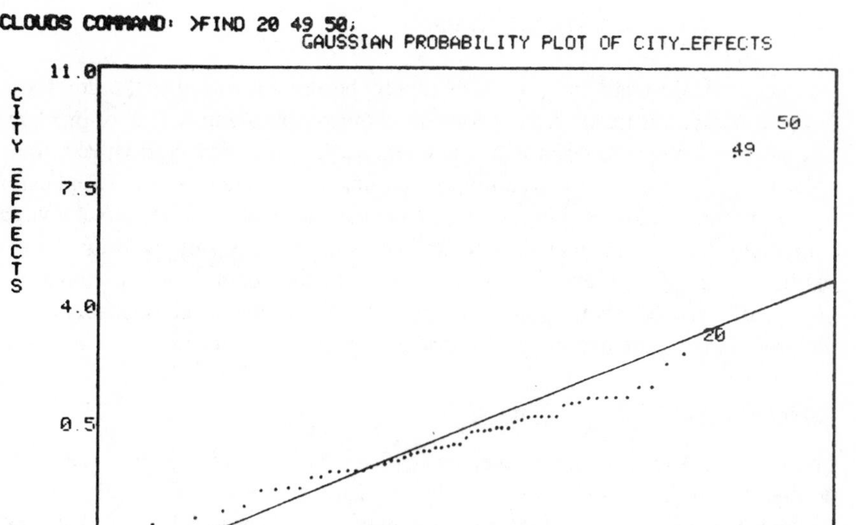

FIGURE 2 Finding points with the CLOUDS command FIND.

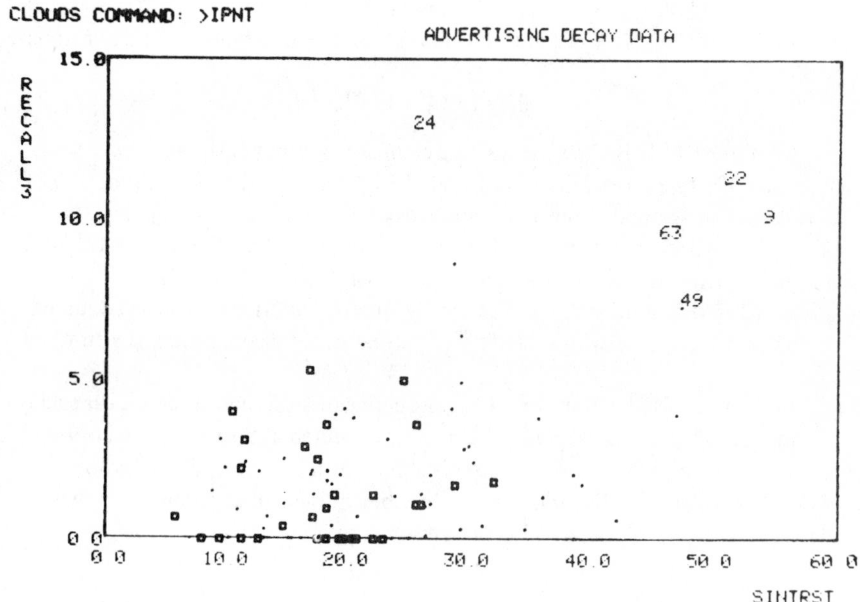

FIGURE 3 Points with a given attribute are given a special symbol using the MARK command in CLOUDS.

graphics applied to data in his area, he rapidly becomes involved in the process of creating pictures for his needs. Whereas many people are unwilling to propose new statistical methods because of the mathematics that might be involved, they are willing to suggest pictures and plots that might help. Many areas of application – psychology, sociology, biology, medicine, management – have more of a visual component than formal mathematics and statistics. People working in these areas can help define what they would like to see about their data. Given the tools to implement what they want, significant progress has been and can be made. In the process, communication and feedback are greatly enhanced.

Complex Models

Important advances in numerical analysis [2; 4] have made reasonably effective nonlinear optimization routines available on many computers. This makes it possible to consider the calibration of nonlinear models using a least-squares loss function or, even better, a loss function based on the philosophy of resistance. Some specialized algorithms have been developed, and are becoming widely available.

A problem in linear models (and an equally serious but often less obvious problem in nonlinear models) is collinearity. A major new attack on this problem has been the concept of ridge regression introduced by Hoerl and Kennard [8]. Consider trying to calibrate the linear model $Y = X\beta + e$. If we assume that X has been scaled so that X^TX is a correlation matrix then one family of ridge estimates for β is

$$\hat{\beta}_k = (X^TX + kI)^{-1} X^TY$$

and we note that $\hat{\beta}_0$ is the least-squares estimate. A number of ways have been proposed [5] to choose k including making a "trace" of the $\hat{\beta}_k$ as a function of k. It is easy to see that if there is collinearity and X^TX is nearly singular, then $(X^TX + kI)$ will be better behaved. The price we pay is that the ridge estimators are biased. There are strong theoretical results [10] and Monte Carlo studies [20] that indicate $\hat{\beta}_k$ is a generally superior alternative to least-squares. At the very least, it provides a family of estimates for examination as a function of k.

Progress has also been made on the problems of selecting variables, especially for linear models. Ridge regression can be interpreted as a form of selection in the principal component coordinate system, and the C_p plots proposed by Mallows [12] are also becoming more feasible and useful as computer power increases and graphic displays become more common.

Large Data Bases

We are learning to first explore data on its own terms in a flexible mode, unencumbered by considerations of probability, significance, and the like. For large data sets graphics is essential, because plots can summarize large numbers of data points while still carrying along enough fine detail where it is needed. Linking clustering techniques with graphics has also proved to be a fruitful way to explore large data bases.

But there is a more basic principle that is being applied on large data sets. When taking a first look at data it is often not necessary to use *all* of the data. Subsamples and subsummaries may tell us a great deal, and they can be obtained more rapidly at less cost. We are beginning to see data analysis systems designed so that getting subsamples and subsummaries is an integral part of the system and not something requiring a great deal of programming effort.

For large discrete multivariate data sets exploratory techniques based on the theory of log–linear models [3] are now used to sort out the pattern of dependencies in a set of discrete variables. When there are only two variables, there are only two basic situations: either the variables are independent of each other, or they are not. When there are more variables, complex conditionally independent relationships can be studied. No longer are we limited to examining just pairwise correlations and linear relationships.

One Pass Processing

Today the emphasis is on an iterative approach to data analysis. If we consider the relation

$$\text{data} = \text{fit} + \text{residual}$$

then iterative, and increasingly interactive, data analysis proceeds by first trying simple fits or perhaps simple calibrated models, and then exploring the residuals, often as if they were a new data set. Successive layers of structure can be peeled away and built into a more complex fit, but the residuals at each step are carefully examined for further structure or indications of poor fit. We are gradually learning how to provide effective summaries, often graphic in nature, at the end of each of these steps to guide us in taking the next step and to provide useful feedback and communication. Interactive computing seems to provide the best framework for this type of data analysis.

Rigid Assumptions

We are now more likely to let the data speak freely and help us to determine realistic assumptions. The careful data analyst is slow to make assumptions before he explores the data. Families of estimators like those associated with ridge regression (and, as described in the next section, robust estimation) are becoming available. These families of estimators often provide ways to explore how various assumptions might be affecting our conclusions and what the consequences might be.

Finally, I think we are searching for data analysis techniques that are good in a variety of situations rather than optimal in a few. If they are good, they will survive even without a formal mathematical proof of their optimality. This also means we shall probably have several good techniques to use on a problem rather than one restricted optimal approach.

Resistance

Notice that the definition of resistance we gave in the last section was based on the data – there was no mention of probability. When we introduce probability and other modeling assumptions into the problem, we often call the techniques robust rather than resistant. Robust methods of data analysis either yield results and conclusions which are relatively unaffected by moderate departures from the assumptions apparently underlying the analysis or, as methods, are highly efficient in the presence of such departures, or (most likely) both.

Various mathematical formulations of robustness have been developed by Hampel [7] and Huber [9]. A major Monte Carlo study of robust estimators was undertaken in [1]. Briefly, we can describe robust calibration in terms of criterion functions as follows. Assume we want to find estimates for θ in a model $f(x_i, \theta)$ (perhaps nonlinear) where x_i denotes the explanatory variables. If y_i is the response variable and σ a scale factor, then we try to find

$$\min_{\theta} \sum_{i=1}^{n} \rho_c\left(\frac{y_i - f(x_i, \theta)}{\sigma}\right) \tag{4.1}$$

and use the values of θ and σ at the minimum as our estimates. One family of criterion functions proposed by Huber is

$$\rho_c(u) = \begin{cases} \frac{1}{2}u^2 & |u| < c \\ c|u| - \frac{c^2}{2} & |u| \geqslant c \end{cases}$$

When $c = \infty$ we have least-squares and, in general, $\rho_c(u)$ is like least-squares in the middle and least absolute deviations for large values of u. Except when $c = \infty$, $\rho_c(u)$ provides robust estimates with the efficiency at the Gaussian error model dependent upon c. (Least-squares is most efficient at the Gaussian but is very inefficient for modest departures from Gaussianity.) Scale has been included in (4.1) because $\rho_c(u)$ is not scale invariant.

Thus we again have created a family of estimators. There are various ways to choose c, and by varying c we can examine how sensitive our estimates are to large residuals. In Figures 1 and 2 we can see the difference between the fit obtained using a least-squares regression equation and a robust regression equation. We should note that there are other forms of robust loss functions, some of which are bounded and therefore give effectively no weight to large residuals. Many efficient routines exist for solving (4.1) and the associated normal equations.

Validity

We have already seen how the ridge and robust estimator families provide ways to examine some of our assumptions and check the stability of our estimates. A basic approach to validation is the jackknife [13; 14].

The data is divided into r groups of approximately equal size. Let $\theta_{(j)}$ be the result of a complex calculation (say calibrating a model) on the portion of the sample that omits the j-th subgroup, i.e. on a pool of $r - l$ subgroups. Let θ_{all} be the corresponding result for the entire sample, and define *pseudovalues* by

$$\theta_{*j} = r\theta_{\text{all}} - (r-1)\theta_{(j)} \qquad j = 1, 2, \ldots, r.$$

The jackknifed value θ_*, the way we combine the pseudovalues, and an estimate s_*^2 of its variance are given by:

$$\theta_* = \frac{1}{r}\sum_{j=1}^{r} \theta_{*j}$$

$$s_*^2 = \frac{1}{r(r-1)} \sum_{j=1}^{r} (\theta_{*j} - \theta_*)^2.$$

The jackknife approach provides us with a form of internal validation and a rough estimate of variability. The basic principle is crucial: by removing a portion of the data and calibrating on the rest we can study the influence of small portions of the data on the resulting estimates. Of course, if possible, we would like to hold some data completely aside for further validation.

Consider how we might use this idea to help determine a good value for k or c in the ridge or robust case. Drop one data point (or some subset), calibrate on the rest, and then predict the omitted value of the dependent variable using the calibrated model. Square the difference between the actual and predicted value and repeat for each data point or group. For a given value of say, c, we have a measure of internal predictive quality. Repeat for several values of c and we have a plot that can help us understand the quality and predictive ability of our fit. Increased computing power and good numerical analysis make such ideas practical today.

Prior Information

Ridge regression can be viewed as a form of Bayesian regression [11] and this allows many possible extensions. In particular we can incorporate certain kinds of prior information in our calibration process. One way to generalize the ridge idea is to modify (4.1) and try to find

$$\min_{\theta} \sum_{i=1}^{n} \rho_c\left(\frac{y_i - f(x_i, \theta)}{\sigma}\right) + k \sum_{j=1}^{p} \lambda_j(\theta_j - \delta_j)^2. \tag{4.2}$$

The δ_j are our prior values of θ_j, and λ_j is a precision or weight saying just how strongly we feel about our prior, δ_j. The parameter k determines the relative importance of our prior information (the second sum) and the data (the first sum). Notice when $k = 0$ we get a standard robust estimate. Again we may wish to vary k and perhaps use the validation ideas discussed above to gain further insight.

A special form of (4.2) which uses a robust and scale invariant loss function is

$$\min_{\theta} \sum_{i=1}^{n} |y_i - f(x_i, \theta)| + \sqrt{k} \sum_{j=1}^{p} \sqrt{\lambda_j}\, |\theta_j - \delta_j|.$$

When $f(x_i, \theta)$ is linear this problem can be solved easily for all k by parametric linear programming methods.

Robust estimates are nonlinear and loosely speaking, unbiased. Ridge type estimates are biased, but linear. Recall that least-squares estimates are often said to be the best linear unbiased estimates. We have relaxed the linear and unbiased assumptions in (4.2) in order to provide more flexible, sensitive and, perhaps, better data analysis tools.

Access to Methods

In the past ten years access to statistical computing has been primarily provided by stand-alone programs like the BMD series or various subroutine libraries. Packages like SPSS, PSTAT and others are also being used increasingly by those who want to concentrate on the analysis and not on the computer complexities.

Unfortunately most of these means of access suffer from portability problems (although much progress has been made) and timeliness – they are generally well behind the state-of-the-art. In fact, it is increasingly important that experimental programs, properly denoted as such, be made available for testing and feedback from various user groups.

Gradually algorithms (and in some cases programs) are being published which, if you have the talent, can then be implemented on a local computer. Publication is slow and validation a serious problem, but it is at least a start.

Perhaps the best hope is through networks associated with regional computing research centers. First, data could be entered via the network or drawn from data banks maintained by public and private organizations. Second, new methods of analysis can be put on the system by arrangement with the research center. In fact the center, probably with government support, would make a constant effort to place new, even highly experimental methods on the system. Third, in certain cases, widely used models or classes of models would also be available for use with different data, calibration, and prediction technologies. Finally, the analyst or researcher could access all this via a local device, most likely a typewriter terminal with an associated CRT display or a CRT terminal with a hard copy device.

This is not enough, however. We need a flexible file system for sharing data and programs, a convenient command language and perhaps above all, a way to combine commands to form macros. These macros make it possible to experiment with many different combinations of data processing components in order to provide good tools for analyzing particular types of data.

If we have the building blocks contributed by a large number of researchers, and a way to put them together, then we have a chance to improve data analysis. In fact, we might hope that after some experience certain macros would survive to become new data processing building blocks (and perhaps then coded for efficiency).

For several years I have been associated with the M.I.T. Center for Computational Research where a group of data analysts, computer scientists, numerical analysts and others have been working to create a research center and computer system like that I described. All of its facilities are available via a local phone call to the M.I.T. network in most major cities in the U.S.A.

The system now operating is called TROLL. A large number of new programs and macros have been made available as TROLL Experimental Programs (TEP [15]). Many of the ideas discussed in earlier parts of this paper are available via TEP and the network.

We cannot forget minicomputers. A lot of data analysis is and will be done on them. Some of these machines (and hopefully more in the future) are a part of distributed systems based on a large host computer. Again networks and regional centers could provide software research and updates with appropriate portions fed to local minicomputers for specialized use at a particular location.

Responsive Data Analysis Research

Here I will draw especially on my own experience. Users of a computer network data analysis system can talk back and they do. In fact, as more tools are made available, the more possibilities users discover to combine and improve, and therefore, more new programs are designed and placed on the system. Most of the comments relate to specific problems the user has – he feels he will be able to investigate his data and model better if certain tools are available or existing tools are modified. It is often a real challenge for the developers of data analysis techniques to respond.

A particular example of this process concerns the graphics system, CLOUDS, described earlier. While the basic system is still used for point clouds, it is far more often used in macros to create special plots and pictures for graphic data analysis. Many of these plots are a result of direct interaction with users in application areas like economics and management science.

5 CHALLENGES

Each of the areas outlined above contains room for research. Some important areas have been omitted completely. When talking about challenges, it is useful to group these topics together in a different way.

Those involved with data analysis must pay particular attention to the following three areas:

1) Developing new methodology *and* communicating it to others.

2) Understanding how data analysis methodology affects the communication channels and information flows in society.

3) Using what others have learned about social processes, communication, and artificial intelligence to examine the data analysis process itself and perhaps "improve" it.

In the first area we face the very real problem of integrating old and new methods. We are only just starting to develop graphic and robust methods. How do we interface these with the large body of existing tools? We know how to find robust estimates, but very little about finding confidence intervals for these estimates. Do we not use robust estimates or can we use what we have combined with approximate confidence intervals based on older methodologies? If so, how do we tell others what this means?

In the second area, a basic problem is multiplicity. Gradually (perhaps too gradually) data analysts are losing their fear of analyzing a body of data in a number of different ways and then viewing the results from different perspectives. As data analysts do change, the results of data analysis will be communicated in different ways. We are going to be living with multiplicity, and we are going to have to face reports that show some contradictory results when different methods are used. It seems better to have this than kidding ourselves that there is a single "best" method.

It will take good people working very hard to figure out ways to summarize multiple analyses for public use. Accountants are starting to face this problem. Physicians are struggling with ways to digest complex observational and clinical studies so that they can communicate with their patients about the risks of various treatments. The computer is fostering an explosion of the possibilities for analyzing data. We data analysts must provide more than just methods, if the new technology is to have a really beneficial impact on society.

In the third area we consider the internal aspects of data analysis. Most data analysis will involve a computer to help process data. But data analysis is more than processing data. John Tukey [18] has used the term data investigation to describe the process of data analysis. If we are to work effectively, we need to interact with the raw data and processed data so that we can control the flow of our investigation. We feel that it is almost never true that a data investigation can be completely prescribed before the data is examined. Rather, the data will give us various indications and we will proceed by choosing alternatives based on these indications.

Our efforts with the TROLL system and TEP have taught us several things. An important one is that if you provide a reasonably flexible system, with good communication among its parts, and a macro language, users will help to create the tools they need to do effective research. But what about the person who wants to investigate data using the existing supply of tools?

We have a long way to go in this area. We need to provide some guidance, after a particular tool has been applied, about where a person might want to go next. A very experienced data analyst has much of this in his head – can we put some of this experience into the software to provide help for less experienced users? Perhaps a menu of possibilities and online or documented examples would help. With so many new tools available, and so much data generated by looking at the original data in many ways, some help is clearly needed. This is the intermediate area between having the data analyzed by a single large data processing operation (hands off) and a collection of many data processing parts with no guidance from the system on how to proceed with the investigation.

Some of us are starting to think about this intermediate course where one could still link the parts of the analysis together in his own way, but where there would be lots of signposts to help. We suspect that we will need all the help we can get from current research in artificial intelligence and cybernetics in order to make progress in this area.

REFERENCES

[1] Andrews, D. F., Bickel, P. J., Hampel, F. R., Huber, P. J., Rogers, W. H., and Tukey, J. W., *Robust Estimates of Location: Survey and Advances,* Princeton, N.J.: Princeton

University Press, 1972.

[2] Becker, R., Kaden, N., and Klema, V., "The singular value analysis in matrix computation", NBER Working Paper 46, 1974.

[3] Bishop, Y. M. M., Fienberg, S. E., Holland, P. W., *Discrete Multivariate Analysis: Theory and Practice*. Cambridge, Mass.: M.I.T. Press, 1974.

[4] Chambers, J. M., "Fitting nonlinear models: numerical techniques", *Biometrika*, **60**: 1–13, 1973.

[5] Efron, B. and Morris, C., "Stein's estimation rule and its competitors – an empirical Bayes approach", *Journal of the Amer. Stat. Assoc.*, **68**: 117–130, 1973.

[6] Gilbert, J., Light, R. and Mosteller, F., "Assessing social innovations: an empirical base for policy", Unpublished Manuscript, Dept. of Statistics, Harvard University 1974.

[7] Hampel, F. R., "Robust estimation: a condensed partial survey", *Z. Wahrscheinlichkeitstheorie*, **27**: 81–104, 1973.

[8] Hoerl, A. E. and Kennard, R. W., "Ridge regression: biased estimation for nonorthogonal problems", *Technometrics*, **12**: 55–68, 1970.

[9] Huber, P. J., "Robust regression: asymptotics, conjectures, and Monte Carlo", *Annals of Statistics*, **1**: 799–821, 1973.

[10] James, W. and Stein, C., "Estimation with quadratic loss", in *Proceedings of the Fourth Berkeley Symposium*, Vol. 1: 91–105, University of California Press, 1961.

[11] Lindley, D. V. and Smith, A. F. M., "Bayes estimates for the linear model", *J.R. Statist. Soc. B*, **34**: 1–41, 1972.

[12] Mallows, C. L., "Some comments on C_p", *Technometrics*, **15**: 661–675, 1973.

[13] Miller, R. G., "A trustworthy jackknife", *Ann. Math. Statist.*, **35**: 1594–1605, 1964.

[14] Mosteller, F. and Tukey, J. W., "Data analysis, including statistics", *Handbook of Social Psychology, 2nd edition*, 80–203, Vol. 2, Lindzey, G. and Aronson, E., (Eds.) Reading, Mass.: Addison-Wesley, 1968.

[15] "TROLL experimental programs", NBER Computer Research Center Documentation Series D0070, M.I.T. Information Processing Services, Cambridge, Mass.

[16] Tukey, J. W., *Exploratory Data Analysis*, New York: Addison-Wesley, 1977.

[17] Tukey, J. W., "Some graphic and semi-graphic displays", in Bancroft, T. A., (Ed.), *Statistical Papers in Honor of George W. Snedecor:* 293–316, Ames, Iowa: Iowa State University Press, 1972.

[18] Tukey, J. W., "Data analysis, computation and mathematics", *Quarterly of Applied Mathematics:* 51–65, 1972.

[19] Welsch, R. E., "Graphics for data analysis", *Computers and Graphics*, **2**: 31–37, 1976.

[20] Wermuth, N., "An empirical comparison of regression methods", Unpublished Doctoral Dissertation, Department of Statistics, Harvard University 1972.

INCENTIVES AND INFORMATIONAL CONTROL

INTRODUCTION

The papers in this section respond to what I believe to be a major revolution in social organization and control, one that is based less on technology than on the way human beings relate to each other.

Traditional forms of organization in business and industry developed chiefly from patriachical forms at the beginning of the industrial revolution. They now consist essentially of hierarchical structures of rational control with a single, formal, goal-setting authority at the top and a great many workers at the bottom. Downward movement through this organizational pyramid is correlated with a decrease in responsibility, with a decrease in the span of control, with a decrease in freedom and, what is probably most important, with a decrease in the satisfaction of work and the feeling of individual worth.

The revolution that is taking place in industrially advanced societies has been called variously "a transition to post-industrial society", "a drift towards industrial or economic democracy", etc. This revolution is also tied to advances in cybernetics. On the one hand, cybernetics has fostered the progressive automation of production with consequent changes in the composition and educational prerequisites of the labor force and the depreciation of menial jobs in particular. On the other hand, with its emphasis on self-government, cybernetics has brought into focus a new concept of man and of work.

Keys to this new concept of man may be found in the increased desire of individuals:

a) to *be informed* about and to comprehend the nature and purpose of their work as opposed to the localization of knowledge at higher levels of the hierarchy, secrecy towards lower levels and the subsequent feelings of arbitrariness and alienation that are found in traditional forms of organization. With this desire, man is seen as seeking a sense of his significance in the context of social organization.

b) to *define their own goals* and to choose among means as opposed to being instructed or programmed by some formal authority. With this desire, man is seen as seeking self-reliance and self-confidence.

c) to *cooperate with others* or to join efforts as opposed to being forced into individual competition, possibly at the expense of others, with the psychological side effects of suspicion, the suboptimal use of skills, the experience of either winning or losing, etc. Here man is seen as a member of a community that offers some security but considers differences and even conflicts to be a part of social reality.

d) to *engage commitments* to work and to other human beings as opposed to competing for individual rewards and avoiding punishments. With this desire, man is seen as motivated to work because it is intrinsically satisfying, because it is fun and stimulating.

e) to *freely communicate* emotions and to appreciate the idiosyncrasies of other human beings as opposed to the restrictions of prescribed formal-rational discourse in traditional organizational settings and the reduction of the individual to a functional commodity. With this desire, man is seen as seeking his own individuality.

Some of these concepts have provided the basis of a recent criticism of the behavioral sciences and organizational theory [1] which comes to the conclusion that current knowledge is underdeveloped if not unable to guide the management of social organizations in any way other than through traditional authoritarian control structures. This unfortunate state of affairs is magnified by the increasing gap between theory and practice which may be unbridgeable for some time to come. In the United States, experiments in social organization can be traced to the depression years and were common during World War II. Recent experiments by Procter & Gamble and by Nabisco have been well publisized. But attention to experiments with the participation of workers in decision making, teamwork, decentralization, profit sharing, etc., has now shifted to Europe where experiment are being conducted on all levels and in many countries – Yugoslavia, Great Britain, and West Germany for example [2]. Problems of collaboration among the members of the European community are also being solved in novel ways. Organizational changes of this kind present a considerable challenge to behavioral scientists, organizational theorists and economists who by and large have not been able to keep up with advances in organizational practice.

It is therefore fortunate to be able to include in this volume a survey by William A. Hetzner of various forms of work reorganization in six European companies. In this paper Hetzner distinguishes and gives separate accounts of four kinds of reorganization activities:

1) attempts to delegate the responsibility for production output to a group rather than to individual workers;

2) attempts to involve workers in decision making processes regarding plannin organization and quality control;

3) attempts to enlarge the task of individual workers in order to combat mon tony and rigid specialization; and

4) attempts to replace the foreman who functions as a formal interface betwe workers and management by team leaders who emerge in the process of work.

Drawing upon a large number of interviews, the author discusses the successes and failures of these organizational changes and links them to socio-technical constraints. Acknowledging the absence of adequate theory, Hetzner calls for a systematic study of the growing number of organizational experiments being conducted in the United States and in Europe.

The other four papers in this section are more formal in nature and attempt to solve some of the conceptual problems created by these changes.

In the first of these papers, Leonid Hurwicz makes a distinction between classical control theory and structural change theory and, considering the latter, identifies information and incentives as its most important concepts. He observes that the behavior prescribed within an organization and the incentives offered to its members may be at odds with each other and may push the organization as a whole in an unanticipated direction or force it to produce suboptimally. Conversely, organizational structures may be described as optimal whenever given incentives move the behavior of that organization in a prescribed direction. This formulation leads Hurwicz to pose several interesting questions, not all of which have been answered. For example, do incentive compatible mechanisms exist that would guarantee optimality (not possible for economies involving public goods). Do equilibrium conditions exist that guarantee truthfulness? Etc.

Hurwicz' work has been a major source of influence on the economic analysis of organizations. He was the keynote speaker for the session of the conference from which these papers stem. Drawing on his work the second and third papers by Paul R. Kleindorfer and Murat R. Sertel respectively, take up separate issues in the design of enterprises from the point of view of incentives and information. These authors investigate several aspects of a formal representation of enterprises which they had previously developed together. Their concept of an enterprise consists of a group of economic agents including workers and an owner or manager, whose joint efforts produce an output, which is then shared among the agents according to some rule. The sharing rule acts then as an incentive scheme for the enterprise. This formal system leads to a variety of theorems with organizational interpretations.

Kleindorfer considers the effects of unequal distribution of knowledge among economic actors, particularly imperfect knowledge on the part of workers concerning production, capital and labor of others. The behavioral and welfare consequences of such imperfections are examined for alternative production relations and different sharing rules are presented, e.g. incentives based on individual inputs vs. rewards based on divisional outputs or the outputs of an enterprise as a whole. Solutions to such formal problems may provide the key to understanding the economic aspects of the participation of workers in decision making and may shed light on informational and incentival conditions for the successful organization of enterprises.

Sertel's paper compares the effects of two incentive schemes, constant wage and output sharing, on the efficiency and equity of individual enterprises. It turns out that, ignoring the costs of monitoring, there exists a wage scheme which is superior to all constant-share incentive schemes from the point of view of the capitalist, though not necessarily from that of the worker. However, considering the costs incurred for information gathering and processing, the superiority of one scheme over the other is not so clear. In his paper, Sertel considers a variety of ways for deciding on incentive schemes, problems of setting the level of employment, etc., and discusses their politico-economic implications.

In the fourth paper, Steven A. Ross takes up another issue that lies at the root of cooperation between unequal parties. He notes that from an economic perspective, decision theory has nearly always assumed that individuals make rational, self-interested choices. And yet there are many social situations in which a decision maker, the agent, possesses but cannot communicate all the information a second party, the principal, would need to solve his problem. The agent is therefore forced to make choices that affect both the principal and himself. Such a relation between parties exists, for example, between a physician and his patient, between a parent and his child, between an expert consultant and the firm which hires him, and between a government and its citizens. Two equally unacceptable conceptualizations of such a situation are that the agent maximizes his own gain or that he puts the welfare of the principal above all. The first is considered unethical and the second unrealistic. Ross' algebraic theory of altruistic choices provides a solution to this agency problem and thus paves the way for an economic interpretation of this social relationship.

In the fifth paper of this section, Henry Tulkens responds to the need for adequate theory guiding decisions within the European Economic Community. The EEC is a historically unprecedented collaboration of autonomous states with little or no theory to understand, predict or guide their bargaining procedures and their outcomes in conflict situations. The specific problem addressed by Tulkens is the pollution of the southern part of the North Sea to which each adjacent country contributes and by which each is affected. The paper introduces "polluting activity" and "environmental quality" parameters into a general equilibrium model of an international economy and finds, naturally, that "Laissez-faire" equilibria are non-optimal in the Pareto sense because the level of polluting activities exceeds the assimilative capacity of the waters. The paper then develops a dynamic model of negotiations between the countries involved which describes a path of joint decisions leading the international economy from a laissez-faire equilibrium to a Pareto optimum. Finally a variety of measures are investigated such as the creation of an agency for fixing the costs of waste discharge into the North Sea. This and other measures are examined from the point of view of each country's interests and the nature of the informational exchanges required for their operation, is analyzed.

Although these papers are quite different in terms of formal rigor, scope, and anticipated practical implications, they all reflect the need to incorporate information, communication, incentives and interests into models of human behavior whether on an individual or organizational level. They thus describe aspects of what I consider to be a new image of man in the context of the economics of social organization.

REFERENCES

[1] Argyris, C., *The Applicability of Organizational Sociology,* Cambridge University Press, 1972.

[2] Jenkins, D., *Job Power: Blue and White Collar Democracy,* Garden City, N.Y. Doubleday, 1973.

ON THE INTERACTION BETWEEN INFORMATION AND INCENTIVES IN ORGANIZATIONS

LEONID HURWICZ
University of Minnesota

Although the title of my paper refers to organizations in general, the reader should be warned that my thinking has evolved primarily in the context of economic organizations – be they whole nations or individual firms. I do believe that my framework, even though developed for dealing with economies, is also relevant to other types of organizations. It is, however, a framework quite distinct from that familiar from control theory. Since the term "control" appears in the title of this conference, as well as of this session, I shall start by showing how I see the relationship of my framework, to be called *structural change framework*, to the *control theory framework*. I shall follow this up by formulating first the informational and then the incentival aspects of structural change in organizations, and then look at the problem of choosing organizational structure given the interactions between these two aspects.

CONTROL THEORY VERSUS STRUCTURAL CHANGE FRAMEWORK

To begin with, there are important features common to the two approaches. Both are *normative* in spirit. That is, they do not accept the status quo, but rather look for modes of intervention that would bring the system as close to optimality as possible. Thus the mode of intervention is the unknown of the problem. But while rejecting a purely passive attitude toward the workings of the system, they also try to avoid the danger of Utopianism by taking into account the constraints to which intervention and its effects are subject.

The two approaches differ, however, with respect to the form of intervention they envisage. This can be seen against the background of a simple dynamic model governing the behavior of some *state variable*, to be denoted by $\mathbf{x}$, usually a vector. Since the model is dynamic, the state variable is a function of time, say

$x(t)$. The laws according to which it varies can be formulated as systems of differential equations, as is customary in much of control theory, or systems of difference equations, as is often convenient in economics. At this point of our presentation, in order to facilitate comparison, we shall use the differential equation formulation.

Using a dot over a symbol to denote the derivative with respect to time, we can represent a purely autonomous dynamic system by the (vectorial) equation

$$\dot{x} = f(x) \tag{1}$$

Assuming $x = (x_1, \ldots, x_n)$, this is equivalent to the system

$$\frac{dx_i}{dt} = f^i(x_1, \ldots, x_n) \quad i = 1, \ldots, n \tag{1'}$$

Now the control theory model introduces another (vectorial) variable, the *control variable*, denoted by **u**, assumed to be subject to our manipulation. Since the control variable influences the behavior of the state variable, the dynamic system represented by (1) above is replaced by

$$\dot{x} = g(x, u) \tag{2}$$

If we think of nonintervention as corresponding to setting the control variable at zero, we may consider (1) to be a special case of (2), with

$$f(x) = g(x, 0) \tag{3}$$

Given a criterion of optimality, the problem is that of finding the appropriate time pattern $u(t)$ to be followed by the control variable. But reality is usually complicated by various perturbations, so that instead of dealing with a system governed by (2) we have

$$\dot{x} = h(x, u, y) \tag{4}$$

where **y** represents the disturbance. Again, (2) may be regarded as a special case of (4) with $y = 0$, so that

$$g(x, u) = h(x, u, 0) \tag{5}$$

Typically, the perturbation **y** can be neither predicted in advance nor even observed after the event, although we may assume it to be a random variable with known probability distribution. It is clear that in such circumstances it would be inadvisable to choose the time pattern $u(t)$ before observing the impact of the disturbance on the state variable. It is possible, however, to determine the optimal *policy* concerning the selection of control variable level as a function of the state variable that will be observed as the process develops. Denoting this function (sometimes called *feedback* controller *synthesis* function) by ψ, we can express the dependence of the control variable on the state variable by

$$u(t) = \psi[x(t)] \tag{6}$$

If this feedback synthesis is adopted as the rule governing the control of the system, the behavior of the system is described by substituting (6) into (4), i.e. by

$$\dot{x} = h[x, \psi(x), y] \tag{7}$$

Thus, although the direct instrument of control is the variable **u**, the unknown of the problem becomes the synthesis function ψ, since – by hypothesis underlying the control theory model – we cannot modify the laws governing the system as expressed by the function **h** and by the probability distribution of the perturbation **y**. It is in this latter respect that the structural change framework is different.

It is, of course, natural to regard the *structure function* **h** in (4) as immutable if it represents a law of nature, say the Newtonian laws of mechanics. But if (4) represents a human system, some of the components h^i of the structure function $h = (h^1, \ldots, h^n)$ represent the behavior patterns and decision rules that are being followed by those operating the system, and these are far from immutable. They can be changed or influenced by laws, regulations, reward and punishment structures, etc. Hence it is possible to consider the function **h** itself as an unknown of the problem, to be selected in an optimal manner. It would, of course, be unrealistic to suppose that behavior can be imposed in any desired manner; hence the unknown function **h** will be assumed to be selected from some family H of *a priori* admissible functions, with H representing the known invariants of human behavior and other constraints. This formulation, involving optimization with respect to the structure function **h** over the *a priori* admissible family H is characteristic of what we have called the structural change framework.

Let h_0 denote the structure function in the absence of any intervention and h_* the changed structure. It is possible to think of the control theory model as a special case of the structural change model in which

$$h_*(x, u, y) = h_0(x, \psi(x), y) \tag{8}$$

i.e. where the change in the structure function is effected by feedback synthesis. An important example arises in connection with influencing an economic system through properly designed reward formulas. To see this we only need to regard the reward, say allocation of goods, as a control variable, with output as a state variable and the reward formula as the feedback synthesis relating the reward to output. If u_i and x_i denote respectively the reward to and the output produced by the i-th individual, the reward (synthesis) formula can be written as

$$u_i = \psi^i(x_i).$$

In economics one often thinks of the component structure functions h^i as expressing either natural or imposed rules of behavior. An example of such a rule is a firm's decision-making behavior based on profit maximization with prices treated parametrically, i.e. regarded as uninfluenced by the firm's decision. Such behavior would be natural for a very small firm in a very large economy. On the other hand, such behavior could be imposed by law in an economy with only a few large firms. Indeed, the Lange–Lerner model of a socialist price-guided economy postulated the imposition of this rule in the absence of increasing returns to scale. It is simple to interpret such a proposal within the structural change framework, with h_0 corresponding to natural behavior of large firms (presumably oligopolistic or monopolistic – which involves nonparametric treatment of prices) and h_* corresponding to the imposed rule of profit maximization with parametrically treated prices. This interpretation disregards the question of enforcement of the imposed rule. If the enforcement is explicitly built into the model, the laws governing the penalties for disobeying the imposed rules could be interpreted along the lines of the control theory framework, with penalties as control variables and the laws as feedback synthesis functions.

We see, therefore, that there is no conflict between the two frameworks and the choice between them is to a large extent a matter of convenience in the context of a particular problem. Although we shall from now on work within the structural change framework, there would be little difficulty in translating the discussion into control theory language.

STRUCTURAL CHANGE IN AN ECONOMIC ADJUSTMENT MODEL

In this section we shall consider problems of structural change against the background of an economic adjustment process model which has certain affinities with the simplest of the above models, viz. that of Eq. (1), in that it lacks both control variables and perturbation factors, although it differs from (1) in that it uses difference rather than differential equations. The basic state variables of the adjustment process model are *messages* transmitted between participants. For the sake of simplicity we treat messages as if they were broadcast to all participants so that only the sender but not an addressee need be specified. We denote by $m^i(t)$ the message sent out by participant i at time t. In order to make the model applicable to a variety of organizational (in particular, economic) structures, we do not restrict the nature of messages; they can be verbal or numerical (scalar or vectorial) and even pictorial (graphs of functions, etc.). The set of signals eligible to be used as messages is called a *language* and is denoted by M. Corresponding to the structure functions f^i in Eq. (1′) above, we encounter in an adjustment process *response functions* (also denoted by f^i) which determine the messages emitted by participants at a point in time, given earlier messages and information acquired by the participants independently of the message exchange process. Assuming we denote the information acquired independently of the message process by the i-th participant to be constant over time and denoting it by z^i, we represent the adjustment process by the difference equation system

$$m^i(t+1) = f^i[m(t); z^i] \qquad t = 1, 2, \ldots; i = 1, 2, \ldots, N \tag{9}$$

where N is the number of participants and $m(t) = [m^1(t), \ldots, m^N(t)]$. We shall further suppose that the process possesses a unique *equilibrium* (stationary) message N-tuple $\overline{m} = (\overline{m}_1, \ldots, \overline{m}_N)$ defined by

$$\overline{m}^i = f^i(\overline{m}; z^i) \qquad i = 1, 2, \ldots, N \tag{10}$$

Furthermore, as is often true in economics, we shall focus on the system's performance at this equilibrium position. (This can in some cases be justified by postulating rapid convergence of the system to its equilibrium values, but such an assumption should be regarded as only one of methodological convenience; it shows how much work still remains to be done in this area.) Now Eq. (10) only specifies the equilibrium position of the message exchange process; a transition must be made to events in the sphere of real phenomena. For the economist, these phenomena are the resource flows between participants (exchange) and other economic actions, including consumption and production. Thus the economic process is not specified until a rule is present determining the resource flows given the outcome of the message exchange process. We therefore complete our model by introducing the *outcome function* ϕ which determines the equilibrium resource flow $\overline{a}$ resulting when the equilibrium message N-tuple is $\overline{m}$, i.e.

$$\overline{a} = \phi(m) \tag{11}$$

[For instance, in the case of pure exchange, the outcome $\overline{a}$ can be thought of as a N by N matrix whose (i, j)-entry is $a_{ij} = (a_{ij1}, \ldots, a_{ijL})$, with a_{ijk} denoting the net flow of commodity k from participant i to participant j in a world with L commodities.]

Adopting the point of view of the structural change framework, we now consider the problem of the designer of the system. The designer is assumed to have an evaluation criterion which he applies to the outcome $\overline{a}$ of the process. For instance, economists often use the criterion of Pareto-optimality of the outcome. (An outcome is defined as Pareto-optimal relative to the possibilities and preferences characterizing an economy if there is no other possible outcome more attractive for some participants and equally attractive for others. A complete specification of possibilities and preferences is called *environment* and denoted by **e**; a class of environments is denoted by E.) Noting that optimality is relative to the environment, one may ask whether a given "*economic mechanism*" (defined by the language M, the response function $f = (f^1, \ldots, f^N)$, and the outcome function ϕ) guarantees the optimality (here and below this term is understood in the Pareto-sense) of outcomes relative to all environments **e** belonging to a specified class E. For instance, the so-called perfectly competitive mechanism has been shown to yield optimal outcomes for what are sometimes referred to as "classical" environments, where the sets and functions characterizing the environment have certain convexity and continuity properties and where externalities and public goods are absent. On the other hand, for nonclassical environments the perfectly competitive mechanism may fail to have equilibrium positions altogether (e.g. due to increasing returns, i.e. to nonconvexity) or may produce nonoptimal equilibrium outcomes (e.g. in the presence of externalities such as pollution).

Within the structural change framework let us look at the problem of a system designer whose evaluation criterion is the Pareto-optimality of equilibria. His situation is very different depending on whether the mechanism is being designed for operation in classical environments only – or for some broader category of environments. When the mechanism is being designed for operation in classical environments only, there is an obvious candidate for a mechanism to be adopted, viz. the perfectly competitive mechanisms, since in classical environments equilibrium positions exist and yield optimal outcomes. But we do not know as yet whether this candidate is eligible in the sense of belonging to the family of feasible mechanisms (the *a priori* admissible family of structure); in fact, we have not as yet formulated any criteria for eligibility for membership in the family of feasible mechanisms (admissible structures). Although one obviously cannot hope for an exhaustive listing, we distinguish two important categories of requirements: informational and incentival. In the following section we focus on the informational requirements, to the exclusion of incentival considerations.

INFORMATIONAL PROPERTIES OF ADJUSTMENT MODELS

As just indicated, the discussion of this section is focused on the problem of designing a mechanism (i.e. finding a language M and functions **f** and ϕ) guaranteeing the existence and optimality of equilibria for a specified class E of environments (whether restricted to classical ones or broader). Since we are at this stage disregarding the issue of incentives, we shall not ask whether the participants would be inclined to obey the rules of the prescribed mechanism, but only whether rules can be found that would – if obeyed – yield optimal outcomes. However, we shall be concerned about the informational characteristics of the rules.

This concern is primarily related to the fact that, typically, no single participant has – at least at the beginning of the adjustment process – the information required to determine which resource flows would be optimal. As we have seen above, optimality depends on possibilities and preferences. In turn, the possibilities of the economy depend on the availability and distribution of productive resources among the participants, to be represented by the *initial endowment* N-tuple $w = (w^1, \ldots, w^N)$, where w^i denotes the resource endowment of the i-th participant, as well as on the *technologies* of the society's producers, to be represented by the technology N-tuple $Y = (Y^1, \ldots, Y^N)$, with Y^i denoting the technology of the i-th producer. Similarly, preferences are represented by the N-tuple $R = (R^1, \ldots, R^N)$, with R^i representing the preference relation (map, pattern) of the i-th participant. Merely to determine whether a given resource flow is possible one would have to know something about each of the initial endowments and technologies; to answer questions of optimality, one would also need to know the various preferences. In practice, different decision makers are likely to have only partial information about these matters. One simple although extreme assumption is that each participant knows his own components of the three N-tuples w, Y, and R but no one else's, i.e. that the i-th participant knows his characteristic $e^i = (w^i, Y^i, R^i)$ of the environment but is completely ignorant of all the other characteristics e^j (j different from i). We shall refer to such a situation as *initial dispersion* of information.

Assuming such initial dispersion of information and (for the sake of analytical simplicity) absence of memory or learning, we may replace in Eqs. (9) and (10) the symbol z^i (representing the information acquired by participant i independently of message exchange) by his characteristic e^i. For this is now the only information – other than messages just received – he is assumed to have available in arriving at the message $m^i(t + 1)$ to be emitted during the given time interval. Such a process, written as

$$m^i(t+1) = f^i(m(t), e^i) \quad (9'); \qquad \bar{m}^i = f^i(\bar{m}, e^i), \quad (10'),$$

is called *privacy-preserving*.

The property of being privacy-preserving partly captures the notion of *informational decentralization* underlying much of the thinking in this area. However, the initial dispersion of information would not be much of an obstacle to the determination of optimality if it was easy to transmit the individual characteristics e^i and other information from one participant to another. For in that case each participant could send, at time $t = 0$, a message $m^i(0)$ containing the complete description of his e^i to (say) the first participant; aside from problems of computational capacity, the first participant could then determine an optimal resource flow pattern and, in turn, inform all participants of appropriate actions to take. Thus to supply an essential element of *informational decentralization* we must express analytically the limitations on the capacity to transmit and/or process information. A natural first step is to assume limits to the complexity of the messages to be transmitted and to the complexity of algorithms implicit in calculating the values of the response functions. We shall concentrate on the messages.

A clue as to these limits is obtained from a study of known mechanisms, in particular the perfectly competitive. So far we had been considering the messages as elements of some arbitrary abstract set M called the *language*. To transmit complete information as to a producer's technology, the language would have to contain as elements sets of a variety of shapes in the L-dimensional space, where L is the number of commodities. The transmittal of preferences might require even a richer language. In any case, even for classical environments, the language M would be larger (in the sense of cardinality) than any finite-dimensional space. I.e. even for a classical economy with specified numbers of participants N and commodities L, there is no finite integer K such that complete information about individual characteristics could be conveyed by messages consisting of K-dimensional vectors. Yet the competitive process provides an algorithm which is privacy-preserving and yields optimal equilibria with a language of dimension $K = N.(L-1)$.

It is also important to note that this is accomplished by the competitive process with smooth (e.g. continuously differentiable) response and outcome functions. (The dimensionality of the message space is not invariant and hence not meaningful without some smoothness restrictions on the functions used by the mechanism; hence a smoothness restriction will be implicit in subsequent references to the dimensionality of the message space.)

The foregoing discussion suggests that without some restriction on the dimensionality of the message space (language M) used the notion of informational decentralization loses much of its interest. For this reason we shall call a mechanism *informationally decentralized for a given class E of environments* if its language M is of finite dimension (given N and L), the response functions are privacy preserving and both the response and outcome functions are smooth. Of course, there is nothing sacred about this definition and, indeed, I have used alternative concepts on other occasions. What seems to me important is that if we at all use the notion of informational decentralization, we should make it rigorous enough to give precision to the question of decentralizability of various classes of environments. We shall be calling a class E of environments *informationally decentralizable* if it is possible to design a mechanism that is informationally decentralized for this class and where equilibria exist and are optimal. Thus, for instance, the class of "classical" environments is informationally decentralizable because the competitive mechanism is privacy preserving, smooth, and – for a given N and L – only requires a language of dimension $N.(L-1)$ to guarantee the optimality of equilibria.

The situation is much less clear with regard to environments that are not "classical", e.g. environments characterized by increasing returns, indivisibilities, and various other nonconvexities or discontinuities. Already in the 1930's Lerner, Hotelling, and Lange developed the marginal cost pricing principle which, under certain assumptions can be viewed as a necessary first order condition for optimality. Under somewhat stronger assumptions one can also design a mechanism whose equilibria are characterized by the equality of marginal costs and prices, with prices treated parametrically, and optimality is likely to result, at least for initial positions close to such an equilibrium. However, from recent research work it appears that even relatively mild departures from the "classical" domain may be sufficient to kill decentralizability.[1] Under such circumstances one will have to settle for something less than decentralization – or something less than full optimality.

INCENTIVE PROPERTIES OF ADJUSTMENT MODELS

We have just seen that there are no informationally decentralized mechanisms guaranteeing optimality for nonclassical as well as classical environments, except in certain narrow categories of cases. On the other hand no such informational difficulties arise when we confine ourselves to classical environments. However, even classical environments lack immunity with regard to incentive problems, unless we restrict ourselves to "atomistic" environments – those where every economic unit (firm, household) can be regarded as infinitesimal in relation to the market in which it operates.[2] To understand the issue we shall first introduce the concept of *incentive compatibility*. Although this concept can be formulated in a very general manner, we shall consider it only in a very simplified setting. We shall suppose that a central authority can impose (and enforce) an outcome function ϕ^* of its choosing and that it also, in addition, attempts to prescribe the response functions to be used by the participants in the process. We shall denote the prescribed response function for the i-th participant by f^{i*}. There would, of course, be no problem of incentives if the prescribed response functions could also be enforced without difficulty. We shall assume, however, that each participant is able to depart from the prescribed behavior, although within certain limits. We shall denote by F^i the class of response functions (called *enforceable domain*) the i-th participant can choose from; naturally this class ordinarily contains the prescribed behavior f^{i*}.

Each participant now must decide whether to follow prescribed behavior or whether to choose one of the alternatives available within F^i and if so – which one. Given the behavior of others, and assuming he knows the outcome function, he can calculate the consequences of the alternative behaviors open to him and hence find that behavior which would maximize his level of satisfaction. But since others are facing the same decision problem, it is far from obvious that he can take the behavior of others as given. We recognize the dilemma as that typical of a noncooperative game and hence we approach it in the spirit of the theory of such games. More specifically, we shall use the concept of *Nash equilibrium.*

To do so, we shall find it convenient initially to define *behavior functions* g^i by the relations

$$g^i(m^1, \ldots, m^N) = f^i(m^1, \ldots, m^N, e^i) \qquad i = 1, \ldots, N \tag{12}$$

That is, a behavior function incorporates both the response function and the characteristic and simply describes what message the i-th participant will emit given the messages he has just received. Now the stationary (equilibrium) message N-tuple $\overline{m} = (\overline{m}^1, \ldots, \overline{m}^N)$ is given by the relations

$$\overline{m}^i = g^i(\overline{m}) \qquad i = 1, \ldots, N$$

or, in vector notation,

$$\overline{m} = g(\overline{m}), \qquad g = (g^1, \ldots, g^N) \tag{13}$$

We see that the stationary message N-tuple $\overline{m}$ is a fixed point of the behavior function **g**. Assuming this fixed point to be unique, we see that it is determined by the behavior function **g**. We may write this as

$$\overline{m} = \lambda(g). \tag{14}$$

Given the outcome function ϕ^* we can use Eq. (14) to express the outcome **a** as a function of **g**, viz.

$$a = \phi^*[\lambda(g)] \tag{14}$$

and, finally, the satisfaction derived by the i-th participant – measured through a utility indicator u^i – as

$$u^i(\phi^*[\lambda(g)]) = P^i(g) \tag{15}$$

The composite function $P^i(g) = P^i(g^1, \ldots, g^N)$ may be regarded as the payoff function in a noncooperative game in which the behavior functions constitute strategy variables. Each participant has available to him as strategy domain G^i the set of all behavior functions obtainable according to (12) with e^i fixed and f^i roaming over the enforceable domain F^i. Having defined the payoff functions and strategy domains for the game, we now apply the definition of a Nash equilibrium as a strategy N-tuple $g^{\#} = (g^{1\#}, \ldots, g^{N\#})$ at which every player is satisfied with his strategy choice provided that others do not abandon theirs. Formally, $g^{\#}$ is a *Nash equilibrium* for the above game if, for every $i = 1, \ldots, N$, the behavior function $g^{i\#}$ belongs to the domain G^i and

$$P^i(g^{\#}) \geqq P^i(g^{1\#}, \ldots, g^{i-1,\#}, g^i, g^{i+1,\#}, \ldots, g^{N\#}) \text{ for all } g^i \text{ in } G^i \tag{16}$$

Although there are well-known controversies in game theory concerning the merits and weaknesses of this solution concept, we shall take it as representing the natural spontaneous behavior of the participants, i.e. the direction in which the incentives draw them. In the light of this interpretation, it is clear what one means by saying that a given prescription for behavior is compatible with incentives: a situation in which the Nash equilibrium coincides with the prescription. Formally, let g^{i*} denote the behavior functions obtained from (12) by using on the right-hand side the prescribed f^{i*} response functions (and the true characteristics e^i). We define a prescribed mechanism (M, f^*, ϕ^*) an *incentive-compatible* (with regard to the domains F^i) if the Nash equilibrium behavior is the same as the prescribed behavior, i.e. if

$$g^{\#} = g^* \tag{17}$$

A justification of this terminology lies in the fact that, provided we accept the hypothesis of "Nash-like" spontaneous behavior, an incentive-compatible mechanism would be self-enforcing.[3]

In economics, the issue of incentive compatibility became prominent in connection with the search for an allocation mechanism that would generate optimal resource allocation of public as well as private goods, since the competitive process was clearly inadequate for the task. An alternative mechanism proposed by Lindahl satisfied the requirements of informational decentralization and would yield optimal allocations provided that the participants would act on the basis of truthful information about their respective preferences. The difficulty, stressed by Samuelson, was that in an informationally decentralized system one could not enforce truthful behavior and, unfortunately, participants could profit by departing from truth. To use our present terminology, the Lindahl mechanism was not incentive-compatible. Samuelson's conjecture (formulated in the mid-1950's) was that we were facing a fundamental difficulty – that one could not design an incentive-compatible mechanism guaranteeing optimality for economies with public goods. It is only recently that his conjecture has been shown correct. Indeed it has also been shown that a similar difficulty arises in "classical" environment economies lacking public goods unless we are in the "atomistic" case.[4]

To understand these results we relate them to the preceding formal framework. We may suppose that the central authority has no knowledge of the *true* characteristic, now to be denoted by $\mathring{e}^i$. Hence it is possible for the participant to adopt as his behavior function any g^i satisfying (12) with the prescribed f^{i*} but using some "strategic" (true or false) characteristics, to be denoted by $\breve{e}^i$.[5]

Since under these circumstances, with f^{i*} kept constant, variations in g^i can only be achieved by manipulating $\breve{e}^i$, we may consider the $\breve{e}^i$ to be the strategic variables (with some domains corresponding to "plausible" even though false values). The utilities can now be regarded as functions of the strategic characteristics and a Nash equilibrium is defined in the spaces of the characteristics. Proceeding as above, let $\breve{e}^{\#}$ denote the Nash equilibrium N-tuple of strategic values of the characteristics; we shall call this equilibrium the *Manipulative Nash Equilibrium*. A prescribed mechanism will be called *incentive-compatible over E in the* *narrow sense* if the Manipulative Nash Equilibrium strategic N-tuple of characteristics $\tilde{e}^{\#}$ is the same as the true one $\mathring{e}$, i.e. if

$$\tilde{e}^{\#} = \mathring{e} \text{ for all } \mathring{e} \text{ in } E. \tag{18}$$

Under our hypothesis of spontaneous Nash-like behavior the interpretation is that in such a mechanism truthfulness is self-enforcing. Samuelson's assertions can be translated into this framework as stating that no mechanism guaranteeing optimali can also guarantee[6] the self-enforcing truthfulness of (18); in particular it follows that the Lindahl mechanism lacks the property of (18).

So far we have followed the tradition of our subject by focusing on the relationship of the equilibria to truth, or, more generally, of spontaneous self-interested behavior to prescribed behavior. But it is possible to take a broader view and simply ask whether spontaneous behavior will lead to an optimum, without raising the question whether spontaneous behavior obeys the edicts of the center. In fact, it has been shown that in certain environments involving publi goods[7] it is possible to design mechanisms whose Manipulative Nash Equilibria do generate optimal allocations, even though the strategic equilibrium characteristics are not, in general truthful, i.e. where $\tilde{e}^{\#} \neq \mathring{e}$.[8]

Unfortunately, the environments in which such optimal allocations are self-enforcing (in the sense of being Manipulative Nash Equilibria) constitute a small class.[9] For broader classes the situation changes for the worse, so that, in general, one [see Hurwicz, 4] cannot hope for self-enforcing optima, even if we abandon our insistence on truthfulness and obedience to prescribed behavior rules. However, we must remind ourselves that these negative results are obtained subject to the requirement of informational decentralization. When this requirement is abandoned, we can imagine inspection systems that, by violating the privacy postulate, might make departures from prescribed behavior impossible. Again, we seem to run into the dilemma of having to sacrifice either optimality or the self-enforcement feature or informational decentralization.[10]

It should be stressed that the phenomenon of nonoptimality of manipulative equilibria arises even in the economist's standard model of pure exchange (Edgeworth Box) and not merely when there are public goods present. It is useful to consider the classroom example where the environment is represented by the Edgeworth Box with two goods and two traders, and the prescribed mechanism is that of perfect competition. We know that if the environment is classical, (i.e. preferences convex, etc.) and the participants follow the prescribed rules, the system does possess positions of equilibrium (the competitive equilibrium) and these positions are optimal. But suppose that the participants' behavior is manipulative. What allocations will be generated by the Manipulative Nash Equilibria? It turns out that there is an infinity of such allocations (indeed a continuum) and that they fill the interior of the set bounded by the so-called offer curves[11] of the two traders. We may note that none of these allocations is optimal, although som of them are arbitrarily close to an optimum. Unfortunately, others are arbitrarily close to the initial allocation, i.e. as far away from an optimum as one can get under a noncoercive mechanism. (See Figure 1.)

There does not as yet exist an approach that would narrow down a range or indeterminacy in such situations, nor is there an agreed upon way of arriving at

ω: Initial endowment

P', P": Pareto-optimal allocations

Γ^1, Γ^2: Offer curves

N: Manipulative Nash allocations (interior of lens-shaped area, not including the point c)

c: True competitive equilibrium

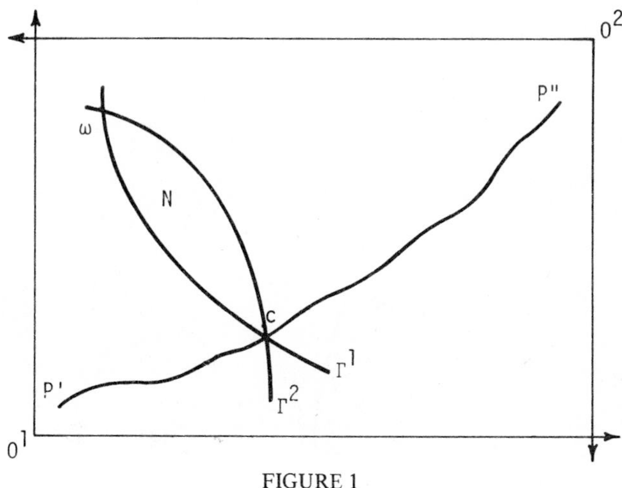

FIGURE 1

some average or minimax measure of inefficiency of such a mechanism. Yet there is an urgent need for progress in this direction since our general theorems tell us that any alternative mechanism will have some of the same problems.

Up to this point we have been dealing with situations in which the participants could manipulate the preference components of their characteristics, i.e. "misrepresent" their preferences through strategic behavior. However, one could imagine an economy in which preferences are centrally known with sufficient accuracy, but there is no central knowledge as to productivity. The problem then arises as to whether such partial central knowledge is sufficient to install a self-enforcing mechanism with optimal equilibria. It turns out, for a change, that the answer is in the affirmative. To indicate the nature of the solution, with its strengths and weaknesses, we shall outline it briefly in the context of two participants, each of whom has preferences and is also a producer. Since preferences are assumed known, the center can adopt a standard procedure for representing these preferences by utility indicators ("canonical" utilities). Let the utilities be denote by u^i and the production possibility sets by Y^i, $i = 1, 2$. Given these, there is a set of feasible aggregate productions and distributions and, consequently, a set – to be denoted by $U(Y^1, Y^2)$ – of the feasible utility vectors (u^1, u^2). The usual

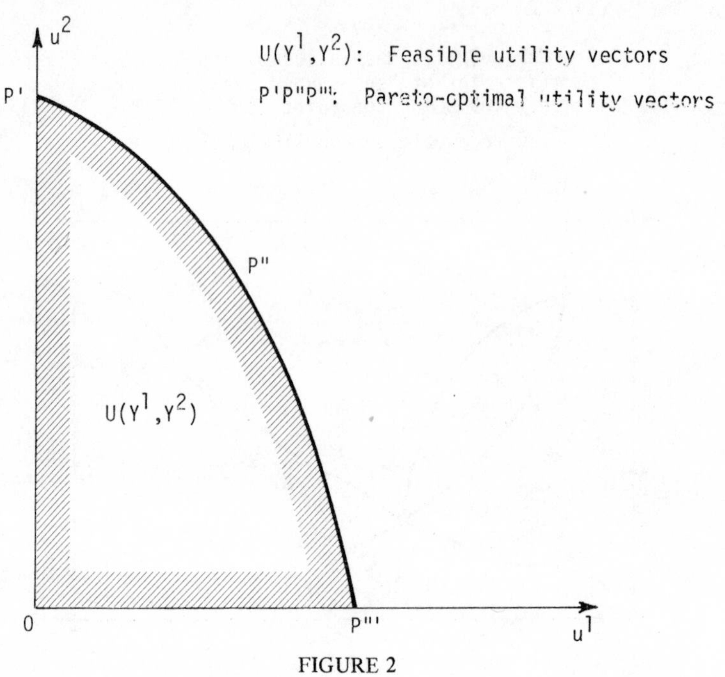

FIGURE 2

diagrammatic representation is in the form of a curvilinear triangle in the (u^1, u^2) – space (see Figure 2). Its "hypotenuse" $P'P''P'''$ represents the various optimal utility combinations.

Now let the allocation rule be formulated as follows. The center fixes the ratio of utilities it regards as desirable, say $u^1 : u^2 = k$. (The desired ratio **k** can, but need not, equal one.) Suppose that the first participant misrepresents his production possibility set and claims some false $\tilde{Y}^1$ instead of the true $\mathring{Y}^1$. We postulate that he cannot with impunity exaggerate the size of his set, so $\tilde{Y}^1 \subset \mathring{Y}^1$. Hence $U(\tilde{Y}^1, \mathring{Y}^2)$ is bound to be smaller than (contained in) $U(\mathring{Y}^1, \mathring{Y}^2)$.

Now the procedure for allocation is to pick out of $U(Y^1, Y^2)$ the best point on the ray corresponding to the desired ratio **k**. It should be clear from Figure 3 that under such circumstances, the first participant cannot improve his utility by misrepresentation and, in general, will worsen it. (The implicit assumptions are such as to make the "hypotenuse" of the curvilinear triangle downward sloping; this is implied by the continuity and monotonicity of preferences.) Furthermore, the situation is the same whether the other participant is truthful or not. Thus "truth is the dominant strategy". But truth generates optimality, since the point

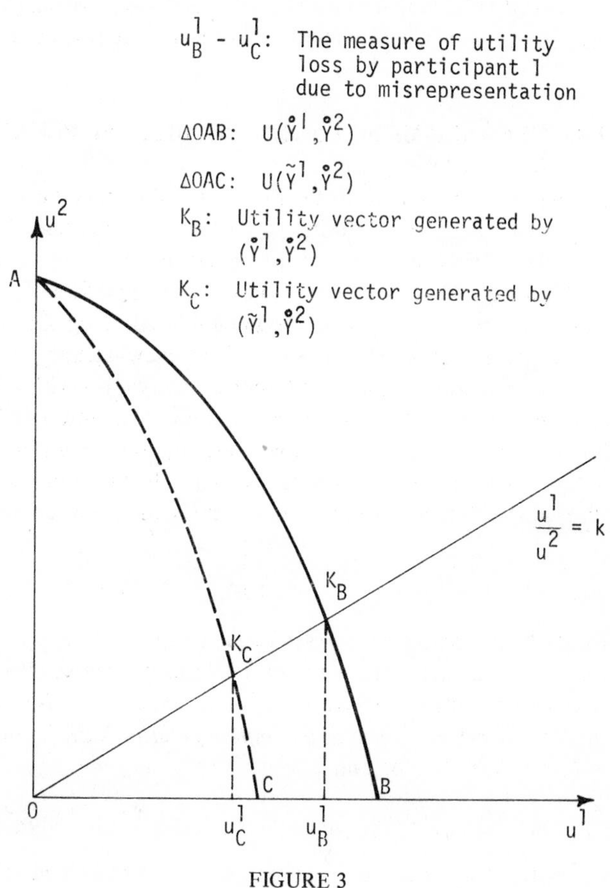

FIGURE 3

chosen is at the intersection of the ray corresponding to **k** with the "hypotenuse", and all points of the "hypotenuse" are optimal.

The preceding example illustrates the interaction, referred to in the title of this paper, between incentives and information. We had seen earlier that there is no self-enforcing mechanism guaranteeing optimality when preferences are unknown to the center (and so preference-manipulation possible). In the example with production we see that the situation changes drastically when preferences are assumed known to the center, even though production sets are still unknown: here we do have a self-enforcing mechanism guaranteeing optimality at the Nash Equilibrium. Suppose, however, that preferences, instead of being centrally known, can become known with a certain expenditure of resources on inspection, etc. Would it be worth acquiring such information? The answer depends on a comparison of the loss of efficiency in the absence of such information (a loss to be expected from our general theorems) with the cost of acquiring the information.

"SYNTHETIC" MECHANISMS FOR NONMANIPULATIVE BEHAVIOR

There are differences of opinion as to the likelihood of manipulative behavior postulated in the preceding section. Although my own inclination is to regard such behavior as very likely under many circumstances, there are situations where nonmanipulative behavior may well prevail. It is therefore of considerable interest to investigate the possibility of designing self-enforcing allocation systems in the absence of manipulation. Here the model is somewhat simpler. We still have the outcome function ϕ which is assumed imposed and enforceable; they constitute the incentive structures. Let the utility function of the i-th individual be written as $u^i(a)$ where a is the resource flow. Since the outcome function makes the resource flow a function of the message N-tuple, the utility becomes a composite function of this N-tuple; we shall denote this composite function by v^i. That is,

$$v^i(m) = u^i(a) = u^i[\phi(m)] \qquad m = (m^1, \ldots, m^N), i = 1, \ldots, N \qquad (19)$$

Now in a world of nonmanipulative behavior each participant uses his true preferences and the i-th message component m^i becomes his strategy variable. Again treating this as a Nash noncooperative game, we regard v^i as the i-th payoff function. Thus an N-tuple m^* is defined as a *Nonmanipulative Nash Equilibrium* if, for every $i = 1, \ldots, N, m^{i*}$ in M, and

$$v^i(m^*) \geqq v^i(m^{1*}, \ldots, m^{i-1,*}, m^i, m^{i+1,*}, \ldots, m^{N*}) \text{ for all } m^i \text{ in } M \qquad (20)$$

Now it turns out that, under certain assumptions of convexity and differentiability, it is not difficult to find outcome functions that guarantee the optimality of resource flows at equilibrium. (A trail blazing example is due to Groves and Ledyard [1] for the case of public goods.) What is perhaps particularly interesting is that one can obtain (necessary) conditions in the form of differential equations defining rather large classes of such outcome functions and then, as it were, manufacture a variety of allocation systems possessing the desired properties by finding functions satisfying these differential equations. We shall illustrate this by the pure exchange (Edgeworth Box) example, partly because of its simplicity – but also to show a "synthetic" alternative to the conventional competitive system.

Let there be two goods **x** and **y** and three traders. The outcome functions (components of ϕ) will be denoted by X^i and Y^i. Thus X^i determines the net trade (increment) in terms of good **x** going to the i-th trader, and similarly for Y^i. We assume the language M to be the real axis; i.e. every message m^i is a real number. We do not provide any interpretation for the possible meanings of these messages. $X^i(m^1, m^2, m^3)$ is the net increment in holdings by **i** of good **x** when the respective messages emitted by the three participants are m^1, m^2, m^3; similarly for $Y^i(m^1, m^2, m^3)$. Now a necessary condition for the optimality of equilibria[12] turns out to be

$$X^1_1(m^*) \,/\, Y^1_1(m^*) = X^2_2(m^*) \,/\, Y^2_2(m^*) = X^3_3(m^*) \,/\, Y^3_3(m^*) \qquad (21)$$

where the subscript **i** indicates partial differentiation with respect to m^i and all derivatives are evaluated at the equilibrium point $m^* = (m^{1*}, m^{2*}, m^{3*})$.

In addition, since we are assuming pure exchange, at equilibrium the (net) trades must add up to zero, i.e.

$$X^1(m^*) + X^2(m^*) + X^3(m^*) = Y^1(m^*) + Y^2(m^*) + Y^3(m^*) = 0 \qquad (22)$$

While these conditions are only necessary, they turn out to be sufficient when second-order derivatives have appropriate signs. Also, although conditions (21) and (22) are only required to hold at an equilibrium, it does not hurt if they hold for all values of **m**, and it turns out easy to find such functions, even among quadratics. As an example of such a "synthetic" system we may give the following

$$\left\{\begin{aligned} Y^i(m) &= m_i - (\tfrac{1}{2})(m_j + m_k) \\ X^i(m) &= -m_i(\tfrac{1}{2}m_1 + m_2 + m_3) + \\ &\quad +(\tfrac{1}{4})(m_j^2 + m_k^2) + 2m_j m_k \end{aligned}\right\} \qquad \begin{aligned} &i = 1, 2, 3; i \neq j \neq k \neq i \\ &j, k = 1, 2, 3 \end{aligned} \qquad (23)$$

(Here we use subscripts on the **m**'s to avoid confusion with algebraic exponents.) It is easily verified that Eqs. (22) are identically[13] satisfied. Also,

$$Y^i_i(m) = 1 \quad \text{and} \quad X^i_i(m) = -(m_1 + m_2 + m_3) \quad \text{for all } \mathbf{m} \text{ and } i = 1, 2, 3 \quad (24)$$

Hence Eq. (21) are also identically satisfied. Finally, second order conditions can be shown to hold also. Therefore, the incentive structure (outcome function) specified by Eq. (23) guarantees the optimality of Nonmanipulative Nash Equilibria defined in (20).

A warning is in order. We do not claim that the above allocation system has all the properties that one might wish for. However, it is interesting to compare it with the competitive system. They both guarantee optimality of equilibria. The "synthetic" system in (23) is probably inferior with regard to the domain[14] of existence of equilibria. Also, unlike the competitive system, the "synthetic" system can place participants at utility levels inferior to initial ones. On the other hand, it happens (although this cannot be regarded as crucial) that the "synthetic" system sometimes uses a message space of lower dimensionality (its dimension is three) than that version of the competitive system whose equilibria can be regarded as Nash Equilibria (the dimension there being four, three quantities and a price). What does seem important is our ability to manufacture such synthetic systems from purely mathematical considerations rather than having to rely on precedents of observed economies.

INCENTIVE STRUCTURES FOR UNILATERAL MAXIMIZATION

In the preceding sections we have used as our optimality concept the Pareto definition which treats all participants' welfare symmetrically. We sometimes encounter problems, however, in which the incentive structure is set up in such a manner as to maximize the degree of attainment of objectives for one of the participants through incentives designed to intensify the efforts of others.

We shall illustrate this by two examples, one from the private sector and one involving public welfare.

The private sector example, in a very simplified form, deals with the problem of a landowner whose land is being worked by a sharecropper. Ignoring inputs other than the laborer's effort **z**, we postulate that his output **y** is a strictly increasing function **f** of effort; his utility function **u** is assumed to decrease with the amount of effort he expends but increase with the reward **r** (his share[15] of output). The owner wishes to maximize his share of output π. Informational assumptions made are the following: the owner can only observe the output **y** but not the effort **z**; the laborer observes both **y** and **z**; also, the functions **f** and **u** are known to the laborer but not to the owner. The owner's only control "variable" is the reward formula ρ, specifying the laborer's reward **r** as a function of his output **y**. Thus the model can be written as

$$\left.\begin{aligned} y &= f(z) \\ U &= u(r, z) \\ y &= r + \pi \\ r &= \rho(y), \end{aligned}\right\} \tag{25}$$

and the owner's problem is to find a reward formula ρ to maximize π.

To solve the problem, one must make some assumptions about the laborer's behavior. We shall postulate that he behaves in a nonmanipulative manner and seeks to maximize his utility level U given the functions f, u, and ρ. Because of the structure of the model, the laborer's utility can be expressed as a function of effort, viz.

$$V = u[\rho(f(z)), z], \tag{26}$$

or, since the production function is invertible, as a function of output, viz.

$$U = u[\rho(y), f^{-1}(y)] \tag{27}$$

Making the appropriate regularity assumptions, we shall find that there is a unique output level $\bar{y}$ maximizing U for a given reward formula ρ, and a corresponding $\bar{z} = f^{-1}(\bar{y})$. We shall write $\bar{y} = Y[\rho; f, u]$ to indicate the dependence of $\bar{y}$ on the functions in brackets. Now the owner's problem is to maximize his share, i.e. to maximize the expression

$$\pi = Y[\rho; f, u] - \rho(Y[\rho; f, u]) \tag{28}$$

with regard to the reward formula ρ which he is free to choose. The difficulty is, of course, that the solution to this maximization problem depends on the functions **f** and **u** which are unknown to the owner. Hence he is faced with a problem in choice (decision-making) under uncertainty.

To cope with this problem in its full generality one would have to select one's preferred principle of decision-making under uncertainty, be it minimax or Bayesian. In the latter case a prior distribution on the space of function pairs (f, u) – representing the landowner's ideas about the likely forms and parameters of production and utility functions – would also have to be supplied; thus we would be postulating partial, although subjective, information concerning these functions. On the other hand, if the minimax approach is adopted, we could stay with the assumption of complete ignorance (within specified function classes) on the part of the landlord.

We shall not attempt here to deal with the general problem. Instead, we shall consider a very special case where production is characterized by constant returns, so that

$$f(z) = cz, \qquad c > 0, \tag{29}$$

and the sharecropper's utility function is linear in the reward and quadratic in effort,

$$u(r, z) = r - (\tfrac{1}{2})bz^2, \qquad b > 0 \tag{30}$$

We may assume that the landlord knows the functional forms (29)–(30), but not the values of the parameters **b**, **c**.

Furthermore, instead of permitting the landlord to choose any arbitrary reward formula ρ, we shall confine him to fixed share ratios, so that

$$\rho(y) = \alpha y \qquad 0 \leqq \alpha \leqq 1 \tag{31}$$

Thus the landlord's problem boils down to finding the best (from his point of view) value of the parameter α. Now it turns out that for any given value of α, treated parametrically by the sharecropper, the corresponding level of output maximizing the sharecropper's utility will be

$$y = \alpha\hat{y} \tag{32}$$

where

$$\hat{y} = c^2/b \tag{33}$$

is the Pareto-optimal level of output for this economy. Therefore, the landlord's share is

$$\overline{\pi} = \overline{y} - \overline{r} = (1 - \alpha)\overline{y} = (1 - \alpha)\alpha\hat{y} \tag{34}$$

Hence, regardless of the (to the landlord unknown) parameter values, **b**, **c**, the landlord's best strategy is to set $\alpha = \frac{1}{2}$. We note that this choice results in a level of output that is only one half of optimal output.

Now there arises the question whether the landlord could increase his gain if he were to abandon the fixed share ratio formulas (31) and consider other functions ρ of output. This problem has not as yet been completely solved, but we have examined it when the criterion used is that of maximizing the minimum (with respect to the unknown $\hat{y}$) level of $\pi/\hat{y}$. It appears that from this point of view the landlord cannot do better (get a higher minimum guarantee regardless of the parameter values) than by offering the sharecropper the fixed share ratio arrangement with $\alpha = \frac{1}{2}$, as above, with the guaranteed ratio $\pi/\hat{y} = (1 - \alpha)\alpha = \frac{1}{4}$. (Here we continue to assume that Eqs. (29)–(30) hold.)

We have so far been assuming that the landlord has no information concerning the values of the parameters **b**, **c**, and, in particular, does not know $\hat{y}$. As before, we are particularly interested in the changes that occur when the information structure is altered. Now it is not too difficult to see that there exist reward functions that would induce the laborer to select an output level arbitrarily close to the optimal y and with the ratio $\pi/\hat{y}$ arbitrarily near (and above) $\frac{1}{2}$. This can be seen as follows. For a continuously differentiable reward function ρ, the laborer's (interior) first-order utility maximization condition is

$$\rho'(\overline{y}) = \overline{y}/\hat{y} \tag{35}$$

Hence (see Figure 4), for the reward function ρ_1 (whose derivative ρ_1' is indicated by the solid line), the equilibrium output would be $\overline{y}_1$. The corresponding share π_1 of landlord's output is measured by the area above ρ_1', to the left of $\overline{y}_1$ below the line[16] $w = 1$, and to the right of the vertical axis. Now the limiting position of ρ' would make it coincide with the ray[17] ϕ', and in this case the ratio $\pi/\hat{y}$ would equal $\frac{1}{2}$. By choosing a reward function ρ_2 whose derivative ρ_2' is indicated by the broken line in Figure 2, the landlord would induce the output $\overline{y}_2$ very near $\hat{y}$ and his ratio $\pi_2/\hat{y}$ would be very near $\frac{1}{2}$. Thus he would be getting arbitrarily close to his maximum, but he could do this only if he knew the value of $\hat{y}$, which implies (partial) information concerning the parameters **b**, **c**. Otherwise the reward function of the type ρ_2 would be very good for some values of the parameters and very bad for others.

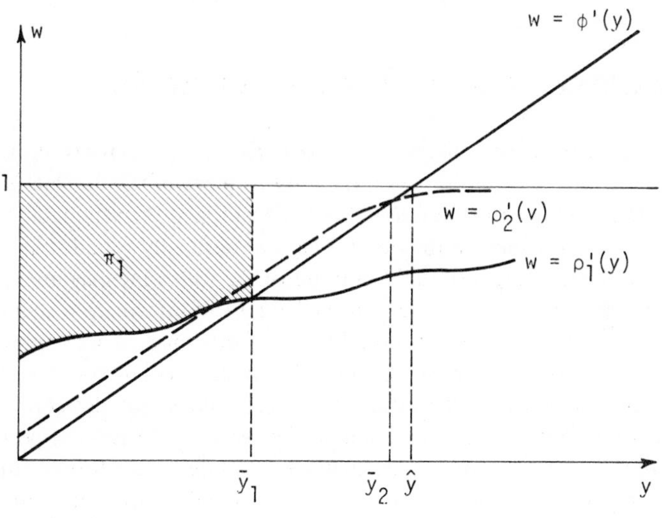

FIGURE 4

Thus we have another illustration of the extent to which the possibility of designing an effective reward structure depends on the state of information available to the designer, in this case the landowner.

At this point we shall redeem an earlier promise to provide another illustration characterized by a similar relationship between information and incentives, but involving issues of public interest. We can think of a community threatened by an outside danger. The community's total output of goods and services **y** (which we shall treat as if it were one-dimensional) must be divided between the defense of the community's existence, to be denoted by π, and its consumption **r**. As individual workers, the members of this community maximize their utility which depends positively on the consumption **r** and negatively on the effort **z**. But as voters, these citizens give priority to the community's survival. Hence they want to design an incentive structure that would maximize the community's survival probability, i.e. maximize π – without regard to the effort involved, but taking into account the utility maximizing behavior of individual workers. Clearly, the mathematical structure of this problem is equivalent to that of the landowner–

laborer example above. In particular, therefore, it is clear that more effort for defense could be extracted through a properly constructed reward structure from the citizenry – if additional information were available with regard to production functions and preferences. Alternatively, given additional information, with the same level of defense, more would be available for consumption **r**. (This is so because total output **y** would be closer to Pareto-optimality.)

INFORMATION COST AND PERFORMANCE OF SYSTEMS

In the course of the above analysis, we have come across instances of "trade-offs" between expenditure of resources on the improvement of the designer's (or the participants') state of information on the one hand, and the attainable level of the system's performance on the other. I have occasionally found it helpful to think of this relationship abstractly in terms of the following diagram (Figure 5). As our abscissa, we use something to be called cost of information acquisition, it being assumed that the resources used to acquire information are deployed in an efficient manner. It should be understood that such costs are in fact multidimensional and include not only the various resources utilized but also certain intangibles, e.g. loss of privacy. As our ordinate, we use some measure of performance of the system, e.g. the maximum attainable level of defense effort in our last example, or (say) a constant plus the negative of the distance from Pareto-optimality in the earlier examples. Denote the abscissa (cost of information) by **c** and the ordinate (maximum attainable performance) by **p**. Now any organizational structure or mechanism, to be denoted by μ, determines a point in the (c, p) space; denoting this point by (c_μ, p_μ), we interpret c_μ as the (minimal) informational cost associated with the mechanism μ and p_μ as the (maximal) performance attainable with this mechanism. A mechanism μ' can be called *more efficient than* μ'' if it has either lower informational cost or higher performance or both; a mechanism can be called *efficient* if no other feasible mechanism is more efficient.

Now let us also suppose that the society's values imply a preference in the (c, p) space, with more **p** and less **c** naturally being preferred. In Figure 5 we have drawn a conventional map of this sort and also indicated by shading the set A of (c, p) points corresponding to *feasible* mechanisms. The North-West boundary B of this set contains the points corresponding to *efficient* mechanisms. Among points of this boundary there is one, labeled T (for tangency) which is *best* among the feasible ones in terms of the preference map.

Given the assumptions made, one would want to choose as the mechanism to be used that which generates the combination (c_T, p_T) corresponding to the point T. If we did not have the preference map, we could not select the point T, but would presumably want to avoid points that are not on the (efficient) boundary

B. The shape of this boundary represents the nature of the unavoidable trade-offs between information cost and performance.

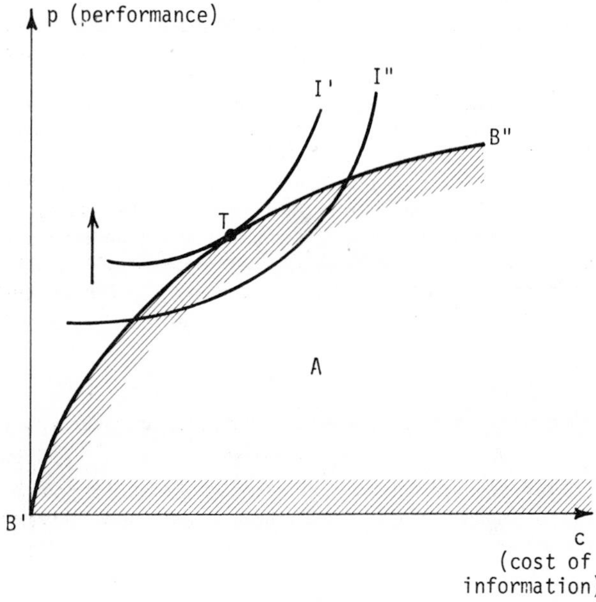

FIGURE 5

Of course, this way of looking at our problems is only schematic. For applications, one must go back to detailed formulations describing the mechanism in terms of language, response and outcome functions, and commit oneself to a specific performance criterion (optimality concept). In addition, a model must be constructed relating the informational costs to the structure of the mechanism and other data. We have made no more than a start by distinguishing such informational properties as privacy-preserving features and the dimension of the message space. We are still far from being able to use the construction of Figure 5 in a concrete way.

ENDNOTES

[1] Note that informationally decentralized mechanisms form a much wider category than the class of mechanisms (of which marginal cost pricing system is a member) that guide decision-making through prices. Hence the claim that a class of environments is not informationally decentralizable is significantly stronger than an assertion that no price mechanism can be used to guarantee the optimality of outcomes.

[2] There is need here for a terminological warning: recent measure-theoretic mathematical economics literature uses the term "nonatomic" to describe what we are calling, in line with more traditional usage, "atomistic".

[3] Also, our definition of incentive compatibility ignores the possibility of collusions, i.e. it would be more accurate to use the term *individual* incentive compatibility. The theory can be extended to cover collusions and coalitions as well.

One could extend the theory by postulating that the central authority, instead of prescribing specific behavior, merely confines the participants to a set of prescribed behaviors, i.e. imposes certain limitations. We could then define incentive compatibility as being present if the Nash equilibrium behavior $g^{\#}$ was an element of the set G^* of permissible behaviors.

[4] See Hurwicz [3].

[5] The assumption that the participants abide by the prescribed response function f^{i*}, although possibly using a false characteristic e^i, can be motivated as follows. We can imagine that there is an enforcement-inspection system which would require that a participant be able to justify his behavior as compatible with his claimed characteristic. Were he to use some "illegal" adjustment function (whether with true or false characteristic), he might generate (through (12)) a behavior function (assumed observed by the inspector) that could not be explained as "legal".

[6] For all $\mathring{e}$ in E.

[7] Those where preferences can be represented by utility functions linear in the private goods.

[8] The fact that such equilibria involve behavior departing from that prescribed may imply that the distributive aspects of the resource allocation are different from those intended by the designers of the system.

[9] See Endnote 7.

[10] A note of warning. The notion of optimality or efficiency implicit in this discussion ignores the resources used to operate the mechanism; it only looks at resources used for production and consumption. We may term this notion "gross" optimality (or efficiency), while the corresponding "net" notion would take into account the resources used to operate the system. Now suppose that, in order to reach "gross" optimality, we decide to sacrifice self-enforcement or informational decentralization. There will then arise a need for additional resource use in enforcement or inspection and it is conceivable that the resulting allocation will be further away from "net" optimality than the original one. (In addition, of course, there is also the "non-economic" loss due to added restrictions on individual freedom.)

[11] An offer curve of the i-th participant is the locus of points he would choose when maximizing utility subject to the budget constraint with the price ratio of the two goods ranging over all positive numbers while his initial endowment and preferences remain fixed.

[12] I.e. the Nonmanipulative Nash Equilibria.

[13] Not only at equilibrium.

[14] In the space of environments.

[15] Here the term "share" means the absolute amount of the good produced given to a participant rather than the fraction of the total.

[16] In Figure 4, w is the symbol for the ordinate.

[17] $\phi(y) = y^2/2\hat{y}$ is the disutility of effort expressed as a function of output. Hence $\phi'(y) = y/\hat{y}$ is represented by a ray from the origin, with $\phi'(\hat{y}) = 1$.

REFERENCES

[1] Groves, T. and Ledyard, J., "An incentive mechanism for efficient resource allocation in general equilibrium with public goods", Center for Mathematical Studies in Economics and Management Science, Discussion Paper No. 119, 1974. A revised version published in *Econometrica*, 45 (1977), 783–809, as "Optimal Allocation of Public Goods: A Solution to the 'Free Rider' Problem."

[2] Hurwicz, L., "Centralization and decentralization in economic processes", in Eckstein, A. (Ed.), *Comparison of Economic Systems;* 79–102, Berkeley: University of California Press, 1971.

[3] Hurwicz, L., "On informationally decentralized systems", in McGuire, C. B. and Radner, R. (Eds.), *Decision and Organization:* 1–29, Amsterdam: North Holland, 1972.

[4] Hurwicz, L., "On the Pareto-optimality of manipulative Nash equilibria", presented at the International Econometric Society, World Congress, Toronto, August, 1975.

[5] Mount, K. and Reiter, S., "The informational size of message spaces", *Journal of Economic Theory,* 8, No. 3: 389–396, 1974.

INFORMATIONAL ASPECTS OF ENTERPRISE DESIGN THROUGH INCENTIVES

PAUL R. KLEINDORFER
University of Pennsylvania, Philadelphia
and International Institute of Management, Berlin

1 INTRODUCTION

There has been a recent flurry of activity in the area of decentralization through workers' control and participative management schemes.[1] Although there is no consensus at this point as to how to design such schemes, it is understood that the structure of the incentives and information available to workers and management are key issues in any such design. This paper will investigate interrelationships between these two basic ingredients of the enterprise design problem.

In discussing incentives it is useful to distinguish three strands of research relevant to this issue. The one, which owes much of its philosophy to Hurwicz,[2] has been concerned primarily with resource allocation mechanisms for entire economies, and has produced results helpful in structuring fundamental informational and computational aspects of economic system (e.g. enterprise) design.

The second strand of relevant research lies in the area of the political economy of the organization of production. Prime examples of this are the recent paper by Alchian and Demsetz [2] on information costs and the structure of the classical firm and the work by Fitzroy [4] on the relationship between the organization of production on the one hand and worker alienation and freedom on the other.

A third area of relevance to the present discussion is organization theory and, in particular, team theory.[3] The classical team theory approach assumes that members of the organization all share a common objective to which each member contributes through his actions. Given no conflict of interest, the classical team design problem is concerned with the design of appropriate informational mechanisms for coordinating decision-making, but no incentive problem is involved. Groves [6] has extended this approach somewhat to allow for conflict of interest among team members. Then again, although Grove's results are interesting in setting the stage, they suffer from two major flaws from the point of view of modeling the interaction of participants in an enterprise. First, technological externalities (in production) between team members' behaviors are absent. Second, the distribution of incentives to organization members is assumed to cost nothing.

The approach taken here will introduce a more realistic model of incentives and technological aspects of the enterprise design problem, but this will be done at the expense of representing information interchange among participants on a far simpler level than Groves. One might therefore view the two approaches as complementary.

We begin our analysis by formally defining an enterprise. Roughly, by an "enterprise" is meant a group of economic agents (or workers) whose joint efforts produce an output which they then share among themselves according to a sharing rule established by the enterprise owner (or central planner). Some basic results of Sertel [13] and Kleindorfer and Sertel [9] are then summarized and certain informational problems are delineated. Specifically, consequences of imperfect knowledge on the part of workers concerning the technology (i.e. the production function) or the level of inputs of other productive factors (e.g. capital or labor of other workers) are studied. Finally, the differential effects of local incentives (e.g. rewards based on divisional output) versus global incentives (rewards based on enterprise output) as well as self-policing and team loyalty aspects of participative management are discussed.

2 PRE-ENTERPRISES AND ENTERPRISES

This section summarizes basic concepts and results of Kleindorfer and Sertel [9]. Our first definition serves also to introduce basic notation and assumptions for the sequel.

2.1 *DEFINITION*: By a *pre-enterprise* we mean an ordered quadruplet

2.1.0 $\Omega = \langle N, X, u, f \rangle$,

where

2.1.1 $N = \{0, \ldots, n\}$ with n a positive integer;

2.1.2 X is the product ΠX_i of a family $\{X_i | i \in N\}$ with each $X_i \subset R$ nonempty, closed and convex (where R is the real line);

2.1.3 $u = \{u_i\colon X_i \times R \to R | i \in N\}$ is a family of continuous real-valued functions u_i, each of which is additively separable[4] in the form $u_i(x_i, r) = w_i(x_i) + r_i (x_i \in X_i, r_i \in R)$ with $w_i\colon X_i \to R$ twice continuously differentiable and strictly concave for each $i \in \{1, \ldots, n\}$;

2.1.4 $f\colon X \to R$ is twice continuously differentiable, and, for each $x_0 \in X_0$, concave on $\{x_0\} \times X_1 \times \ldots \times X_n$.

To interpret the elements of Ω, the set N represents the "personnel", of which agent 0 will be called the *capitalist*. With each member $i \in N$ is associated an abstract *behavior space* X_i and preferences represented by the *utility function* u_i. The *production function* of the enterprise is f, where $f(x)$ represents the monetary value of the output resulting from the collective behavior $x = (x_0, x_1, \ldots, x_n)$.

Let $\overline{N} = \{1, 2, \ldots, n\}$ and define

$$\Lambda = \{\lambda \in R^n: 0 \leqslant \lambda_i \leqslant 1, \lambda_1 + \ldots + \lambda_n \leqslant 1\}. \tag{1}$$

Given Ω, each $\lambda \in \Lambda$ determines a family $\{\lambda_i f: X \to R | i \in N\}$ of incentive functions whereby agent i receives a constant share (λ_i) of the product $(f(x))$ resulting from the collective inputs (x) of the organization members. The residual $\lambda_0 f(x) = (1 - \Sigma\lambda_i)f(x)$ may be viewed as the *enterprise profit*. We assume that the capitalist is interested in setting $\lambda \in \Lambda$ and $x_0 \in X_0$ so as to maximize enterprise profit.

Given the above class of incentive schemes we see that, once $(\lambda, x_0) \in \Lambda \times X_0$ is set by agent 0, i's effective preferences are represented by the utility function $v_i(\cdot\,; \lambda_i, x_0): X \to R$ defined by

$$v_i(\overline{x}; \lambda_i, x_0) = w_i(x_i) + \lambda_i f(x_0, \overline{x}), \qquad \overline{x} \in \overline{X}, \text{ where } \overline{X} = \Pi\{X_i | i \in \overline{N}\}. \tag{2}$$

Using these effective preferences we can define an adjustment process predicated on the assumption that all the agents $i \in \overline{N}$ behave (noncooperatively) so as to maximize their own well-being.

For each $i \in \overline{N}$ define the *i-exclusive behavior space* $\overline{X}^i = \Pi\{X_i: i \in \overline{N} \backslash \{i\}\}$ (e.g. $\overline{X}^1 = X_2 \times \ldots \times X_n$), with generic element $\overline{x}^i \in \overline{X}^i$. Given any $\overline{x} \in \overline{X}$ consider, for each $i \in \overline{N}$, the problem

$$\underset{X_i}{\text{Maximize}}\; v_i(\cdot\,, \overline{x}^i; \lambda_i, x_0). \tag{3}$$

We note in passing that if agent i is to solve (3) he must know f and be able to observe $\overline{x}^i$ (as well as λ_i and x_0). We assume this to be so for the moment. *Under this assumption* we see that if, for each $i \in \overline{N}$, the problem (3) has a unique solution $\alpha_i[\lambda_i, x_0](\overline{x}^i) \in X_i$, then a family of n reaction maps $\alpha_i[\lambda_i, x_0]: \overline{X}^i \to X_i$ is determined from which we define the *adjustment process* $\alpha[\lambda, x_0]: \overline{X} \to \overline{X}$ as

$$\alpha[\lambda, x_0](\overline{x}) = \{\alpha_i[\lambda_i, x_0](\overline{x}^i): i \in \overline{N}\} \qquad (\overline{x} \in \overline{X}). \tag{4}$$

The process so defined is noncooperative in the sense that each $i \in \overline{N}$ takes the behavior of each $j \in \overline{N}^i = \overline{N} \backslash \{i\}$ to be given in reacting to the current collective behavior. For the moment we rule out coalitions and bargaining among the members of $\overline{N}$.

We will always work under assumptions[5] insuring the existence and uniqueness of solutions to the problem (3) for each $i \in \overline{N}, x \in X, \lambda \in \Lambda$. This being so, once $(\lambda, x_0) \in \Lambda \times X_0$ has been fixed, a well-defined adjustment process $\alpha[\lambda, x_0]$ is determined. Given a pre-enterprise Ω we will call the restriction $\Omega(\lambda, x_0)$ of Ω to a fixed $(\lambda, x_0) \in \Lambda \times X_0$ an *enterprise*.

2.2 *DEFINITION*: Given a pre-enterprise Ω, let $\Omega(\lambda, x_0)$ be an enterprise and $\alpha[\lambda, x_0]$ its adjustment process. By an *equilibrium* of $\Omega(\lambda, x_0)$ we mean a fixed point of $\alpha[\lambda, x_0]$, i.e. a point $\tilde{x} = \alpha[\lambda, x_0](\tilde{x}) \in \overline{X}$.

If $\tilde{x} \in \overline{X}$ is an equilibrium of $\Omega(\lambda, x_0)$, then by definition of $\alpha[\lambda, x_0]$

$$v_i(\tilde{x}_i, \tilde{x}^i; \lambda, x_0) = \operatorname*{Max}_{X_i} v_i(\cdot, \tilde{x}^i; \lambda, x_0), \qquad i \in \overline{N}. \tag{5}$$

In Kleindorfer and Sertel [9] weak conditions (e.g. compactness of X) are given under which an equilibrium $\tilde{x}(\lambda, x_0)$ exists for every $(\lambda, x_0) \in \Lambda \times X_0$, and furthermore under which the point $\tilde{x}(\lambda, x_0)$ varies continuously in (λ, x_0). In this case, we define the *equilibrium map* $\tilde{x}$: $\Lambda \times X_0 \to \overline{X}$ of a given pre-enterprise Ω through $\tilde{x}(\lambda, x_0) = \alpha[\lambda, x_0](\tilde{x}(\lambda, x_0))$. Pre-enterprises having a well-defined and continuous equilibrium map will be called *completely regular*. Such pre-enterprises enjoy many regularity properties including, for example, the fact [see 9] that agent i's equilibrium behavior $\tilde{x}_i(\lambda, x_0)$ is monotonically nondecreasing in his share λ_i. Stability of equilibrium can also be shown under weak regularity assumptions.[6] Altogether then, the picture emerging from these results is the following. If each agent $i \in \overline{N}$ has perfect knowledge of f and can observe the behavior x^i of all other agents without distortion, then the *enterprise design problem* may be stated succinctly as

$$\operatorname*{Maximize}_{\Lambda \times X_0} \left(1 - \sum_{\overline{N}} \lambda_i\right) f(\tilde{x}(\lambda, x_0)), \tag{6}$$

a problem which has a solution if, for example, $\tilde{x}$ is continuous and $\Lambda \times X_0$ compact.

We now propose to study the effects of relaxing the information requirements implicit in (3). We will do this only for two specific forms of pre-enterprise.

2.3 *DEFINITION*: We will say that a pre-enterprise $\Omega = \langle N, X, f, u \rangle$ is a *Multiplicative Cobb-Douglas* (MCD) pre-enterprise if it satisfies:

2.3.1 $X_i = \{x_i \in R | x_i \geqslant 0\}, \qquad i \in N;$

2.3.2 $f(x) = \prod_N x_i^{\delta_i}$ where $\delta_i \geqslant 0$ and $\sum_N \delta_i \leqslant 1;$

2.3.3 $w_i(x_i) = -b_i x_i^{\beta_i}$ where $b_i > 0, \beta_i > 1, \qquad i \in N.$

Ω will be said to be an *Additive Cobb-Douglas* (ACD) pre-enterprise if it satisfies 2.3.1, 2.3.3, and

2.3.4 $f(x) = x_0^{\delta_0}\left(\sum_{\overline{N}} x_i\right)^{\delta_1}$ where $\delta_0, \delta_1 \geqslant 0, \delta_0 + \delta_1 \leqslant 1.$

An important difference between MCD and ACD pre-enterprises is that in the former $\partial^2 f/\partial x_i \partial x_j \geqslant 0$, for all i, j for which $i \neq j$, whereas in the latter $\partial^2 f/\partial x_i \partial x_j \leqslant 0$ for all i, j. The effects of this in the MCD (ACD) case can be expected to be that as others work harder there is more (less) incentive for agent i to work since increased efforts by others will increase (decrease) his marginal productivity and, therefore, also his marginal work incentive $(=\lambda_i \partial f/\partial x_i)$.

The first-order conditions for equilibrium (see (5)) for the MCD and ACD problems are seen to be, respectively,

$$\text{MCD:} \quad b_i \beta_i \tilde{x}_i^{\beta_i - 1} = \frac{\delta_i \lambda_i}{\tilde{x}_i} x_0^{\delta_0} \prod_{\overline{N}} \tilde{x}_i^{\delta_i} \qquad i \in \overline{N}; \tag{7}$$

$$\text{ACD:} \quad b_i \beta_i \tilde{x}_i^{\beta_i - 1} = \delta_1 \lambda_1 x_0^{\delta_0} \left(\sum_{\overline{N}} \tilde{x}_j \right)^{\delta_1 - 1} \qquad i \in \overline{N}. \tag{8}$$

Multiplying both sides of (7) by $\tilde{x}_i$ and both sides of (8) by $\sum_{\overline{N}} \tilde{x}_j$, we obtain

$$\text{MCD:} \quad \frac{b_i \beta_i \tilde{x}_i^{\beta_i}}{\delta_i \lambda_i} = f(x_0, \tilde{x}), \qquad i \in \overline{N}; \tag{9}$$

$$\text{ACD:} \quad \frac{b_i \beta_i \tilde{x}_i^{\beta_i - 1}}{\delta_i \lambda_i} \left(\sum_{\overline{N}} \tilde{x}_j \right) = f(x_0, \tilde{x}), \qquad i \in \overline{N}. \tag{10}$$

Combining (9) (resp. (10)) with the requirement that $f(x_0, \tilde{x}) = x_0^{\delta_0} \prod_{\overline{N}} \tilde{x}_i^{\delta_i}$ (resp. $f(x_0, \tilde{x}) = x_0^{\delta_0} \left(\sum_{\overline{N}} \tilde{x}_i)^{\delta_1} \right)$ the equilibrium is characterized for MCD (ACD) pre-enterprises. We now analyze the effects of informational imperfections on equilibrium behavior and production in MCD and ACD pre-enterprises.

3 THE MULTIPLICATIVE CASE

We first note that not all possible informational imperfections are reasonable. In particular, we shall require the *consistency condition at equilibrium*, that the perception by each agent $i \in \overline{N}$ of the monetary value of enterprise production coincides with the actual value $f(x)$. This seems a reasonable requirement since each agent i receives a constant and known share λ_i of this monetary value, so lasting distortions seem unlikely.

Now suppose that an agent $i \in \overline{N}$ has a distorted perception of the production function f or of the i-exclusive behavior $x^i = (x_0, \ldots, x_{i-1}, x_{i+1} \ldots x_n)$ of other agents. Let $\hat{v}_i \colon \Lambda \times X \to R$ represent agent i's *perceived effective utility function*, defined by

$$\hat{v}_i(\bar{x}; \lambda, x_0) = w_i(x_i) + \lambda_i \hat{f}^i(x), \tag{11}$$

where $w_i(x_i)$ is given by 2.3.1 and where agent i's *perceived production function* $\hat{f}^i: X \to R$ is defined by

$$\hat{f}^i(x) = x_i^{d_i} \psi_i(x^i) \tag{12}$$

where d_i represents agent i's perception of δ_i and $\psi_i: X^i \to R$ is some nonnegative function representing i's perception of the effect of other productive factors $j \in N^i$ on enterprise output.

We see that if agent i solves (3) (using $\hat{v}_i$ instead of v_i) then, for any $x^i \in X^i$, $\lambda \in \Lambda$, the first order conditions are

$$b_i \beta_i x_i^{\beta_i - 1} = \lambda_i d_i x_i^{d_i - 1} \psi_i(x^i), \quad x^i \in X^i. \tag{13}$$

Solving (13) for x_i we have, in analogy with (4), the *imperfect-information adjustment process* $\hat{\alpha}[\lambda, x_0] : \bar{X} \to \bar{X}$ determined by

$$\hat{\alpha}[\lambda, x_0](\bar{x}) = \{\hat{\alpha}_i[\lambda_i, x_0](\bar{x}^i): i \in \bar{N}\} \quad (\bar{x} \in \bar{X}), \tag{14}$$

where, from (13)

$$\hat{\alpha}_i[\lambda_i, x_0](\bar{x}^i) = \left[\frac{\lambda_i d_i}{b_i \beta_i} \psi_i(x^i)\right]^{\frac{1}{\beta_i - d_i}} \quad (\bar{x}^i \in \bar{X}^i, i \in \bar{N}). \tag{15}$$

If we define an *imperfect-information equilibrium* analogous to 2.2, but using the imperfect-information adjustment process (14) in place of (4), we obtain the conditions for an imperfect-information equilibrium $\hat{x}(\lambda, x_0)$ to be

$$\hat{v}_i(\hat{x}_i, \hat{x}^i; \lambda_i, x_0) = \text{Max}\ \hat{v}_i(\cdot, \hat{x}^i; \lambda_i, x_0), \qquad i \in \bar{N}, \tag{16}$$

or, using (13) simultaneously[7] for all $i \in \bar{N}$,

$$\frac{b_i \beta_i \hat{x}_i^{\beta_i}}{\lambda_i d_i} = \hat{x}_i^{d_i} \psi_i(\hat{x}^i), \qquad i \in \bar{N}. \tag{17}$$

Now, using the consistency condition at the imperfect-information equilibrium[8] $\hat{x}(\lambda, x_0)$, we must have $\hat{f}^i(x_0, \hat{x}) = f(x_0, \hat{x})$; $i \in \bar{N}$. Using (12) and (17), this yields

$$\frac{b_i \beta_i \hat{x}_i^{\beta_i}}{\lambda_i d_i} = f(x_0, \hat{x}), \qquad i \in \bar{N}. \tag{18}$$

One may rewrite the left-hand side of (18) as

$$\frac{b_i \beta_i \hat{x}_i^{\beta_i}}{\lambda_i d_i} = \frac{b_i \beta_i \hat{x}_i^{\beta_i}}{\hat{\lambda}_i \delta_i}, \tag{19}$$

where $\hat{\lambda}_i = \lambda_i d_i / \delta_i$. Thus, comparing (18)–(19) with the equilibrium conditions (9) for the perfect information case, we see that the effect of all information imperfections is as if agent i's perceived share were changed in proportion to d_i/δ_i. That is, $\hat{x}(\lambda, x_0) = x(\hat{\lambda}, x_0)$ with $\hat{\lambda}_i$ as given above. Note that any distortions agent i might entertain with respect to the productivity or behavior of others are washed out by the consistency condition.

Finally, as noted in the last section, the equilibrium behavior $\breve{x}_i(\lambda, x_0)$ is monotonically increasing in λ_i. Thus, the above indicates that, *ceteris paribus*, if $d_i < \delta_i (d_i > \delta_i)$, then agent i's equilibrium contribution will be less (greater) than under perfect information. Since $\partial^2 f/\partial x_i \partial x_j \geqslant 0$ in the MCD case, any decrease (increase) in agent i's equilibrium behavior will decrease (increase) agent j's ($i \neq j$) equilibrium behavior as well.[9] We see that the MCD case is straightforward. Increases or decreases in agent i's perceived (own) productivity lead directly to increases or decreases in everyone's individual equilibrium contribution and therefore also in enterprise (equilibrium) output. We summarize these results in the following Theorem.

3.1 *THEOREM*: *Let* Ω *be an MCD pre-enterprise* (*as specified in* 2.3) *with informational imperfections as described by* (11) *and* (12). *Then, for each* $(\lambda, x_0) \in \Lambda \times X_0$, *there exists an imperfect-information equilibrium* $\hat{x}(\lambda, x_0)$ *satisfying the consistency condition. Moreover, if the elasticity of output with respect to an agent i's contribution is perceived by that agent to be less than* (*respectively, greater than*) *the true elasticity, i.e. if* $d_i < \delta_i$ (*respectively*, $d_i > \delta_i$), *then* (*ceteris paribus*) *for each* $(\lambda, x_0) \in \Lambda \times X$ *and for every* $j \in \overline{N}$ *the resulting* (*imperfect-information*) *equilibrium contribution* $\hat{x}_j(\lambda, x_0)$ *will be less than* (*respectively, greater than*) *the equilibrium contribution* $x_j(\lambda, x_0)$ *that would obtain under perfect information. Consequently, if* $d_i > \delta_i$ (*respectively*, $d_i < \delta_i$) *for some* $i \in \overline{N}$, *then the equilibrium output* $f(x_0, \hat{x}(\lambda, x_0))$ *with informational imperfections will be greater than* (*respectively, less than*) *the equilibrium output* $f(x_0, x(\lambda, x_0))$ *under perfect information.*

Note that the above Theorem holds regardless of the particular form, captured by $\{\psi_i : i \in \overline{N}\}$, of the informational distortions present. The effect of the consistency condition is to negate all imperfections except those due to misperceived factor elasticities.

4 THE ADDITIVE CASE

We now consider an ACD pre-enterprise (see 2.3) and we suppose that agent i's perceived effective utility function $\hat{v}_i$: $\Lambda \times X \to R$ is given by

$$\hat{v}_i(\bar{x}; \lambda_i, x_0) = w_i(x_i) + \lambda_i \hat{f}^i(x), \qquad x \in X, \tag{20}$$

where $w_i(x_i)$ is again given by 2.3.1 and where agent i's perceived production function $\hat{f}^i: X \to R$ is defined by

$$\hat{f}^i(x) = x^{d_{i0}}(x_i + \psi_i(\bar{x}^i))^{d_{i1}}, x \in X, \qquad i \in \bar{N}, \tag{21}$$

with $d_{i0}, d_{i1} > 0, d_{i0} + d_{i1} \leqslant 1$, and $\psi_i: \bar{X}^i \to R$ a nonnegative real-valued function. A simple example of ψ_i would be a linear function of $\bar{x}_i$, as would be the case where i multiplies the input x_j of the worker closest to him by the number of other workers employed in the enterprise.

Our first result will be that if no agent $i \in \bar{N}$ grossly overestimates the contribution of others (i.e. of $j \in \bar{N}^i$) then an imperfect-information equilibrium exists in this case.

4.1 *THEOREM: Let Ω be an Additive Cobb-Douglas pre-enterprise. Suppose for each $i \in \bar{N}$ that the function ψ_i in (21) is continuous and satisfies*

$$\psi_i(\bar{x}^i) \leqslant (1/\delta_i)\left(\sum_{\bar{N}} {}_i x_j\right), \qquad \bar{x}^i \in \bar{X}^i. \tag{22}$$

Then, for each $(\lambda, x_0) \in \Lambda \times X_0$, there exists an imperfect-information equilibrium $\hat{x}(\lambda, x_0)$ satisfying the consistency condition.

Proof: As in (16) any imperfect-information equilibrium $\hat{x}(\lambda, x_0)$ must satisfy

$$\hat{v}_i(\hat{x}_1, \hat{x}^i; \lambda, x_0) = \operatorname*{Max}_{X_i} \hat{v}_i(\cdot, \hat{x}^i; \lambda, x_0), \qquad i \in \bar{N}. \tag{23}$$

Taking first-order conditions, (23) implies

$$b_i\beta_i\hat{x}_i^{\beta_i - 1} = d_{i1}\lambda_i x_0^{d_{i0}}(\hat{x}_i + \psi_i(\hat{x}^i))^{d_{i1}-1}, \qquad i \in \bar{N}. \tag{24}$$

Multiplying both sides of (24) by $\hat{x}_i + \psi_i(\hat{x}^i)$ and using (21), we obtain

$$b_i\beta_i\hat{x}_i^{\beta_i - 1}(\hat{x}_i + \psi_i(\hat{x}^i)) = d_{i1}\lambda_i\hat{f}^i(x_0, \hat{x}), \qquad i \in \bar{N}. \tag{25}$$

Applying the consistency condition (that $\hat{f}^i(x_0, \hat{x}) = f(x_0, \hat{x}), i \in \bar{N}$), equation (25) implies[10]

$$\frac{b_i\beta_i\hat{x}_i^{\beta_i-1}}{d_{i1}\lambda_i}(\hat{x}_i + \psi_i(\hat{x}^i)) = f(x_0, \hat{x}), \qquad i \in \bar{N}, \tag{26}$$

where $f(x_0, \hat{x}) = x_0^{\delta_0}(\hat{x}_1 + \ldots + \hat{x}_n)^{\delta_1}$ from 2.3.4.

Our Theorem asserts that, for each $(\lambda, x_0) \in \Lambda \times X_0$, there exists an $\hat{x} = \hat{x}(\lambda, x_0) \in \bar{X}$ satisfying (26). This we now show. To begin with, take any $(\lambda, x_0) \in \Lambda \times X$ and, for each $i \in \bar{N}$, define the imperfect-information adjustment process $\hat{\alpha}_i[\lambda_i, x_0] : \bar{X}^i \to X_i$ by[11]

$$\frac{b_i\beta_i\hat{\alpha}_i(\bar{x}^i)^{\beta_i-1}}{d_{i1}\lambda_i}(\hat{\alpha}_i(\bar{x}^i) + \psi_i(\bar{x}^i)) = f(x_0, \hat{\alpha}_i(\bar{x}^i), \bar{x}^i), \qquad (\bar{x}^i \in \bar{X}^i), \tag{27}$$

where we have suppressed the arguments (λ_i, x_0) of $\hat{\alpha}_i$.

The remainder of the proof is structured as follows. We first show that for each $i \in \overline{N}$, $\hat{\alpha}_i$ is well-defined, i.e. that for each $\overline{x}^i \in \overline{X}^i$, a unique $\hat{\alpha}_i(\overline{x}^i) \in X_i$ is determined through (27). We then produce a compact, convex set $\overline{Y} \in \overline{X}$ such that $\hat{\alpha}[\lambda, x_0] : \overline{Y} \to \overline{Y}$ where $\hat{\alpha}[\lambda, x_0] : \overline{X} \to \overline{X}$ is defined through

$$\hat{\alpha}[\lambda, x_0](\overline{x}) = \{\hat{\alpha}_i[\lambda_i, x_0](\overline{x}^i) : i \in \overline{N}\}, \qquad \overline{x} \in \overline{X} \tag{28}$$

We then show that $\hat{\alpha}[\lambda, x_0]$ is continuous so that Brouwer's Fixed Point Theorem implies the existence of an $\hat{x} \in \overline{Y}$ with $\hat{\alpha}[\lambda, x_0](\hat{x}) = \hat{x}$. From (26)–(28) this $\hat{x}$ is the desired imperfect-information equilibrium.

To show that $\hat{\alpha}_i[\lambda_i, x_0]$ is well-defined as a map, we note from (27) that

$$\frac{\partial}{\partial \hat{\alpha}_i}\left[\frac{f(x_0, \hat{\alpha}_i, \overline{x}^i)}{(\hat{\alpha}_i + \psi_i(\overline{x}^i))}\right] = \frac{(\hat{\alpha}_i + \psi_i(\overline{x}^i))\partial f/\partial \hat{\alpha}_i - f(x_0, \hat{\alpha}_i, \overline{x}^i)}{(\hat{\alpha}_i + \psi_i(\overline{x}^i))^2}, \tag{29}$$

where

$$\partial f/\partial \hat{\alpha}_i = \partial f/\partial x_i|_{x_i = \hat{\alpha}_i} = \delta_1 x_0^{\delta_0}(\hat{\alpha}_i + \sum_{\overline{N}^i} x_j)^{\delta_1 - 1} \tag{30}$$

But, using the hypothesis (22) and $\delta_1 < 1$, we see that

$$\delta_1(\hat{\alpha}_i + \psi_i(\overline{x}^i)) \leqslant (\hat{\alpha}_i + \sum_{\overline{N}^i} x_j), \tag{31}$$

so (30)–(31) coupled with (29) and (2.3.4) imply

$$\frac{\partial}{\partial \hat{\alpha}_i}\left[\frac{f(x_0, \hat{\alpha}_i, \overline{X}^i)}{(\partial_i + \psi_i(\overline{x}^i))}\right] \leqslant 0, \qquad \overline{x}^i \in \overline{X}^i \tag{32}$$

We now rewrite (27) as

$$\frac{b_i \beta_i \hat{\alpha}_i^{\beta_i - 1}}{d_{i1}\lambda_i} = \frac{f(x_0, \hat{\alpha}_i, \overline{x}^i)}{(\hat{\alpha}_i + \psi_i(\overline{x}^i))} \qquad (\overline{x}^i \in \overline{X}^i) \tag{33}$$

From (33), (32) and the fact (*N.B.* $\beta_i > 1$) that the left-hand side of (33) is strictly increasing in $\hat{\alpha}_i$, we see that there exists a unique[12] $\hat{\alpha}_i(\overline{x}^i) \in X_i$ solving (33) (and therefore also (27)) for each $\overline{x}^i \in \overline{X}^i$.

We now verify that there exists a compact, convex set $\overline{Y} \subset \overline{X}$ such that $\hat{\alpha}$: $\overline{Y} \to \overline{Y}$. In particular, we show for each $i \in \overline{N}$ that

$$\hat{\alpha}_i(\overline{x}^i) \leqslant M \triangleq \text{Max}\left\{\underset{j \in \overline{N}}{\text{Max}}\,(nC_j)^{(1/\beta_j - 1)}, 1\right\} \tag{34}$$

whenever $x_j \leqslant M$ for all $j \in \overline{N}$, where

$$C_j = \frac{d_{j1}\lambda_j x_0^{\delta_0}}{b_j \beta_j}, \quad j \in \overline{N} \tag{35}$$

We start by fixing some $i \in \overline{N}$. We then note from the definition (27) of $\hat{\alpha}_i$

$= \hat{\alpha}_i[\lambda_i, x_0](\bar{x}^i)$ that

$$\hat{\alpha}_i^{\beta_i} + \psi_i(x^i)\,\hat{\alpha}_i^{\beta_i - 1} = C_i\left(\hat{\alpha}_i + \sum_{\bar{N}^i} x_j\right)^{\delta_1} \tag{36}$$

Since $\hat{\alpha}_i$ and ψ_i are nonnegative, (36) implies

$$\hat{\alpha}_i^{\beta_i} \leqslant C_i\left(\alpha_i + \sum_{\bar{N}^i} x_j\right)^{\delta_1} \tag{37}$$

Now there are just two possibilities: (i) $\hat{\alpha}_i + \sum_{\bar{N}^i} x_j \leqslant 1$ and (ii) $\hat{\alpha}_i + \sum_{\bar{N}^i} x_j > 1$.

In case (i), (37) implies $\hat{\alpha}_i^{\beta_i} \leqslant C_i$. If $C_i < 1$, then, since $M \geqslant 1$ by (34), we have $\hat{\alpha}_i \leqslant M$, as desired. Otherwise ($C_i \geqslant 1$), we have from (34) $n \geqslant 1$, and $\beta_i > 1$

$$C_i^{1/\beta_i} \leqslant C_i^{1/(\beta_i - 1)} \leqslant (nC_i)^{1/(\beta_i - 1)} \leqslant M \tag{38}$$

so that $\hat{\alpha}_i^{\beta_i} \leqslant C_i$ implies $\hat{\alpha}_i \leqslant M$ again.

In case (ii), (37) and $\delta_1 < 1$ imply

$$\hat{\alpha}_i^{\beta_i} < C_i\left(\hat{\alpha}_i + \sum_{\bar{N}^i} x_j\right) \tag{39}$$

Suppose first that $M = 1$. Then, from (34) and properties of the power function, we have

$$nC_i \leqslant M = 1 \tag{40}$$

Thus, if we assume that $x_j \leqslant M$ for every $j \in \bar{N}^i$ (39)–(40) yield

$$\hat{\alpha}_i^{\beta_i} < C_i\left(\hat{\alpha}_i + \sum_{\bar{N}^i} x_j\right) \leqslant \frac{1}{n}(\hat{\alpha}_i + n - 1) \tag{41}$$

Now at $\hat{\alpha}_i = 1$ we have

$$\hat{\alpha}_i^{\beta_i} = \frac{1}{n}(\hat{\alpha}_i + n - 1), \tag{42}$$

so (41)–(42) and $\beta_i > 1$ imply $\hat{\alpha}_i \leqslant 1 = M$ as desired.[13]

Now suppose that $M > 1$. Then (34) implies $C_i \leqslant (M^{\beta_i - 1})/n$, so that, assuming as before that $x_j \leqslant M$ for every $j \in \bar{N}$, (39) implies

$$\hat{\alpha}_i^{\beta_i} \leqslant (M^{\beta_i - 1}/n)(\hat{\alpha}_i + (n-1)M) \tag{43}$$

But when $\hat{\alpha}_i = M$, the inequality in (43) holds as an equality. Therefore, an $\hat{\alpha}_i$ solving (39) (or (36)) must satisfy $\hat{\alpha}_i \leqslant M$.

Since $i \in \bar{N}$ was arbitrary in the above argument we have shown that $\hat{\alpha}_i[\lambda, x_0]: \bar{Y} \to \bar{Y}$ with $\bar{Y} = \{x \in \bar{X}: x_i \leqslant M, i \in \bar{N}\}$. $\bar{Y}$ is clearly compact and convex.

Finally, we show that $\hat{\alpha}_i$ is continuous on Y. To that end, note that the functions f and ψ_i defining $\hat{\alpha}_i$ in (27) are continuous, so the graph of the restriction of $\hat{\alpha}_i$ to $\bar{Y}$ is closed. The compactness of $\bar{Y}$ now implies the continuity of $\hat{\alpha}_i$ on $\bar{Y}$

(see Dugundji [1966], XI.2.7, p. 228). With the remarks following (27), this completes the proof.

4.2 *THEOREM: Consider an ACD pre-enterprise with informational imperfections as described by* (20) *and* (21). *Suppose* ψ *in* (21) *is continuous and satisfies* (22). *If for every* $i \in \overline{N}$ *and* $\overline{x} \in \overline{X}$, $d_{i1} \leqslant \delta_1$ *and* $\psi_i(\overline{x}^i) \geqslant \sum_{\overline{N}^i} \overline{x}_j$ *(resp.* $d_{i1} \geqslant \delta_1$ *and* $\psi_i(\overline{x}^i) \leqslant \sum_{\overline{N}^i} x_j$*), then for any* $(\lambda, x_0) \in \Lambda \times X_0$ *the actual equilibrium output* $f(x_0, \hat{x}(\lambda, x_0))$ *will be less (resp. greater) than or equal to the corresponding equilibrium output* $f(x_0, \tilde{x}(\lambda, x_0))$ *under perfect information.*

Proof: We show only the unbracketed assertion, the other being analogous. Thus, fix $(\lambda, x_0) \in \Lambda \times X_0$ and suppose that $d_{i1} \leqslant \delta_1$ and $\psi_i(\overline{x}^i) \geqslant \sum_{\overline{N}i} x_j$ for each $i \in \overline{N}$ and $x \in \overline{X}$. Then from (10) and (26), $\tilde{x}$ and $\hat{x}$ in the statement of our Theorem must satisfy

$$\frac{b_i\beta_i\tilde{x}_i^{\beta_i-1}}{\delta_1\lambda_1} = x_0^{\delta_0}\left(\sum_{\overline{N}} x_j\right)^{\delta_1-1}, \qquad i \in \overline{N}; \tag{44}$$

and

$$\frac{b_i\beta_i\hat{x}_i^{\beta_i-1}}{d_{i1}\lambda_i}(\hat{x}_i + \psi_i(\hat{x}^i)) = x_0^{\delta_0}\left(\sum_{\overline{N}} \hat{x}_j\right)^{\delta_1}, \qquad i \in \overline{N} \tag{45}$$

Suppose now that $f(x_0, \hat{x}) > f(x_0, \tilde{x})$ or equivalently (assuming $x_0 > 0$)

$$\sum_{\overline{N}} \hat{x}_j > \sum_{\overline{N}} \tilde{x}_j \tag{46}$$

Then $\delta_1 < 1$ implies from (44)

$$\frac{b_i\beta_i x_i^{\beta_i-1}}{\delta_1\lambda_i} > x_0^{\delta_0}\left(\sum_{\overline{N}} \hat{x}_j\right)^{\delta_1-1}, \qquad i \in \overline{N} \tag{57}$$

Now divide both sides of (47) by the corresponding side of (45) and use $d_{i1} \leqslant \delta_1$, obtaining

$$\left(\frac{\tilde{x}_i}{\hat{x}_i}\right)^{\beta_i-1} > \frac{\hat{x}_i + \psi_i(\hat{x}^i)}{\hat{x}_i + \sum_{\overline{N}i} \hat{x}_j}, \qquad i \in \overline{N} \tag{48}$$

By the hypothesis on ψ_i, the r.h.s. of (48) is no less than 1. Since $\beta_i > 1$, this implies $\tilde{x}_i \geqslant \hat{x}_i$, $i \in \overline{N}$, which contradicts (46). Therefore, $f(x_0, \hat{x}) \leqslant f(x_0, \tilde{x})$ as asserted.

4.3 *THEOREM*: *For an ACD pre-enterprise with informational imperfections, assume for all* $i \in \bar{N}$ *that* $\psi_i(\bar{x}^i) = \sum_{\bar{N}} x_j, \bar{x} \in \bar{X}$ (i.e. *the only distortions are possibly that* $d_{i0} \neq \delta_0, d_{i1} \neq \delta_1$). *If* $d_{i1} > \delta_1$ ($d_{i1} < \delta_1$) *then, ceteris paribus,* $\hat{x}_i(\lambda, x_0) \geqslant \tilde{x}_i(\lambda, x_0)$ (*resp.* $\hat{x}_i(\lambda, x_0) \leqslant \tilde{x}_i(\lambda, x_0)$) *and* $\hat{x}_j(\lambda, x_0) \leqslant x_j(\lambda, x_0), i \neq j$ (*resp.* $\hat{x}(\lambda, x_0) \geqslant x_j(\lambda, x_0), i \neq j$), $(\lambda, x_0) \in \Lambda \times X_0$.

Proof: We simply note that, in this case, (26) becomes

$$\frac{b_i\beta_i\hat{x}_i^{\beta_i-1}}{d_{i1}\lambda_i} = x_0^{\delta_0}\left(\sum_{\bar{N}} \hat{x}_j\right)^{\delta_1-1}, \qquad i \in \bar{N} \tag{49}$$

Proceeding as with the MCD case, we see that (49) can be rewritten as

$$\frac{b_i\beta_i\hat{x}_i^{\beta_i-1}}{\delta_1\hat{\lambda}_i} = x_0^{\delta_0}\left(\sum_{\bar{N}} \hat{x}_j\right)^{\delta_1-1}, \qquad i \in N, \tag{50}$$

where $\hat{\lambda}_i = d_{i1}\lambda_i/\delta_1$. This is clearly equivalent to the equilibrium conditions (44) for the perfect information case with λ_i changed to $\hat{\lambda}_i$. The reasoning now is the same as in the MCD case (see equation (19) and following remarks). This completes the proof.

5 DISCUSSION AND EXTENSIONS

An interesting aspect of the above results is their agreement with the analysis of Fitzroy [4] to the effect that the agent (or group) which has the authority to set the incentive scheme would find it in its interest to foster certain informational distortions on the part of workers. For example, the above results indicate that the Capitalist (or Central Planner) would find it desirable for workers to overestimate their own productivity. The workers may also gain from such informational imperfections, since, as equilibrium output increases, so also do their rewards. An analysis of the gains or losses in the welfare of the workers is outside the scope of the present paper, however. Such an analysis would be particularly interesting at the optimal incentive scheme and capital input, i.e. at the (λ^*, x_0^*) solving (6). As an example, it can be shown[14] that the optimal incentive scheme for an MCD pre-enterprise is $\lambda_i^* = \delta_i/\beta_i (i \in \bar{N})$, independent of x_0. But we may rewrite the equilibrium conditions (18) for an MCD pre-enterprise as

$$\frac{\hat{b}_i\beta_i\hat{x}_i^{\beta_i}}{\lambda_i\delta_i} = f(x_0, \hat{x}), \qquad i \in \bar{N}, \tag{51}$$

where $\hat{b}_i = b_i\delta_i/d_i$. From this and the result just stated ($\lambda_i^* = \delta_i/\beta_i$) we see, comparing (9) and (51), that the optimal solution to (6) with $\hat{x}$ substituted for $\tilde{x}$ is unchanged since only b_i has changed and λ_i^* is independent of b_i. Such a simple result is not likely to obtain where inputs are substitutes for one another as in the ACD case.

Another interesting topic not treated here is the information base for the incentives themselves. We have assumed above that only incentives based on output are allowed and we have, in fact, only studied the constant-share sub-class of those. There is clearly a trade-off to be made between the scope and precision of the information base of feasible reward schemes and the equilibrium output obtained. Sertel [14] shows that, neglecting the cost of monitoring, a wage scheme (based on measuring individual inputs) exists which is superior to all constant-share incentive schemes from the point of view of the capitalist (although workers' welfare may decrease under the wage scheme in question). Such results remind one of Taylor's experiments and the later Hawthorne experiments which demonstrated the (declining) effectiveness of group incentives as a function of group size.[15] When these results on the trade-off between monitoring-computation costs and performance are combined with those of this paper on the effects of informational distortions, a much clearer picture of the costs and benefits of participative management seems to emerge.

A final caveat on the nature of these results is in order. A key assumption in this paper and other analyses of the economic aspects of participation is that workers' behavior is "noncooperative" (see § 2 above). Yet it can be expected that participation could lead to quite different behavior. For one thing, informational distortions would tend to be minimized through interaction and discussion.[16] For another, "team loyalty" and other coalition effects could substantially increase enterprise output over what might have been predicted from a noncooperative model.[17] In principle, these issues can be treated by assuming some "natural" coalition structure (e.g. one determined by the organizational-technological structure of the pre-enterprise) and then applying a modified version of the noncooperative analysis pursued here, wherein the basic actors would be the coalitions instead of individual workers. This approach, however, awaits both theory and empirical evidence on coalition decision-making in response to specified group reward structures such as product-sharing and collective wage schemes.

ENDNOTES

[1] See, e.g. Vanek [16] and references therein for an introduction to published studies of this issue. See also the bibliography assembled in Fitzroy [5].

[2] See, e.g. Hurwicz [7], [8].

[3] See Marshak and Radnor [10] and Groves [6].

[4] The additive separability of u_i is not essential to the development, and is used primarily for ease of exposition. More generally, one would simply require u_i to be monotonically increasing in r_i and jointly concave in x_i and r_i, as in Sertel [13].

[5] Several sets of sufficient conditions can be found in Kleindorfer and Sertel [1976]. Compactness of X is one such condition.

[6] See Kleindorfer and Sertel [9] for details. The pre-enterprises introduced below all have (locally) stable enterprise equilibrium (at least when information distortions are neglected).

[7] N.B. When $\lambda_i = 0$, clearly $x_i = 0$ is the maximizing behavior for agent i. In the case at hand, where $f(x) = \Pi x_i^{\delta_i}$, this leads to $f = 0$, which causes all other agents to produce nothing also. Thus, agent 0 gives positive shares λ_i to all agents $i \in \overline{N}$.

[8] One may expect some form of consistency conditions to hold at points other than the imperfect-information equilibrium as well. For example, the perceived adjustment process $\hat{\alpha}$ might be defined through

$$(*)\ b_i\beta_i(\hat{\alpha}_i[\lambda, x_0](\bar{x}))^{\beta_i - 1} = \lambda_i d_i f(x_0, \bar{x})/(\hat{\alpha}_i[\lambda, x_0](\bar{x})), \bar{x} \in \overline{X}, i \in \overline{N}.$$

This expression, like (13)–(15), equates marginal disutility of effort to marginal utility of income. In this case, however, the latter is computed using the proportionality of marginal product to average product (i.e. $f/\hat{\alpha}_i$) in the Cobb-Douglas case. Moreover, in (*), average product is computed, applying the consistency condition, using actual observed output. If (*) were used in defining $\hat{\alpha}$ instead of (13), the (imperfect-information) equilibria of $\hat{\alpha}$ would, of course, be the same. However, properties like stability of equilibrium clearly depend on the adjustment process itself.

[9] The intuitively appealing fact that $\partial\tilde{x}_i(\lambda, x_0)/\partial\lambda_j \geqslant 0$ for all $i, j \in \overline{N}$ is established in Kleindorfer and Sertel [9] by showing that the equilibrium map $\tilde{x}_i(\lambda, x_0)$ for MCD pre-enterprises satisfies

$$\log \tilde{x}_i(\lambda, x_0) = (1/\beta_i\epsilon)\left[\delta_0 \log x_0 + (\epsilon + (\delta_i/\beta_i)) \log C_i + \sum_{\overline{N}^i}(\delta_j/\beta_j) \log C_j\right]$$

where $\epsilon = \left[1 - \sum_{\overline{N}}(\delta_j/\beta_j)\right]$ and $C_i = (\delta_i\lambda_i/b_i\beta_i), i \in \overline{N}$.

[10] If either $\lambda_i = 0$ or $x_0 = 0$, then (23) implies $\hat{x}_i = 0$. We assume henceforth that $x_0 > 0$ and that $\lambda_i > 0$ for each $i \in \overline{N}$.

[11] Another imperfect-information adjustment process, having the same equilibria as that determined through (27), would be obtained by substituting $f(x_0, x)$ for the r.h.s. of (27). This would actually be more realistic as an adjustment process than that defined through (27) as it would not require any knowledge of f (as the current (27) implicitly does) in computing $\hat{\alpha}_i$. Only f itself would have to be observed. See also endnote 8 above.

[12] This is not strictly true, for when $\bar{x}^i = 0$ (so that by (22) also $\psi_i(x^i) = 0$) then $f(x_0, \hat{\alpha}_i, \bar{x}^i) = x_0^{\delta_0}\hat{\alpha}_i^{\delta_1}$. So both $\hat{\alpha}_i = 0$ and $\hat{\alpha}_i' = (x_0^{\delta_0} d_{i1}/b_i\beta_i)^{1/(\beta_i - \delta_1)}$ solve (27). It is clearly $\hat{\alpha}_i' > 0$ which is the solution to (33), however, so $\hat{\alpha}_i(x^i)$ is in any case single valued.

[13] The reader can verify for any real constants $C > 0$, $\beta > 1$, and d, that the graphs of $\hat{\alpha}_i^\beta$ and $C(\hat{\alpha}_i + d)$ only cross once. Before this cross point $\hat{\alpha}_i^\beta < C(\hat{\alpha}_i + d)$ and after it $\hat{\alpha}_i^\beta > C(\hat{\alpha}_i + d)$.

[14] See Sertel [14] in this volume.

[15] See Taylor [15], pp. 74–77, and Parsons [12] for a discussion of these findings. See

also Vroom [17] in this regard. Such group size effects are, of course, related to similar phenomena in the provision of public goods, see e.g. Olson [11].

[16] See Adizes [1] for several illustrations of consensus effects through participation.

[17] A possible reaction rule for a coalition $B \subset \bar{N}$ which would "explain" such increased productivity would be that, for any collective behavior $x^B \in X^B = \{X_j | j \in N \backslash B\}$ of non-coalition members, coalition B reacted by solving the problem Max $\{f(x_B, x^B) | x_B \in X_B, u_i(x_i, \lambda_i f(x_B, x^B)) \geqslant u_i(\tilde{x}_{i,B}(\lambda_B, x_0), \lambda_i f(x_B, x^B)), i \in B\}$ where, for $i \in B$, $\tilde{x}_{i,B}(\lambda_B, x_0)$ represents the Nash equilibrium (given x^B) within coalition B. This would have coalitions maximize physical product subject to each coalition member's being at least as well off as he would be under a unilateral, noncooperative adjustment process within the coalition. It seems clear that equilibrium output would increase (over the noncooperative case) were such an adjustment process to be followed by each coalition. The case where the entire work force decides on its collective behavior cooperatively is analyzed in Reference [9].

REFERENCES

[1] Adizes, I., *Industrial Democracy: Yugoslav Style,* New York: The Free Press, 1971.

[2] Alchian, A. A. and Demsetz, H., "Production, information costs, and economic organization", *The American Economic Review,* **62:** 777–795, 1972.

[3] Dugundji, J., *Topology,* Boston: Allyn & Bacon, 1966.

[4] Fitzroy, F. R., "Economic organization and human capital", Discussion Paper No. 32, Alfred-Weber-Institute, University of Heidelberg (Germany), April, 1973.

[5] Fitzroy, F. R., "Alienation, freedom, and economic organization: a review note", Alfred-Weber-Institute, University of Heidelberg (Germany), July, 1974.

[6] Groves, T., "Incentives in teams", *Econometrica,* **41:** 701–720, 1973.

[7] Hurwicz, L., "On information decentralized systems", Ch. 14 in McGuire, C. B., and Radner, R. (Eds.), *Decision and Organization,* Amsterdam: North Holland, 1972.

[8] Hurwicz, L., "The design of mechanisms for resource allocation", *The American Economic Review,* **63:** 1–30, 1973.

[9] Kleindorfer, P. R. and Sertel, M. R., "Enterprise design through incentives", International Institute of Management, West Berlin, 1976.

[10] Marschak, J. and Radner, R., *Economic Theory of Teams,* New Haven: Cowles Foundation, 1972.

[11] Olson, M., Jr., *The Logic of Collective Action,* Cambridge, Mass.: Harvard University Press, 1965.

[12] Parsons, H. M., "What happened at Hawthorne", *Science,* **183:** 922–932, 1974.

[13] Sertel, M. R., "Elements of equilibrium methods for social analysis", unpublished Ph.D. dissertation, M.I.T., January, 1971.

[14] Sertel, M. R., "A medley on shares vs. wages", in this volume.

[15] Taylor, F. W., *Scientific Management,* New York: Harper and Brothers, 1911.

[16] Vanek, J., *Self-Management: Economic Liberation of Man,* Harmondsworth: Penguin, 1974.

[17] Vroom, V. H., *Work and Motivation,* New York: Wiley, 1964.

[18] Wilson, R., "The theory of syndicates", *Econometrica,* **36:** 119–132, 1968.

A MEDLEY ON SHARES VERSUS WAGES

MURAT R. SERTEL
International Institute of Management, Berlin

0 A MEDLEY OF TWO TALES

This medley tells two tales, each relating a set of selected results emanating from a separate series of ongoing explorations into the design of economic systems.[1]

The first tale has emerged as part of some work by P. R. Kleindorfer and this author [3] on the economics of enterprise design with special emphasis on output-sharing. It computes profit-maximizing schemes of output-sharing and factor-pricing, solving two rather salient classes of enterprise design problems in the presence of perfect information about the production technology and the preferences of the participating agents.[2] It then compares the outcomes under these two incentive schemes, to find that in capitalist-managed enterprises pricing schemes are always superior, from the viewpoint of maximizing enterprise profits, to sharing schemes, so long as agents' factor contributions to the productive process are readily (costlessly) observable. At the same time it explicitly obtains various measures of the "share-price gap" thus exhibited. These measures of the share-price gap then identify the maximal extent to which costs of measuring agents' factor contributions (costs of supervision) may be incurred before the gap reverses itself.[3]

In contrast to the focus on the economic design of the individual enterprise in our first tale, the theme of our second tale concerns design features of a political economy in a macro-analytic mode and form of specification. The results reported in this second half of our medley derive from a study by Y. M. I. Dirickx and the present author [1]. Here we begin to investigate social systems whose basic behavioral relations are determined, in the standard fashion of economic theory, by its agents' utility- and profit-maximization, but certain rules and parameters affecting the behavior of these agents are set through institutional mechanisms such as voting. The study is motivated by a multitude of interrelated questions, some concerning the foundations and consequences of politico-economic syndromes such as "democratic capitalism", some the formation and deformation of classes and the nature of class conflict under such syndromes, and some the multitude of associated issues of "fairness".

Now the two original studies on which the present medley is based differ in their focus and mode of analysis. The first study (Kleindorfer and Sertel, [3]) focuses on efficiency in economic design at the enterprise level, while the second (Dirickx and Sertel, [1]) concerns itself with fairness and the distribution of income at the level of an entire (aggregate) political economy. There is, however, a common theme of class conflict that arises in each of these investigations. The form taken by this conflict as it arises in the second study has already been suggested above. In the first study two types of conflict emerge. One concerns what form of incentive (e.g. wage-based or output-sharing) schemes are to be instituted in an enterprise. The other type of conflict here concerns which economic agent(s), (e.g. the capitalist or the workers) will be vested with the authority to choose and implement an incentive scheme from the established class of feasible incentive schemes.

Besides exhibiting certain common elements of class conflict in micro- and macro-economic design problems, this medley juxtaposes the issues of efficiency and equity for individual enterprises and for the economic system within which such enterprises are embedded. This is due to a hunch that the design of efficient and fair economic systems can be achieved by addressing the issue of efficiency primarily at the enterprise level with the issues of equity and fairness separated out and addressed at the level of the overall political economy. Demonstrating the conditions under which such a recipe would work is the task of a set of propositions worthy of the stature of "fundamental theorems" of economic system design. Aspiring to much humbler goals, this essay is meant only as a skeletal outline of some of the issues which such an endeavor will likely encounter.

1 SHARES VERSUS WAGES IN ENTERPRISE DESIGN[4]

Consider a set of economic agents whose joint contribution to a productive process yields an output which they all consider to be a good, but each of whom derives only an ever-decreasing extra (marginal) pleasure from increasing his contribution. More specifically, observe the following

1.0 *DEFINITION*[5]: By a *pre-enterprise* we mean an ordered quadruplet

1.0.0 $\Omega = \langle N, X, u, f \rangle$

where

1.0.1 $N = \{0, \ldots, n\}$ with n a positive integer,

1.0.2 $X = \prod_N X_i$ is the product of a family $\{X_i \subset R \mid i \in N\}$ of nonempty, closed and convex subsets of R, the real line,

1.0.3 $u = \{u_i: X_i \times R \to R \mid i \in N\}$ is a family of continuous functions additively separable in the form

$$u_i = w_i(x_i) + r_i \qquad (x_i \in X_i, r_i \in R)$$

with w_i twice continuously differentiable and strictly concave,

1.0.4 $f: X \to R$ is twice continuously differentiable and, for each $x_0 \in X_0$, concave on $\{x_0\} \times X_1 \times \ldots \times X_n$.

Here we interpret N as a set of agents, x_i as the contribution of agent $i \in N$, and u_i as his utility as dependent directly on x_i through his utility w_i of contributing, and on his receipts or remuneration r_i in terms of the output $f(x)$ yielded, via a production function f, by a joint contribution $x = (x_0, \ldots, x_n)$.

Essentially, an *enterprise* is obtained from a pre-enterprise by equipping it with an *incentive scheme*, i.e. a rule tying the remuneration r_i of each agent to the collective contribution $x = (x_0, \ldots, x_n)$. Here we consider and compare two ways of doing this, namely, by way of either a sharing scheme or a pricing scheme. Let us see precisely what these are and how they give rise to enterprises when coupled with a pre-enterprise. Toward this let us once and forever single out the agent 0 to give him the role of enterprise designer. (One may think of this agent as a central planner, a capitalist or a landowner, interpreting his contribution x_0 to the productive process, accordingly, as some mixture of the services of capital and land.) In his role of enterprise designer, given a pre-enterprise Ω as in 1.0, agent 0 is to decide, not only on the size of his contribution x_0, but also on the incentive scheme to institute.

Now a sharing scheme is defined through a real vector $\lambda = (\lambda_1, \ldots, \lambda_n)$ assigning a nonnegative share λ_i of the output f to each agent $i \in \bar{N} = \{1, \ldots, n\}$ in such a way that $\sum_{\bar{N}} \lambda_i f(x) \leqslant f(x)$ for each $x \in X$, i.e. such that $\sum_{\bar{N}} \lambda_i \leqslant 1$. Thus, the set of possible sharing schemes arises from the simplex

$$\Lambda = \left\{\lambda \in R_+^n \mid \sum_{\bar{N}} \lambda_i \leqslant 1\right\},$$

where R_+^n stands for the nonnegative orthant of n-dimensional Euclidean space. Instituting a sharing scheme means taking an element $\lambda \in \Lambda$ and setting $r_i = \lambda_i f$ for each agent $i \in \bar{N}$, so that the remuneration of each such agent is tied to the joint contribution x via $r_i(x) = \lambda_i f(x)$ as a constant share λ_i of the output $f(x)$ determined by that joint contribution x.

On the other hand, instituting a pricing (or wage) scheme means choosing a nonnegative vector $p = (p_1, \ldots, p_n) \in R_+^n$ of prices (or wages) p_i and setting the remuneration r_i of each agent $i \in \bar{N}$ through the formula $r_i(x_i) = p_i x_i$. In contrast to a sharing scheme, a pricing scheme thus remunerates each agent $i \in \bar{N}$ (each "worker") as a function of his own contribution, and *independently* of the contributions of other agents in N.

Now we look at the behavior of the "workers" $i \in \bar{N}$ when agent 0 puts in a contribution x_0 and institutes either a sharing scheme $\lambda \in \Lambda$ or a pricing scheme $p \in R_+^n$. In the case of a pricing scheme, say $p \in R_+^n$, this is quite straightforward. Each agent $i \in \bar{N}$ computes his optimal behavior $\dot{\alpha}_i[p_i] \in X_i$, so that

$$w_i(\dot{\alpha}_i[p_i]) + p_i\dot{\alpha}_i[p_i] = \operatorname*{Max}_{X_i} \; [w_i(x_i) + p_i x_i],$$

and thus the behavior of the agents exhibits the *equilibrium* value $\dot{\mathbf{x}}[p]$ defined by

$$\dot{\mathbf{x}}[p] = \{\alpha_i[p_i]\}_{\bar{N}}.$$

Defining (somewhat repetitiously)

$$\dot{\alpha}[p] = \{\alpha_i[p_i]\}_{\bar{N}}$$

for each $p \in R_+^n$, $\dot{\alpha}[p]$ can be considered as a mapping of $\bar{\mathbf{X}} = \prod_{\bar{N}} X_i$ into $\bar{\mathbf{X}}$, each $\dot{\alpha}[p]$ having a constant value $\dot{\mathbf{x}}[p]$ depending on p above.

On the other hand, when agent 0 institutes a sharing scheme $\lambda \in \Lambda$ instead of a pricing system, a typical agent $i \in \bar{N}$ can compute his optimal behavior only as a function of the i-exclusive behavior

$$x^i = \{x_j\}_{N^i}, \text{ where } N^i = N \setminus \{i\}.$$

In particular, given i-exclusive behavior $x^i \in X^i = \prod_{\bar{N}^i} X_j$ and a sharing system assigning the share $\lambda_i f(x_i, x^i)$ of output to agent $i \in \bar{N}$, the set of optimal behaviors of i consists of the behaviors $\tilde{\alpha}_i[\lambda_i, x_0]\,(\mathbf{x}^i) \in X_i$ maximizing $w_i(\cdot) + \lambda_i f(x_i, x^i)$ where $\mathbf{x}^i = \{x_j\}_{N^i}$ with $\bar{N}^i = \bar{N} \setminus \{i\}$.

For each $i \in \bar{N}$ this defines, in general, a mapping $\tilde{\alpha}_i[\lambda_i, x_0]$ of $\mathbf{X}^i = \prod_{\bar{N}^i} X_j$ into the space of nonempty-closed-and-convex subsets of X_i. When the Cartesian product

$$\tilde{\alpha}[\lambda, x_0] = \prod_{\bar{N}} \tilde{\alpha}_i[\lambda_i, x_0]$$

defined through

$$\tilde{\alpha}[\lambda, x_0](\bar{\mathbf{x}}) = \prod_{\bar{N}} \tilde{\alpha}_i[\lambda_i, x_0](\bar{\mathbf{x}}^i)$$

is singleton-valued for each $\bar{\mathbf{x}} \in \bar{\mathbf{X}}$, we treat $\tilde{\alpha}[\lambda, x_0]$ as a transformation of $\bar{\mathbf{X}}$ into itself and refer to it as the (noncooperative) *adjustment process* determined by (λ, x_0), saying that (λ, x_0) has determined a *sharecropping enterprise.*

Similarly, an ordered pair $(p, x_0) \in R_+^n \times X_0$ is called an *enterprise design*, in particular a *pricing design*, and it is said to determine a *pricing enterprise* with *adjustment process* $\dot{\alpha}[p]: \overline{\mathbf{X}} \to \overline{\mathbf{X}}$.

To define the enterprise design problems that interest us, we agree to call a pre-enterprise *sharecropping-regular* if each sharecropping design $(\lambda, x_0) \in \Lambda \times X_0$ determines an enterprise with a unique *sharecropping equilibrium* $\tilde{\mathrm{x}}[\lambda, x_0]$, i.e. with a unique fixed point of the adjustment process $\tilde{\alpha}[\lambda, x_0]: \overline{\mathbf{X}} \to \overline{\mathbf{X}}$. Note that every pre-enterprise is *pricing-regular*, in the sense that every pricing design $(p, x_0) \in R_+^n \times X_0$ determines an enterprise with an adjustment process $\dot{\alpha}[p]$: $\overline{\mathbf{X}} \to \overline{\mathbf{X}}$ having a unique *pricing equilibrium* $\dot{\mathrm{x}}[p]$, i.e. a unique fixed point.

Now the *sharecropping enterprise design problem* is that of choosing a sharecropping (enterprise) design $(\lambda, x_0) \in \Lambda \times X_0$ so as to maximize the utility (i.e. the net profit)

$$w_0(x_0) + \lambda_0 \tilde{f}(\lambda, x_0)$$

of agent 0, where $\lambda_0 = 1 - \sum_{\overline{N}} \lambda_i$ and

$$\tilde{f}(\lambda, x_0) = f(x_0, \tilde{\mathrm{x}}(\lambda, x_0))$$

The *pricing enterprise design problem*, on the other hand, is that of choosing a pricing (enterprise) design $(p, x_0) \in R_+^n \times X_0$ so as to maximize the utility (i.e. the net profit)

$$w_0(x_0) + \dot{f}(p, x_0) - \Sigma p_i \dot{\mathrm{x}}_i(p, x_0),$$

where

$$\dot{f}(p, x_0) = f(x_0, \dot{\mathrm{x}}[p]).$$

In the case of a log-linear pre-enterprise, as defined below, both of these design problems have unique solutions and we will present these:

1.1 *DEFINITION*: A pre-enterprise Ω will be said to be *log-linear* if it fits the form:

1.1.0 $X_i = R_+ \qquad (i \in N)$;

1.1.1 $u_i(x_i, r_i) = -b_i x_i^{\beta_i} + r_i$ with $b_i > 0\ (i \in N)$ and $\beta_i > 1 \qquad (i \in \overline{N})$;

1.1.2 $f(x) = \prod_N x_i^{\delta_i}$ with $\delta_i \in R_+ (i \in N)$ and $\sum_N \delta_i \leqslant 1$.

N.B. These are the pre-enterprises called multiplicative Cobb-Douglas in Definition 2.3 of Kleindorfer [2].

Now we present the sharecropping equilibrium map for log-linear pre-enterprises and then, for each $x_0 \in X_0$, we exhibit the sharecropping enterprise design problem for such pre-enterprises. Proofs of all theorems in this section are omitted, since they are to be found in Kleindorfer and Sertel [3] and often require material therein or are lengthy.

1.2 *THEOREM: Let Ω be a log-linear pre-enterprise. Then Ω is sharecropping regular. In fact, the sharecropping equilibrium map $\tilde{x}$: $\Lambda \times X_0 \to \overline{X}$ has the form*

$$1.2.0 \quad \tilde{x}_i(\lambda, x_0) = [(C_i\lambda_i)^{\gamma_0} x_0^{\delta_0} \prod_{\overline{N}} (C_j\lambda_j)^{\delta_j/\beta_j}]^{1/\gamma_i} \qquad (i \in \mathrm{N}),$$

where

$$1.2.1 \quad C_j = \frac{\delta_j}{b_j\beta_j}, \gamma_j = \beta_j\gamma_0 \qquad (j \in \overline{\mathrm{N}}) \text{ with } \gamma_0 = 1 - \sum_{\overline{N}} \frac{\delta_j}{\beta_j}$$

and so is continuous, continuously differentiable in the interior of $\Lambda \times X_0$, and monotonically increasing, each equilibrium $\tilde{x}(\lambda, x_0)$ being stable. The sharecropping equilibrium product $\tilde{f}$: $\Lambda \times X_0 \to R$ has the form

$$1.2.2 \quad \tilde{f}(\lambda, x_0) = f(x_0, \tilde{x}(\lambda, x_0)) = K\left(x_0^{\delta_0} \prod_{\overline{N}} \lambda_i^{\delta_i/\beta_i}\right)^{1/\gamma_0}$$

where, using the notation of 1.2.1,

$$1.2.3 \quad K = \prod_{\overline{N}} C_i^{\delta_i/\gamma_i}.$$

1.3 *THEOREM: Let Ω be a log-linear pre-enterprise. For each $x_0 \in X_0$, the sharecropping enterprise design problem constrained to $\Lambda \times \{x_0\}$ is solved by the design (λ^*, x_0) with shares $\lambda_i^* = \delta_i/\beta_i$ (independent of x_0)($i \in \mathrm{N}$).*

Note that, although x_0 is fixed in 1.3, λ^* is actually independent of x_0. To determine the optimal sharecropping design fully, we need also to determine the optimal input x_0^*. But this is independent of λ^* and, when x_0 is capital input, essentially a matter of the cost of capital. We treat this case in Kleindorfer and Sertel [3]. We now present the pricing equilibrium map for log-linear pre-enterprises, and then for each $x_0 \in X_0$ we exhibit the solution to the pricing enterprise design problem for such pre-enterprises.

1.4 *THEOREM: Let Ω be a log-linear pre-enterprise. Then Ω is pricing-regular. In fact, the pricing equilibrium map $\dot{x}$: $R_+^n \times X_0 \to \overline{X}$ has the form*

$$1.4.0 \quad \dot{x}_i(p, x_0) = \left(\frac{p_i}{\beta_i b_i}\right)^{\frac{1}{\beta_i - 1}} \qquad (i \in \overline{\mathrm{N}})$$

and so is continuous, continuously differentiable on the interior of $R_+^n \times X_0$, and *monotonically increasing in p with each equilibrium globally stable. The pricing equilibrium product* $\dot{f}: R^n_+ \times X_0 \to R$ *has the form*

1.4.1 $$\dot{f}(p, x_0) = x_0^{\delta_0} \prod_{\bar{N}} \left(\frac{p_i}{\beta_i b_i} \right)^{\frac{\delta_0}{\beta_i - 1}}$$

1.5 *THEOREM: Let* Ω *be a log-linear pre-enterprise. For any* $x_0 \in X_0$, *the pricing enterprise design problem constrained to* $R^n_+ \times \{x_0\}$ *is solved uniquely by the design* (p^*, x_0) *with prices*

1.5.0 $$p_i^* = \frac{\delta_i}{\beta_i} \left[\frac{\beta_i^2 b_i}{\delta_i} \right]^{1/\beta_i} \left[x_0^{\delta_0} \prod_{\bar{N}} \left(\frac{\delta_j}{\beta_j^2 b_j} \right)^{\delta_j / \beta_j} \right]^{\frac{\beta_i - 1}{\gamma_i}} \qquad (i \in \bar{N})$$

Having computed the optimal sharing scheme from the viewpoint of an (enterprise-designing) agent 0 contributing any given quantity of x_0, we are now able, at least for log-linear pre-enterprises, to compare the efficiency of the outcomes under these two alternative profit-maximizing schemes. We present a basic comparison in the following

1.6 *THEOREM (SMALL SHARE-PRICE GAP): Let* Ω *be a log-linear pre-enterprise and take any* $x_0 \in X_0$. *Table 1 below compares the outcome of instituting the optimal sharing scheme* λ^* (see 1.3) *with that of instituting the optimal pricing scheme* p^* (see 1.5). *With*

$$G = \frac{\tilde{G}}{\dot{G}} = \left(\prod_{\bar{N}} \delta_i^{\delta_i / \beta_i} \right)^{1/\gamma_0} < 1,$$

this gives the conclusion that, for each $i \in \bar{N}$,

$$G = \frac{\tilde{f}^*[x_0]}{\dot{f}^*[x_0]} = \frac{\tilde{r}_0^*[x_0]}{\dot{r}_0^*[x_0]} = \frac{\lambda_i^*}{p_i^*[x_0]} < 1$$

and

$$\frac{\tilde{x}_i[x_0]}{\dot{x}_i[x_0]} = (\delta_i G)^{1/\beta_i} < 1$$

and, finally,

$$\frac{\tilde{u}_i[x_0]}{\dot{u}_i^*[x_0]} = \frac{\beta_i - \delta_i}{\beta_i - 1} G,$$

where starred quantities superscripted by a tilda [resp. a dot] are measured at optimal sharing schemes [resp. at optimal pricing schemes]. In particular, for every $x_0 \in X_0$, *the pricing enterprise design* $(p^*[x_0], x_0)$ *is strictly superior to the sharecropping enterprise design* (λ^*, x_0) *from the viewpoint of the Capitalist:*

TABLE 1: EFFICIENCY COMPARISON OF SHARECROPPING AND PRICING AT ANY $x_o \varepsilon X_o$ $(i \varepsilon \bar{N})$

SHARECROPPING	PRICING
$\lambda_i^* = \frac{\delta_i}{\beta_i}$	$p_i^*[x_o] = (\beta_i b_i)^{\frac{1}{\beta_i}} (\frac{\delta_i}{\beta_i} \dot{G})^{\frac{\beta_i - 1}{\beta_i}} x_o^{(\frac{\delta_o}{\gamma_o} \frac{\beta_i}{\beta}}$
$\tilde{x}_i^*[x_o] = (\frac{\delta_i^2 \tilde{G}}{\beta_i^2 b_i})^{\frac{1}{\beta_i}} x_o^{\frac{\delta_o}{\gamma_i}}$	$\dot{x}_i^*[x_o] = (\frac{\delta_i \dot{G}}{\beta_i^2 b_i})^{\frac{1}{\beta_i}} x_o^{\frac{\delta_o}{\gamma_i}}$
$\tilde{f}^*[x_o] = \tilde{G} x_o^{\frac{\delta_o}{\gamma_o}}$	$\dot{f}^*[x_o] = \dot{G} x_o^{\frac{\delta_o}{\gamma_o}}$
$\tilde{r}_o^*[x_o] = \gamma_o \tilde{G} x_o^{\frac{\delta_o}{\gamma_o}}$	$\dot{r}_o^*[x_o] = \gamma_o \dot{G} x_o^{\frac{\delta_o}{\gamma_o}}$
$\tilde{r}_i^*[x_o] = \frac{\delta_i}{\beta_i} \tilde{G} x_o^{\frac{\delta_o}{\gamma_o}}$	$\dot{r}_i^*[x_o] = \frac{\delta_i}{\beta_i} \dot{G} x_o^{\frac{\delta_o}{\gamma_o}}$
$\tilde{u}_i^*[x_o] = (1 - \frac{\delta_i}{\beta_i}) \frac{\delta_i}{\beta_i} \tilde{G} x_o^{\frac{\delta_o}{\gamma_o}}$ $(i \varepsilon \bar{N})$	$\dot{u}_i^*[x_o] = (1 - \frac{1}{\beta_i}) \frac{\delta_i}{\beta_i} \dot{G} x_o^{\frac{\delta_o}{\gamma_o}}$ $(i \varepsilon \bar{N})$
$\tilde{G} = \left(\prod_N \frac{\delta_i^2}{\beta_i^2 b_i}\right)^{\delta_i/\gamma_i}$	$\dot{G} = \left(\prod_N \frac{\delta_i}{\beta_i^2 b_i}\right)^{\delta_i/\gamma_i}$

One fact made evident by 1.6 is that the profit-maximizing pricing scheme imparts a strictly greater affluence to all participants – capitalist and workers alike – than does the profit-maximizing sharing scheme (i.e. $\tilde{r}_i^*[x_0]/\dot{r}_i^*[x_0] < 1$). Turning to utilities of enterprise participants, we see that agent 0 always prefers the optimal pricing scheme to the optimal sharing scheme. For agents $i \in \bar{N}$ the utility comparison provided in 1.6 may be written as

$$\frac{\tilde{u}_i^*[x_0]}{\dot{u}_i^*[x_0]} = \frac{\beta_i - \delta_i}{\beta_i - 1} G = \frac{\beta_i - \delta_i}{\beta_i - 1} \left(\prod_{\bar{N}} \delta_i^{\delta_i/\beta_i} \right)^{1/\gamma_0} \qquad (i \in \bar{N}).$$

Since $\gamma_0 = 1 - \sum_{\bar{N}} \delta_i/\beta_i$, the reader may check from this that the ratio $\tilde{u}_i^*[x_0]/\dot{u}_i^*[x_0]$ exceeds unity as β_i approaches unity. [This may be easily verified by noting that $\delta_i^{\delta_i}$ always exceeds $(1/e)^{1/e} \cong .692$ ($e \cong 2.718$) on $[0, 1]$, as can be checked by considering the function $\log \delta_i^{\delta_i} = \delta_i \log \delta_i$.] The point of this is that, with the capitalist at the helm, some or all of the workers may prefer a different class of incentive schemes (output-sharing) to that preferred by the capitalist pricing).

An important extension of the above, described in detail in Kleindorfer and Sertel [3], is to the "labor-managed enterprise[7] design problem", in which the authority to choose and set the incentive scheme is vested in the workers instead of the capitalist. A basic result obtained there is that, for every level of capital input $x_0 \in X_0$, there exists a sharecropping (resp. pricing) scheme which, together with appropriate lump-sum transfers of income among enterprise participants, is Pareto superior – from the point of view of capitalist and workers alike – to the profit-maximizing sharecropping (resp. pricing) scheme. Thus, the authority to institute the incentive scheme for an enterprise is always a focus of class conflict.

The above development should suffice to give a flavor of the research directions of Kleindorfer and Sertel [3], which is the main purpose of relating the tale up to here. Now we pass on to our second tale. At the end we will have something to say on possibly desirable extensions of both underlying studies (Kleindorfer and Sertel [3] and Dirickx and Sertel [1]), as well as comments to indicate their various shortcomings.

2 SHARES VERSUS WAGES IN DEMOCRATIC CAPITALISM[8,9]

Here we consider a political economy where capital goods are owned by a profit-maximizing capitalist class who are responsible for employment decisions in the economy, but where the wage level is determined "democratically". We refer to this syndrome as *democratic capitalism*. In particular, we take $\Omega = [0, 1]$ as the set of agents consisting of a capitalist class $C = [0, c]$, with $0 < c < \frac{1}{2}$, and a labor class $L = [c, 1]$.

Denoting Lebesque measure by μ and the set of Borel subsets of R_+ (the non-negative half-line) by $B(R_+)$, we take a scalar-valued production function F: $B(R_+) \times B(R_+) \to R_+$ with $F(K, M) = f(k, m)$, where $k = \mu(K)$, $m = \mu(M)$ and f is of the Cobb-Douglas form

$$f(k, m) = k^{1-\alpha} m^{\alpha} \quad \text{with} \quad \alpha \in (0, 1)$$

Thus, given a Borel set K of capital goods and a Borel set M of laborers employed, F produces a communal income depending on the (Lebesque) measure of the input sets K (capital) and M (labor).

Throughout our present analysis we will work with a Borel set $K \subset R_+$ of capital goods with fixed (Lebesque) measure $\mu(K) = k$ whose ownership is shared uniformly by the capitalists, i.e. by the agents $\omega \in C$. In general, this class owns the "means of production" and employs some Borel set $M \subset L$ of laborers. The level $\mu(M) = m$ of employment at any given wage $w \in R_+$ yields the profit

$$\pi(m) = f(k, m) - wm$$

and there is a proportional profit tax (the only tax in the economy) at a rate $t \in (0, 1)$ whose proceeds $t\pi$ are distributed uniformly over the labor class L, the profit after taxes (i.e. $(1 - t)\pi$) being uniformly distributed over the capitalist class C.

Associated with each agent $\omega \in \Omega$ is a utility function u_ω: $R \to R$ depending only on his income x and of the form

$$u_\omega(x) = \text{sign}\,(x)(|x|)^{\gamma(\omega)}$$

with γ a positive measurable real-valued function on Ω. Thus, given a wage w, each capitalist $\omega \in C$ seeks to maximize his share $\frac{1}{c}(1 - t)\pi$, or equivalently π, and the capitalist class is thus unanimous in the choice of the (profit-maximizing) level of employment.

We take a very simplistic labor market in that we assume each laborer $\omega \in L$ offered employment at a nonnegative wage to automatically accept. Thus, the capitalists' demand for labor determines the employment level m. As profit $\pi(m)$ as a function of m at any wage $w \in R_+$ is concave, first order conditions in maximizing profit determine the (profit-maximizing) employment level m^* as a function of wage in the following fashion:

$$m^*(w) = \begin{cases} 1 - c & \text{for } w \leqslant \overline{w} \\ \alpha^{1/1-\alpha} k w^{1/\alpha-1} & \text{for } w \geqslant \overline{w}, \end{cases}$$

where

$$\overline{w} = \alpha k^{1-\alpha}(1 - c)^{\alpha-1}$$

is the highest wage at which *full employment* ($m = 1 - c$) obtains.

Now we turn to some of the more political aspects of our political economy, indeed to the voting that is to determine the wage level. We consider an open ballot (OB) and two sorts of secret ballot (SB) as voting systems. In all cases, however, there is universal suffrage, one man having one vote to cast for one wage bill $w \in R_+$, and the majority rule obtains. By this rule we mean that a wage bill $w \in R_+$ is majority winner, and gets instituted, if the set $\Omega(w)$ of agents voting for that wage contains a closed set of measure no less than $\frac{1}{2}$.

No matter which of the voting systems is in operation, voting takes place in full knowledge of the employment function m^* above: tables of the employment level as a function of the wage w to be instituted are published and distributed to all the agents. Each agent votes for a wage maximizing his expected utility of income. For the capitalists this clearly means voting for $w = 0$. The analysis of the voting behavior of the workers, i.e. the agents $\omega \in L$, is somewhat more complex; it depends on the voting system and the vote-influencing behavior of the capitalists.

In this part of our analysis we posit a particular coalition structure for our polity. Namely, we assume that the capitalists are perfectly organized and the laborers perfectly unorganized: formally, the admissible coalitions are the capitalist class C and all sets of zero measure consisting of laborers. [This evidently fails to reflect polities where labor parties or unions have a strong voice (for labor), but it may approximate features of early capitalism and of fascism.]

Our next assumption, at this point, that affects the voting outcome concerns the attitudes of workers toward risk. We assume that the interval $[c, c + \frac{1}{2}]$ of workers do not like risk, i.e. that $\gamma(\omega) \leqslant 1$ for every $\omega \in [c, c + \frac{1}{2}]$.

Our objective from here on is to determine voting behavior and the majority-winning wage bill under various voting systems and, finally, to present a welfare and fairness analysis of the outcomes, viewing the distribution of income, under these systems.

The general rule guiding workers' voting behavior is that of expected utility maximization. At any wage w to be instituted, each worker knows the probability $m^*(w)/(1 - c)$ that he will be employed, in which case he earns w on top of other incomes, such as his share $\frac{1}{1 - c} t\pi$ in taxes, and the remaining probability $(1 - m^*(w))/(1 - c)$ that he will be unemployed, foregoing w as income.

The case of the open ballot (OB) is perhaps the easiest to study. As laborers are perfectly unorganized, each of them is willing to sell his vote for an arbitrarily small positive price – as each labor coalition has zero measure, its vote has no effect on the voting outcome and any price obtained for voting one way or another is pure gain in their necessarily noncooperative computations of benefit. Thus, for an arbitrarily small (positive) price, the capitalist class is able – and certainly willing – to purchase enough votes from the workers to secure $w = 0$ as the majority-winning outcome of the OB voting process.

In the case of the secret ballot (SB), it is not possible to effect a genuine purchase of votes, for there is no way of checking whether "the goods" are delivered. Therefore, to affect the voting outcome, one must make it in the interest of a voter to vote in the fashion to be effected. Furthermore, it turns out that, if left alone, the workers in $[c, c + \frac{1}{2}]$ all find $\bar{w}$, the largest wage-level securing full-employment, to be expected-utility-maximizing. Thus, if left alone, the workers in this interval vote for $w = \bar{w}$, whereby $\bar{w}$ is instituted as the majority-winning wage bill. What can be done to perturb this outcome? The capitalist class may *donate* sufficiently large shares in its own interests to a sufficiently large set of workers, in an attempt to create a situation where "what is in the interests of GM (nickname for C)" is in the interests of these workers, who then vote according to those interests, i.e. for $w = 0$, and bring it into majority-winning position.

We consider two types of ruling regarding such donations in the SB case. A *donation*[10] is to be understood as a simple function $\delta: L \to R_+$, where $\delta(\omega)$ is a share in after-tax profit, $(1 - t)\pi$, to obtain (after the voting) at whatever wage gets instituted through that (SB) voting. The two cases we look at are the nondiscriminatory secret ballot (NDSB), where donations are required to be uniform (i.e. constant functions), and the discriminatory secret ballot (DSB), where any simple function $\delta: L \to R_+$ is permitted as a donation.

It turns out that the least costly workers whose voting behavior is to be perturbed by the capitalist class are those who are not risk-loving, i.e. those $\omega \in [c, c + \frac{1}{2}]$, and that the critical (minimal) share switching the vote of such workers (from $w = \bar{w}$ to $w = 0$) is $(1/(1 - c)) + \epsilon$ where ϵ is an arbitrarily small positive real number. Actually, it suffices to switch the votes of the set $[c, \frac{1}{2}]$ to secure $w = 0$ as majority winner. Now in the case of the DSB it pays for C to do this, giving the rest of L a zero donation, if $\alpha > (1/2 - c)/(1 - c)$. In the case of the NDSB, the best donation, for C, is no donation: $\delta \equiv 0$.

Accordingly, we obtain the welfare and fairness analysis depicted in Tables 2 an[d] below, concluding this study. *Fairness* here is to be understood as the poorest being as well off as possible. In Table 2 the income densities resulting from the voting schemes OB, NDSB, and DSB are shown for the three classes of interest: the capitalists, the workers who receive donations under DSB (class $D = [c, \frac{1}{2}]$), and the remaining workers $P = [\frac{1}{2}, 1]$ whom we call the "Proletariat". In Table 2 we denote π evaluated at $m^*(w)$ as $\pi^*(w)$. In particular, $\pi^*(0) = \pi(1 - c) = \bar{f}$ and $\pi^*(\bar{w}) = \pi(m^*(\bar{w})) = (1 - \alpha)\bar{f}$, where $\bar{f}$ is the full employment communal income.

Straightforward calculations show, from Table 2, that the different classes have different preferences regarding the equilibrium outcomes of the various voting schemes. The relative ranking of each class is presented in Table 3 (where $\sim$ denotes indifference and $<$ denotes strict preference). Note that there is, in general, class conflict concerning the best voting scheme.

Finally, the fair tax rate under the separate voting schemes is given in Table 2. By substituting t^* into the expressions for income densities in Table 2 one com-

TABLE 2: INCOME DENSITIES AND FAIR TAX RATES

(N.B. $\bar{f} = k^{1-\alpha}(1-c)^{\alpha}$ is full-employment communal income.)

<table>
<tr><th rowspan="2"></th><th rowspan="2">OB</th><th rowspan="2">NDSB</th><th colspan="2">DSB</th></tr>
<tr><th>$\alpha \leq \frac{1/2-c}{1-c}$</th><th>$\alpha > \frac{1/2-c}{1-c}$</th></tr>
<tr><td>C = [O, C)</td><td>$\frac{1}{c}(1-t)\pi^*(O)$
$= \frac{1}{c}(1-t)\bar{f}$</td><td>$\frac{1}{c}(1-t)\pi^*(\bar{w})$
$= \frac{1}{c}(1-t)(1-\alpha)\bar{f}$</td><td>$\frac{1}{c}(1-t)\pi^*(\bar{w})$
$= \frac{1}{c}(1-t)(1-\alpha)\bar{f}$</td><td>$\frac{1/2}{c(1-c)}(1-t)\pi^*(O)$
$= \frac{1/2}{c(1-c)}(1-t)\bar{f}$</td></tr>
<tr><td>D = [C, 1/2]</td><td rowspan="2">$\frac{1}{1-c}t\pi^*(O)$
$= \frac{t}{1-c}\bar{f}$</td><td rowspan="2">$\bar{w} + \frac{1}{1-c}t\pi^*(\bar{w})$
$= \frac{\alpha+t(1-\alpha)}{1-c}\bar{f}$</td><td rowspan="2">$\bar{w} + \frac{1}{1-c}t\pi^*(\bar{w})$
$= \frac{\alpha+t(1-\alpha)}{1-c}\bar{f}$</td><td>$\frac{1}{1-c}\pi^*(O)$
$= \frac{1}{1-c}\bar{f}$</td></tr>
<tr><td>P = (1/2, 1]</td><td>$\frac{1}{1-c}t\pi^*(O)$
$= \frac{t}{1-c}\bar{f}$</td></tr>
<tr><td>Fair Tax Rate t*</td><td>$1 - c$</td><td>$1 - \frac{c}{1-\alpha}$</td><td>$1 - \frac{c}{1-\alpha}$</td><td>$\frac{1}{1+2c}$</td></tr>
</table>

TABLE 3: CLASS CONFLICT

<table>
<tr><th></th><th>$\alpha \leq \frac{1/2-c}{1-c}$</th><th>$\alpha > \frac{1/2-c}{1-c}$</th></tr>
<tr><td>C = [0, C)</td><td>OB > DSB ~ NDSB</td><td>OB > DSB > NDSB</td></tr>
<tr><td>D = [0, 1/2]</td><td rowspan="2">OB < DSB ~ NDSB</td><td>OB < NDSB < DSB</td></tr>
<tr><td>P = (1/2, 1]</td><td>OB ~ DSB < NDSB</td></tr>
</table>

putes that they are equal (to $\bar{f}$) across all classes for OB, NDSB, and, as long as $\alpha \leqslant (.5-c)/(1-c)$, DSB. When $\alpha > (.5-c)/(1-c)$ the fair tax rate under DSB leads to equal income densities for C and P with a higher income density accruing to D. When the tax rate is not set at its fair value, of course, income densities may vary more substantially among classes with resulting further class conflicts. Additional results on this issue are obtainable through [1].

3 FINAL NOTES

Here we briefly remark on a few points relating to the two tales just told.

Note first that both tales fail to mention a capital market. The study by Dirickx and Sertel [1] has to be extended to account for a capital market and price determination therein. The study by Kleindorfer and Sertel [3] on which the first tale is based actually does continue, treating the case where the enterprise-designer (agent 0) reckons with an externally given cost of capital, i.e. with a perfectly competitive capital market or, in the case where the enterprise-designer is a central planner, a shadow price of capital. We find that the share-pric gap in efficiency reported in 1.5 is only widened (to a "large share-price gap") when agent 0 chooses the optimal pricing design $(p^*[\dot{x}_0^*], \dot{x}_0^*)$ instead of the optimal sharecropping design $(\lambda^*, \tilde{x}_0^*)$. The gap may be interpreted as due to a lack in supervision or supervisability of the contributions of the "workers" $i \in \bar{N}$ by agent 0.

Second, one may look at the labor market in the two tales told. In the first, labor is "geographically" fixed, i.e. already assigned to enterprises, but has to be motivated to work. In contrast, labor in the second tale needs no motivation to work and is perfectly mobile and malleable in use. It appears worthwhile to incorporate utility for leisure into the second model and to obtain a labor supply that is sensitive to wage, determining the employment level at a given wage as the infimum of (this wage-sensitive) supply and demand.

Again with regard to the second tale, one may remark as to the nature in which a *so-called* "people's capitalism" arises here. This syndrome is that where a portion (L\P) of the labor class "defects" from the proletariat, i.e. the rest of the labor class, joining hands with the capitalist class against whose interests the interests of the labor class lie. In this so-called "people's capitalism", it is only a minority of the labor class who actually have sufficiently large capitalist interests (shares in $(1-t)\pi$): it is only a small class who own large shares in the "stock market".

The final note of this medley is a question: How to combine the above two aspects of politico-economic system design? How can the first (the second) be aggregated (disaggregated) to fit with the second (first)?

ENDNOTES

[1] The author is thankful to his separate research partners, Professors P. R. Kleindorfer (University of Pennsylvania) and Y. M. I. Dirickx (International Institute of Management, West Berlin; Katholieke Universiteit Leuven) for permitting him to report, in what follows, their joint findings with him prior to publication elsewhere. Along with Professor L. Hurwicz (University of Minnesota), they are also thanked for helpful comments on a draft of this paper.

[2] Interesting consequences of certain typical informational imperfections in this regard are demonstrated by Kleindorfer [2].

[3] A not so uncommon real-world scenario against the background of which this first tale may be heard is that of absentee large landowners resorting to sharecropping. Consider the case of a large landowner residing in Istanbul to whom the local supervision of the cultivation of his land by his peasants in rural Anatolia is such a nuisance that, despite a share-price gap, he settles for a traditional sharecropping agreement which requires only short visits to "the old homestead" around harvest time to determine the size of the crop to be shared.

[4] The results presented in this section are summarized from Kleindorfer and Sertel [3].

[5] See also 2.1 of [2].

[6] See endnote 4 of [2].

[7] As defined, for example, in [4].

[8] This section summarizes certain results of Dirickx and Sertel [9].

[9] The notation of this section differs in general from that of the last section.

[10] The reader is to imagine a donation as the mailing of shares $\delta(\omega)$ in $(1-t)\pi$ on the day before the voting takes place, entitling the typical recipient $\omega \in L$ of the donated shares to $\delta(\omega)\,(1-t)\pi(w)$ as extra income on top of $\frac{1}{1-c}t\pi(w)$ and, with probability $m^*(w)$, the wage w to be instituted.

REFERENCES

[1] Dirickx, Y. M. and Sertel, M. R., "Class conflict and fairness in 'democratic capitalism', West Berlin", International Institute of Management Preprint Series, IIM, 1975. (*Public Choice*, forthcoming 1978)

[2] Kleindorfer, P. R., "Informational aspects of enterprise design through incentives", in this volume.

[3] Kleindorfer, P. R. and Sertel, M. R., "Enterprise design through incentives, West Berlin", International Institute of Management Preprint Series, IIM, 1975.

[4] Vanek, J., *The General Theory of Labor Managed Market Economies*, Ithaca and London: Cornell University Press, 1970.

AN ECONOMIC THEORY OF ALTRUISM[1]

STEPHEN A. ROSS
Yale University, New Haven

The study of altruism is an old issue in the literature of philosophy. Debates on the possibility and the meaning of altruistic choice predate the development of economic choice theory by centuries.[2] But it is somewhat surprising that, having developed a theory of rational self interested choice for the individual, economists and decision theorists have not attempted to develop a similar theory for altruistic choice. There are, I think, two primary reasons for this.

First, the economist has a healthy suspicion of altruism in positive models. The success of self interest in modern consumer theory has led to a reluctance to abandon it even in situations where it is patently inapplicable. This, in turn, has led to a tautological view of choice that denies the possibility of altruistic choice. The act of giving, for example, without any material recompense gives satisfaction to the giver, and the selfish maximization of this satisfaction becomes the raison d'être for the act. I do not wish to become involved in such semantic issues which really stem from the beneficent content of the word altruism; I merely want to point out that a problem and area of study may remain despite attempts to define it away. Second, the problem itself has not been well posed and is easily confused with the more general issue of social choice. The problem of establishing an axiom system for altruistic choice may be wrongly thought of as being the same as the problem of choosing criteria to guide social choice. There is, however, an important distinction which will be exploited in this paper.

The altruistic choice problem is a special form of an agency relationship [10; 11]. An agency relationship arises when one individual, the agent, makes choices which affect not only himself, but also a second party, the principal. The relationship between the physician and the patient is a good example of the sort I wish to study. The physician cannot fully communicate to the patient all of the information which he possesses about the domain of decision. As a consequence, the physician makes choices for the patient and to the extent to which he does so he can be considered to be an agent of the patient. The guiding preference structure for these decisions, however, is unclear. At one extreme, taking a traditional view, the physician may act solely to maximize his own welfare. The fee will then be the maximum amount of surplus that the physician can extract from the patient. If it is less, then this can be explained by assuming that the physician is engaged in an ongoing relationship or a supergame in which he is wary of signalling his avarice to potential patients. Alternatively, we could always appeal to institutionalized contractual arrangements as the constraint on the physician, but this explanation would seem to be jeopardized by the costliness of enforcement.

An alternative view of the situation is that the relationship itself is embedded in a social structure which specifies that the physician must act in an ethical fashion. Of course, the incentive structure which induces the physician to accept the code may be purely selfish. However, to the extent to which the physician has internalized this ethical code there is a gain in efficiency from the consequent absence of the requirement that the code be enforced and the physician monitored. By internalizing the ethical code, the physician now derives personal satisfaction from curing the patient, or, perhaps, from simply being faithful to the code. Whether or not this is termed altruism is the semantic issue which we have no need to delve into. Why or what inducement leads the physician to accept such a code also makes no difference for what follows. The deciding point for definitional purposes concerns the content of the code itself. I will assume that choice is altruistic, in an economic sense, if the code is not paternalistic, that is, if the code respects the sovereignty of the patient. Ideally, in such a case the physician, as an agent, will be choosing as the patient, as a principal, would have him do. The object of this paper is the problem of designing a proper code, ignoring questions of the extent to which different codes can be internalized. In other words, assuming that the agent wishes to choose for the principal in an altruistic fashion respecting the principal's preferences, how should he go about doing so? Why the agent might wish to do so will not concern us here. Finally, the distinction between the problem as posed and the social choice problem should be clear; we are concerned with a situation where an agent chooses for a single principal not for a polity. While the formal tools of the argument are the analysis of social choice theory, the acceptance of the postulates rests on a quite different conceptual basis.

In the first section we will examine the problem of the altruistic physician in sufficient detail to bring out the central features of the general problem of altruistic choice. In the second section we present and argue for a particular formal system for making such choices. The second section is also of some independent interest since it provides a very simple proof of a result of Harsanyi [4]. The third section discusses the relationship between the formal system and the relevant literature and concludes the paper.

I The data for the physician's problem is illustrated in Figure 1.

We have let Π_A and Π_B denote the respective subjective probabilities held by the physician about the patient's preferences for treatment a or b. In column A,

	A	B
a	1	0
b	0	1
	Π_A	Π_B

FIGURE 1

for example, the 1 in the first entry and the 0 in the second simply indicate that with preferences A, treatment a is preferred to b. The construction of the subjective probabilities is validated through an axiom system such as that of Savage or Raiffa. What is important is that they represent the beliefs held by the agent, the physician, and not by the patient. Given this array, a variety of choice techniques is available. To illustrate, one decision criteria is for the physician to simply choose the treatment that is most probably preferred by the patient. Thus, a or b would be chosen as $\Pi_A \gtrless \Pi_B$. Whatever appeal this procedure may have, however, disappears when we introduce a third treatment, c, as in Figure 2.

In Figure 2, numbers are again used to represent ordinal preferences. By the procedure of choosing the one with the maximal probability, a would be chosen. Treatment c, however, is preferred to a under both preferences B and preferences C, i.e. with probability .6. In fact, it is easy to see that just as in Condercet's cyclic example in voting theory, each choice is dominated by another with probability at least .6.

	A	B	C
a	2	0	1
b	1	2	0
c	0	1	2
	$\Pi_A =$	$\Pi_B =$	$\Pi_C =$
	.4	.3	.3

FIGURE 2

Furthermore, as in voting procedures, it is clear that procedures of this sort take as their raw material the ordinal preference scales represented by A, B, and C and leave out entirely questions of intensity of preference. While this may be desirable (or unavoidable depending on the viewpoint) in social choice problems, in this problem of altruistic choice A, B, and C represent different preferences for the same individual, the patient. In our original choice between a and b, for example, the physician might assign nearly equal probabilities to a or b being best, but might feel that if a is the favorite it will be very intensively favored relative to b, but that if b is the favorite the preference is much less intensive. This mode of reasoning would support the choice of treatment *a*.

An alternative way to represent the data that the physician possesses is given in Figure 3.

	A	B	C	$E\{U\}$
a	.5	.4	.3	.41
b	.3	.8	.1	.39
c	0	.5	.9	.42
	Π_A $= .4$	Π_B $= .3$	Π_C $= .3$	

FIGURE 3

Now each treatment *a*, *b*, or *c*, is conceived of as a lottery giving prizes lying in some space of alternatives, Ω, that the physician uses as a scale for the patient. The numerical quantities in the tableau represent the von-Neumann Morgenstern utilities that the physician believes the patient would assign to choices in Ω. For example, consider the choice of *b*, and for convenience let Ω represent a linearly ordered space of monetary rewards. Suppose that the physician can choose two reference rewards labeled 0 and 1, so that with probability one the physician feels that the best and worst consequences of the treatments would be neither preferred nor inferior to 0 and 1. In other words, each consequence can be assigned a value in Ω and treatments are, generally, lotteries on Ω. Now, under the Marschak system of axioms [7] we can construct a utility function on Ω taking values in (0, 1), by assigning to any consequence a utility, u, with the property that the consequence as viewed by the patient is indifferent to a reference lottery

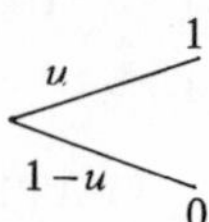

For treatment *b*, for example, we are assuming that preferences *A* represent a case where the patient is indifferent between receiving *b* or a lottery that offers a .3 chance of winning $1. Equivalently, the physician might find it easier to directly assess for the patient the *A* certainty equivalent, say $.20 of receiving treatment *b*. If U is the utility scale of the patient on Ω, then $U(.20) = .3$. Each choice can now be evaluated in expected utility terms, as is shown in the last column of the tableau in Figure 3. As we can see, by taking account of intensities, *c* emerges as the choice followed by *a* and *b*.

The problem becomes considerably more difficult, though, when the agent is unsure not only of the Ω value of each choice, but also of the principal's attitudes towards risk. In the above analysis we used the same utility scale, U, on Ω. Each choice was simply scaled on Ω as a random variable. From the physician's point of view, for example, there was a .4 probability that the principal, the patient, would be indifferent between receiving \$.20 for sure or treatment b. The physician then evaluated the choices by using the patient's utility function, U, over monetary rewards, Ω. As is appropriate, the physician's risk attitudes played no role in the analysis. But, of course, the physician could only use his subjective view of what the patient's risk attitudes are, and, in general, at best the principal could assign a subjective probability distribution on risk attitudes. Notice, too, that with alternative risk attitudes the pattern of normalization becomes important; with a single risk attitude, as in the above example, the choice is insensitive to the utility scaling. The next section proposes a formal axiom system for dealing with these problems.

II The system of postulates presented below is motivated by the desire to develop a rational economic theory of altruistic choice. A desirata of such a system would be that altruistic choice for a particular principal should be independent of the preferences of the agent. It is too much to require that all agents should make the same choice for a given principal, since different agents will, in general, possess different subjective information on the same principal. If they do agree on the same information, though, we will require that they make the same choices.

To begin with, let A denote the space of feasible alternatives, a, among which the agent must choose for a given principal. Notice that A is not meant to exhaustively list all *conceivable* alternatives. The first two postulates describe the data and setting of the problem as viewed by the agent.

Postulate 1 There is some space, Ω, of conceivable rewards such that the agent can assign subjective probabilities Π_i, $i = 1, \ldots, n$, to the principal having a von-Neumann Morgenstern utility index, U_i, on Ω.

Postulate 2 The agent can assign to each choice $a \in A$ a probability distribution, γ_a on Ω, such that given U^i the agent perceives the principal as being indifferent between having the consequences of act a, or having the γ_a lottery on Ω.

Postulate 2 is somewhat less familiar than a number of its special cases, but its generality is of value. Consider, for example, a situation where the agent is going to invest wealth for a principal whose attitudes towards risk he does not know. It seems natural to let Ω be the space of dollar rewards (say $[0, \infty]$). An action $a \in A$ is now a portfolio of investment options, and A is a subset of the set of lotteries on Ω. Alternatively, though, the law in awarding monetary compensation in civil suits is identifying an action, a, such as an injury incurred in an accident, with an equivalent amount of monetary compensation. Thus, if Ω is a dollar reward space, A may bear no direct relationship to it as an object space.

Notice, too, that γ_a is independent of the principal's perceived risk attitudes, U^i. Even though the principal may hold opinions about the probability distribution of returns from a and even though those could alter with U^i, the agent is free to ignore these subjective probability attitudes in altruistic choice, provided that he views γ_a as his own subjective distribution. In using the agent's probability distribution we are really taking an ex post view. Ex ante, if the principal's probability assessments differ from those of the agent, even with a single risk attitude, U, the principal might be quite unhappy with the agent's choices. Nevertheless, to a great extent agency relationships arise because of a presumption that the agent possesses superior information and this is what we are capturing by using the agent's subjective beliefs. Of course, to the extent that the agent is informed of the principal's probability assessments, they will influence his own assessments.

From Postulates 1 and 2, the agent can form a subjective probability distribution over the ways in which to order the alternatives in A. With probability Π^i the agent feels that the principal, given γ_a, would rank alternatives in A by the functional criteria

$$_{\Omega}E\{U^i \mid \gamma_a\} \equiv \int_{\Omega} U^i(x)d\gamma_a$$

We will denote these induced orderings on A by the symbol R^i; i.e. given $a, b \in A$ aR^ib if and only if $\underset{\Omega}{E}\{U^i \mid \gamma_a\} \geqslant \underset{\Omega}{E}\{U^i \mid \gamma_b\}$. Strict preference is denoted by $P^i(aP^ib$ if aR^ib and not $bR^ia)$, indifference by $I^i(aI^ib$ if aR^ib and $bR^ia)$.

Since there will exist a $C_a^i \in \Omega$, such that

$$U^i(C_a^i) = \underset{\Omega}{E}\{U^i \mid \gamma_a\},$$

there is a certain choice, $C_a^i \in \Omega$ that fulfills the role of γ_a in Postulate 2. The reason for not stating Postulate 2 directly in this seemingly simpler form is provided by the investment advisor's example. The advisor assesses the equivalent γ_a directly and computes C_a^i only as a derivative consequence. In the case of compensation for monetary loss, though, it might be easier to assess C_a^i directly.

We will now require that the agent choose amongst alternatives in A by means of a complete ordering on A that is consonant with the orderings R^i. In social choice theory Postulate 3 is referred to as the positive responsiveness condition or, simply, the Pareto condition.

Postulate 3 If for $a, b, \in A$,

$$aP^ib \text{ all } i,$$

then we must have

$$aPb.$$

Postulates 1, 2, and 3 are sufficient to enable us to validate the choice procedure chosen for the example in Section II. When there is a single $U^i(\Pi_j = 0; j \neq i)$, then Postulates 1 and 2 enable the construction of R^i as in the last column of the tableau in Figure 3, and Postulate 3 simply instructs the agent to order A by R^i as the principal would do.

Our technique for dealing with the general case will be to follow the analysis developed by Harsanyi for the social choice problem where R^i denotes the preference ordering of the i-th member of society. We will find, though, that the required postulates are on a stronger footing when applied to the altruistic choice model, as opposed to questions of social choice. Following Harsanyi let us assume then that the agent obeys Marschak's axioms in the altruistic choice problem. Beginning with the least controversial postulate we have:

Postulate 4 The agent chooses altruistically amongst sure alternatives in Ω by means of a complete ordering, $\succsim$, on Ω. Furthermore, $\#\Omega \geqslant 4$.[3]

In what follows, too, the analysis is greatly simplified at no real loss in generality by specializing Ω to be a finite set and we shall do so, setting $\#\Omega = m < \infty$.

Postulate 5 If x, y, and $z \in \Omega$ and

$$x \succsim y \succsim z,$$

then there exists a subjective probability p such that

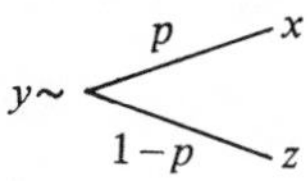

$$y \sim \left\langle \begin{array}{l} \overset{p}{\longrightarrow} x \\ \underset{1-p}{\longrightarrow} z \end{array} \right.$$

Postulate 6 If $x, y \in \Omega$ with $x \sim y$, then for any $z \in \Omega$ and any subjective probability p

$$\left\langle \begin{array}{l} \overset{p}{\longrightarrow} x \\ \underset{1-p}{\longrightarrow} z \end{array} \right. \sim \left\langle \begin{array}{l} \overset{p}{\longrightarrow} y \\ \underset{1-p}{\longrightarrow} z \end{array} \right.$$

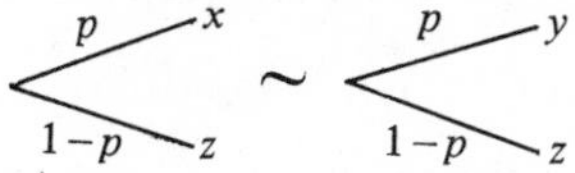

Postulate 5 is the continuity axiom and Postulate 6 is the substitution principle found by Samuelson and Malinvaud to be implicit in the original von-Neumann Morgenstern axioms, and made explicit by Marschak. The application of Postulate 5 to problems of social choice has received little criticism in the social choice literature, authors such as Diamond [3] and Sen [14], preferring instead to focus on the substitution principle. In section III we will examine in detail the criticisms of these postulates that have been advanced in the social choice literature, but for the moment we will simply be content to motivate them and derive their implications.

From Postulates 1 through 6 we can prove a basic result due to Harsanyi [4]. In our context Harsanyi's result implies that the altruist's preference ordering on Ω is simply a weighted sum of the subjectively perceived preferences that may be held by the principal. Instead of assuming the positive responsiveness condition, Postulate 3, to prove this result, Harsanyi required the postulate that if $x, y \in \Omega$ and for all $i, xI^i y$, then xIy for society. He then had to impose positive responsiveness as a separate condition. Unlike ours, his also required the existence of n certain prospects in Ω such that the U^i could be scaled so as to give the unit vectors when applied to these prospects.[4] The following proof, then, is of interest in its own right.

Theorem 1 Given any normalization of the U^i's, there exist constants, α_i, such that the altruistic ordering, $\mathfrak{L}$, on Ω is given by

$$W(U^1, \ldots, U^n; \Pi_1, \ldots, \Pi_n) = \sum_{i=1}^{n} \alpha_i(\Pi_1, \ldots, \Pi_n) U^i; \alpha > 0$$

Notice that the weights α_i depend in general on Π as well as on the normalization.

Proof: Let

$$U_{ij} \equiv U^i(X_j) \text{ for } X_j \in \Omega; j = 1, \ldots, m,$$

and let

$$U \equiv U_{ij}$$

Let p denote the vector of probabilities of a lottery with prizes in Ω. Since the ordering, $\mathfrak{L}$, of the altruist satisfies Marschak's postulates, there exists a utility function W on Ω which reflects this ordering for the purposes of choice under uncertainty. Representing W by the vector $\omega = (\omega_1, \ldots, \omega_m)$, where

$$\omega_i \equiv W(X_i); X_i \in \Omega,$$

the positive responsiveness condition, Postulate 3, requires that for any p, p^* probability vectors

$$Up \gg Up^*$$

implies that

$$\omega p > \omega p^*$$

This is equivalent to requiring that for all vectors η

$$\sum_i \eta_i = 0,$$

and

$$U\eta \gg 0,$$

imply that

$$\omega\eta > 0.$$

It follows immediately that the ω is a linear combination of the constant vector, and the columns of U, where the latter weights must be nonnegative and not all 0. Since we can scale the constant to be zero we are finished. *Q.E.D.*

Given W, the agent chooses amongst alternatives in A by the induced ordering on A generated by W. Knowing the form of W is a major step in the development of an altruistic choice procedure, but unless we can derive the dependence of the weights α on the probability vector, Π, we are in no position to make choice. To enable us to obtain a complete ordering on A, we will impose the following additional postulates.

Postulate 7 (Symmetry) There exists a normalization of the U^i's, $i = 1, \ldots, n$, such that for all sequences $\{U_1, \ldots, U_s\}$ drawn from $\{U^1, \ldots, U^n\}$ (i.e. $U_i \in \{U^1, \ldots, U^n\}$),

$$W\left(U_1, \ldots, U_s; \frac{1}{s}, \ldots, \frac{1}{s}\right) = \frac{1}{s}U_1 + \ldots + \frac{1}{s}U_s$$

Postulate 7 is, in some sense, the most controversial one in the system. In particular, we are suggesting a special normalization in the space Ω. We will consider Postulate 7 in detail in Section III.

The remaining two postulates in our system are probably much more acceptable.

Postulate 8 (Pooling) If $U^i = U^j$ (when appropriately scaled), then

$$W(U^1, \ldots, U^n; \Pi_1, \ldots, \Pi_i, \ldots, \Pi_j, \ldots, \Pi_n) = (\alpha_i + \alpha_j)U^i + \sum_{k \neq i,j} \alpha_k U^k$$

$$= W(U^1, \ldots, U^n; \Pi_1, \ldots, \Pi_i + \Pi_j, \ldots, \Pi_n).$$

In other words, if two risk attitudes are identical we can simply pool their associated probabilities. Finally, we impose a regularity condition on W. Postulate 9 states that sufficiently similar situations lead to similar orderings.

Postulate 9 (Continuity) $W(U^1, \ldots, U^n; \Pi_1, \ldots, \Pi_n)$ is continuous in Π.

Theorem 2 Under the normalization of Postulate 7,

$$W(U^1, \ldots, U^n; \Pi_1, \ldots, \Pi_n) = \Sigma \Pi_i U^i.$$

Proof: Consider a $(\Pi_1, \ldots, \Pi_n)$ vector. Since the rationals are dense on the unit interval there exists a sequence of integer ratios

$$\frac{q_i^s}{r_i^s} \to \Pi_i.$$

Hence, the sequence

$$\Pi^s \equiv \left(\frac{K_i^s}{K^s}\right) \to \Pi$$

where

$$K_i^s = q_i^s \prod_{j \neq i} r_j^s$$

and

$$K^s = \prod_j r_j^s$$

Now, applying the normalization of Postulate 7

$$\begin{aligned} &W(U^1, \ldots, U^n; \Pi^s) \\ &= W\left(U^1, \ldots, U^n; \frac{K_1^s}{K^s}, \ldots, \frac{K_n^s}{K^s}\right) \\ &= W\left(U^1, \ldots, U^1, U^2, \ldots, U^2, \ldots, U^n, \ldots, U^n; \frac{1}{K^s}, \ldots, \frac{1}{K^s}\right) \\ &= \frac{K_1^s}{K^s} U^1 + \ldots + \frac{K_n^s}{K^s} U^n \\ &\to \Pi_1 U^1 + \ldots + \Pi_n U^n, \end{aligned}$$

where we have made use of Postulate 8 and Postulate 7 in the second and third steps. Postulate 9 now establishes the result. *Q.E.D.*

Theorem 2 provides the decision criteria to be used by the altruistic agent.[5] In Section III we will discuss the normalization procedure that yielded it and the conceptual basis of the controversial postulates. Accepting these postulates, though, we must conclude that the altruistic agent should choose for the principal by using as a criteria the expected von-Neumann Morgenstern utility function obtained by weighting the appropriately normalized subjective utility functions by the respective probabilities that the principal holds for them.

III This section discusses the individual postulates presented in Section II and briefly considers both the implementation of the criteria as a normative approach and its positivistic implications.

Postulates 1 and 2 have already been discussed in Section II and are probably not difficult to accept, but they do pose the altruistic choice problem in a fashion that may seem a bit strange. It would have been more straightforward to have simply identified the alternative space, Ω, and the reward space A and have dealt with orderings on only a single space. On a purely formal level, too, there is nothing in the postulates to exclude the possibility that $A \subseteq \Omega$ (or $A = \Omega$). Nevertheless, something is lost by such a procedure. In a practical choice problem, the space Ω can be specified by the agent prior to any knowledge of what the feasible class of acts, A, will be. For example, if the agent is able to treat Ω as simply dollar rewards (or even indirectly do so through the indirect cardinal utility function on a price-income space), then a wide array of action sets A, quite distinct from Ω, can be accommodated.

Postulate 3 is a strong form of the Pareto principle. It requires that if one alternative, a, strictly dominates another, b, in all probable preference structures the principal might possess, then the altruistic agent should also observe that strict preference. It is cast in this form rather than, say, a form that observed weak preference so as to avoid problems of the sort that arise most clearly when the agent knows the principal's preferences. The weak form would allow the agent to be indifferent between a and b even when the principal is known to prefer a to b. As Sen points out even the Pareto principle is not free from criticism in the social choice literature, but it would seem to be on stronger ground in the theory of altruistic choice. The objections in the social choice literature are largely motivated by situations where one would want to discount the preferences of some individual as being irrelevant or, in some sense, improper. In our context such a judgment would be a form of paternalism. In effect, the agent would be making the judgment that "he knows better" than the principal what is good for the principal even when they both agree on the same information.

Let us consider this argument further since, certainly in situations like that of Section II, it would seem that paternalism would be in order. To introduce yet another example, an argument could be made that even though a child would prefer to eat ice cream for supper rather than vegetables, the parent should impose his preferences on the child and insist on vegetables. This is, I believe, the canonical example of what might be thought of as paternalism, but it need not be an imposition of preferences or in violation of the postulates for altruistic choice. Presumably, the parent is making that choice because of a greater awareness of the consequences of the child's choice. If the ice cream leads to dental problems, then the child, unaware of this or unable to conceive of it, would no doubt prefer the parent's choice as an adult.[6] In this fashion, much of what seems like paternalism may, in fact, be a form of altruistic choice. The only sense in which the agent imposes attitudes on the principal comes through the use of the agent's subjective probabilities.

Postulate 4 states that the agent can always choose in A, and does not seem to be terribly controversial, but one can construct examples that would cast some doubt upon it. The physician, for example, is admonished not to practice medicine on loved ones, presumably because objectivity of choice would be difficult. At the extreme, the agent might simply not be able to choose in such situations, and the resulting ordering would be rendered incomplete.

Postulates 5 and 6, though, embody the heart of the application of the axioms for rational individual choice under uncertainty to the altruistic choice problem, and they deserve some comment. Postulate 5, the continuity requirement, appears to be more easily accepted for general choice problems than Postulate 6, but even it is not without its troubles. As Luce and Raiffa [5] point out and Sen [14] demonstrates by example, Postulate 5 omits the possibility of special, i.e. lexicographic attitudes towards gambling. In the altruistic choice problem, this seems of little consequence; it is precisely the effect of such attitudes on the part of the agent that we wish to exclude. If the principal possesses such attitudes, then this would deny that he had a von-Neumann Morgenstern utility index on Ω. If the altruist feels, perhaps on moral grounds, that such attitudes ought to be imposed on the principal's choices; this, too, is excluded by Postulate 5.

Criticism of the use of the Marschak axioms for social choice has focused on Postulate 6. Consider the following paradox constructed by Diamond [3]. On grounds of equity two social states, $(U^1 = 1, U^2 = 0)$ and $(U^1 = 0, U^2 = 1)$, where U^i is the utility of individual i, might well be considered socially indifferent. Applying the substitution principle, we would then conclude that the lottery

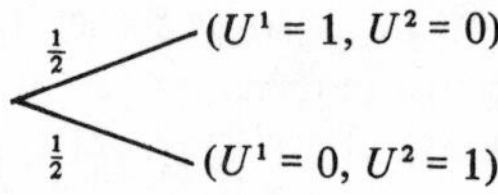

would be indifferent to the sure state $(U^1 = 1, U^2 = 0)$. Yet, this appears to violate a basic concept of fairness. Sen discusses two counterarguments to this paradox. First, that the result is not paradoxical at all, since ex post after the lottery one of the sure states will be chosen so that the mechanism of choice is really irrelevant. Second, and more telling, as Sen observes, the appeal of the Diamond paradox rests sensitively on the particular normalization of the utility indices. Adding one unit to U^2 might make the sure thing appear fairer than the lottery.[7]

Where does all this leave us in the altruistic choice problem? The arguments advanced against the substitution principle have stressed examples where it leads to an ethically inferior or unfair choice being made. Fairness and equity, though, have no role in the altruistic choice problem. Suppose that the probabilities that the principal has either U^1 or U^2 as the cardinal utility index are each $\frac{1}{2}$. The indifference between the lottery and an alternative that gives $U^1 = 1$ if the utility index is U^1 and $U^2 = 0$ if it is U^2, seems quite reasonable. The lottery itself is really a compound lottery,

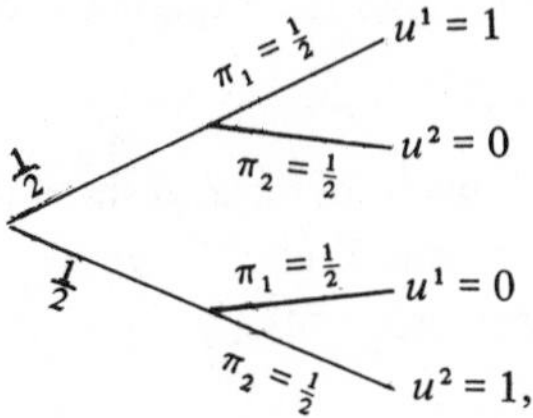

and setting it indifferent to the lottery

$$\pi_1 = \tfrac{1}{2} \quad u^1 = 1$$
$$\pi_2 = \tfrac{1}{2} \quad u^2 = 0,$$

is as innocent as asserting the equivalence of compound lotteries; in either case the agent is selecting a lottery with a probability of $\frac{1}{2}$ that the principal will get unit utility. Of course, this interpretation rests heavily on the acceptance of the normalization that makes the agent indifferent between the principal having $U^1 = 1$ or $U^2 = 1$ if he turns out to be of type 1 or type 2. The existence of such a normalization is assumed by Postulate 7. Postulate 7 does not, however, give the altruist a specific formula for constructing the normalization. Furthermore, Postulate 7 can be easily weakened.

A somewhat more circuitous, but more transparent, route to establishing Theorem 2 comes from breaking Postulate 7 down into two constituent postulates. For convenience we will suppose that there are only two types, U^1 and U^2.

Postulate 7′ Reduction of compound lotteries.

Postulate 7″ Suppose a and b to be two conceivable alternatives such that, without loss of generality

$$U^1(a) = 1,\ U^2(a) = 0$$

and

$$U^1(b) = 0,\ U^2(b) = 1.$$

Under this normalization, in all compound lotteries on a and b the prizes

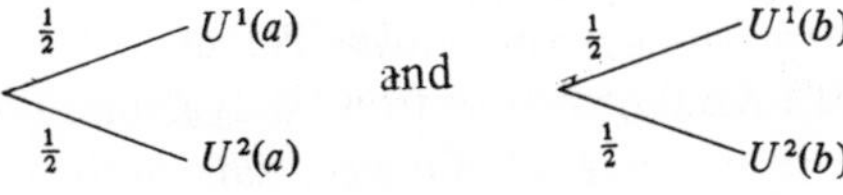

may be substituted, retaining indifference of the resulting lotteries.

Theorem 2 now becomes Theorem 2′.

Theorem 2′ Same as Theorem 2 with Postulates 7′ and 7″ substituted for Postulate 7.

Proof: To prove Theorem 2′ we only have to show that the new postulate system implies Postulate 7. (Postulates 7′ and 7″, however, give us a direct proof quite easily.) Consider any sequence of the sort in Postulate 7 drawn from $\{U^1, U^2\}$. Let

$$W\left(U_1, \ldots, U_S; \frac{1}{S}, \ldots, \frac{1}{S}\right) = \sum_{i=1}^{S_1} \alpha_i U^1 + \sum_{i=s_1+1}^{S} \alpha_i U^2;$$

where we have grouped the U^1's at the beginning. Now, consider the pair (U^1, U^2) where U^1 comes from the i-th entry, $i \leqslant s_1$, and U^2 from, say, the s-th entry. By compounding the lotteries and substitution

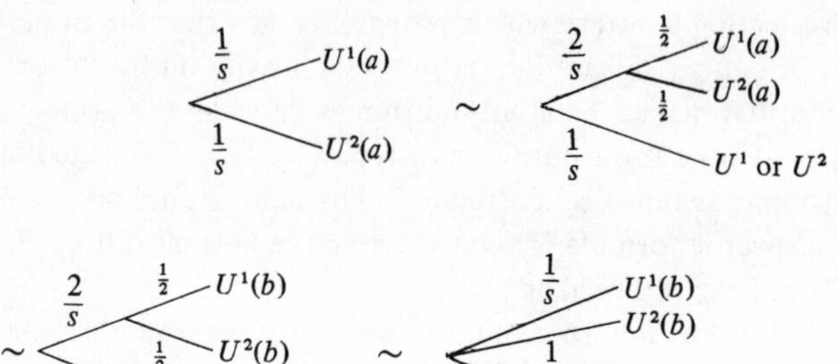

Hence,

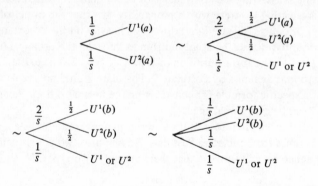

$$\alpha_i U^1(b) + \alpha_s U^2(b) = \alpha_i U^1(a) + \alpha_s U^2(a)$$

or

$$\alpha_i = \alpha_s.$$

Similarly, for all $j \geqslant s_1 + 1$

$$\alpha_j = \alpha_1,$$

and we can normalize to get Postulate 7. *Q.E.D.*

Postulate 7′ is, perhaps, easily accepted, but Postulate 7″ is in direct violation of the Independence of Irrelevant Alternatives (IIA) assumption of social choice theory. Deleting a or b from the set Ω of *conceivable* alternatives will, in general, alter choices within the set, A, of *feasible* alternatives. Arrow [1] argues for the adoption of IIA on the grounds that choice should not be affected by the presence of infeasible alternatives. For example, the choice of a president in the next election should not be affected by the presence or absence of Abraham Lincoln. However, while the preferences between the two candidates should not be affected by the *feasibility* of choosing Lincoln, equally clearly it may be affected by the conception of a Lincoln presidency. In effect, we are agreeing to norm choices by a Lincoln. As the feasible set, A, varies the choice between alternatives in A does satisfy IIA. The ordering of alternatives in $B \subset A$ is unchanged if the feasible set changes to $A' \supset B$. It is only when the *conceivable* set, Ω, is altered that IIA is violated.

Postulates 7′ and 7″, unlike Postulate 7, also suggest a way to carry out the normalization, but how in practice is the altruist to find acts a and b? One way to proceed would be to select the best $x^i \in \Omega$ for U^i and the worst $x_i \in \Omega$ for U^i and set

$$\sup_{x \in \Omega} U^i(x) = U^i(x^i) = 1,$$

and

$$\inf_{x \in \Omega} U^i(x) = U^i(x_i) = 0,$$

on the grounds that in choosing for the *same* individual the intensity of preference and displeasure should be identical at the extremes of the conceivable scale. There are no natural scales or metrics on Ω other than the ones induced by the U^i orderings, and it is only with the best and the worst *conceivable* choices that comparable alternatives can be found. Since irrespective of risk attitudes the choice is being made for the same individual, such a normalization is equivalent to making the agent indifferent between alternatives a and b which are each best or worst with probabilities $\frac{1}{2}$, $\frac{1}{2}$, although when a is best b is worst and conversely. To put it slightly differently, if we let E be the event, "the individual has risk attitudes U_1" and "not E" for U_2, then the probability of E is $\frac{1}{2}$ and "not E" is $\frac{1}{2}$. The normalization requires that the lotteries

$a \equiv$ (E → "best"; not E → "worst") and $b \equiv$ (F → "worst"; not E → "best"),

are indifferent.

IV In conclusion, an axiomatic system for making altruistic choices has been developed above. Whether or not this particular scheme proves acceptable, the applicability of such a decision procedure should be clear. For one thing, it would facilitate economic analysis of the ethical codes of behavior which govern both market and nonmarket transactions.

ENDNOTES

[1] This work was supported by the National Science Foundation Grant No. SOC 74–20292.

[2] Along with the writings of Jeremy Bentham, Butler and Hume also dealt extensively with the topic. A critical secondary source is [9].

[3] Formally, alternatives lie in A, but by a "sure alternative in Ω" we will mean the certainty equivalent of an action in A. Choice in A is by means of the ordering induced on A by $\succsim$.

[4] Camacho and Sonstelie [2] have also noted and corrected this aspect of Harsanyi's proof, but they, too, require the separate imposition of positive responsiveness. Furthermore, as with Harsanyi, their proof adopts a particular normalization of the U^i, which is irrelevant to the proposition.

[5] It is important to understand the sense in which Theorem 2, by the use of Postulates 7, 8, and 9 definitively restricts the form of the criteria. Theorem 1 only specifies that for *each* Π, there is some set of weights $\langle\alpha_i\rangle$. Now, for any particular choice of a normalization of the U^i's without further postulates we cannot infer that $\alpha_i = \Pi_i$. Of course, it is trivially true that if we adjust the U^i normalization for each Π_i we can set the "weights" at Π_i, but this is no restriction on the *functional* relation between Π and α.

[6] There are several possible misinterpretations of this statement that I should like to avoid. I am not suggesting here that the parent chooses for the child according to a minimizing regret notion. What is important is *not* that after the child has cavities then he will regret the ice cream, but rather that when the child is a fully reasoning adult, he will have wished the parent to have chosen vegetables even if the exact dental consequences were somehow withheld. I also do not wish to suggest that the altruist ought to apply some notion of extended sympathy along the lines proposed by Arrow [1]. What the parent might have wished were done for him is not necessarily what is best for the child. In practice, though, such extended sympathy is probably a prevalent mode, by which, at the least, agents form estimates of principals' preferences. Finally, the parent–child example is probably not a good one for this section since, the child is, in some sense, not a fully rational being. I suspect that this presumption is not as clear as it might at first appear, but I wish to avoid this issue, and the reader should feel free to alter the example accordingly.

[7] In a sense, though, this line of reasoning misses the point. An argument can be made that to be fair we should ignore utility considerations and concentrate on goods distributions alone. If there is one unit of a good to be distributed (the single life giving coconut on the desert island), then Diamond's paradox [3] with the good substituted for utility levels seems much stronger. The difficulty with this approach, though, is that if we agree to treat symmetric goods allocations and lotteries such as (a, b) and (b, a) as indifferent, we will, in general, be in conflict with the Pareto principle. It is easy, in fact, to construct examples where two states are viewed as indifferent by all individuals, but where one is Pareto preferred to the symmetric analogue of the other.

REFERENCES

[1] Arrow, K. J., *Social Choice and Individual Values,* New York: Wiley, 1951; 2nd ed. 1963.

[2] Camacho, A. and Sonstelie, J., "Cardinal welfare, individualistic ethics, and interpersonal comparisons of utility: a note", *Journal of Political Economy,* **82,** No. 3: 607–611, May/June, 1974.

[3] Diamond, P., "Cardinal welfare, individualistic ethics, and interpersonal comparisons of utility: a comment", *Journal of Political Economy,* **75**: 765–766, 1967.

[4] Harsanyi, J. C., "Cardinal welfare, individualistic ethics, and interpersonal comparisons of utility", *Journal of Political Economy,* **63,** 309–321, 1955.

[5] Luce, R. D. and Raiffa, H., *Games and Decisions,* New York: Wiley, 1957.

[6] Malinvaud, E., "Note on von Neumann-Morganstern strong independence axion", *Econometrica* **20:** 679, 1952.

[7] Marschak, J., "Rational behavior, uncertain prospects, and measurable utility", *Econometrica,* **18:** 111–141, 1950.

[8] Raiffa, H., *Decision Analysis,* Reading, Mass., and London: Addison-Wesley, 1968.

[9] Roberts, T. A., *The Concept of Benevolence,* New York: MacMillan, 1973.

[10] Ross, S. A., "The economic theory of agency: the principal's problem", *American Economic Review,* **63:** 134–144, May 1973.

[11] Ross, S. A., "On the economic theory of agency and the principle of similarity, in Balch, M. D. (Ed.), *Essays on Economic Behavior under Uncertainty:* 215–237, Amsterdam: North-Holland, 1974.

[12] Samuelson, P. A., "Probability, utility and the independence axiom", *Econometrica,* **20:** 670–678, 1952.

[13] Savage, L. J., *The Foundations of Statistics,* New York: Wiley, 1954.

[14] Sen, A. K., *Collective Choice and Social Welfare,* San Francisco: Holden-Day, 1970.

AN ECONOMIC MODEL OF INTERNATIONAL NEGOTIATIONS RELATING TO TRANSFRONTIER POLLUTION

HENRY TULKENS
C.O.R.E., Université Catholique de Louvain, Belgium

I MOTIVATION AND PLAN OF THE PAPER[1]

This paper attempts to formulate a mathematical model of negotiations taking place between governmental delegates of countries whose citizens share a common resource, viz. the "Southern Bight" (i.e. the southern part of the North Sea) and are concerned about its quality for various and possibly conflicting reasons.

Let us briefly describe the problem: in each of the countries involved (U.K., France, Belgium, Netherlands, Germany, Denmark and Norway) various industries and municipalities use the North Sea for discharging residuals, both off-shore and near their national coasts; such discharges have increased considerably over the last decades. Today, however, ecologists warn that the amount of discharges in that region has reached a point which is beyond that of the so-called assimilative capacity of the waters; even the ecologically unsophisticated vacationers along the coasts complain increasingly about the quality of the North Sea waters. Thus, if we approach the situation from the individualistic viewpoint of each of the seven countries involved, it appears that each of them both *benefits* and *suffers* from the prevailing state of affairs: the discharges of residuals take place in the sea, quite probably because this is a (nationally) cheaper way to dispose of them than any kind of in-land treatment; on the other hand, each country has to put up with the nuisance effects of the aggregate amount of the discharges, mainly from the scenic and health points of view. Of course, within each country, beneficiaries of discharges and sufferers from pollution are not necessarily the same persons; but this is only a side issue in the present investigation. More important is the fact that the relative advantages from discharging on the one hand, and disadvantages from pollution, on the other hand, vary considerably from one country to another (e.g. discharges from the industrial area of Dunquerque are carried north-eastwards by marine currents, and therefore affect French waters only for a short time, while they affect Belgium and the Netherlands more permanently).

The paper addresses itself to the following question: if it is felt (or shown) that the water quality in the Southern Bight is nonoptimal (in a sense to be made precise below), can a *collective decision mechanism* be defined that would

(*a*) determine changes in the countries' discharges at sea such that the sea's quality level be brought to an "optimal" level; and
(*b*) be specified in such a way that for each country taken individually, benefits from improvements in the quality of the sea would be larger (or at least equal to) the costs imputed to it by the decision mechanism?

The propositions 1 and 2 proved below (*section III*) establish that the dynamic process presented in *section II* of this paper indeed possesses these two properties.

The process is formulated in terms of gradual reductions over time of the physical discharges that each country allows its national economic agents to make: a hydrodynamic and ecological model is then called upon for expressing the estimated effects on the quality of the sea of such gradual reductions; finally, compensatory transfer payments among countries are defined, which are supposed to accompany the discharge reductions; these are shown to cover the costs of the reductions, and at the same time to give the process property (*b*) defined above. In particular, if such compensatory transfer payments were managed by a specially designed international agency, it is shown that, under the proposed process, the budget of the agency would break even.

In concluding *section IV*, the question is approached (IV.1) whether the model proposed here also enjoys another and much stronger property, namely assuming that each country has the choice between either participating, or not participating in the negotiation (as formalized here), does participation always benefit all participants more than nonparticipation? Some conditions for a positive answer are suggested, but the idea remains a conjectural one at this stage. Finally (IV.2), the compensatory payments scheme implied by the process is confronted with the currently popular "polluters pay" principle.

II THE BASIC MODEL

II.1 *Notation; Agents and Commodities; Ecological and Behavioral Relations*

Let $\mathscr{I} = \{i | i = 1, \ldots, I\}$ be the set of countries involved in the pollution problems of the Southern Bight, either as polluters, or as victims of its polluted waters, or both. Let $\mathscr{B} = \{\beta | \beta = 1, \ldots, B\}$ denote the set of pollutants (i.e. material substances characterized by their physical and chemical properties) subject to negotiations between countries i (or a subset of the latter). Let $\mathscr{D} = \{d | d = 1, \ldots, D\}$ be the set of geographical points, inland or on the sea surface, where pollutant discharges or dumpings take (or may take) place; let finally $\mathscr{L} = \{\ell | \ell = 1, \ldots, L\}$ be the set of geographical points on the sea surface where water is sampled and its quality technically evaluated, to serve as a basis for sanitary and ecological policy decisions. All the above sets are assumed to be finite.

Environmental variables (typically the amounts per unit of time of the various pollutants present in the waters of the Southern Bight) are written in vector form $R = (R_1, \ldots, R_\alpha, \ldots, R_A)$, where $\alpha \in \mathscr{B} \times \mathscr{L} = \{\alpha | \alpha = 1, 2, \ldots, A; A = BL\} \equiv \mathscr{A}$; thus, each component R_α refers to an amount of some pollutant β observed at some sampling location ℓ in the sea waters. Similarly, *discharge variables* (i.e. amounts of each pollutant discharged by each country per unit of time) are written as vectors $S_i = (S_{i1}, \ldots, S_{i\gamma}, \ldots, S_{iG}), i \in \mathscr{I}$, where $\gamma \in \mathscr{B} \times \mathscr{D} = \{\gamma | \gamma = 1, 2, \ldots, G; G = BD\} \equiv \mathscr{G}$. By convention, $R_\alpha \leqq 0$ and $S_{i\gamma} \leqq 0$ for all α, i and γ.

We shall assume that the relations between environmental and discharge variables can be determined by a predictive model such as, e.g. the one called "Math. Modelsea" [3], whose general form reads:

$$R_\alpha = f_\alpha(S_1, \ldots, S_G) \qquad \alpha \in \mathscr{A} \tag{1}$$

with

$$S_\gamma = \sum_{i \in \mathscr{I}} S_{i\gamma} \qquad \gamma \in \mathscr{G}. \tag{2}$$

The functions f_α, assumed nondecreasing in all their arguments, are called transfer functions. Only their first-order partial derivatives $\partial f_\alpha / \partial S_\gamma$ are supposed to be known in the negotiation model developed below.

As far as the *economic variables* are concerned, we shall denote by two vectors x_i and y_i the quantities of all commodities respectively consumed and produced per unit of time in country i. For simplification purposes, we shall consider that each vector x_i has only two components $x_{i1} \geqq 0$, denoting a physical aggregate of all commodities consumed by households in country i, and $x_{i2} \leqq 0$ denoting an aggregate of all factors of production supplied by households of country i (including, typically, labour). Similarly, let $y_i = (y_{i1}, y_{i2})$ where $y_{i1} \geqq 0$ denotes the aggregate (physical) output of commodities in country i and $y_{i2} \leqq 0$ the aggregate factors used in production (including labour).

Moreover we shall assume that outputs from production (y_{i1}) can be traded among countries, whereas factors of production (y_{i2}) cannot. This last qualification implies that $x_{i2} = y_{i2}$ for every country i.

The reduction to R^2 of the space of economic variables, together with the assumption of nontradability of commodity 2, amounts to exclude from the analysis a detailed account of international trade phenomena (except for the possibility of transfers of commodity 1 between countries), and especially of price effects resulting from anti-pollution decisions eventually reached by the countries involved. On the one hand, such exclusion allows a substantial expository simplification for the pollution negotiation process. On the other hand, the consideration of a model embodying international trade effects raises rather serious additional problems of both theoretical and applied character, which would deserve separate treatment. The presently limited model nevertheless keeps its relevance from the environmental point of view.

With every country i, let us finally associate two basic concepts whose purpose is to describe economic behaviors, namely an (aggregate) *preference function* $W_i(x_i, R)$, defined on a consumption set $\mathscr{X}_i \subset \mathscr{R}^{2+A}$, and assumed to be monotonously and strictly increasing in all its arguments, strictly quasi-concave, continuous and twice differentiable, and an (aggregate) *production function* $F_i(y_i, S_i) = 0$, defined on a bounded production set $\mathscr{Y}_i \subset \mathscr{R}^{2+G}$, and assumed to be strictly increasing in all its arguments, convex, continuous and twice differentiable.

Notice that each production set $\mathscr{Y}_i$ is determined only by the variables y_i and S_i, which are specific to country i; production conditions in i are thus assumed to be independent of the discharges S_j made by the other countries $j \neq i$, and of the quality parameters of the sea, R_α, $\alpha \in A$, as well. This assumption amounts to ignore in our model the inter-country externalities that arise at the production level; we concentrate only on the "producer-consumer" externality aspect of the problem. The reason for this limitation comes from the nonconvexity problems that may arise when production externalities are taken into account. The mathematical methodology that will be used below for converging to an optimum is no longer applicable in such cases, and other methods should be sought for.

II.2 *States of the International Economy and Behavioral Assumptions*

Definition 1 The *feasible states* of the economic-ecological system constituted by the Southern Bight and the surrounding countries are the vectors $\{x_i, y_i, R, S_i; i \in \mathscr{I}\}$ such that (1) and (2) hold, $\sum_i x_{i1} = \sum_i y_{i1}$, and for every i, $x_{i2} = y_{i2}$, $(x_i, R) \in \mathscr{X}_i$ and $(y_i, S_i) \in \mathscr{Y}_i$. Let Ω denote the set of feasible states.

Definition 2 *Optimal states* of the system are the feasible states $\{x_i^+, y_i^+, R^+, S_i^+; i \in \mathscr{I}\}$ such that there exists no alternative feasible state $\{x_i^0, y_i^0, R^0, S_i^0\}$ for which $W_i(x_i^0, R^0) \geq W_i(x_i^+, R^+)$, $i \in \mathscr{I}$, with a strict inequality for at least one i.

Using standard Lagrangean techniques, it can be shown that, given our assumptions (and ignoring corner solutions), necessary and sufficient conditions for a feasible state to be an optimum are:

$$\sum_{j \in \mathscr{I}} \sum_{\alpha \in \mathscr{A}} \pi_{j\alpha} \eta_\gamma^\alpha - \phi_{i\gamma} = 0 \qquad i \in \mathscr{I}, \gamma \in \mathscr{G}; \tag{3}$$

$$\pi_{i2} - \phi_{i2} = 0 \qquad i \in \mathscr{I}, \tag{4}$$

where the notation is as follows:

$$\pi_{i\alpha} \stackrel{\text{def}}{=} \frac{\partial W_i/\partial R_\alpha}{\partial W_i/\partial x_{i1}}; \qquad \phi_{i\gamma} \stackrel{\text{def}}{=} \frac{\partial F_i/\partial S_{i\gamma}}{\partial F_i/\partial y_{i1}};$$

$$\pi_{i2} \stackrel{\text{def}}{=} \frac{\partial W_i/\partial x_{i2}}{\partial W_i/\partial x_{i1}}; \qquad \phi_{i2} \stackrel{\text{def}}{=} \frac{\partial F_i/\partial y_{i2}}{\partial F_i/\partial y_{i1}}; \quad \text{and} \quad \eta_\gamma^\alpha \stackrel{\text{def}}{=} \frac{\partial f_\alpha}{\partial S_\gamma}.$$

The marginal rate of substitution $\pi_{i\alpha}$ measures the amount of commodity 1 (the numeraire) that country i is willing to contribute to obtain a unit reduction of pollutant α (at some location ℓ); similarly, the marginal rate of transformation $\phi_{i\gamma}$ represents the cost for country i, measured in terms of foregone production of commodity 1, of a unit reduction of its discharges of pollutant γ (at some location d).

Thus by (3), an optimum is characterized by the fact that the cost of reducing any discharge $S_{i\gamma}$ should be equal to the sum of what all countries are together willing to contribute to avoid the effects across the ecosystem of such discharge.

The mere definition of an optimum does not imply that such a situation is likely to occur. On the contrary, the nature of pollution phenomena is precisely such that the individual behaviors of each economic agent (in our case, of each country i) lead the economic-ecological system to a state of affairs which is *not* an optimum in the sense defined above. This is why current concern about "transfrontier" pollution is so relevant, and why the latter deserves special corrective action, organized through specifically designed cooperation schemes.

In order to make this claim more precise, we need to state explicitly at this point how one assumes that each country "behaves" individually, or, in other words, we need to specify the economic (x_i and y_i) and environmental (S_i, R) choices that each country is likely to make *in absence of environmental cooperation* with the other countries. This we shall do by means of the following concept and assumption:

Definition 3 An "*environmentally nationalistic equilibrium (E.N.E.) for country i, relative to given environmental behaviors $\bar{S}_j, j \in \mathcal{I}, j \neq i$, of the other countries*" is a set of economic and environmental choices $\tilde{x}_i, \tilde{y}_i, \tilde{S}_i$ and $\tilde{R}$ such that $(\tilde{x}_i, \tilde{y}_i, \tilde{S}_i, \tilde{R})$ maximizes $W_i(x_i, R)$ on $\mathcal{X}_i$ subject to (i) $x_i = y_i$; (ii) $F_i(y_i, S_i) = 0$; and (iii) $R_\alpha = f_\alpha(S_1, \ldots, S_G)$, $\alpha \in \mathcal{A}$, where $S_\gamma = \sum_{\substack{j \in \mathcal{I} \\ j \neq i}} \bar{S}_{j\gamma} + S_{i\gamma}$, $\gamma \in \mathcal{G}$.

The justification for the term "equilibrium" lies, as usual in economics, in the notion of (constrained) individual maximizing behavior implied by this concept. On the other hand, the reason for the adjectives "environmentally nationalistic" is best seen by noting the following *property* of such an equilibrium [which is easily derived from the first order conditions for a maximum of $W_i(\cdot)$]:

$$\sum_\alpha \pi_{i\alpha}\eta_\gamma^\alpha - \phi_{i\gamma} = 0 \qquad \gamma \in \mathcal{G} \tag{5}$$

Thus, at an E.N.E., each country i makes choices such that its own cost of unit reduction in discharges $S_{i\gamma}$ is equal to what *itself* is willing to contribute to avoid the effects of those discharges on the ecosystem; in other words, no account is taken by country i of the effects of its discharges on other countries $j \in \mathscr{I}, j \neq i$.

Behavioral assumption 1 In absence of international agreement on pollution, each country is supposed to make its own economic and environmental choices in such a way as to find itself at an E.N.E., taking as given the behavior of the other countries.

If we now apply this assumption to all countries, a precise description of the economic-environmental situation prevailing in the Southern Bight and in the surrounding countries *when no cooperation is taking place* among the latter, is provided by the following concept:

Definition 4 A "*noncooperative equilibrium*" (N.C.E.) of the system is a set of national economic and environmental choices $\{\tilde{x}_i, \tilde{y}_i, \tilde{S}_i, \tilde{R}\}$ for every $i \in \mathscr{I}$, such that (i) $\tilde{R}_\alpha = f_\alpha(\tilde{S}_1, \ldots, \tilde{S}_i, \ldots, \tilde{S}_G)$, $\alpha \in \mathscr{A}$, and (ii) every country i is at an "environmentally nationalistic equilibrium relative to the behaviors of the other countries".

It is easy to see that a N.C.E. is a feasible state, and that (5) and (4) hold for every i at such an equilibrium. Comparing this with (3), it is now clear that actions taken in isolation by the various countries lead the system to a state *that cannot be an optimum*.

This is so perfectly in agreement with common sense that it might seem trivial; what is less so is that we now have an explicit description of the nonoptimal outcome of the absence of international cooperation (and thus a well-defined starting point for an analysis of negotiations), as well as a precise expression for the discrepancy between this starting point and the optimum we want to reach through negotiations.

II.3 *Statement of a Negotiation Process*

From now on, all variables in the model are functions of time, i.e. $R_\alpha(t)$, $S_{i\gamma}(t)$, $x_{ih}(t)$, $y_{ih}(t)$; most often, however, the argument t is omitted. A dot on top of a variable denotes the operator d/dt.

Our interest lies in analyzing the move of the economic-ecological system from a state of absence of cooperation to an optimum. Given behavioral assumption 1, it is natural to consider that, at time $t = 0$, the system is at a N.C.E., to be denoted as $\{x_i(0), y_i(0), S_i(0), R(0)\}$, $i \in \mathscr{I}$; these values of the variables are to be called later on "initial values of the solution".

Now, the main effect of the negotiation process is to impose gradually on each country i various upper bounds on their allowed discharges $S_{i\gamma}$, and hence on the quality variables R_α of the ecosystem, say $\bar{S}_{i\gamma}(t)$ and $\bar{R}_\alpha(t)$, for any $t>0$. When this is the case, the countries can no longer be assumed to behave as defined by the E.N.E. concept. We need to characterize otherwise their remaining economic choices; to this effect we introduce:

Definition 5 A "*conditional equilibrium for country i*" (i.e. conditional upon given environmental choices internationally made, and upon the economic behaviors of the other countries) is a set of economic choices $\tilde{\tilde{x}}_i, \tilde{\tilde{y}}_i$, and a set of constants $\bar{S}_j (j \in \mathscr{I}), \bar{x}_{j1}, \bar{y}_{j1} (j \in \mathscr{I}; j \neq i)$ such that $(\tilde{\tilde{x}}_i, \tilde{\tilde{y}}_i)$ maximizes $W_i(x_i, \bar{R})$ subject to (i) $x_{i1} - y_{i1} = \sum_{j \neq i} \bar{y}_{i1} - \sum_{j \neq i} \bar{x}_{i1}$; (ii) $F_i(y_i, \bar{S}_i) = 0$; and (iii) $f_\alpha(\bar{S}_1, \ldots, \bar{S}_i, \bar{S}_j, \ldots, \bar{S}_I) - \bar{R}_\alpha = 0, \alpha \in \mathscr{A}$. Such a conditional equilibrium is characterized by the conditions $\pi_{i2} - \phi_{i2} = 0, i \in \mathscr{I}$. Accordingly, we also introduce:

Behavioral assumption 2 Every country i participating in the negotiation process defined below is assumed to make its economic choices in such a way as to find itself, at all t, at a conditional equilibrium in the sense of definition 5.

Consider now the following system of differential equations[2]:

$$\dot{S}_{i\gamma} = a\left(\sum_j \sum_\alpha \pi_{j\alpha} \eta_\gamma^\alpha - \phi_{i\gamma}\right) \quad i \in \mathscr{I}, \gamma \in \mathscr{G}, \tag{6}$$

$$\dot{R}_\alpha = \sum_\gamma \eta_\gamma^\alpha \sum_i \dot{S}_{i\gamma} \qquad \alpha \in \mathscr{A} \tag{7}$$

$$\dot{x}_{i2} = \dot{y}_{i2} \qquad i \in \mathscr{I} \tag{8}$$

$$\dot{x}_{i1} = \dot{y}_{i1} - \sum_\alpha \pi_{i\alpha} \dot{R}_\alpha + \sum_\gamma \phi_{i\gamma} \dot{S}_{i\gamma} + \delta_i a \sum_{j \in \mathscr{I}} \sum_\gamma \dot{S}_{j\gamma}^2 \qquad i \in \mathscr{I}, \tag{9}$$

where a is an arbitrary positive scalar (assumed = 1 below, and henceforth ignored without loss of generality) and

$$0 < \delta_i < 1, \sum_i \delta_i = 1. \tag{10}$$

A *solution* for this system consists of a vector-valued function Z: $[0, +\infty] \to \Omega$; $t \to Z[t; Z(0)]$, whose derivative with respect to time exists and is equal to (6)–(9), and defined by

$$Z[t; Z(0)] = \{S_1[Z(0)], \ldots, S_I[Z(0)]; R_1[Z(0)], \ldots, R_A[Z(0)];$$
$$x_1[Z(0)], \ldots, x_I[Z(0)]; y_1[Z(0)], \ldots, y_I[Z(0)]\}$$

where $Z(0) = [S_1(0), \ldots, S_I(0); R_1(0), \ldots, R_A(0); x_1(0), \ldots, x_I(0); y_1(0), \ldots, y_I(0)]$ is the initial value of the solution. We shall assume, by convention, that such initial value is a noncooperative equilibrium of the economic-ecological system, in the sense of definition 4.

Equilibrium values of the solution are values of the variables S_i, R, x_i and y_i, $\forall i \in \mathscr{I}$ at some time $\overset{+}{t}$ such that the right-hand side of the system (6)–(9) is equal to zero. The system is said to *converge* if there exists a solution for the equations such that for $t \to \infty$ the values of the variables tend to equilibrium values of the solution.

It is easily seen that equilibrium values of a solution constitute an optimum in the sense defined in section II above, since the necessary and sufficient conditions (3) are met for such values. Thus, if the process so defined converges at all, it converges to an optimum.

II.4 *Interpretation*

The interpretation of the system (6)–(9) as a "negotiation process" among countries concerning their discharges $S_{i\gamma}$ goes as follows: each country is supposed to be represented by one delegate, who essentially determines at every t with his colleagues the magnitudes $\dot{S}_{i\gamma}(t)$ and $\dot{x}_{i1}(t) - \dot{y}_{i1}(t)$, according to their respective definitions in (6) and (9); this may be seen as taking place in a sequence of steps:

(i) each delegate makes known to the conference the coefficients $\pi_{i\alpha}(t)$ and $\phi_{i\gamma}(t)$, as they are evaluated within his own country i;

(ii) ecological experts provide the conference with the matrix of transfer coefficients $\eta_{\gamma}^{\alpha}(t)$;

(iii) then, reductions of each country's discharges of the various pollutants β at all locations d, $\dot{S}_{i\gamma}(t)$, are specified according to formula (6), and their resulting effects across the entire ecosystem are estimated as magnitudes $\dot{R}_{\alpha}(t)$, computed according to formula (7);

(iv) finally, net transfers $\dot{x}_{i1}(t) - \dot{y}_{i1}(t)$ of commodity 1, as specified by equation (9), are collected from – or distributed to – each of the I countries. To be more specific, let us imagine that an international agency has been set up by the I countries, prior to the negotiation, which is in charge of managing these transfers *as they are decided by the negotiators*, as the process develops over time.

The net transfer of each country i may then be seen as the algebraic sum of three successive payments (for clarity, we consider here the case where $\dot{S}_{i\gamma}$ and $\dot{R}_{\alpha} > 0$ for all α's and γ's; however, since the model does not imply any sign restriction on these variables, the payments described below are reversed when $\dot{S}$ and $\dot{R}$ are negative): first, *country i pays to the agency* an amount $-\sum_{\alpha} \pi_{i\alpha}\dot{R}_{\alpha} < 0$, i.e. a payment based on what country i has announced it is willing to contribute ($\pi_{i\alpha}$) per unit change in each of the R_{α}'s times the actual changes $\dot{R}_{\alpha}$; second, *the agency pays to country i* an amount $\sum_{\gamma} \phi_{i\gamma}\dot{S}_{i\gamma} > 0$, i.e. the cost to that country of reducing its discharges by the specified amounts $\dot{S}_{i\gamma}$, given that it has announced a unit cost $\phi_{i\gamma}$ for each of them; third, *the agency pays to country i* an amount $\delta_i \sum_{j \in \mathscr{I}} \sum_{\gamma} \dot{S}_{j\gamma}^2 > 0$, which can be shown (see Proposition 1(*a*) in Section III) to be a fraction of the surplus that would remain in the hands of the agency if only the first two payments just mentioned were made.

For each i, the algebraic sum of these payments may be positive, negative, or zero, depending upon the relative magnitudes of the coefficients $\pi_{i\alpha}$ and $\phi_{i\gamma}$, on the one hand, and of the variables $\dot{S}_{i\gamma}$ and $\dot{R}_\alpha$ on the other. In the aggregate however, the sum of all payments and receipts made by the agency can be shown to break even (see proposition 1(*b*) in section III). Thus, the negotiation process appears to be driven by repeated cost-benefit analyses, which determine the direction of decisions at each step. The key information on which it rests is the set of coefficients $\pi_{i\alpha}$, $\phi_{i\gamma}$ and η_γ^α, all duly updated as the process develops.

III PROPERTIES OF THE PROCESS

III.1 *Individual Rationality and Feasibility*

Proposition 1 (*a*) Along the process (6)–(9), $\dot{W}_i > 0$ for every $i \in \mathscr{I}$;
(*b*) $\sum_i \dot{x}_{i1} = \sum_i \dot{y}_{i1}$ for all $t \geqq 0$.

Proof: (*a*) Along the negotiation process, one has at every t, and for every i:

$$\dot{W}_i = \frac{\partial W_i}{\partial x_{i1}} \dot{x}_{i1} + \frac{\partial W_i}{\partial x_{i2}} \dot{x}_{i2} + \sum_\alpha \frac{\partial W_i}{\partial R_\alpha} \dot{R}_\alpha$$

$$= \frac{\partial W_i}{\partial x_{i1}} [\dot{x}_{i1} + \pi_{i2}\dot{x}_{i2} + \sum_\alpha \pi_{i\alpha}\dot{R}_\alpha].$$

Replacing $\dot{x}_{i1}$ in this expression by its value in (9) leads to:

$$\dot{W}_i = \frac{\partial W_i}{\partial x_{i1}} [\dot{y}_{i1} + \sum_\gamma \phi_{i\gamma}\dot{S}_{i\gamma} + \pi_{i2}\dot{x}_{i2} + \delta_i \sum_j \sum_\gamma \dot{S}_{j\gamma}^2];$$

in addition, by (8), $\dot{x}_{i2} = \dot{y}_{i2}$ along the process, and $\pi_{i2} = \phi_{i2}$ in view of behavioral assumption 2; hence,

$$\dot{W}_i = \frac{\partial W_i}{\partial x_{i1}} [\dot{y}_{i1} + \sum_\gamma \phi_{i\gamma}\dot{S}_{i\gamma} + \phi_{i2}\dot{y}_{i2} + \delta_i \sum_j \sum_\gamma \dot{S}_{j\gamma}^2]. \qquad (11)$$

Now, differentiating the production function $F_i(\,.\,) = 0$ yields

$$\frac{\partial F_i}{\partial y_{i1}} \dot{y}_{i1} + \frac{\partial F_i}{\partial y_{i2}} \dot{y}_{i2} + \sum_\gamma \frac{\partial F_i}{\partial S_i} \dot{S}_{i\gamma} = 0,$$

or, since $\partial F_i/\partial y_{i1} > 0$,

$$\dot{y}_{i1} = -\phi_{i2}\dot{y}_{i2} - \sum_\gamma \phi_{i\gamma}\dot{S}_{i\gamma}.$$

Replacing $\dot{y}_{i1}$ in (11) by this value gives

$$\frac{\partial W_i}{\partial x_{i1}} [\delta_i \sum_j \sum_\gamma \dot{S}_{j\gamma}^2] > 0,$$

which is positive since W_i is a strictly increasing function of all its arguments, $\delta_i > 0$, and all terms within brackets are positive squares.

(*b*) Starting from equation (9) of the process, we have

$$\sum_i \dot{x}_{i1} = \sum_i \dot{y}_{i1} + \left[\sum_i \left(\sum_\gamma \phi_{i\gamma}\dot{S}_{i\gamma} - \sum_\alpha \pi_{i\alpha}\dot{R}_\alpha + \delta_i \sum_i \sum_\gamma \dot{S}_{i\gamma}^2\right)\right].$$

We have to show that the expression within brackets is equal to zero, or equivalently that

$$\sum_i \left(\sum_\gamma \phi_{i\gamma}\dot{S}_{i\gamma} - \sum_\alpha \pi_{i\alpha}\dot{R}_\alpha\right) = -\sum_i \sum_\gamma \dot{S}_{i\gamma}^2,$$

where use has been made of the fact that $\sum_i \delta_i = 1$. Using now (7) we have

$$-\sum_i \sum_\gamma \dot{S}_{i\gamma}^2 = \sum_i \left[\sum_\gamma \phi_{i\gamma}\dot{S}_{i\gamma} - \sum_\alpha \pi_{i\alpha} \left(\sum_\gamma \eta_\gamma^\alpha \sum_j \dot{S}_{j\gamma}\right)\right]$$

and

$$-\sum_i \sum_\gamma \dot{S}_{i\gamma}^2 = \sum_i \left[\sum_\gamma \phi_{i\gamma}\dot{S}_{i\gamma} - \sum_\gamma \sum_\alpha \pi_{i\alpha}\eta_\gamma^\alpha \sum_j \dot{S}_{j\gamma}\right]$$

$$= \sum_i \left[\sum_\gamma \phi_{i\gamma}\dot{S}_{i\gamma} - \sum_\gamma \sum_j \dot{S}_{j\gamma} \sum_\alpha \pi_{i\alpha}\eta_\gamma^\alpha\right]$$

$$= -\sum_i \sum_\gamma \dot{S}_{i\gamma}\left[\sum_i \sum_\alpha \pi_{i\alpha}\eta_\gamma^\alpha - \phi_{i\gamma}\right],$$

where the term within brackets equals $\dot{S}_{i\gamma}$ in view of (6). *Q.E.D.*

The main implication of part (*a*) of this proposition is that all countries do benefit from the procedure, no matter whether they are polluting, polluted, or both. The relationship between this "individual rationality" property and the "polluters pay" principle is considered in section IV below. Note also that the importance of the "benefit" $\dot{W}_i$ for each country i is proportional to the coefficient δ_i which determines the fraction it receives of the surplus $\sum_j \sum_\gamma \dot{S}_{j\gamma}^2$ distributed by the agency as the "third" payment mentioned above (p. 207). On the other hand, together with equation (8) of the process, part (*b*) of this proposition shows that along the procedure, the economic-ecological system is led *through a sequence of feasible states* (in the sense of definition 1 above), from the initial N.C.E. to the final optimum. In our terminology, a deficit or a surplus of the agency managing the transfers would imply that the negotiation process leads the system into an unfeasible state, as it happens to be the case with other schemes proposed in the literature [4].

III.2 *Existence of a Solution and Convergence to an Optimum*

Assumption 3 The system (6)–(9) is Lipschitzian.

While this could be proved as a lemma, given our previous assumptions on the functions W_i and F_i, on the sets $\mathscr{X}_i$ and $\mathscr{Y}_i$ and given Proposition 1, we state it here as an assumption in order to avoid lengthy technical developments in this essentially expository paper. For an explicit treatment of the Lipschitzian property of processes of this type, see TULKENS and ZAMIR [5, Section IV].

Proposition 2 Under assumption 3 and those stated earlier on the functions W_i, F_i and f_α, the process (6)–(9) has a unique solution which is globally stable; it converges to an optimum.

Proof: (1) We use first the quasi-stability theory of UZAWA [7, pp. 619–620]: (*a*) In view of assumption 3, existence, uniqueness and continuity of a solution $Z[t; Z(0)]$ is guaranteed; (*b*) moreover, the solution is contained in the compact set of feasible states (see definition 1); (*c*) a "modified Lyapunov function" is provided by $W(t) = -\sum_i W_i(t)$, which is shown to be monotonous and nonincreasing by proposition 1. Hence, by theorem 1 of [7] the process (6)–(9) is quasi-stable.

(2) Every limit point of the process is an optimum, since it verifies conditions (3).

(3) Given continuity and monotonicity of W_i, and compactness of the set of feasible states, one has that

$$\forall i, \lim_{\nu \to \infty} W_i[z(t_\nu); z(0)] = W_i^+$$

is unique. This in turn implies that the limit point z^+ (an optimum in view of (2) above) is also unique: indeed, if there were two limit points, say z^+ and z^{++}, strict quasi-concavity of W_i would imply, for every i,

$$W_i[(z^+ + z^{++})/2] > W_i(z^+) = W_i(z^{++}),$$

contradicting the optimality of z^+ and z^{++}. Hence, the set of equilibrium values of the solution is finite, implying equivalence between quasi-stability and stability (see [7, p. 619]). Q.E.D.

IV CONCLUDING REMARKS

IV.1 *Is it Worth Participating in such a Process for Every Country?*

If for some reason some country j decided not to join in the negotiation, knowing that the other ones will nevertheless proceed according to the above model, would such an attitude be beneficial or detrimental to such country? The answer is easily provided by computing

$$\hat{W}_j = \frac{\partial W_j}{\partial x_{j1}} \dot{x}_{j1} + \frac{\partial W_j}{\partial x_{j2}} \dot{x}_{j2} + \sum_\alpha \frac{\partial W_j}{\partial R_\alpha} \hat{R}_\alpha,$$

where, under such "isolationist" circumstances,

$$\hat{R}_\alpha = \sum_\gamma \eta_\gamma^\alpha \left(\sum_{\substack{i \in \\ i \neq j}} \dot{S}_{i\gamma} + \dot{S}_{j\gamma} \right),$$

$$\dot{x}_{jh} = \dot{y}_{jh}, \qquad h = 1, 2, \tag{12}$$

and

$$\frac{\partial F_j}{\partial y_{j1}} \dot{y}_{j1} + \frac{\partial F_j}{\partial y_{j2}} \dot{y}_{j2} + \sum_\gamma \frac{\partial F_j}{\partial S_{j\gamma}} \dot{S}_{j\gamma} = 0.$$

As in part (a) of proposition 1 above, we may rewrite

$$\hat{W}_j = \frac{\partial W_j}{\partial x_{j1}} \left[\dot{x}_{j1} + \pi_{j2} \dot{x}_{j2} + \sum_\alpha \pi_{j\alpha} \sum_\gamma \eta_\gamma^\alpha \left(\sum_{\substack{i \in \mathscr{I} \\ i \neq j}} \dot{S}_{i\gamma} + \dot{S}_{j\gamma} \right) \right],$$

and derive, from the differentiation of $F_j(\,.\,) = 0$,

$$\dot{y}_{j1} = -\phi_{j2} \dot{y}_{j2} - \sum_\gamma \phi_{j\gamma} S_{j\gamma}.$$

Using (12) and recalling the first order conditions of an E.N.E. for j, the expression for $\hat{W}_j$ reduces to

$$\hat{W}_j = \frac{\partial W_j}{\partial x_{j1}} \left[\sum_\alpha \pi_{j\alpha} \sum_{i \neq j} \sum_\gamma \eta_\gamma^\alpha \dot{S}_{i\gamma\,;\,\mathscr{I} - \{j\}} \right], \tag{13}$$

where

$$\dot{S}_{i\gamma,\,\mathscr{I} - \{l\}} = \sum_{\substack{i \in \mathscr{I} \\ i \neq j}} \pi_{i\alpha} \eta_\gamma^\alpha - \phi_{i\gamma}.$$

If the factor within brackets in (13) is strictly positive, i.e. if the other countries do take anti-pollution decisions, $\hat{W}_j$ is positive. Clearly there is thus an incentive for (any) country j *not* to participate in a transfrontier pollution agreement, as soon as the other ones start doing something. This is not a phenomenon genuine to the process developed above: rather, it is an intrinsic characteristic of the public good character of our transfrontier pollution problem.

It has been shown above, however, that if country j were to participate in the process, its benefit $\dot{W}_j$ would depend upon the value of coefficient δ_j assigned to it for distributing the surplus emerging from the transfer payments scheme.

Thus, the question whether for a given country j it is worth participating in the above negotiation process boils down to the following: is the surplus large enough to be shared among participants in such a way that each of them benefits more from participating than from not participating? Formally, the problem is to find values $\overset{+}{\delta}_j$ such that for every $i \in \mathscr{I}$,

$$\dot{W}_j = \frac{\partial W_j}{\partial x_{j1}} \overset{+}{\delta}_j \sum_{i \in j} \sum_{\gamma} \dot{S}_{i\gamma}^2 \geqslant \widehat{W}_j = \frac{\partial W_j}{\partial x_{j1}} \sum_{\alpha} \pi_{j\alpha} \sum_{i \neq j} \sum_{\gamma} \eta_{\gamma}^{\alpha} \dot{S}_{i\gamma}, \mathscr{I} - \{j\}. \tag{14}$$

Preliminary investigation of this problem by means of game theoretic tools leads to the conclusion that there may not exist, in general, such values $\overset{+}{\delta}_j$ satisfying also $\sum_j \overset{+}{\delta}_j = 1$ (feasibility). In special cases, however (which depend upon the structure of the preference functions W_i) they may exist. On the other hand, if (14) cannot be satisfied, one may want to look for $\overset{++}{\delta}_j$'s which would minimize $\widehat{W}_j - \dot{W}_j$, i.e. the incentive not to participate.

IV.2 *Do the "Polluters Pay" under this Process?*

It is currently impossible to propose any decision process in the pollution field without taking a stand with respect to the so-called "polluters pay" principle, which has much popularity in both official and unofficial circles. Clearly, the present process does *not* derive from that rule; on the contrary one could say that its characteristic is that, to every polluter (viz. each country i) all countries (*including i itself*) do pay a transfer the total of which exceeds the cost of reducing its discharges. On the other hand, there is also the important feature that all countries do benefit from the outcome.

Essentially, there appears to be here a shift in emphasis, from the question of "who should pay"? to that of "is it worthwhile for everybody to combat pollution"? A rather strong rationale can be called upon for such a shift. Indeed, the essence of the "principle" is that it is a normative one, whose application requires an authority empowered to impose burdens on some economic agents, *without* (or with only little) *compensation* (thus, actions such that $\dot{W}_i < 0$). Now, at the international level such an authority is non existent; and it seems quite unlikely that any country would voluntarily sign an agreement establishing such an authority. Therefore, if the "principle" may deserve consideration at the *national* level (for an analysis of its various forms, see [5]), it cannot have much relevance as regards the substance of *international* environmental decisions; here, the only policy proposals susceptible of entailing voluntary agreement seem to be those which are shown to be clearly beneficial to each party taken individually. Thus, if the objective of the analysis (and of the negotiations) is to help reaching an optimum, agreements motivated by and based on the self-interest of the parties involved are probably – but also alas ! – a stronger means to get things done, than quoting "principles", no matter how attractive they may be from other viewpoints.

ENDNOTES

[1] This is a revised version of a paper presented at the Conference on "Communications and Control in Social Processes", sponsored by the American Society for Cybernetics and the University of Pennsylvania, Philadelphia, October 31–November 2, 1974, and based on reports CB1/5 and CB1/7 written under contract with the Belgian Ministry of Scientific Policy under its "Programme National de Recherche et de Développement sur l'Environnement". This revision has benefitted from support of the Ford Foundation (Program of Research on International Economic Order). Thanks are due to Professors Theodore BERGSTROM, François de DONNEA, Jacques NIHOUL, Robert ROSENTHAL, Yves SMEERS, Robert WILSON, and especially Maurice MARCHAND.

[2] The reader who is familiar with the theory of public goods will have recognized that we are dealing here with an economy with "international public goods"; the present dynamic process is an adaptation to our environmental problem of similar procedures recently proposed in [1] and [2].

REFERENCES

[1] Drèze, J. and de la Vallée Poussin, D., "A tâtonnement process for public goods", *Review of Economic Studies,* **38**: 133–150, 1971.

[2] Malinvaud, E., "Prices for individual consumption, quantity indicators for collective consumption", *Review of Economic Studies,* **39**: 385–406, 1972.

[3] Nihoul, J. C. J., "Mathematical model of continental seas: preliminary results concerning the Southern Bight", Institut de Mathématique, Université de Liège, Belgium; mimeo, 1971.

[4] Smets, H., "Le principe de la compensation réciproque: un instrument économique pour la solution de certains problèmes de pollution transfrontière", Note du Secrétariat, Comité de l'Environnement, O.C.D.E., Paris, Septembre, 1973.

[5] Tulkens, H. and Schoumaker, F., "Stability analysis of an effluent charge and the 'polluters pay' principle", *Journal of Public Economics,* **4**: 245–269, 1975.

[6] Tulkens, H. and Zamir, S., "Local games in dynamic exchange processes", CORE Discussion Paper Nr. 7606, April 1976, to appear in *Review of Economic Studies,* 1978.

[7] Uzawa, M., "The stability of dynamic processes", *Econometrica,* **29**: 617–631, 1961.

THE ORGANIZATION OF WORK AND THE CHALLENGE TO MANAGEMENT AND ENGINEERING

WILLIAM A. HETZNER
IIT Research Institute, Chicago

INTRODUCTION

Over the past decade, industrialized countries have shown an increased interest in altering traditional patterns in the organization of production systems to more effectively and efficiently allocate human and material resources. Indicators of change are reflected in the proliferation of programs involving job enlargement, job enrichment, participative management and industrial democracy.

The initial impetus of this revision of the organization of production systems in the United States can be traced to the depression years of the thirties [5] and the war years of the forties [4]. The locus of work reorganization activities in recent years seems to have shifted to Western Europe where numerous experiments have been introduced and innovative studies associated with work reorganization undertaken. This paper deals with some apparent trends in the organization of work in five selected European companies. Implications of these trends on scientists, practitioners and workers are discussed in terms of areas of further research and development activities.

The literature on work organization has expanded greatly in recent years, reflecting scholarly interest in this field. As in any emerging field of study, many of the concepts are not yet clearly defined and empirical evidence is greatly lacking. With the apparent increase of work reorganization activities in industry, there is both an opportunity and a need for systematic study of these activities.

The study described herein is an exploratory examination of work reorganization activities in five companies. In terms of the literature, the data suggests that a clear distinction must be made between approaches to work organization (change processes) and the resultant forms of work organization. Changes in task structure (job enlargement) and authority structure (job enrichment, participative management, industrial democracy) were evident in many instances of work reorganization. In many cases the actual forms of these changes in task and authorit structure were not well defined before the fact, but emerged "naturally" from the change processes initiated in the companies. On an *à posteriori* basis, the data seemed to suggest that successful change process in the companies surveyed explicitly or implicitly followed the so-called socio-technical approach first described by Trist [6]. This approach with its emphasis on the interaction betwee technological and socio-psychological considerations in the design of work organization is described and its importance indicated in subsequent sections of this paper.

DATA COLLECTION METHODOLOGY

Selection of companies, respondents and data collection methodology at this stage of the study were not totally systematic. Companies were selected partially on the basis of their experience or interest in work reorganization and on the extent of prior contact with our group. Data collection involved an open-ended interview schedule and review of articles and brochures concerning activities of the participating companies.

Three of the five companies in the study represent the automative or related industries, one the electronics industry and one the office machine industry. The degree to which these companies are involved in work reorganization activities varies considerably. At one end extreme, two companies have very little experience and involvement. At the other extreme, two companies had much experience and large programs in the area.

Thirteen representatives from the five European companies visited were interviewed. Although over sixteen separate experiments or experiences with work reorganization were described in the course of the interviews, no one respondent appeared to be aware of all possible experiences within their companies. One respondent noted, "there are activities (in the company) that we do not know about, because they are initiated at the foreman level". Due to the limited sample of respondents and companies and the unstructured nature of the data collection, the following discussion is presented for the purpose of providing candidate areas for further research.

TRENDS IN WORK ORGANIZATION

Reorganization Due to Special Circumstances

Work reorganization activities in the companies surveyed were discussed primarily at changes in assembly and other production processes characterized by a rigid division of labor. Emphasis on this area is generally attributed to high incidence of labor difficulties (high turnover and absenteeism), union pressures to increase wages and to improving working conditions, government pressures to improve working conditions, or a combination of these factors. The presence of these "special circumstances" which initiated a search to traditional patterns of work organization for alternatives is consistent with the findings of Reuhl [4].

Criteria to Assess the Success or Failure of Reorganization Activities

Since it was anticipated that the reasons for the introduction of work reorganization activities would vary from company to company, the development of a generalized measure of success for these activities did not seem to be possible. For example, the criteria of success for programs designed to reduce turnover and absenteeism were not expected to be the same as for programs designed to reduce union or government pressures to improve working conditions. For the purposes of this study, an *á priori* decision was made to define success in terms of the degree to which a particular program was perceived as achieving its objectives, with the expectation that these objectives would vary greatly.

As it turned out, there were varying objectives and criteria for success in the cases of work reorganization surveyed. It was also evident, however, that for a majority ($N = 6$) of ten cases perceived as successfully achieving their objectives, decreased worker absenteeism and turnover were considered. In about four of these cases decreased worker turnover and absenteeism were explicit objectives of the programs. In four of approximately six cases of "failure", increases or no change in worker turnover and absenteeism were cited as criteria for failure.

Other criteria used to assess success or failure included quality of output, productivity and labor satisfaction. Of these additional factors, increase in quality ($N = 4$) was most often mentioned as resulting from work reorganization activities. Productivity was cited in two cases. In one, decreases in downtime of machines offsetting increases in wage payments to the worker was utilized as the criterion of success. In another, output per labor hour decreased, but the project was still considered as successful for reasons of increased quality. Labor satisfaction was mentioned in a number of cases, but it was not directly measured. Instead, labor turnover and absenteeism were typically considered as surrogates of satisfaction.

Characteristics of Work Reorganization

The characteristics of successful work reorganization activities that seemed to be relatively consistent between and among the companies are as follows:

1. *Grouping of Workers:* Responsibility for production output delegated to a group of workers rather than an individual worker. Groups consisted of from three to twenty-five workers.

2. *Decision Making:* Varying amounts of decision-making, authority, in terms of day-to-day planning, organization and quality control vested in the workers.
3. *Job Enlargement:* Increase in the number of tasks carried out by an individual worker.
4. *Team Leader:* A member of the work group rather than a traditional foreman responsible for interfacing with management and other production work groups.

Table 1 reflects distribution of these characteristics in companies participating in the study.

From this table it appears that the characteristics are highly interrelated rather than independent. The tendency towards work groups as the unit of production and job enlargement activities is equally present in both successful and unsuccessful instances of work reorganization. These factors must be considered along with the decision-making authority and form of supervision (team leader versus foreman) before a scenario of success emerges.

Caution should be used in making strong inferences from data presented in Table 1. There is a great deal of difference as to the extent of grouping of workers, decision-making authority, job enlargement, and the role of the team leader across cases surveyed. These differences will be more fully indicated in subsequent sections of this paper. It is important to note at this time that instances of work reorganization having a similar profile of characteristics according to Table 1 are not necessarily similar in actual practice.

GROUPING OF WORKERS

The development of work groups was evidenced in seven of the ten successful cases of work reorganization and in four of the six unsuccessful cases. Work groups were operationally defined as consisting of from three to twenty-five workers, where the group rather than its individual members were assigned responsibility for achieving required production output.

The formation of work groups can be viewed from both a socio-psychological and technological viewpoint. Criteria were not designed in advance to determine which viewpoint was relatively more important in the cases studied. However, the indications are that successful work reorganization activities did explicitly or implicitly consider grouping of workers from both perspectives.

This is consistent with the approach of Trist and his colleagues at the Tavistock Institute of Social Research, who suggest that effective work design must consider the mutual interdependence of the technological system (machines, material and space) and the social system (relationships between the workers carrying out

TABLE 1

Characteristics of work reorganization activities

	COMPANY	CASE NO.	WORK GROUP	DECISION MAKING	JOB ENLARGEMENT	TEAM LEADER
SUCCESSFUL CASES	Company A	1			X	
		2	X	X	X	X
	Company B	3	X		X	X
		4			X	
		5	X	X	X	X
	Company C	6	X	X	X	X
		7	X	X	X	X
	Company D	8	X	X	X	X
		9			X	X
		10	X		X	X
	TOTAL		7	5	10	8
UNSUCCESSFUL CASES	Company A	11	X		X	
	Company B	12	X	X	X	X
		13			X	X
	Company D	14			X	
		15	X	X	X	
	Company E	16	X		X	
	TOTAL		4	2	6	2

the necessary tasks). Because of this mutual interdependence, work organization is considered in terms of a socio-technical systems. Social scientists, they contend, have too often considered only isolated elements of the technological system, such as the repetitive and piecemeal nature of tasks; while industrial and manufacturing engineers have not fully considered the effects of the technological system on the social system.

Research at Tavistock on deep seam coal mines [6], textile mills [3] and by other researchers in the U.S. has indicated that for any technological system there often exists a choice of alternative forms of work organizations that can effectively operate the technology. The range of alternative forms of work organization is constrained only by the tasks and task interrelationships required by the technological system. While there are generally a number of social systems that can operate a technology with some degree of success, ultimate success depends upon the adequacy with which the social system is able to cope with the requirements of the technology.

From a socio-technical perspective, one can argue that the technological systems and their accompanying tasks and task interrelationships have tended to undergo change in recent years for assembly and other production processes characterized by a rigid division of labor. Production processes based on a rigid division of labor and well established lines of reporting, control and supervision were initially developed to produce products with little or no variation in their characteristics. The superiority of the technology of the assembly lines in other rigid production processes to produce products with few model variations more than offset any problems, technical or social, caused by their relative rigidity. Tasks were well defined and task interrelationships fixed so that they could be effectively dealt with by the assembly line itself as well as the established lines of reporting, control and supervision.

Increasing demands for product variation has been recently experienced by many of the investigated companies and has placed a great deal of stress on production processes based on a rigid division of labor. In one case, recent technological developments have altered the very nature of the companies products and the accompanying technological systems. Such significant changes in demand and technology have led to increasing variable task requirements and task interrelationships in production processes.

Production processes characterized by rigid division of labor, Trist contends, are not able to cope with changing conditions such as those experienced by the investigated companies except at the expense of increasing the stress placed on workers, sacrificing smooth cycle progress or drawing more labor onto the line. Work organization based on task groups, tend to be more flexible and, therefore, more efficient for variable task requirements caused by changing conditions. This efficiency is manifested not only in the increased ability of work groups to cope with the tasks involved, but also in the satisfaction of the workers' personal and interpersonal needs through task performance and their interdependence with other workers.

The socio-technical approach does not suggest that work groups are appropriate in all production processes or settings. Rather the requirements of the technological system determine if grouping is possible and efficient for a particular process or setting. This seemed, at an impressionistic level, to be borne out by the experience of the companies surveyed. Of the eight instances of work reorganization involving assembly lines, the introduction of work groups was considered successful in six cases, four involved multi-model products. Two of the three unsuccessful cases involved products with very little variation in models.

There are, however, other reasons for success or failure that are potentially important. One of these factors which is typically considered as a necessary part of grouping workers is decision-making authority which is discussed in the following section.

DECISION-MAKING AUTHORITY

The delegation of decision-making authority to work groups is generally considered as a requisite of the grouping of workers. The data collected in this study tends to confirm the necessity of some decision-making authority being delegated to work groups. However, the evidence is not significant enough to indicate it as essential for success. Decision-making authority delegated to workers seemed to vary greatly in the cases surveyed, both in terms of type and degree. Some groups had been delegated varying degrees of authority over planning, organizing and inspecting their own work. Other groups were delegated varying degrees of authority over one or two of these factors. In five of the seven cases of successful work groups and two of the four unsuccessful cases some degree of decision-making was delegated to the groups. In the remaining two successful and two unsuccessful cases of work groups there was no *apparent* delegation of decision-making authority.

A possible explanation for the varying degree and type of decision-making authority again may be found by viewing work organization in terms of a socio-technical system. From this perspective it would seem that the degree and type of decision-making authority delegated to work groups is a function of the degree of flexibility required by tasks and task interrelationships. The more the flexibility required, the greater the degree and comprehensiveness of decision-making authority necessary to achieve flexibility. It would seem evident that too little or too much decision-making authority can be contributing factors to the failure of work groups to cope with the technological system.

Whether work groups resulted from the change process or not, worker involvement in planning work reorganization activities seemed to be important in all cases surveyed. The data suggest that success is based on the degree of involvement of the workers affected by work reorganization in the planning process. This point was emphasized by all respondents in the study. The time frame of this planning process often spans two or more years, beginning with the involvement of unions in the preplanning stages, up to the establishment of worker-management-engineering planning teams. As one respondent stated, "this is not something you jump into without a great deal of thought". Once a company has involved unions and workers in the planning process a precedence has been established which is difficult, perhaps even impossible, to reverse.

In some cases, involvement of the workers and unions in the planning process resulted in only minor modifications to work organization not totally anticipated by management. In other cases, major modification in work organization occurred. Regardless of the level of change, involvement in planning was perceived, according to one respondent, "as an indication that management was concerned with the lot of the worker, and willing to listen to their complaints". In a number of instances, the planning teams responsible for recommending or initiating changes have been retained to "reinforce the positive benefits of initial experience" and to suggest further modifications to work organization. This indicates to some extent, that groups may be as important in the planning of change as is the outcome of the change process.

JOB ENLARGEMENT

In all cases ($N = 16$) of work reorganization surveyed, job enlargement activities were in evidence. Job enlargement for this study was defined as a worker performing a greater number or variety of tasks as a result of work reorganization. In some cases, tasks were enlarged to permit a worker to perform an entire unit of work. In others, enlargement meant that operating personnel now also carried out maintenance and repair activities.

Successful job enlargement in terms of increases in the number of similar tasks carried out by workers generally occurred where the production process was relatively labor-intensive and tended to be accompanied by grouping of workers, the delegation of some decision-making authority, and a team leader form of supervision. Enlargement by itself was successful only in cases where the production process was highly mechanized. Only in one or two cases was enlargement of mechanized production processes accompanied by grouping of workers and a task leader form of supervision. Decision-making authority was not delegated in any of the successful cases of enlargement of highly mechanized production processes. In one unsuccessful case where decision-making authority was delegated to workers involved with a highly mechanized process, machine downtime and rejects increased.

From a socio-technical perspective, grouping of workers, decision-making authority and job enlargement go hand-in-hand. The findings of this study seemed to confirm this perspective and lead further to the formulation of a hypotheses that job enlargement is a function of tasks and task interrelationships defined by the technological system. It should be pointed out, however, that for highly mechanized production processes, job enlargement alone may be feasible without the grouping of workers or the delegation of decision-making authority.

TEAM LEADER FORM OF SUPERVISION

Various "team leader" forms of supervision were evident in ten instances of work reorganization (eight successful and two unsuccessful). Team leaders varied from members of a working group assigned as a liaison between other operating groups, management and engineering to foremen trained in "democratic leadership" methods. The role of the team leader was generally a function of the degree of work group autonomy. Team leaders as "linking pins" were generally found in highly autonomous work groups. "Democratic" foremen were generally found in groups in which the worker had little or no autonomy.

SUMMARY OF TRENDS IN WORK ORGANIZATIONS

The preceding discussions seem to indicate that successful work reorganization activities are a function of the ability of organization relationships among workers (social system) to cope with the technology, and the ability of the technology to cope with environmental demands. The form of organizational relationship best able to operate a particular technology depends upon the complexity and flexibility of tasks and task interrelationships defined by the technological system. In highly complex, changing task and task relationships, work groups having a high degree of autonomy and a team leader form of supervision seem to be an efficient form of work organization. For less complex, rigid tasks and task requirements, a relatively rigid division of labor and well established lines of reporting, control and supervision appears to be effective and efficient.

FURTHER AREAS OF POTENTIAL RESEARCH

The development of autonomous work groups presents some interesting and important areas of concern both to practitioners and researchers. A primary area of concern apparent from the interviews with the six European companies is the "integration" of work groups to achieve organizational goals and objectives.

The establishment of work groups tends to facilitate the flow of information and work among group members, enabling the group to deal with complex and changing tasks and task interdependencies. This very same process, however, tends to create barriers to the flow of information and work between work groups. The differentiation of tasks into groups results not only in task differences among these groups but also in differences in group goals, structures, style of operations, physical and organizational locations, and the cognitive and emotional orientations of group members. These differences can create barriers to the flow of information and work. As Lawrence and Lorsch [2] contend the requirements of *differentiation* to effectively carry out specific tasks and the requirements of *integration*, to effectively accomplish the organization's purposes, are "essentially antagonistic and one can be achieved only at the expense of the other". While it is not clear that organizations cannot develop means of both effectively carrying out tasks and integrating these tasks together, it is certainly true that achieving effective differentiation of tasks makes effective integration more difficult.

There seem to be two major areas of "integration" difficulties in the development of autonomous work groups. First, there is the problem of integration of work groups with management and engineering. This can be considered as vertical integration. Second, there is the problem of integration among work units or horizontal integration.

Of the two problem areas, vertical integration seems to be the most prevalent and difficult. Work groups have in general assumed some of the day-to-day decision-making authority held by management and engineering.

Management and engineering no longer can simply tell the worker what to do in all cases. This loss of decision-making authority is perceived as quite threatening and has caused a great deal of apprehension on the part of management and engineering groups with direct interfaces with production workers.

Decision making authority and control over information are two major sources of power (the degree of influence of one individual or group over another) in an organization [1]. The threat of loss of decision-making authority by management and engineering tends to create a reluctance on their part to provide information to work groups. This lack of information can affect the quality of work group decisions.

One method of solving the problem of vertical integration that has met with some success, is the development of planning teams. These teams, typically made up of members from the work group, engineering and management staff, investigate the problems of layout and design of work and facilities for the purpose of providing management and engineering input on the day-to-day decisions made by the group. In addition, these teams also provide feed back on group and company performance, which creates "a sense of belonging" to the larger organization.

Horizontal integration involves the flow of information and work between groups. Communication is an important aspect in decision-making. Therefore, it is essential that information be supplied to the work group in a prompt, timely and accurate manner. Decisions made on the basis of data supplied affect not only matters concerning work activities but also accomplishes a more efficient work flow process.

The extent of the problem of horizontal integration seems to be a function of the degree of work group autonomy. The greater the autonomy of the work groups involved, the less effective is information transfer between them. Autonomy, therefore, can have an adverse effect on decision-making and work flow.

In the companies surveyed, one can see the effect of autonomy on work flow in the large storage areas of buffers that have sometimes been used between work groups. Large buffers, however, increase the cost of in-process inventories and cannot guaranteee that a particular group will receive a subassembly or part when they need it.

In addition to buffers, personal communications (face-to-face or remote) between team leaders have been utilized to integrate work groups. The success of personal communications, however, seems to be effective for relatively long-range decisions. This suggests the possibility of utilizing remote techniques for communications, such as interactive computer terminals, across work groups as an aid to short-range decision-making. Computers are currently used to assist management in keeping track of work-in-progress. These same computers could be used interactively to provide information to work groups in order to plan, organize and schedule their daily activities. For example, a work group could interrogate the computer by giving it a range of production possibilities or single estimates of production output. The computer compiling information on in-process inventories and the activities of other groups could respond with the results of various alternatives, a "preferred" daily output or both. On the basis of this response or further interrogations the group could then schedule their activities.

Of course, the use of computers as a decision aid is a function of the degree of group autonomy. The proposition suggested there is that the computer, if properly used, is not a threat to group autonomy, and yet can provide prompt, timely, accurate communications between work groups.

The potential use of computers and the entire area of the horizontal integration of work groups is an area in which social scientists and cybernaticians can make a significant contribution. The problem of intergroup communication is critical to the future success of work reorganization activities. Future research and experimentation can provide solutions to what appears to be an increasingly challenging problem.

CONCLUSION

The title of this paper, "Organization of Work" is posed as a challenge to the field of management and engineering. Evidence points out that practice in this area has tended to precede theory building and will continue to do so at least in the near future. Special circumstances, such as high turnover and absenteeism among workers will force engineers and managers to search for alternative solutions to traditional work organization.

While the primary challenge in work organization is at the practitioner level, the importance of the role of scientists in this process is also quite clear. With the apparent increase in work reorganization activities, there is a clear need for scientists to initiate systematic studies of the varying experiences among companies and programs, to begin to provide the theories and principles necessary for future activities. In addition, as this paper indicates, there are numerous areas of research within work reorganization activities that can have an impact on the efficiency and effectiveness of these new forms of organization.

REFERENCES

[1] Leavitt, H. J., "Applied organization change in industry: structural technological, and humanistic approaches", in March, J. G. (Ed.), *Handbook of Organization,* Skokie, Illinois: Rand McNally, 1965.

[2] Lawrence, P. R. and Lorsch, J. W., "Differentiation and integration in complex organizations", *Administrative Science Quarterly,* **12:** 1–47, 1967.

[3] Rice, A. K., *Productivity and Social Organization: The Almedabad Experience,* London: Tavistock Institute, 1958.

[4] Ruehl, G., "Workstructuring, Part 1", *Journal of Industrial Engineering,* **32:** 32–37, 1974.

[5] Schleicher, W. F., "Let's Put Joe Back into the Pyramid", paper presented at the 11th Annual Meeting and Technical Conference on the Numerical Control Society, Toronto, 1974.

[6] Trist, E. L., "On socio-technical systems", in Bennis, W. G., Benne, K. O. and Chin, R. (Eds). *The Planning of Change,* 2nd ed., New York: Holt, Rinehart and Winston, 1969.

TOWARDS SOCIAL CONTROL THEORY

INTRODUCTION

A good theory of social control does not yet exist. This is rather unfortunate because social control penetrates all spheres of human life. Social institutions from the nuclear family to the national government control their members. The history of social conflict and revolution shows that shifts in control from one group to another or from one mechanism to another lie at the root of almost all radical social change. Social control is intimately linked to human values, beliefs and ideologies, which are manifest primarily in the constraints they impose upon individual behavior. Law enforcement, social work, management, education, advertising, etc., provide a living for many people, all of whom are engaged in some form of social control. The very idea of engineering and of planning implies control over social resources. While many writers would regard the concept of social control as a fundamental ingredient in any theory of society it is surprising that knowledge about this process is so meager.

On the constraining side, it is sometimes argued that the conspicuous absence of valid theories of social control stems from the tendency of power elites to suppress all knowledge that would diminish their power. From this follows that those subjected to the power of others have no access to the experience that a theory of social control would describe. There also exists the fear that knowledge about social control will ultimately be placed into the wrong hands, where it will be transformed into techniques for brainwashing, repression by totalitarian regimes or into a life such as that depicted in Orwell's *1984*. While there might be some truth in these contentions, they both are based upon an acceptance of the *status quo*.

On the facilitating side, it is argued that the unchecked development of technology, with its systematic exploitation of natural resources and consequent pollution of the environment, as well as the exponential increase in human population have created conditions of instability, of runaway growth intrinsically destructive of mankind. This is not the place to develop threatening scenarios. The issue here is to get on with the job of understanding those processes of interaction among individuals, social groups and institutions that determine or at least influence the transformation of the surface of the earth and the fabric of society, with the ultimate aim of creating a better life for all of us.

I believe that the reason for the lack of progress in this area lies in the complexity and in the large variety of phenomena designated by the theme "social control". There does not even exist a consensus regarding the goals of theories of social control, much less regarding their content. All that can be accomplished in this introduction, therefore, is to outline some of the possible goals of a theory of social control and place them into the context of some existing conceptualizations.

In entirely general terms, control theory is concerned with models or with portions of the real world that can be represented diagrammatically and algebraic by:

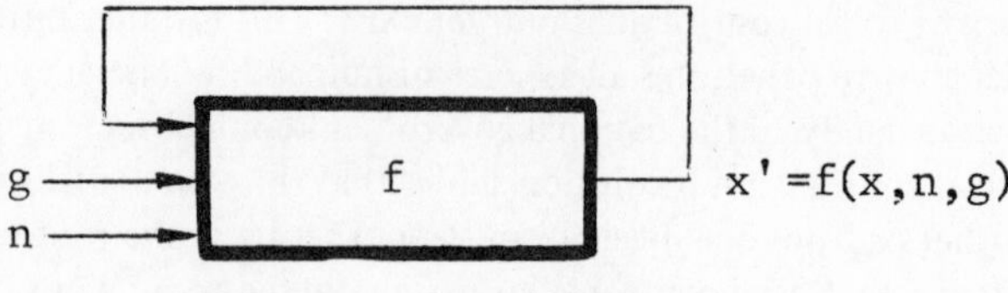

FIGURE 1 Control Theory

Where x indicates a system's observable state, n represents unaccountable noise whether it occurs in form of external disturbances or internal unreliabilities (entropy), g designates external guidance variables and f represents the structure of the system and describes the behavior $x' \to x'' \to x''' \to \ldots$ as a function of the interplay of internal variables x, noise n and guidance variables g. Since it is the guidance variables through which one may influence the behavior of the systems, g must be free to vary at will while n is considered independent with x to be controlled or to be kept within desirable limits, the goal.

Control is manifest in a system's ability to keep x within specifiable limits despite the impact of noise (from outside disturbances or from inside unreliabilities, decay and breakdown).

This general form includes complete *exogenous control* as special case in which the system has no dynamic of its own and simply follows the external variables:

FIGURE 2 Exogenous Control

exogenous control requires g to be compensatory of the effects n has on x.

It includes complete *endogenous control* in which the system is governed by its own purposes and closed to guidance variables:

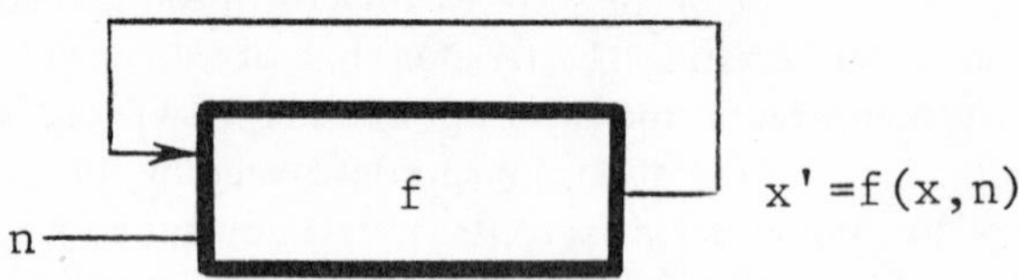

FIGURE 3 Endogenous Control

Endogenous control requires that the structure of the system resist the effects n has on x.

The general form also includes the special case of the linear feedback regulator which is basic to control theory in engineering and of which the above are generalizations. Again, diagrammatically and algebraically:

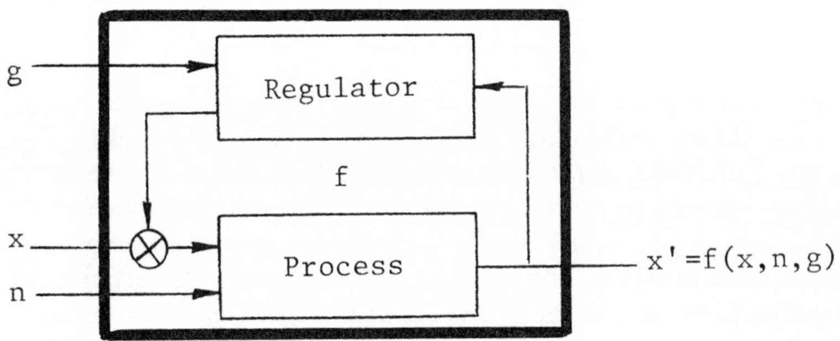

FIGURE 4 Linear Feedback Control

where the guidance variable g becomes the externally set goal *or* the reference signal, against which the results of a disturbed process, x', may be matched. The regulator assures that the input to the process keeps x' within limits specified by g despite existing external or internal noises n.

Conant and Ashby [1] have shown that every successful regulator of a process must be a model of that process, i.e. contain a suitable representation. This is quite consistent with William T. Power's view in this section that control systems must be "capable of stabilizing against disturbance an internal representation of an external state of affairs" to which one might add that such a representation might have to be stabilized against the effects of internal unreliabilities, breakdown or conflicts as well. The success of regulation, the ability to control is thereby in fundamental ways related to available information.

Accordingly, we can distinguish three principal aims of control theory:

(1) The first and most simple aim is to predict where a system comes to rest, the goal it seeks by itself without the alteration of g. In this case the system is open only to noise and otherwise governed by internal purposes only. With such an aim in mind, the scientist appears in the role of an external observer who, without exerting any influence of his own, attempts to understand where, in the face of discernible noise, the interactions and mutual influences within the system are leading.

(2) The second aim of control theory is to identify those conditions in g that would, within available time and in the face of discernible noise, make a system

reach a given goal or keep some variable x within specifiable limits. It presumes the possibility of influencing the setting of a goal from the outside of the system. The scientist becomes knowingly obtrusive relative to the system under study. Only with reference to a known (independently stated) goal can one speak of deviations, of errors and of deviation reducing (or amplifying) feedback. It is this a conservative aim, because the structure of the system is taken for granted, only the variability of the goal is assumed.

(3) The third aim of control theory is to create (design or influence the growth of) new structures within a system, as reflected in the function f, through which the system would endogeneously move to an exogeneously defined goal. While this is the principal aim when it comes to the design of new technology, its role in social science is fairly limited because individuals cannot be rearranged entirely at will. However, this difficulty should not be construed as a rejection of purposeful structural change in society. In fact structural change might be the only way to overcome presently unstable social conditions. (See Pergler and Buckley's contribution in this section.)

For lack of better terminology, let me call a theory that accomplishes the first goal: a theory of *self-control* (whereby intentions, explicit goal definition and even their awareness by those involved in the process is neither required nor need it be known from the outside), a theory that accomplishes the second goal: a theory of *strategic control*, and a theory that accomplishes the third goal: a theory of *structural control*.

While I cannot point to a social theory that would accomplish any one of the above stated aims, there are at least five research traditions which deserve a brief mention, in part to describe conceptualizations relevant to a future social control theory and in part to provide a context within which the contributions to this section may be appreciated. These are identified by their focus on social control mechanisms, socialization processes, social influence, collective behavior and management respectively.

One of the chief omissions in theories of social control is an appreciation of the dynamics involved even though the term "process" has been available in sociology at least since the work of C. H. Cooley [5]. Perhaps for this very reason most theories reduce control to a situation in which the behavior of a system (individual or social group) follows a guidance variable or can in some respect be *unidirectionally* constrained. This is already evident in the work of E. A. Ross, who published a series of articles on the subject, beginning in 1895 with the first volume of the *American Journal of Sociology* [13] and culminating in a book on *Social Control* [14] published in 1901. Basic to this work is the distinction between the control of society *by individuals* and the control of individuals *by society*. In either case, control flows in only one direction and the emphasis is on strategical considerations. Ross' main contribution lies in his identification of *social control mechanisms* such as public opinion, law, education, religion, rituals, art and the like through which individual behavior is coordinated and in his concern for the social purposes these control mechanisms serve.

Social control mechanisms of the kind studied by Ross are assumed to be effective only if adherence to or deviation from the norm they specify leads to sanctions, i.e. positive sanctions or rewards and negative sanctions or punishments. The elaboration of these concepts are found in G. C. Homan's theory of Social behavior as exchange [7], in T. Parsons theory of Social Systems [e.g. 11] and many others. While the concept of feedback is implicit in these authors' descriptions of control, it is not employed to account for the dynamics of this process.

Building upon this one-sided concept ot control, with a relatively fixed society as the controlling agent, Cooley's work on *Social Organization* [4] and *Social Process* [5] introduces the idea of *socialization*, though not by this name. He focuses attention on the conceptions that individuals acquire in the course of social participation, and he explains how the ability to empathize and the need for social participation account for the development of specific concepts of self and of social reality, including an individual's personality and ideals. In the course of time these concepts become so interrelated that the individual becomes an integrated member of society. Socialization, as this process is now called, is a form of individual-structural control through which external conditions are internalized by an individual and transformed into *values*, *beliefs* and *social stereotypes* which, after they have been acquired, reduce the need for continuous social guidance and give the illusion of self-control. The internalization of social control is a concept that has been used and further developed by numerous social scientists, among them Max Weber [17], and George H. Mead [9]. Thus, socialization essentially builds up and maintains within individuals the equivalent of Ashby's "model of a process" or Power's "internal representation" which is moreover assumed to be compatible with the functional requirements of a society.

In addition to the study of social control mechanisms and of socialization there exists by now a voluminous literature on social influence, basically on influence among individuals. At least since Weber's analysis of bureaucracy, we have known that an individual cannot exert social control without some form of legitimization, without being given some form of authority. According to Weber, an authority may be *traditional* as when a subject perceives orders by a superior to be justified on the grounds that this is the way things have always been done; *rational-legal* as when a subject perceives orders to be justified on the grounds of their consistency with his own abstract rules about how things work or are to be done; and *charismatic*, as when a subject identifies with a superior or leader and considers orders to be justified on the grounds of the superior's personality [17].

Much of social psychological literature seems to assume that influence operates through the distribution and exchange of some *valuable resource* such as "wealth, prestige, skill, information, physical strength and the ability to gratify the 'ego needs' that people have for such intangibles as recognition, affection, respect and accomplishment". Social psychological experiments with small groups has shed light on how perceptions change when such resources are used for gratification or punishment and how the allocation of these resources affects the behavior of the group as a whole [2]. I cannot review here the contribution made by studies of propaganda, mass media effects, political decision making and the like, except to say that these studies share the conception of an unidirectional flow of influence, mediated by the exchange or dissemination of resources, and based on the mutual perception of intentions, and capabilities.

A fourth research tradition relevant to social control theory may be found in the study of *collective behavior*: crowds, fads, collective reactions to crises, social movements, etc. This school of thought has broken with at least two assumptions underlying the social control mechanism, the socialization process and the social influence approach. The first of these is the dichotomy between society and the individual and the implied contention that one controls the other. For example, R. E. Park and E. W. Burgess [10], who contributed to the study of collective behavior, identify control as a mutual constraint that individuals impose upon each other in the course of ongoing interactions. The second assumption, rejected by this approach, is the association of rationality with the process of control. The assumption of rationality – explicitly stated goals and consciously manipulated means – is already manifest in Ross' distinction between "influence without aims and purposes" (e.g. fads, fashion) and "social control" proper, (e.g. law, administrative rules), and it runs all the way from the organismic approach of O. Spengler [15] to Parsons' social systems theory [11]. Park and Burgess, on the other hand, contend that individuals in group situations always respond to a perception of the motives and interests of others and that out of such interactions, common behavioral orientations inevitably emerge, whether or not they become dominant within a group.

Naturally, such orientations are not entirely stable. They are more vulnerable to change the less they are rooted in the history of the group. Individuals in interaction are always *self-organizing*; and, in the course of mutually constraining each other, they develop structure, acquire a history, and take on norms, values, and goals which, because of the diffused causality of the underlying process, give the appearance of belonging to something other than the individuals that create it. The inevitability of this process has been shown elsewhere [8]. It follows that the sharing of objectives among parts is not a necessary prerequisite in order for a whole to converge to an apparent goal, i.e. to exhibit purposive behavior. With these behavioral orientations in mind, the tradition of research on collective behavior is likely to make further contributions to social control theory, particular to an understanding of the process of *self-control*.

The final and not at all the least important source of contributions is the practice and science of management and decision theory in particular. Here the rational approach – externally stated goals and the manipulation of a system as a means to achieve them – dominates. Analogies to the feedback control cycle are frequent: the manager acts on a portion of the organization he is in charge of, obtains information resulting from his actions, compares this with his stated objectives and then decides on a course of action. Beer [1], Vickers [16] and many others including Ackoff, have elaborated and contributed to this approach to social control theory. Among the important implications of this rational approach to management decisions is the conception of organizational structure as *hierarchical* with controlling actions on one level setting the goals on the subordinate level (except for the relatively autonomous top management), defining spans of controls, assuming a logically consistent system of values, etc.

In terms of the three aims set forth above research in either of these five traditions has yet to provide a satisfactory theory of social control. But research in these traditions concurs that it would be inadequate simply to extend the conception of control through linear negative feedback from engineering into the social domain. People are not linked by wires with predetermined effects, as is the case with mechanical components. They always have the options of following a command, of ignoring a command, or of questioning the legitimacy of the authority to provide guidance. Unlike the thermostat, people have abstract goals of their own. Social control is mostly nondeterministic and always symbolic in nature. The symbolic nature of social control is particularly manifest in the final stage of socialization processes, at which deviance is reduced without continual guidance. Moreover, human beings and the relations between them are extremely complex and the ability to transmit and process information is a scarce resource. The phenomenon of collective behavior and the problems of management science both serve to demonstrate that strategic and structural control is not always possible.

The papers in this section all shed some light on the problem of theory development in the area of social control. The first paper by Russell L. Ackoff, is interesting; for it unquestionably relies on the rational control paradigm, while pointing out two ways in which this model fails in practice. Both are subsumed by Ackoff under the aesthetic aspects of management. The first inadequacy stems from the fact that managers, and I suppose all rational human beings, do not manipulate means solely to achieve certain ends, rather, choices among means may also be a reflection of individual preferences for particular actions. Ackoff links these intrinsic or stylistic values of means (as opposed to their extrinsic or instrumental values) to notions of the quality of life, a concept that is emerging with considerable force though it is far from being clearly understood.

The second inadequacy of the rational control paradigm is one that has been experienced by many operations researchers who find that their carefully worked out proposals are just not implemented, lead to outright rejection, dissatisfaction or ineffective results. The traditional rational paradigm responds to such failures either by accusing the opponents of such proposals of being "irrational" or by suggesting that the developers of such proposals were not well enough informed about the goals to be achieved or the constraints to be considered. Recognizing that a management consultant cannot possibly be aware of all the political intricacies of a social organization he is working for, the stylistic preferences of those subjected to change and the side-effects of proposed changes, Ackoff's paper espouses that a consultant make the search for practical solutions as participatory as possible. In particular the consultant should involve all those who will control, those controlled or otherwise affected by control. In his paper Ackoff suggests that this second inadequacy derives from "a lack of sense of progress" which may be overcome when participative planning is based on what he calls the "*idealized design* of a system", an approach the paper considers in some detail.

An aesthetics of managerial control thus appears to have two faces: style and ideals. Ackoffs idealized design seems to offer a way of overcoming the inadequacies of the traditional rational control paradigm in management situations.

In the second paper in this section, Erwin Laszlo considers some other limitations of the rational management-decision-making paradigm when the subject of control is a global one. Taking the need of controlling the world system as given, he views such a system as having essentially two kinds of components: *organizations* and *communities*. These components are controllable (steerable) in quite different ways. Because of the clear delineation of their constituent parts and operating functions and because of their underlying rationality, organizations are more readily amenable to steering than communities. One characteristic of communities is a high degree of indeterminacy, which makes modeling and subsequent rational guidance difficult. It would follow that very large socio-cultural systems of the kind that constitute the world system are best controlled by means of their component organizations. Communities may be influenced only indirectly insofar as they adjust to planned changes in the surrounding organizations. Laszlo suggests that the cybernetic modeling of global systems be restricted to the organizational components of such systems. He believes, leaving communities out is most likely to yield manageable results. However, the modeling of large communities seems to be an important challenge as well. If the extend of self-control exceeds that of strategic or structural control, this in itself would be an important finding.

The third paper in this section is a report of ongoing work. Premysl Pergler, a control engineer, and Walter Buckley, a sociologist, are collaborating with others in an effort to develop a conceptual framework for social systems within which social control processes play an important role. The paper proposes several definitions which are intended to shed light on problems of organizational control structures. In their contribution, the two most central concepts are *morphostasis* and *morphogenesis*. In Morphostasis according to the author, the goal of regulation is to maintain the structure of a system, while morphogenesis refers to structural changes that result in the improvement of better goal attainment or regulation of a system.

While the use of these terms is idiosyncratic to the authors and "considerable difficulties of this approach" are freely admitted, the paper stresses the necessity to conceive morphostasis to cooccur in virtually all social systems and the subsequent requirement to evaluate the impact of any stimulus on both processes. Underlying the paper is the hope that an understanding of these phenomena may transform the currently cyclical processes:

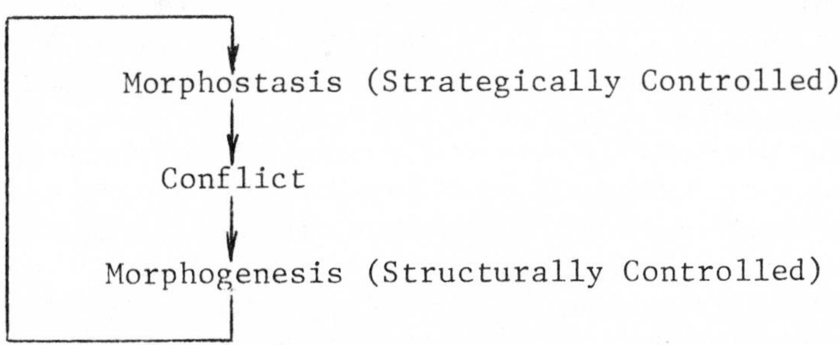

into one of *continuous morphogenesis*.

The fourth and fifth papers of this section present somewhat more specific statements about social control. This is due in part to the fact that both authors have been working for some time on separate theories of communication and control.

With a reorientation of psychological theory in mind, William T. Powers worked out elsewhere [12], a negative feedback theory of human perception, which has widely been seen as attack on and a way out of the confines of behaviorism. The central building block of this theory is a system that automatically adjusts the effects of its output on its environment so as to maintain some perception of that environment according to a given reference signal. (A perception is understood here as an input that is recognized and processed by the system). An important feature of the theory is that these elementary building blocks of feedback control can be combined hierarchically so as to account for more complex and higher order perceptions. A block that is higher up in such a hierarchy controls one or more below by providing the reference signals of the latter. It is fair to assume with Powers, that an organism's behavior is indeed the result of control hierarchies such as those for which his theory offers explanatory constructs. It might be noted here in passing that it is this assumption that traditional behaviorism was unable to consider as a possibility. In the paper included here, Powers extends this theory of brainlike and nervous activities to small groups and societies which, when organized in terms of control hierarchies, can be assumed to have similar properties. In particular, it follows from his theory that the number of options available decreases as the number of feedback blocks within the hierarchy increases. Moreover, if there are too many higher order feedbacks blocks, conflicts become inevitable. Powers suggests that this might increasingly be true in industrial society. He concludes with a warning that the growth in population and the increase in organizational forms may have exhausted the degree of freedom individuals once enjoyed. The current social control structure may be a source of conflict that is harmful for society as a whole. His contribution does not develop specifically social concepts of control but shows a very useful way of studying strategic control within hierarchical systems such as characterize many aspects of society.

In the final paper of this section, Anatol Holt addresses himself to "information as a system-relative concept". With a background in chemistry, in linguistics, and as a trained computer scientist who has made many contributions to parallel processing, Holt is now working on a formal structural control theory which centers around the social organization of work. He recognizes that the coordination of activities performed by different individuals is a crucial goal of any organization. This is accomplished through communication, which he conceives of as a highly specific interaction that must occur within a specific organizational setting and, most importantly, at an appropriate moment in time. To assess communication within social organizations requires a formal language that describes both the organizational structure – the circuitry of social roles – and the desired process. Holt has proposed such a language elsewhere [6]. In this volume, his paper discusses the requirements a communication theory or an information theory must satisfy in order to serve social organizational ends. It concludes therefore with a discussion of how the theories of Shannon and Wiener fail.

The papers of Powers and Holt are somewhat removed from the social science traditions reviewed above. Powers derives his notion of control from the coordinating activities of the brain, which is probably more centralized than most social organizations. Holt derives his notion of control from the logically consistent behavior of the computer, which is probably more formal than any real social organization. However, it is this kind of contribution which challenges existing conceptions of control in the social sciences.

REFERENCES

[1] Beer, S., *Decision and Control: the Meaning of Operational Research and Management Cybernetics,* London, New York: Wiley, 1966.

[2] Cartwright, D., "Power and influence in groups: introduction", in Cartwright, D. and Zander, A. (Eds.), *Group Dynamics; Research and Theory* 3rd ed.: 215–235, New York: Harper & Row, 1968.

[3] Conant, R. C. and Ashby, W. R., "Every good regulator of a system must be a model of that system", *International Journal of Systems Science,* **1**, No. 2: 89–97, 1970.
[4] Cooley, C. H., *Social Organization, A Study of the Larger Mind,* New York: Scribner, 1909.
[5] Cooley, C. H., *Social Process,* New York: Scribner, 1918.
[6] Holt, A. W., *et al., Information System Theory Project,* Rome, N.Y.: Applied Data Research, Inc., Air Force Systems Command, Griffiss Air Force, Report No. RADC–TR–68–305, 1968.
[7] Homans, G. C., "Social behavior as exchange", *American Journal of Sociology,* **63**: 597–606, 1958.
[8] Krippendorff, K., "Communication and the genesis of structure", *General Systems,* **16**: 171–185, 1971.
[9] Mead, G. H., *Mind, Self, and Society from the Standpoint of a Social Behaviorist,* Morris, C. W. (Ed.), Chicago: University of Chicago Press, 1962.
[10] Park, R. E. and Burgess, E. W., *Introduction to the Science of Sociology,* 3rd ed., Chicago: University of Chicago Press, 1969.
[11] Parsons, T., *The Social System,* Glencoe, Ill.: Free Press, 1951.
[12] Powers, W. T., *Behavior: The Control of Perception,* Chicago: Aldine, 1973.
[13] Ross, E. A., "Social control", *American Journal of Sociology,* **1**: 513–535, 1895; 753–770, 1896.
[14] Ross, E. A., *Social Control. A Survey of the Foundations of Order,* New York: MacMillan, 1901.
[15] Spengler, O., *The Decline of the West: Form and Actuality,* London: Allen and Unwin, 1926–28.
[16] Vickers, *Sir* G., *Towards a Sociology of Management,* New York: Basic Books, 1967.
[17] Weber, M., *The Theory of Social and Economic Organization,* Parsons, T. (Ed.), New York: Oxford University Press, 1947.

THE AESTHETICS OF MANAGEMENT

RUSSELL L. ACKOFF
University of Pennsylvania, Philadelphia

The science of control took its name, *cybernetics*, from the ancient Greeks. Therefore, it is not completely inappropriate to apply other concepts borrowed from Grecian scholars to this contemporary science.

The philosophers of Ancient Greece divided the pursuits of man into four categories:

(1) the *scientific* – the pursuit of *truth*,

(2) the *political-economic* – the pursuit of *plenty*,

(3) the *ethical-moral* – the pursuit of *goodness* or *virtue*, and

(4) the *aesthetic* – the pursuit of *beauty*.

These categories were refined out of the philosophical thought of centuries; they were not the product of a deliberate effort to divide man's activities into an exclusive and exhaustive set of categories. Obviously they are not mutually exclusive since man clearly can pursue several of these ideals at the same time, through the same acts. Nevertheless, I believe these categories are exhaustive for reasons that will become evident in the course of this paper.

My principal purpose is to redress the imbalance with which these categories have been used in our efforts to understand control, particularly that kind of control exercised by purposeful subsystems over the systems of which they are part; a type of control we call *management*.

In the study and practice of control and management we have been concerned with their scientific aspects, their political economic aspects, and their ethical-moral aspects, but not with their aesthetic aspects. The expression "aesthetics of management" does not even convey meaning to most students or practitioners of management. To be sure, we refer to the *art* of management but in so doing we do not signify the aesthetic aspects of this activity, but our inability to understand all of it scientifically. Art in this sense is the ability to control without understanding all that is involved in so doing. This is not the sense in which I will use the concept. My concern is with the art *in*, not *of*, management.

Failure to reflect on or understand the meaning of aesthetics of management is not surprising in view of the fact that over the last twenty-five centuries very few philosophers have been able to incorporate aesthetics into a comprehensive philosophical system. There has been little systematic development of aesthetics since the Greeks. On the other hand, "aestheticians" tend to give the other three categories of man's activities little attention. As a result, we understand aesthetics much less than science and technology, politics and economics, or ethics and morality. It is safe to say that most of us have some idea as to how each of these three activities relates to the others but somehow or other the idea that aesthetics is antithetic, if not downright hostile, to the other three has gained widespread acceptance.

Thoughtful men would agree that considerable progress has been made in science and technology. Some would agree that progress has also been made in the domain of political economy and in ethics and morality. But one would be surprised to hear it argued that mankind has made aesthetic progress. Indeed, we must admit that little evidence exists on which to base a claim that contemporary man has a greater ability to produce or enjoy beauty than his predecessors.

Nevertheless, in the last two decades aesthetic concerns have been emerging as a major preoccupation of affluent people. But it is doing so in a way that makes it difficult to recognize as a matter of aesthetics. It is emerging in the guise of a concern with the *quality of life*.

It has become traditional for affluent people to separate work from pleasure and play. They have been conscious of aesthetics – or at least beauty, play, and pleasure – in their homes and in their recreational and social activities. But attitudes toward business and work have been dominated in the Western World by the Protestant or, more precisely, the Puritan ethic. This ethic separates work from play and conceptualizes it as an *ascetic*, not an *aesthetic*, activity. Work is still widely thought of as necessary and necessarily unpleasant. The dissatisfaction or suffering it has imposed has been justified or rationalized by many moralists of the Industrial Revolution; it was accepted, if not embraced, as a kind of earthly purgatory in which sin is expiated and virtue is gradually accumulated.

Post-Renaissance man came to conceptualize the universe as a machine created by God to do His work, to serve His purposes. Man already believed he had been created in the image of God. Therefore, it was natural for him to develop machines that would do *his* work, serve *his* purposes. This effort gave rise to the Industrial Revolution.

Consider this revolution. It was carried out by analyzing work into its simplest elements; mechanizing those that could be mechanized; and inserting human beings into those elements of work which were impossible or uneconomical to mechanize. The men and machines used in this way were organized into the production line that became the spine of the modern factory. This process had an ironic consequence: it imposed its machine-like character on man. Man was reduced to behaving like a machine. Work was almost completely dehumanized.

Industrial societies succeeded in this program of dehumanization so long as workers were economically and educationally deprived. But the Industrial Revolution also brought affluence with it, and with affluence came greater economic security and more and better education. Secure and educated workers are increasingly dissatisfied with, and alienated from, machine-like work [4]. Discontent with the quality of work, now widespread in developed countries, is reaching crisis proportions. It is hardly necessary to point out that this discontent is part of a more general dissatisfaction with the quality of life.

The Industrial Revolution also accelerated urbanization and produced a fundamental qualitative change in the nature of man's environment. Most of the environment in which "developed" men spend their time is now man-made. Initially, man-made environments brought great comfort and convenience but in recent years they have been deteriorating rapidly. This deterioration is exacerbated by inaccessibility to untouched Nature and has further sharpened dissatisfaction with the quality of life.

The meaning of "quality of life" is far from clear, but it is clear that it has less to do with what and how much we have in the way of material goods than with the conditions under which we acquire and use them. Quality of life – ordinary life, corporate life, work life, academic life, all kinds of life – is not a matter of products but of processes.

Even to specialists in decision making it is apparent that aesthetic considerations often dominate our activity. Ordinary people are never in doubt about it. I mention but a few examples drawn from my own corporate and governmental experiences.

A small company owned by its three executives wanted to diversify so that they could be challenged more and become more involved in running their business which, to their embarrassment, ran itself rather successfully. In their words, they wanted "to get more *fun* out of it". Fun is a matter of aesthetics, not science, technology, politics, economics, ethics, or morality.

Another example. A major corporation's profits have been suffering recently because of its commitment to producing only the highest quality products in its field. Its cost of materials is now inflating more rapidly than are its competitors, but it refuses to abort its products or abbreviate the processes by which they are made, as its competitors do. To degrade its product would significantly reduce the satisfaction its managers derive from their work. This too is a matter of aesthetics.

Finally, there are two districts of the Federal Reserve banking system both of which have exactly the same functions but which differ significantly in the way they are organized and carry out their functions. The atmospheres in these two banks are completely different. The difference cannot be explained in scientific, technological, political, economic, ethical, or moral terms; it is a matter of aesthetics, a matter of *style*.

The moral of these short stories is clear. Style, which is an important aspect of aesthetics, is also an important aspect of decision making and control. Therefore, let us focus on style for a moment.

Decision making in the face of doubt is problem solving. Through science we have developed rich and useful models of decision making and problem solving. Decision models usually have four components: (1) the decision maker, (2) alternative possible courses of action defined by controllable variables, (3) alternative possible outcomes, and (4) the environment defined by those uncontrolled variables that together with controlled variables can affect the outcome. The components are interrelated by three types of parameter: (1) probability of choice – that is, the probability that the decision maker will select a particular course of action; (2) efficiency of choice – the probability that a course of action if selected will produce a particular outcome in the decision environment; and (3) the relative values of the outcomes to the decision maker in the decision environment. The sum of the products of these measures over all the possible courses of action and outcomes is the *expected relative value* of the choice situation.

Choosing that course of action which maximizes expected relative value is what many economists mean by "rationality". This, it seems to me, is an irrational concept of rationality because it omits a major type of value. Consider the following type of choice situation: only one of a set of exclusive and exhaustive courses of action can be chosen, each of which is certain to produce the same desired outcome. If maximization of expected relative value were the only appropriate criterion of choice, the rational decision maker would be completely indifferent to the alternatives in this situation. But this is seldom the case and for a good reason. *We have preferences for means as well as ends*, for we know that means and ends are relative concepts. We buy a book in order to read; we read in order to learn; we learn in order to earn; and so on. Every end is a means to a further end and every means is an *end-in-itself*. Therefore, means as well as ends have value to us. Means have two kinds of value: *extrinsic* or *instrumental*, and *intrinsic* or *stylistic*. The extrinsic value of a means has to do with its efficiency relative to an end, and intrinsic value of a means has to do with the satisfaction its use produces independently of its outcome.

Consider a few simple examples. Brown shoes may be just as efficient for walking or for dress as black ones. We may nevertheless prefer one to the other because it pleases us more. My preference for black ink over blue, blue over green, and green over purple exists even though I am aware that each works as well as the others for the purposes for which I use them. That set of preferences which each of us has that are independent of considerations of efficiency constitutes our style. Our individuality, our uniqueness, lies as much in our style as it does in what we value as ends and the efficiency with which we pursue them. Style has to do with the satisfaction we derive from what we do rather than what we do it for.

Desired outcomes are performance objectives that impart extrinsic value to the means we employ to pursue them. Uses of preferred means are stylistic objectives that have intrinsic value, value that is independent of the outcomes they bring about. If and when our theories of decision making, problem solving, management, and control do not take style, intrinsic values, into account, they are seriously deficient.

There is another important aspect of aesthetics: the pursuit of ideals. Let me explain.

I turn for a moment to the history of ethics and to the philosphers' search for the one universal desire that might define an *ultimate Good.* Their motive was the belief that only through such a desire might all men and women, past, present, and future, be unified into what we call "mankind". They hoped to demonstrate that mankind is a system whose parts are integrated by pursuit of a common ideal. How ironical that the search for such an ideal failed because those who engaged in it were too sophisticated. Again let me explain.

Once upon a time a young man was granted three wishes. We all know that with the first two of them he managed to get himself into such difficulties that he had to use up his third wish to get back to the condition from which he started. On hearing any one of the many versions of this story most bright children tell us they can do better than the hapless hero with only one wish; they would wish that all their wishes would come true. My teacher, the much-too-little recognized American philosopher Edgar Arthur Singer, Jr. [3] systematized this childlike wisdom by identifying a desire so universal that it unifies all men at all times. It is the desire to be able to satisfy our desires whatever they might be, even if we desire nothing, Nirvana. It is in the nature of purposeful systems – and human beings are purposeful systems – to desire; and there can be no desire not accompanied by the desire to be able to satisfy it. Hence an ideal that can be shared by all men at all times is *omnipotence*, the ability to satisfy any desire. The ideal character of omnipotence is reflected in the fact that virtually every religion of Western culture ascribes it to deity.

Omnipotence is an ideal which, if attained, would assure the fulfillment of all desires, hence the attainment of all other ideals. Therefore, it is what might be called a "meta ideal" for mankind. Like any ideal, however, it is not attainable though it can be continuously approached.

Now consider the four conditions that are necessary and sufficient for continuous progress for every person toward omnipotence. Doing so will return us to the four-fold nature of human activity as conceived by the Greeks.

First, such progress requires a continuous increase in the efficiency of the means by which we can pursue our ends and, therefore, a continuous increase of our knowledge and understanding, in our grasp of *truth*. To provide it is the function of *science*; to enable us to use it effectively is the function of *technology*.

Second, pursuit of omnipotence requires a continuous increase in the availability and accessibility of those resources needed to employ the most efficient means available and, therefore, increasingly efficient production, distribution, and maintenance of wealth. To provide *plenty* is the function of the *political economy.*

Third, it requires continuous reduction of conflict within and between individuals because conflict means that the satisfaction of one (or one's) desire precludes the satisfaction of another (or an other's) desire. Therefore, we pursue both peace of mind and peace on earth: the state of *virtue*. To provide it is the function of *ethics* and *morality*.

Fourth, it requires continuous attention to the aesthetic life. To demonstrate this will be the burden of much of the rest of my paper.

If man is to continuously pursue the ideal of omnipotence he must never be willing to settle for anything less. He must never be either permanently discouraged or completely satisfied. Therefore, whenever he attains one objective he must seek another that he wants more than the previous one and the attainment of which requires a further increase in his power. Thus he must always be able to find new possibilities for improvement and satisfaction.

Singer showed that it is the function of the aesthetic life to *inspire* us; to create the creator of visions of the better and give this creature the *courage* to pursue his vision no matter what short-run sacrifices are demanded. Inspiration and aspiration go hand-in-hand. Inspiring art consists of the works of man that are capable of stimulating new aspirations and a commitment to their pursuit.

In the *Republic*, Plato conceived of art as a stimulant that was potentially dangerous to society because it could threaten its stability. His perception of its function was the same as that put forward here, but his conception of utopia as a *stable* state is not consistent with the four conditions of omnipotence set forth above. These imply a dynamic idealized polity, one that is a *process* rather than a stasis.

In the ideal state as I have presented it man would not have solved all his problems or attained all his objectives. There is at least as much satisfaction to be derived from the pursuit of ends as in attaining them. Therefore, the ideal society, as I conceive it, is not problem-free, but problem-full though power-full.

In contrast to Plato, Aristotle viewed art as a cathartic, a palliative for dissatisfaction, a producer of stability and contentment. While Plato saw art as producing the dissatisfaction with the present that leads to efforts to create the future, Aristotle saw it as producing satisfaction with what has already been accomplished, not as creative, but as *recreative*. Plato and Aristotle were concerned with two aspects of the same thing. Art is both creative and recreative. These aspects of it can be viewed and discussed separately but they cannot be separated. Recreation is the effort to extract pleasure from the here and now, a reward for past effort. It provides "the pause that refreshes" and thus recreates the creator. Art is also inspiration, it drives us to further efforts. Art both pulls us from the past and pushes us into the future. Without art we would falter on the way in our continuous pursuit of ideals.

Note that the only end that can have purely intrinsic value is an end which cannot be a means to a more ultimate end. Hence, only ultimate ends, ideals, can have purely intrinsic value. On the other hand I have argued that the ultimate ideal of mankind must be omnipotence, the ability to attain any objective. The value of this ideal is therefore purely extrinsic. Thus in dialectical fashion we find the complete synthesis of intrinsic and extrinsic value in this ideal.

I have tried to show that decision making, hence control, has two important aesthetic aspects: (1) the style of decision makers, which has to do with immediate sources of satisfaction derived from doing rather from the consequences of doing, and (2) the ideals of decision makers, the outcomes which, if attained, would be completely satisfying to them. Unless we understand the role of style and ideals in decision making and control we cannot develop satisfactory control systems.

Style is at least as important in decision making and control as is efficiency and effectiveness. It has often been observed, for example, that individuals enjoy power – which is the ability to control – for its own sake. This means that control can bring its own satisfactions. Control becomes an end-in-itself as well as a means to other ends. A control system, even a self-control system, that does not match the style of the controller will either not be accepted or, if accepted, will lead to dissatisfaction and ultimate replacement. Thus, the designer or planner of control systems must understand the style and ideals of not only those who are to be in control but also those who are to be controlled if he is to design or plan a system that will work effectively.

Style is multidimensional. Every personality trait is a dimension of it. G. W. Allport and H. S. Odbert identified 17,953 trait names in English [2]. These include ascendant-submissive, introverted-extroverted, aggressive, sociable, charitable, courageous, apathetic, and so on and on. An introvert, for example, seeks to get along with as little information as possible; an extrovert with as much as possible. The introvert prefers to have others manipulate the environment; the extrovert prefers to do so himself. The introvert prefers to work alone; the extrovert with as many others as possible. No one management or control system will satisfy both equally. But determination of the traits possessed by the controllers and the controlled which are relevant in designing a management system is not easy.

There are not enough relevantly knowledgeable behavioral scientists to engage in all the control-system design that is required. Another way out is called for. There is such a way: *involvement of those who will control, be controlled, or otherwise be affected by the system being designed, in the design of that system.* A system designer who is aware of the relevance of style can learn about the stylistic preferences of stakeholders in the system by making its redesign *as parti-*

cipative as possible. Participants in the design process cannot help but put their stylistic preferences into their designs. Nor can they refrain from incorporating their ideals into these designs.

The kind of participative planning that best reveals relevant styles and ideals is based on the *idealized* design of a system. This is a design (or redesign) of a system "from scratch" with all but two constraints removed. First, the design must be technologically feasible though it need not be practical. For example, one could not assume the availability of a direct transfer of the content of one mind to another by telepathy. This technological constraint does not preclude consideration of technological innovations – for example, the picture phone or facsimile transmission – but it restricts them to what is currently believed to be possible. On the other hand, all considerations of financial or political feasibility are removed.

The second constraint is that the system design must be viable; that is, capable of working and surviving if it were implemented.

In brief, then, an idealized design is an explicit formulation of the designer's conception of the system he would create now if he could create any system he wanted.

Since any such design is unintentionally constrained by the designers' lack of information, knowledge, understanding, wisdom, and imagination, the designed system should be capable of learning from its own experience and adapting to internal and external changes. It should be flexible and capable of rapidly improving its own performance.

Such an idealized design is *not* utopian precisely because it is capable of being improved and of improving itself. It is the best we can think of *now* but its design unlike that of a utopia, is based on recognition of the fact that no ideal state can remain ideal for long.

An idealized system differs from a utopia in another important and related respect. Its designers need not pretend to have the final answer to every question. Where they do not have answers they can incorporate into their design experiments that are directed toward finding them. Thus not every issue needs to be resolved in an idealized design. Such a design is not absolute, fixed, or final. It is subject to continuous change in light of information, knowledge, understanding, and motivation acquired after the design is completed.

At my research home, the Busch Center of The Wharton School, we have been and are currently involved in the idealized design and redesign of a number of social systems. Under the leadership of Professor Hasan Ozbekhan we recently collaborated with a number of intellectual and political leaders in France in such a redesign of Paris. Currently we are involved with a number of executives, managers, staff personnel, and users of the Federal Reserve banking system in an idealized redesign. A short while ago we prepared such a redesign of our nation's juvenile justice system working together with a wide variety of participants in that system. We have been and are involved in such redesigns of several foreign and domestic corporations. At the moment we are engaged in a similar effort applied to the national scientific communication and technology transfer system. My own personal idealizations of a number of social institutions appear in my book, *Redesigning the Future*, which has just been published [1].

The redesign of Paris was directed at producing the kind of Paris that the French want, such desire being interpreted in the broadest aesthetic sense. It yielded a conception of a possible Paris that not only incorporated many of the intrinsic values the French have, but it also is *a work of art* in itself because it has inspired many Frenchmen with a determination to try to bring it about.

The idealization process provides an effective way of obtaining meaningful participation of those who will use or be affected by the system. It tends to generate consensus among those engaged in the process. Since most disagreements arise from considerations of efficiency, not value, and since idealized design is preoccupied with intrinsic value, not efficiency, it tends to breed agreement among participants in the process. It facilitates participation because *it is fun* and requires no special skills to be able to engage in and contribute to it.

The effective design of management systems can help managers, those managed, and those otherwise affected become more aware of their aesthetic (intrinsic) values: their styles and ideals. By so doing management systems can be developed which enable men to enjoy their work more and to work closer to the limits of their capabilities. It can thus enable them to reunite work and play, to make work recreative, fun.

By making those involved in management systems extend themselves and reach toward their ideals such a system can also facilitate the combining of work and learning and thus make work a more creative, as well as recreative, experience. The management system designer, the bearer of the Postindustrial Revolution, can therefore bring together what the Industrial Revolution put asunder: work, play, and learning.

A management system which does so is a work of art.

REFERENCES

[1] Ackoff, R. L., *Redesigning the Future,* New York: John Wiley & Sons, 1974.

[2] Allport, G. W. and Odbert, H. S., "Trait-names: a psycholexical study", *Psychological Monographs,* No. 211, 1936.

[3] Singer, E. A. Jr., *In Search of a Way of Life,* New York: Columbia University Press, 1948.

[4] *Work in America: Report of a Special Task Force to the Secretary of Health, Education and Welfare,* Cambridge, Mass. and London: The MIT Press, 1973.

USES AND LIMITATIONS OF THE CYBERNETIC MODELING OF SOCIAL SYSTEMS

ERVIN LASZLO
United Nations Institute for Training and Research (UNITAR)

In this paper I shall discuss some of the applications and limitations of the cybernetic modeling of social systems.

Such modeling raises first of all, the issue of warrant and justification. Are we merely interested in gaining more adequate knowledge of the phenomena studied, in this case, of social systems? Or does the choice of a cybernetic model imply that subsequent to modeling we shall have possibilities of applying the models for controlling and guiding the modeled phenomena? I believe that the choice of a cybernetic model implies the underlying value of operationalization. Multi-person organizations, as other phenomena, can be modeled by different methods, ranging from the phenomenological and *verstehen* approach to the economic, political, anthropological, and psychological constructivist schools, embracing behaviorism and functionalism on the one hand and econometry and sociometry on the other. The choice of a specifically cybernetic model is not dictated by the nature of the phenomenon: as we know, all empirical phenomena are underdetermined by data and permit the construction of an indefinite number of theories (the so-called Mach-Duhem-Poincaré hypothesis). We would look in vain for data which would automatically select toward a cybernetic model; the very data we consider already implies a selection, and such selection in turn implies a theory, at least in implicit form. Only a naive justificationist would argue that certain types of organizations call for cybernetic modeling by their very nature, and he would be guilty of circular reasoning; of having his premises disguised as his conclusions. No: if we are to justify the use of cybernetic modeling, in this or any other empirical domain, we must turn our attention not to our data, but to our *objectives*. Can cybernetic modeling provide us with results that are preferable, by some criteria, to the results of alternative modelings?

Here I believe there are instances where we can clearly answer, yes. Cybernetics was defined by Wiener as the "science of communication and control" [23], and cybernetic models usually permit concrete application, due to the fact that their concepts are purely operational. Any meaning other than that of flows of events with determinate directions and controlled transformations are extraneous to cybernetic models and represent meta-interpretations, traceable to separate assumptions on the part of the investigator. This *ab ovo* emphasis, which is the hallmark of cybernetics, renders it uniquely qualified to provide grounds for controlling, and not merely understanding, the phenomena modeled. This, I suggest, we should accept as the main warrant for the choice of cybernetic models. If we do so, we shall be naturally led to inquire concerning the limits and possibilities of their concrete applications.

My argument, in a nutshell, is the following. First, I suggest that we have a pronounced need for steering our social systems; that we can no longer content ourselves with producing but descriptive models for scholarly contemplation. Second, I argue that there are strict limitations to the direct application of cybernetic models to realworld social systems for purposes of steering and control. Third, I offer some suggestions concerning ways and means of overcoming these limitations by steering larger social systems indirectly, through the impact of more steerable subsystems. Let me elaborate on each of these points in turn.

I THE CONTEMPORARY NEED FOR STEERING SOCIAL SYSTEMS

Marx said, in his *Theses on Feuerbach* [18], that philosophers hitherto have attempted only to explain the world; the task, however, is to change it. He was struck by the dangers inherent in the unfolding of the First Industrial Revolution and hoped by his theories to help to bring the process under control. Today we are no less struck with the dangers inherent in the coming Post-Industrial Revolution – the cybernetic and electronic revolution in the interdetermined, overcrowded, highly polluted and resource-depleted planet – and many among us call for radical changes. To attempt to change any phenomenon is a tall order and Marx, as well as contemporary radicals, may be guilty of exaggeration. But we can no longer contest that there is a need to steer *some* of the determinant processes of our world, if we are to avoid major breakdowns and suffering.

Steering is a cybernetic concept. It derives from the original meaning of *kybernetes* which in classical Greek means guide or pilot. To steer a system is not to change it (which may be arbitrary, and in any case not always possible), but to enable it to cope with its internal and external exigencies. A system is in need of steering if (a) the probability of its continued persistence is consistently reduced by the extension of its actual modes of behavior, and (b) its behavior results in increasing stress on its environment, including other systems, some of which may be subsystems and others suprasystems in relation to it.

Steering a system means exercising adaptive control over it. More exactly, we may follow the definitions of Beer [1] in considering control as self-regulatory activity essential for system stability and survival, and view it as the establishment of the limits on resource expenditures within which the system remains stable. But I wish to add to the criterion of system stability the second, and in my view equally important, criterion of system-produced *environmental stress*. Control then becomes the function of defining the limits within which the resource expenditures of the system are adequate to maintain the system in its environment without degrading or deteriorating that environment. In the sense of this definition, steering is the ongoing exercise of control.

A general statement which proclaims the need for steering with respect to *all* contemporary social systems can be upheld due to the fact that if the highest-level social suprasystem needs steering, so do all its lower-level subsystems. The highest-level system is what Beer called a "recursive" system, and which I, among others, have since termed a "multilevel hierarchical system" [16; 22; 24]. (In Mesarovič' mathematical model it appears as a multiechelon system [21].) The salient feature of such a system is that it is not an additive heap but a constitutive whole; in other words, that its functions and characteristics are determined not only by the functions and characteristics of all its parts separately, but by the interrelations of the parts within the whole. Nonadditivity has led investigators to speak of the irreducibility of such systems and has involved occasional excursions into metaphysics. Yet for our present purposes it will suffice to note that if the functions and characteristics of the whole hierarchical system are the product of all its subsystem functions and characteristics in view of their precise interrelations, than any change in the subsystems and their relations will entail changes in the whole system, and vice versa. Hence if we postulate that the whole system needs steering, i.e. a modification of the parameters of its resource expenditures, by that same token we mean that all its subsystems need corresponding steering (since any change in the whole entails changes in all its parts). The severity of this mathematical postulate is softened in practice by the possibility that some of the entailed subsystem changes may be minor and, for all practical purposes, may be ignored. To paraphrase a popular saying, all social systems need steering, but some need steering more than others.

This conclusion holds true if the total social system on this planet is actually in need of steering. This is the "world system" [12], the "system of world order" [9], the crew of the "space-ship earth" [14], or "the global village" [19]. By our definition, this system needs steering if its existing patterns of resource expenditures are inadequate either to maintain itself in its environment, and/or to maintain the essential qualities of the environment. A large number of "doomsday theories" and "world order scenarios" argue these days that if current resource expenditures are continued, the extension of growth curves into the next few decades provokes major catastrophes [e.g. 15; 20]. Apart from the specific assumptions of the models, we may note that if certain parameters of the world system continue to evolve at their present rates, breakdowns are in fact likely to occur. (This, we should note, is not a prediction, or even a forecast, but a simple conditional hypothesis.) The parameters in question include population growth, pollution generation, industrialization, food, raw-material usage rate (the five Forrester-Meadows "system levels"), combined with energy consumption rates, ratios of wealth between developed and underdeveloped nations, growth of nuclear destruction capacity in terms of new technologies and existing stockpiles, disparity rates of educational level and world view between peoples in different geopolitical regions, drug usage rates, the growth of mystical and "drop-out" cults, the progressive depersonalization of advanced urban environments, and disparities in governmental perceptions of national interests. The dominant characteristic of the eventual breakdown would be determined by the relative growth curves of the parameters: we can envisage catastrophes due primarily to overpopulation, overpollution, the overexploitation of urban-industrial regions, widespread starvation, the failure of essential technologies due to scarcity of raw-materials and energy sources, war triggered by the widening gap between the condominium of superpowers and the rest of the world, nuclear war triggered by conflict among the superpowers or possibly by a series of accidents, discontent among the peoples of the underdeveloped world, the conflict of regional world views, breakdowns of urban and medical technologies and of law and order due to spreading drug and drop-out cults and the indifference of populations who no longer perceive meaning in life [cf. 13], and a series of squirmishes among national governments eroding the texture and stability of civilization. Because any one of these breakdowns is a real possibility, and the occurrence of one or some combination of them is almost a certainty unless the limits of existing patterns of resource expenditures are redefined, it is clear that, on all rational humanistic grounds, the contemporary world system needs corrective steering.

II LIMITATIONS OF THE DIRECT APPLICATION OF CYBERNETIC MODELS FOR PURPOSES OF STEERING

The thesis I am suggesting is that only one variety of social system is well adapted to being steered, and that whatever steering we should exercise over other social systems we must accomplish indirectly, through the impact of the steerable systems, some of which we will have to design specifically for that purpose.

I shall define the steerable systems as those of which the members act in a public, social or professional capacity, and are compensated for their actions by the allocation of values (e.g. monies, prestige, or some other desired commodity). I shall contrast with this type of system those in which the members act in a nonprofessional or private capacity, and are not specifically remunerated for their actions. I shall call one type of system "organizations" and the other "communities".

Social systems encompass both organizations and communities. For purposes of analysis, we can point out that social systems minus communities gives us organizations, and social systems minus organizations give us communities. Organizations can be isolated more easily if they do not directly involve the private lives of their members (even if they affect them in some ways). Most organizations are of this kind. Exceptions are paternalistic corporations, such as those occurring in Japan, military and quasi-military organizations, religious and church organizations, tightly organized underworld groups, and boarding schools and universities. Communities can be conceptually isolated when they exist outside and independently of organizations (e.g. as suburban bedroom communities).

My thesis is that social systems can be steered through the *direct* steering of their *organizational* components and *indirect* steering of their *community* components. There are two principal reasons for limiting direct steering to organizations. First, communities tend to be too indeterminate to permit adequate cybernetic modeling (or conversely, the models tend to become so complex and fuzzy as to be practically useless). Second, communities tend not to respond to model-generated policies, especially if they are counterintuitive. By contrast, organizations can be relatively unambiguously modeled, and they tend to respond to policy decisions regardless of whether these mesh with the members' private views and values.

Let me enlarge on these points in more detail. Cybernetic analysis is premised on the definability of specific system components and functions, and the identification of the defined components and functions with observations made or inferred of the system. In order to identify functions of information transfer (communication), storage, retrieval, and processing, and the system components responsible for them, the phenomena in question must be adequately ordered. For example, cybernetic analysis requires that points where information or energy enters and leaves the system be identified. If these points are irregularly distributed in the observed system along its boundaries, and if, moreover, the rate of input and filtering at given points varies in a random or irregular fashion, the cybernetic model becomes fuzzy, heavily stochastic, and highly complex. Its practical usefulness is greatly reduced, and we may legitimately question whether cybernetic modeling is altogether justified. Cybernetic analysis also calls for the identification of decision-centers, where transformations of the energies and information passing through the system has an appreciable effect on its states as well as on its input-output relations with the environment. If in an observed system energies and information are transformed at an indeterminate number of points with random or irregularly varying consequences on system states and environment, the cybernetic model likewise becomes "jammed". Analogous remarks could be made with respect to other cybernetic functions: monitoring, comparing, information storing and retrieving, and so on. Cybernetic modeling requires sufficient order in the observed phenomena to justify (on epistemological grounds) the postulation of sensors, effectors, regulators, controllers, comparators, monitors and memory-stores, with determinate information and energy-flows. In the absence of such order, cybernetic modeling loses its validity on epistemological, and its warrant on methodological grounds (the resulting models become largely unoperationalizable).

Realworld systems that are too "noisy" or too complexly ordered, select toward noncybernetic models not by the brute force of the data, but in view of the eminent justification of cybernetic modeling by operational control. A cybernetic model with innumerable loops, points of randomness and low levels of probability fails to provide grounds for operational control, and the investigator must ask himself whether in the circumstances some other theoretical scheme would not be more indicated. The question is, whether entire social systems come into the category of cybernetically overcomplex and underdetermined phenomena.

I suggest that social systems, conceived as suprasystems formed by organizations and communities in interrelation, are indeed too complex for adequate cybernetic modeling. Attempts to model them in this way have come up against the objections just mentioned [cf. 3]. Even the modeling of certain organizational aspects of societies, such as their political systems, had to proceed with much care and faced serious problems due to the fuzziness and complexity of data [cf. 6; 7]. Coping with complexity is a problem that besets not only the social scientist, but also his natural scientist colleagues, especially in the fields of ecosystem analysis, population biology, biophysics, and neurophysiology. It is not a cause for despair but for the proper consideration and selection of theoretical tools. In the social sciences this means limiting the use of cybernetic models mainly to organizations and using, at least for the time being, more broadly conceived if less rigorous models for entire societies (e.g. functionalist, systems theoretic, structuralist, action-theoretical, social-psychological, communication-theoretical, or still other models).

Let me address now the second reason I gave for restricting the direct application of cybernetic models to organizations, *to wit*, the tendency of the members of community-type systems to remain impervious to model-generated policies, especially if they prove to be counter-intuitive in the light of their private values and world views. Recent experience with cybernetic and systems models of multi-person systems has shown that the points at which such systems can be steered are usually contrary to common-sense expectations, and that the measures indicated by the models for effective steering are very different from, and often conflict with, expectations based on intuitive mental models or hunches (see especially Forrester, [10; 11; 12]. There is a nonnegligible probability that policy guidelines based on cybernetic or systems models prove to be sufficiently different from ordinary expectations to appear misleading or nonsensical. Under such circumstances they will only find acceptance under one of two conditions (or both): if the system's members are willing to (a) follow policies despite a conflict with their own common sense; (b) suspend their common-sense judgments in favor of the model-generated policies.

Considerable evidence accumulates in recent social science literature to the effect that public policies are usually impotent to change people's basic beliefs and values [cf. 2; 5; interpretation in 8]. This is especially true of the private sphere, where citizens usually think of themselves as on a par with the experts and receive no direct incentive to toe to the official or public line. On the other hand, within organizations members are rewarded for their task-fulfillments, and find themselves in a strongly hierarchical situation where higher levels of command are generally coupled with higher prestige in matters of decision-making (the managers "know").

In the absence of sufficient willingness to follow non- or counter-intuitive policies, and with little in the way of incentives to prompt citizens to suspend, or to change their mind about, their own judgments, community-type systems remain highly unsteerable. They lack the agreement between the two flows of signals, namely the downward flow (from government to people) and the upward flow (from people to government). The downward flow represents societal control which is only effective if it can draw on a corresponding – or at least non-conflicting – upwards flow: the process of consensus-building. In community systems, unlike in organizations, the downward flow alone is unable to generate agreement with the upward flow, as the majority of people persist in their own judgements and values despite governmental proclamations. This renders policies generated by reference to theoretical models largely ineffective with regard to community-type systems. Not so, however, for organizations where incentives and prestige go hand in hand in creating a climate of acceptance even for counter-intuitive policy-decisions.

III POSSIBILITIES FOR THE INDIRECT STEERING OF SOCIAL SYSTEMS

To recapitulate: I have argued that (i) all contemporary social systems are currently in need of purposive steering, yet (ii) only organizations, but not community-type systems, prove to be adequately steerable by reference to theoretical models. The problem before us is whether the direct steering of organizations may suffice for the indirect but nevertheless effective steering of the community-type systems as well – i.e. whether the cybernetic modeling and designing of some societal subsystems is sufficient for steering entire social systems.

To steer a system we must have criteria for the definition of goals, and an understanding of the system dynamics in terms of the nature and relations of the functional subsystems. Criteria for social good-definition are implicit in the suggested definition of the need for steering: a system requires steering if the limits of its resource expenditures must be redefined in order to maintain the system in its environment and maintain as well the quality of that environment. A social goal-state, then, is one where social systems persist in harmony with their social and natural milieu. The problem I raise is whether the direct steerability of organizations suffices to indirectly steer entire social systems to achieve stability coupled with the conservation of the environment.

To deal with this problem we must next elucidate the relevant subsystem relations: the relations of the highly steerable organizations to the relatively non-steerable communities. This may be conceived in terms of superposition. In general, we may say that community systems form an existential foundation for the members. This is partially structured by the network of organizations with which it is overlaid. The distinction between the two types of systems corresponds to the private-public categories in sociology, and is more or less sharp, according to the type of organizational and political-ideological structure we are dealing with. It is least sharp in military and quasi-military organizations, monastery type religious organizations, and highly organized underworld systems. It is sharpest in the case of bureaucracies and business corporations that demand a routine nine-to-five role-fulfillment of their members and leave the details of their private lives untouched (except for determining the limits of their possibilities through the size of the paychecks). The distinction likewise varies with the world view and political ideology dominating a given society. Strongly religious societies tend to collapse the public and private spheres of existence by subordinating both to the dictates of their belief system; and societies dominated by the Communist ideology do this on the strength of their secular economic, social and political doctrines.

Notwithstanding pronounced variations in the degree of distinctness between organizations and communities, it is possible to keep them separate for purposes of analysis and to establish relations between them that hold true regardless of variations. Organizations impact on communities in multiple ways (even when the distinction between them is neat): they tend to determine the educational, social and cultural level of the community members and therewith their values and world views. There is a nontrivial correlation between the social role of individuals in organizations and their personal role within communities. This correlation was stronger in past centuries, when community-roles tended to predispose the availability of social roles, for example, through the class-structure. Being born into a certain class in the community would automatically entail the availability of corresponding roles in the organizational structures, with mobility restricted in both upward and downward directions. However, with the advent of democratic social organization, both on the Western and on the Socialist pattern, the converse influence began to manifest itself: organizational roles now tend to determine community standing, and allow the same degree of social mobility that is built into the organizational system. Although determination of public role by membership in social class is not extinct (it is especially prevalent in the third world), its scope is sufficiently reduced in our own societies to permit us to concentrate on the organizational structure as the main influence on community structure. In western democracies leaders in organizations tend to be leaders in communities, but persons born to leading families and classes in communities no longer have automatic access to positions of organizational leadership.

The point of delineating the general relationships between organizations and communities is to show that the adaptive steering of organizations has as its consequence the indirect steering of communities. Because the behavior of communities determines many of the vectors operating in the world system – such as population growth rates, the demand for energy, food, and drugs, the spread of mysticism and drop-out cults, etc. – it is imperative to steer the normally recalcitrant community systems.

The problem is complex and, I shall address some general features of it by outlining a three phase strategy which I have presented in detail elsewhere [17]. Briefly, the strategy involves using existing adaptive organizations to initiate a consensus-building information-flow which, in a subsequent phase, serves as the support for the creation and effective operation of a supranational functional actor entrusted with the task of steering the decisive aspects of the world system.

The first phases of information-flow are premised on the assumption that individuals can effectively contribute to general societal goals only if they have some knowledge of such goals as well as of the effect of their behavior with respect to achieving them. To this end it is necessary to communicate to the widest masses of people the dangers and the possibilities inherent in choosing alternative social futures. On the one hand they must realize that persistence in certain existing patterns of behavior can lead with a high degree of probability to societal breakdown and suffering on a historically unprecedented scale. On the other hand they must recognize that people collectively can choose more humanistic pathways of societal development suited to the new realities of technology, consumption, ecological balance, and population size. Information on this score constitutes what I termed "ecofeedback".

"Ecofeedback" (where the root "eco", derived from the Greek word *oikos* for house, suggesting that the feedback concerns man's terrestrial house or estate) must divulge to people the global consequences of their personal lifestyles in the areas of energy consumption, pollution, population growth, urbanization, product longevity, resource depletion, and environmental quality.

Existing organizations, especially political and public information institutions, could be adapted to produce a flow of information on the difference between the desirable and the current trends in the world system, and could make plain the particular contribution of individual lifestyles to reducing the gaps between them. To this end two sets of time-series analyses would have to be produced: one set to show the desirable curve of the relevant parameters, and the other the current curve. The discrepancies between them would have to be analyzed in readily understandable language, and traced to personal and organizational choices. For example, the difference between tolerable and intolerable pollution-levels may be generated by the decision of industrial managers to install sophisticated emission-control devices; by the actions of legislators in setting and enforcing adequate emission standards, and by the choices of individuals not to drive large private cars to work (as 65% of Americans still do) but use more efficient cars, motor pools, or public transportation, not to use polluting engines for their leisure-time activities but engage in nonpolluting diversions, and so on. If the effect of each major type of choice is made clear to those responsible for making them, an adaptive shift in values and world-views can be elicited. Few individuals persist in anti-social and anti-humanistic behaviors when they recognize them for what they are. Such modicum of public morality must in any case be premised if we are to grant that humanity could survive by its own morality rather than the brute enforcement of coercive policies.

Social steering could start with the systems most accessible to control, and proceed from there to progressively less steerable systems. The early phases of the strategy make use of the political and institutional subsystems to initiate an information flow designed to inform members of less steerable systems of the consequences of their actions with reference to the stated goal. The primary targets of the information feedback are nonpolitical organizations, such as national and multinational corporations, educational and scientific organizations and mental and physical health agencies. The secondary targets are community systems, in order to make citizens aware that personal lifestyle choices have determinable major impact on the future of their societies. Communities would be reached through the primary-target organizations, as the latter adapt their behaviors to reduce existing differentials between the desirable and the current time-series projections and thus also influence the community systems.

Cybernetic modeling could play a constructive role not only in modeling existing organizations preparatory to steering, but in *designing* special-purpose organizations. Functional organizations to correct social imbalances can of course be designed at any time, but they remain mere utopias in the absence of a consensus permitting their realization. The strategy I am outlining aims at bringing about a consensus among wide strata of the population in its early phases, and at establishing the necessary climate of opinion for the acceptance of concrete policy decisions in the later stage. If the first phases are successful, the specialized organizations which implement the decisions can be translated from the drawing boards of social cyberneticians to reality. In my view the greatest single need in this regard is for designing and implementing a supranational functional actor that could integrate the now fragmented functions of the many specialized United Nations agencies, and have sufficient power and authority to enforce its decisions *vis-à-vis* chauvinistic and short-sighted national governments. Such an actor would be explicitly oriented toward steering the world system by acting upon its national and regional subsystems. Its norms could be premised on the values of cultural and ideological diversity, and its leaders could perceive in the plurality of modes of social and political organization the safeguard which genetic diversity offers for biological development. The supranational functional actor could back up emerging consensus in national political and nonpolitical organizations and in communities, with concrete policies designed to steer each system in a manner conducive to achieving and sustaining equilibrium on the global level. I have described such an actor under the somewhat fanciful title, "The World Homeostat System" [17], and suggested that its operation would testify to the wisdom of world society, much as the homeostasis of the warm-blooded biological organism testifies, in Cannon's [4] words, to the wisdom of the body.

In the views I have presented here, the cybernetic modeling of social systems is placed within the context of a broad humanistic social-steering enterprise, aiming at the achievement of a stable and viable world system. Cybernetic modeling is not an end in itself, but an instrument. The immanent professional objective of cybernetic modeling remains the improvement of the accuracy and completeness of its models of organizations, and their operationalization. However, the transcendent, humanistic objective of cybernetic modeling is its instrumental use for steering our social systems to achieve and maintain stable goal-states in which they can persist without depressing either the quality of life of their members, or the quality of their social and natural environment.

REFERENCES

[1] Beer, S., *Cybernetics and Management,* New York: Wiley, 1959.

[2] Biderman, A. D., "The image of 'brainwashing'," *Public Opinion Quarterly.* **26:** 547–563, 1962.

[3] Buckley, W., *Sociology and Modern Systems Theory,* Englewood Cliffs, N.J.: Prentice-Hall, 1967.

[4] Cannon, W. B., *The Wisdom of the Body,* New York: W. W. Norton, 1932.

[5] Cumming, E. and Cumming J., *Closed Ranks – An Experiment in Mental Health Education,* Cambridge, Mass.: Harvard University Press, 1957.

[6] Deutsch, K. W., *The Nerves of Government,* New York: The Free Press, 1966.

[7] Easton, D., *A Systems Analysis of Political Life,* New York: Wiley, 1965.

[8] Etzioni, A., *Genetic Fix,* New York: MacMillan, 1973.

[9] Falk, R. A., *This Endangered Planet,* New York: Random House, 1971.

[10] Forrester, J. W., *Industrial Dynamics,* Cambridge, Mass.: Wright-Allen Press, 1961.

[11] Forrester, J. W., *Urban Dynamics,* Cambridge, Mass.: Wright-Allen Press, 1969.

[12] Forrester, J. W., *World Dynamics,* Cambridge, Mass.: Wright-Allen Press, 1971.

[13] Frankl, V. E., "Reductionism and nihilism", *in* Koestler, A. and Smythies, J. R. (Eds.), *Beyond Reductionism,* New York: MacMillan, 1969.

[14] Fuller, B., *Operating Manual for Spaceship Earth,* Carbondale, Ill.: University of Southern Illinois Press, 1970.

[15] Goldsmith, E., Allen, R., Allaby, M., Davoll, J., and Lawrence, S., "A blueprint for survival", *The Ecologist,* 2, No. 1, January 1972.

[16] Laszlo, E., *The Systems View of the World,* New York: Braziller, 1972.

[17] Laszlo, E., *A Strategy for the Future: The Systems Approach to World Order.* New York: Braziller, 1974.

[18] Marx, K., "Theses on Feuerbach, appendix", *in* Marx, K. and Engels, F., *Ludwig Feuerbach and The Outcome of the Classical German Philosophy,* Dutt, C. P. (Ed.), New York: International Publisher, 1941.

[19] McLuhan, M., *The Gutenberg Galaxy,* Toronto and New York: Signet New American Library, 1969.

[20] Meadows, Donella H., Meadows, D. L., Randers, J. and Behrens, W. W. III, *The Limits to Growth,* New York: Universe Books, 1972.

[21] Mesarovic, M. D., *Theory of Hierarchical Multilevel Systems,* New York: Academic Press, 1970.

[22] Pattee, H. H. (Ed.), *Hierarchy Theory: The Challenge of Complex Systems,* New York: Braziller, 1973.

[23] Wiener, N., *Cybernetics: or Control and Communication in the Animal and the Machine*. 2nd. ed., New York: M.I.T. Press, 1961.

[24] Whyte, L. L., "Structural hierarchies: a challenging class of physical and biological problems", *in* Whyte, L. L., Wilson, A. G., and Wilson, D. (Eds.) *Hierarchical Structures,* New York: American Elsevier, 1969.

A SYSTEMS FRAMEWORK FOR SOCIAL CONTROL PROCESSES[1]

PREMYSL PERGLER
Ottawa, Canada

and

WALTER BUCKLEY
University of New Hampshire, Durham

SOCIAL MORPHOGENESIS

Recent efforts in developing system models of social control processes has brought home clearly both the promise of such an approach and the many difficulties in applying it to the empirical world. It may be that the "promise" is generated primarily by a feeling of the dire necessity of providing some beginning to more systematic attacks on the understanding and improvement of current societal difficulties. The complexity is discouraging, but any such effort must be evaluated, not against some ideal theory and methodology, but against the improvised, only vaguely conceived, models that underlie current societal policy decision-making, problem solving, and social system regulation generally. From this point of view, almost any serious modeling must be better than what is now applied.

Rather than discuss the promises and problems directly, we will sketch in non-technical terms some basic aspects of our approach and then provide some idea of the semi-formal systems methodology underlying it, in the hope that you will see some promise in it as well as the obvious difficulties.

Let's start with the view that social systems, as continuations of biological systems, have evolved over time, and – like biological systems – their self-regulating structures and processes have evolved, with varying degrees of success in maintaining the viability of the system. Although small, primitive societies might be seen as quite effective self-regulating systems, it can be argued that large-scale societies, especially our industrialized-urbanized complex societies, are only just beginning to evolve the self-regulating structures and processes demanded for longer-run viability in the face of the internal system complexities and external environmental challenges. From a cybernetics point of view, modern societies are rather poor self-regulators, especially in terms of longer run goals and global viability. The "muddling through" approach of many complex societies has led to outcomes neither intended nor desired by most member individuals or groups: urban blight, squandered resources, plundered environment, runaway technology, moral decay of political and economic institutions, paranoid bellicose orientations toward other nations, etc.

Given the problems of regulation in sociocultural systems, we can argue, however, that the historical shift from monarchical or oligarchical types of control or regulating system to the more or less democratic types can be seen as an important step in the evolution of a better self-regulating social system (in a sense to be defined), though it introduces new system problems of its own. (For example, in a democratic system there is potentially fuller information flow, both about internal states of the system and about external disturbances. Also there are potentially more levels or meta-levels of regulation, down to the level of the individual and subgroup, with greater autonomy at lower hierarchical levels and fuller feedback between levels in each direction. As Herbert Simon and others have argued, the more complex the system, the greater the number of hierarchical levels of subsystem it tends to develop, as if complexity at any one level can only become so great if viability is to be promoted. Also, decision-making in more democratic systems has a greater breadth of knowledge and strategy alternatives to draw on. And so forth.)

It is next argued that two of the essential regulative processes in living systems are morphostasis (MS) and morphogenesis (MG). If homeostasis is defined as the regulation of critical system variables to within certain limits, morphostasis – as implied literally – may refer to regulation of the structure – especially the control structure – of the system to within certain limits. In other words, MS refers to processes tending to prevent significant changes in the organization of the system, or the *status quo*, if you wish. Any system must maintain the integrity of its structure with some minimal time invariance, depending on the challenges of the environment.

But given severe enough challenges of the external or internal environment, the viability of a system depends on its ability to change its operating structure – to adapt so as to improve goal attainment or regulate against "disturbances". (This is comparable to the process of species evolution.) We refer to this as morphogenesis, again using the word in its more literal sense.

But now we must make a distinction between at least two types of MG. Throughout the history and prehistory of sociocultural systems we take it that most of the significant structural changes, i.e. MG, have taken place along with a high level of destructive conflict: civil war, revolution, conquest. We will call this "morphogenesis via destructive conflict", or MG_1. This is a high risk type of structural change, increasingly so as social systems became more complex and more powerful.

In the past (and especially in recent historical times) a different kind of institutionalized regulation of the structure has begun to emerge promoting structural changes based on a competition of ideas and joint decision-making. We may call this MG_2. In this type of system, institutional principles incorporate *explicit procedures* for self-modification (e.g., amendments) and hence for change of the structure of the system based on such principles. Such a system, however, still has a long way to go to evolve into its full MG_2 potential.

We may hypothesize that as larger-scale, more complex socio-culture systems evolved, morphostatic mechanism became more thoroughly institutionalized in them compared to MG_2 mechanisms. (This could be argued in terms of the (non-equilibrial) thermodynamics of sociocultural systems: the greater input of information or negentropy to generate the meta-levels of inter-personal coupling and societal integration required for MG_2 regulation.) Modern societies are thus seriously unbalanced in terms of MS *vs* MG_2 processes. The result is to drive the system toward more and more crises and the generation of destructive MG_2 processes.

Part of the research goal is to investigate the additional institutionalized structures and processes needed to generate morphogenesis (MG_2), and make it compatible with the morphostasis. This involves, among other things, (1) more explicit and effective procedures for defining and operationalizing collective goals; (2) procedures for adequately defining the state of the system at any time and its probable trajectories; (3) organizations for generating scientifically informed alternative policy strategies, effective decision-making procedures, and implementation of the chosen strategy; (4) effective citizen participation; (5) better mechanisms for feedback of goal-deviating information. All of this implies the development of much better models of the structure and operation of complex social systems; contemporary social science has hardly made a dent in this, and research support at least the size of that for high-energy physics is demanded.

THE FRAMEWORK

The foregoing is a very informal sketch of the main thrust of our work. During the analysis (and during preparations for practical application) it was found necessary to specify relatively precisely the framework being used. Most of the specification is performed on the conceptual epistemological level with extensions to the mathematical and plain-language levels. To keep the framework both internally consistent and applicable, it was found useful to partially redefine some concepts. The framework (and its comparison with other frameworks) presently exists in the form of working papers. Some of you may be interested in it, and we would certainly welcome critical comment. For pedagogical and epistemological reasons, the starting point is the development of the basic set of operationalized concepts, the more complex ones being defined in terms of the simpler ones. A more or less standard input-output block symbolism is used. The basic unit is the "element", defined as a model of an object of the external world (or occasionally of a process or of a relation). The element has one or more input and output variables, and a "characteristic", or transfer function, defining the relationship between the two – whether continuous or discrete, deterministic or probabilistic. In social systems, most characteristics are probabilistic. In our framework they are defined (both on the mathematical and conceptual epistemological level) as associations in the sense used information theory, i.e. with the condition that the knowledge of the input decreases the uncertainty of the estimate of the output. "Causation" or causal relation is an association of variables whereby a change of the input variable is followed by a change of the output, the change being described again by the characteristic of the element. "Control" is defined using two causal elements; the input to one, the "controlling element", is a "goal" input. Its output is a connecting variable which is identical to the input of the second element, the "controlled element", and constitutes a "structural linkage" between them. If the output variable shows an association with or a matching of the goal input when the structural linkage is intact but fails to do so when the latter is broken, we speak of a "control channel".

A "regulation" system is a control system in which the goal can be achieved by different trajectories of the connecting variables, in response to "disturbance" inputs to the system. In "feedback regulation", disturbances are inputs into the controlled element and some of the output is fed back into the control element. In "regulation via measurement of the disturbance", disturbances are inputs into the control element directly. In both cases the quality of regulation is measured by the degree of association between the output and the goal input.

A "system" is thus a model consisting of a universe of elements with their characteristics and inputs and outputs, and a structure consisting of the structural linkages as defined above.

"Adaptation" of a system becomes a change in the overall characteristic of a regulator that increases the quality of regulation of the system compared to that without the change. If the regulator is decomposed into its system elements and structure, adaptation may be found to involve a change in (a) the characteristics of one or more of the elements, and/or (b) the structure of the system, and/or (c) the universe of elements of the system.

"Homeostasis" is defined as the regulation of critical conditions necessary to the continued functioning of the system. "Morphostasis" refers to homeostatic regulation in which the very structure of the system itself is taken as the goal of regulation. That is, the structure tends to be maintained unchanged despite forces or "disturbances" tending to change it.

Finally, "morphogenesis" is defined as adaptive change of the structure of the system; that is, a change of structure that results in better goal attainment or regulation of the system.

In applying such a framework to social control or regulatory processes, the models are decomposed into generalized "decision-making subsystems". Our decision-making subsystem (or decision channel) is intended as a model of a person or more usually a social organization of persons. It incorporates the several stages in the decision process accepted more or less in modern decision theory. The input to the decision channel is sensory data from the external world, especially linguistic messages. The first element in the channel selects certain of these data to attend to and outputs these to an element that selects a problem requiring decision. (All of these elements are interconnected with an accumulated "knowledge" element and an element representing the goals and values (of the modeled individual or group). The third element selects a set of relevant strategies and a fourth predicts the probable outcomes of each. The fifth evaluates the outcomes in terms of the decision-maker's goal-space, the sixth selects the resulting best strategy and the last selects the initial action for implementing that strategy. The output of the decision-channel is thus that initial action.

The mentioned model of the decision channel was used for two cases of analysis. The first case was decision-making of a person. The analysis was oriented on (a) role of knowledge and of uncertainty, (b) goals and values, and (c) decisions about the society. The partial results have been applied in the second case, more closely related to application.

APPLICATION

The discussion of several alternative processes of morphogenesis, together with the discussion of the decision-making process of a person in the society, led to identification of one channel, which seems to contribute relatively heavily to difficulties of the morphogenesis (i.e. adaptation of the structure) in the contemporary complex society. This channel corresponds to integration of policies, as a component of the process of elaboration of policies. From a comparison of a "cynical" and "naive" versions of models of this channel, and from experiences with policy elaboration in selected concrete cases, a method for policy integration has been developed, that uses the system approach as its general background. The method is being prepared for two possible concrete applications. It is the opinion of the authors, that concurrent work on (a) analysis of selected problems, (b) conceptual framework and (c) concrete application, brings some benefits that may compensate the considerable difficulties of this approach.

The work is supposed to continue. It might be useful to develop alternative models of contemporary American society according to a "conservative" version, a "liberal version", or a "radical version", depending on how those theories visualize the structure of the regulatory system, the universe of elements, and their characteristics. In this way we may promote a constructive scientific critique of various models of actual society and of models of how it might be improved as a viable system. It is our opinion that, especially in the context of the current eco-system movement, there is already a good deal of activity moving in this direction The work outlined here might be seen as a modest exploratory step in the direction of the development of a systems methodology for promoting this movement. Science itself is a morphogenic system and thrives on constructive critique.

ENDNOTE

[1] In order to accommodate the time constraints and to fit into the general direction of the conference, the text of this paper was not read. Rather the presentation concentrated on problems of concrete application of the described framework for policy elaboration and on some related epistomological questions.

DEGREES OF FREEDOM IN SOCIAL INTERACTIONS

WILLIAM T. POWERS
1138 Whitfield Road, Northbrook, Illinois

INTRODUCTION

Freedom is well-known to be a relative term; one is not free to speak lies to a Grand Jury; one is not free to worship a god that demands human sacrifices; one is not free to publish facsimilies of U.S. currency; and one is not free to choose his domicile from those owned and occupied by others. We are all familiar with the fact that there are different degrees of freedom, none of them absolute.

The term "degrees of freedom" has another meaning, which turns out also to be relevant in this discussion. A physical system is said to have *n* degrees of freedom if *n* variables have to be given specific values in order to describe completely the state of the system at a given moment. An object in space has at least six degrees of freedom: three which relate to its location in a three-dimensional coordinate system, and three which relate to its orientation relative to the directions of the coordinate axes. If a system has *n* degrees of freedom, and all but one of them are specified by being given numerical values, there is only one way left in which the system can change. If the location of an object on a flat surface is specified in terms of an X–Y coordinate system, the only remaining way the object can move is up and down. If the X-position is specified and the height above the plane is specified, the only remaining degree of freedom is in the Y-direction.

In a model of a behaving organism, at least as I have approached the problem of modeling, these two senses of "degrees of freedom" turn out to be nearly the same in meaning. We are not absolutely free to indulge in certain behaviors because, I propose, neither we nor the environments with which we effectively deal possesses an infinite number of degrees of freedom in the mathematical-physical sense. The limitation on freedom of behavior due to limits on degrees of freedom of organization can be seen in an individual, and in a society made up of individuals who interact with one another. As will be shown, mere masses of people do not create correspondingly large numbers of degrees of freedom in any sense; in fact, quite the opposite can occur.

CONTROL HIERARCHIES

To begin at the beginning, let us consider a model of behavioral organization derived from Norbert Wiener's original insights [8], the later contributions of W. Ross Ashby [1], a seven-year collaboration with R. K. Clark [7], and the general literature of control theory and early engineering psychology. From these sources, and no doubt many others equally deserving of mention but too diffuse to recover, a picture has emerged of organisms as hierarchies of negative feedback control systems. This model is compatible with the thrust of Donald T. Campbell's long-term development of blind variation and selective retention, which has lately come close to a general control model of evolutionary and learning processes [3], but we will not here consider the origins of the organizations involved in fully-developed adult behavior. The main task here is to develop some general concepts which emerge from the control model, and see what implications they may have with regard to individual organization, and then social organizations.

A control system is defined in this model as any system capable of stabilizing against disturbance an inner representation of an external state of affairs. For all but practicing control engineers, that definition will require some elaboration.

First of all, what does it mean to say that something is stabilized against disturbance? This can only mean that this something, which we will call a *controlled quantity*, is affected both by independent influences and by the actions of the system itself, and that the system's actions systematically oppose the effects of disturbances on the controlled quantity. It was discovered, early in the game of control theory, that if a system is to stabilize some quantity it must sense that quantity, and furthermore it must have an internal standard against which to compare the outcome of that sensing process – a reference with respect to which the sensed quantity can be judged as too little, just right, or too much. The action of the system is based on that judgement, not on the sensed quantity itself nor on the reference itself nor on the disturbance. Departures of the controlled quantity from the reference level are what lead to the actions that limit those departures to a small or even negligible size.

Of especial interest to a control theoretician's thinking about behavior are some new interpretations of Piaget's work with children; von Glasersfeld [4] in particular seems to have found in Piaget an epistemological position that recognizes what I term "internal representation of an external state of affairs" as a central factor in behavioral organization, and furthermore recognizes that these internal representations, which we often term perceptions, are in all likelihood *arbitrary* transformations of the external state of affairs – "reality". In a control model of behavior this epistemological position thrusts itself upon one; there seems to be no other choice. A control system controls what it senses, and what it senses is the result of applying a continuous transformation process to the elementary sensory inputs to the nervous system.

One simple way to put this is to say that a control system controls some particular *aspect* of its environment. How many aspects might there be of a given environment? As many aspects as there are different ways of combining elementary sensory stimuli, and that is a very large number.

It is not, however, an infinite number. In fact, for a single organism, it is probably not even in the realm of what we call today large numbers. The number of degrees of freedom in the perceptual world is limited, of course, by the number of degrees of freedom in the physical universe outside, but it is much more severely limited than that: it is limited by the number of different aspects of the environment that a given organism is prepared to sense at a given time.

That is getting ahead of the story. At the moment, the important notion is that each control system, or subsystem within the organism, *constructs* a quantity by means of sensors and perceptual computations, and acts to stabilize that quantity, a perception. It acts to maintain that quantity matching an internal reference. If the reference remains constant, actions will simply oppose the effects of disturbances in the controlled quantity. If the reference varies, actions will automatically undergo further changes which cause the perception to *track* the changing reference. Control, therefore, may involve maintaining a static or a dynamic condition, protecting it against arbitrary influences from outside.

Already we have a picture of behavior that is considerably different from the traditional one. Instead of being caused to act by external stimuli, the organism under this control model acts purposively, the purpose being to maintain some perception, some aspect of the sensed environment, at a reference level specified inside the organism. There is much more to say about the model itself [6], but for the present we will simply adopt it, and elaborate on it to make it serve the purpose at hand.

A hierarchical control model can be constructed. It is possible to define a set of first-order control systems, corresponding primarily to spinal reflexes, which control very elementary aspects of the environment and musculature. There are probably hundreds of such systems, all at the same level. Each such system requires a signal from further inside the organism which will set its reference level – tell it how much of its perception to produce and maintain. The controlled perception, together with many other perceptions, controlled and uncontrolled, is available to systems further inside the organism. Thus the systems further inside have second-order sensory information about the external state of affairs, and can act to change those second-order perceptions by adjusting the reference-levels of the first-order systems which actually create muscle forces in the process of their own control actions.

In this way a hierarchy is built up, a system at a given level acting by adjusting the reference-levels for lower-level systems, and acting for the purpose of stabilizing its own representation of the external state of affairs.

Now we can begin to see how the degrees of freedom of action might be constrained to a relatively small number at any given time. By "action", we must mean in general not motor behaviors, but selection or specification of reference-levels for control systems. If there are n independent control systems operating at a given level in this hierarchy, then there are just n reference-levels that can be independently specified – at most n, assuming no limitations introduced at lower levels.

It is not necessary to think of one lower-level system having its reference specified by just one higher-level system. Much more generally, and probably more realistically, we can think of the set of all reference-inputs to lower-level systems as defining the degrees of freedom for action, and we can let any higher-order system act by contributing to the reference-settings of many lower-order systems at once. In this way we could accommodate n higher-order systems acting at once on a large subset of m lower-order systems, any one lower-order system receiving a net reference-setting that is the sum of influences from many – perhaps even all – of the systems in the level above. As long as the n higher-level systems distribute their contributions to the lower-level reference settings properly, no conflicts will be created (no paradoxes), and each of the n higher-order systems can act independently to control its own perception.

The crucial word is "properly". The proper distribution of effects can be found only by solving a set of simultaneous equations, and if there are n systems cooperatively making use of n control subsystems, there is one and only one possible solution to the system of equations.

When control systems are involved in situations like this the problem is in some regards not so difficult to solve as it would be if all the systems operated blindly. In fact, the requirements on the distribution of actions by higher on lower systems are much relaxed; some considerable degree of interaction is permitted, the actions of one system disturbing control processes in other systems of the same level. The ability of control systems to maintain control in the presence of disturbances also permits them to work when coupled together – not too tightly – at their outputs.

But the restrictions are still severe enough. When control is involved, the requirement shifts to the *input* side; now it is necessary that each higher-level system sense a different aspect of the lower-environment (consisting of lower-order perceptual representations). By different aspect we mean, of course, an independent aspect; one which could freely be varied no matter what states were being maintained for other aspects of the same environment. So now we are talking about degrees of freedom of perception instead of action, but the basic requirement has not been fundamentally changed: the set of simultaneous equations must still, somehow, be solved, if the n higher-order systems are all to be capable of independent action.

Perceptions are neither entirely consistent nor entirely free of random variations; furthermore, the human brain does not become organized through systematic solution of equations, but through variations (blind or otherwise) and retention of resulting organizations that work to the organism's advantage. It is thus highly unlikely that any hierarchically-organized organism would prove able to employ all n degrees of freedom inherent in an assemblage of n control systems at the same level, even assuming that levels are neatly separated in nature. In order for successful control organizations to emerge, it is most likely the case that the number of available degrees of freedom must be far larger than the number ever simultaneously employed.

The basic reason why these "equations" have to be "solved" is that control organizations cannot work under conditions of direct conflict; the better the quality of control, the worse the consequences of conflict. Failure to maintain independence of the aspects of the environment under control shows up as an attempt by two systems to control the same quantity relative to two different references, which is impossible. Failure to solve the equations thus means loss of control. Loss of control means lowering of survival potential, so we can expect that evolutionary processes will have selected against significant amounts of internal conflict.

Now we have a general picture of a hierarchically-organized organism based on control principles, and we can see that at each level in the hierarchy there is a problem of degrees of freedom any time that many higher-order systems act simultaneously on and through the same set of lower-order systems. The problem is simply that of avoiding internal conflict and losing control altogether. We can now put this aspect of "degrees of freedom" together with the other aspect, the idea that freedom is never absolute, in the less technical usage of the term. Then we will be ready to look at social systems.

RELATIVE FREEDOM AND RELATIVE PURPOSE

When it is said that we are not free to shout "Fire!" in a crowded theater, there is an unspoken assumption in the background. We are not free to do this *if* we are aware of and reject the alternatives: causing a fatal panic or going to jail. Of course if we are willing to accept the consequences, or are unaware of any, we are perfectly free to shout what we like where we like. It is not any law of nature that limits our freedom in such circumstances; the only limit is imposed by the organization of the person in question.

More specifically, the limitation of freedom is imposed by the fact that doing certain acts that satisfy one set of goals or purposes (or in the more noncommittal language of control theory, reference-levels) can cause other controlled perceptions to depart from their reference levels. Conflict can be created, depending on one's structure of perceptions and the reference-levels that go with them. If a person wishes strongly to avoid going to jail and also wishes strongly to shout "Fire!" in a crowded theater, he has a problem. It is not a problem of sequencing; it is a direct conflict, in that satisfying both reference-levels (matching present-time perceptions to them) is essentially impossible. The impossibility is what limits freedom.

Suppose that someone possesses n independent control systems at some level in his hierarchy. This means that in principle, this person could set reference-levels for each of the n perceptions in any way he liked. But now suppose that he has, for whatever reasons, already specified $n - 1$ reference levels, thus implying that $n - 1$ aspects of the lower-order environment (seen from this level) are being controlled in specific states. How much freedom is left for setting the remaining reference level?

From one point of view the remaining goal could be chosen with complete independence; the system setting that goal is not constrained. But there is only one value that can be specified for the remaining reference level that will not create direct conflict. Direct conflict will destroy the ability to control, or severely impair the control abilities of all the systems involved. This will not directly affect the system that erroneously assumed freedom to set the remaining reference level, but there will be a definite indirect effect, if we are talking about a hierarchical control organization. The higher-level systems set lower-level references as their means of control; if the lower systems come into conflict, the result will be an impairment of higher-order control processes. Thus there is a purely *mathematical* limitation on the freedom of the whole system to set simultaneous reference-levels at a given level in the hierarchy. Attempting to violate that limitation destroys control.

Reference-levels constitute intended perceptions: purposes. In a hierarchical system, any level of purpose is at the same time a means for achieving higher-level purposes. The runner adjusts the speed with which he intends to run as a means for controlling the place in which he will finish the race; if it is impossible to come in first, a different choice of running-speeds may make second place possible, and a poor choice will assure not finishing the race at all. Purposes generally have this quantitative aspect in addition to their qualitative natures. Clearly, one is not free to select purposes at a given level if those purposes are also means for achieving higher-order purposes.

In fact, when there are higher-order purposes, external circumstances can be as important a determinant of selection of purposes as any internal determinism at higher levels. If one intends to enter a house, the selection of lower-order controlled perceptions that is made will be strongly influenced by the state of the front door: open or closed, locked or unlocked. A successful control system adjusts its actions to oppose external circumstances that would tend to disturb the controlled quantity – which means that it must allow external circumstances, to a great degree, to control its actions, its selection of lower-level goals.

When freedom and purpose are examined in the light of this hierarchical control model, we can see that no simple slogan can unravel the complexities that come up. "Free will" is a phrase that would be used only by someone who has not really thought about the whole problem. Freedom is relative; sometimes it is impossible. Our chief freedom, it seems, may be the freedom to seek the state in which we suffer the least internal conflict, and thus remain capable of acting on the environment in the way that lets us continue functioning according to our own inner requirements, whatever the basic requirements may be. I would not rule out of this set of basic requirements, by the way, concepts such as orderliness, beauty, elegance, or progress. We have not yet read the entire message in the genes, nor are we in a position to put limits on what it might imply.

FREEDOM IN SOCIAL INTERACTIONS

We have been discussing so far only an individual person. The principles developed, however, should apply at least in part to a social organization, because we have dealt with the individual as a collection of control systems that must work in harmony in order for the whole to function properly. This certainly suggests a parallel treatment of a collection of individuals, each of whom acts as a control system in any given circumstance, and all of whom must learn to live in harmony to avoid the dangers of social conflict.

Before developing these parallels, I want to warn against one tempting extension of the control model. In a human being, it is possible to identify each part of a control system with some part of the person – his sensory equipment, cerebral processes, motor organization. Every function that has to be performed to make up a control system has a place to live, a place where it can be embodied in tissue. When the same organization is extended to a social system, such as an army, one can see many counterparts, but on close inspection these counterparts to the components of a control system are no more than metaphors. An army does not have a perceptual organization, a way of making comparisons, a motor output system. Only the people in the army have such things. An attempt to transfer a control model of an individual directly to a social organization violates the kind of model-making spirit that demands a relationship to the physical world. The kind of model that is needed to represent the interactions among the individuals who make up an army may prove to include feedback phenomena and many other features of a control model – but an army will never be a control system in the same sense that an individual is.

What can we say about social interactions without depending on loose metaphors? And also, I should ask, without having done the many lines of research implied by this approach? Whatever is said will obviously have to be general and tentative, but there are some useful implications to consider.

The existence of conflict between individuals is of theoretical interest if only because of epistemological implications. Although each person might be acting primarily to control aspects of a constructed perceptual reality, the fact that interpersonal conflict can exist and persist indicates that there are regular objective consequences of control behavior. Even though the world of controlled quantities is primarily subjective, it seems that control actions entail producing regular effects in the outside world, regular enough that the control behaviors of two persons may prove mutually exclusive. For one person to maintain control of his perceptions is for another to lose control of his. That is the essence of interpersonal conflict.

We live in a society in which competition is praised as a spur to greater efforts and a higher quality of production. The free enterprise system in principle permits each person to look after his own interests, advancing himself relative to others by increasing the value of his work to others. Again in principle this process should lead to the state in which each person has found a niche which maximizes the benefit to himself while at the same time maximizing his contribution to the well-being of all other persons. Our educational procedures, through competitive sports, grading on a competitive and comparative basis, and at the higher levels acceptance into schools on the basis of relative accomplishments, emphasize and nurture this competitive concept of social evolution.

When sufficient degrees of freedom exist, this design for a social system would seem quite feasible. It is, in fact, a heuristic that leads to the minimization rather than the emphasis of interpersonal conflict. If the system really worked as it is imagined by its proponents to work, *direct* conflict would result in a winner and a loser, and the loser would turn to some other endeavor. Eventually each person would find a position in the society in which his own structure of reference-levels could be satisfied without excessive effort; since the principal cause of excessive effort is direct conflict with other controlling organisms, the result would be a minimization of direct conflict, and in effect a solution to the system of equations describing each organism's requirements and the set of all available means for meeting those requirements. Thus the result of optimal functioning of a society ostensibly based on the principle of free competition should be the reduction of direct competition to the minimum that is actually achievable [Kuhn, 5: 216 ff].

There are at least two important factors which prevent this system from operating to reach a low level of conflict. The first is the fact that a person in a society can interact with the society only as he understands it, not as it actually is. It is not generally understood that the final outcome of free competition ought to be a minimization of competition. Instead, the way we train ourselves to live with competition has resulted in a glorification of conflict itself. Rather than trying to maximize the useful product of our labors, through reduction of conflicts which produce mutually-opposing efforts, we have come to make the opposition of one person by another into a source of reward and prestige. The achievement of such rewards and such prestige becomes a goal accepted as important by large numbers of people, even though few of them have any realistic hope of joining the small circle of winners. Entering into conflicts is said to strengthen us, although in most such cases, such as the example of professional football, it would be hard to say how the resulting kind of strength would be of use to us except in the conflict situation itself. In a society that accepts conflicts as something to be sought, little value is placed on the cleverness with which we find social solutions that avoid direct conflict. Each person's understanding of the nature of his society and the goals implicit in that society organizes all his lower-level purposes. If those purposes do not include the minimization of direct conflict, it becomes rather unlikely that we will find a peaceable solution to our social problems.

The fact that our society apparently seeks conflict situations is not, in the long run, the most important impediment to getting rid of debilitating and dangerous conflicts. Whatever theoretical notions a person may have about his society, he is not going to continue entering into conflicts deliberately if the result is consistently against his interests. There will be at least a strong bias toward avoiding conflicts in which one runs a high risk of emerging the loser. The older one gets, the more evident it becomes that the thrill of victory is ephemeral; while the agony and consequences of defeat are cumulative. One cannot afford to go on forever using his efforts merely to oppose the efforts of others; at least a major part of one's effort ought to be directed toward assuring one's own survival. One cannot take care of himself or his family if the major part of his effort is cancelled out by the efforts of others trying to counteract his disturbances. Sooner or later, one looks for the path of least, not most, resistance.

Then the question becomes: does such a path exist? The answer depends on degrees of freedom.

The number of degrees of freedom in the physical environment is, according to what physics would say, inexhaustible. But it is not a physical model of the environment with which we normally interact and within which we choose our purposes. It is a perceptual model within which we find our goals. The worlds we attempt to control relative to our goals, and the goals themselves, are made up of automobiles and hamburgers, jobs and vacations, bowling and cross-country skiing, passing algebra and plying ladies with gifts in the effort to overcome resistance. It is almost entirely a manufactured world, a world divided into familiar perceptual categories and familiar examples of each category. It is a rather small world, the smaller to the extent that we come to share more and more classes and examples of perceptions, rather than creating our own categories and examples.

This is the other side of civilization. On the one hand, by banding together and pooling our efforts, we can achieve for all of us what none could achieve for himself. On the other hand, by banding together and creating a shared reality, we reduce the size of the universe in which we live, narrowing the choices of goals and the actions recognized as means toward goal achievement. The more of us there are, and the more closely-knit the society we perceive and accept, the fewer become the unused degrees of freedom and the higher becomes the liklihood of direct conflict. The final result can only be a society in which for each person there is one and only one conflict-free set of goals possible, at every level of his organization. All freedom of choice vanishes.

Within an individual, as already mentioned, the hierarchy of systems comes into being through variations and retention of profitable results, not through systematic and efficient design processes. The same holds true for a society. And just as inside one person, in a social system it is not feasible to match the number of personal goals to the available degrees of freedom. The world in which we live – the effective world, the one we perceive – must have far more degrees of freedom than we have goals, if we are to hope to approach something like a minimum level of conflict. If all degrees of freedom but one were exhausted, it is not likely that the remaining one would ever be discovered. Our goal structures must be such that there are many actions that would serve to satisfy any given goal; the richer the store of alternatives, the more likely we are to be able to minimize conflict and maintain control.

The implications are obvious. The more standardized a society becomes, the fewer become the individual goals and the means for achieving them. The more people there are, the fewer degrees of freedom remain. Long before actual exhaustion of degrees of freedom occurs, the level of conflict within a growing and increasingly standardized society must begin a rapid ascent. Failure of an individual to find a unique set of goals and an unopposed means to achieve them forces that individual to compete with others for means and to select goals which can only be met if someone else fails to maintain control. Finding the unique set becomes difficult long before the last degree of freedom is used up. Overpopulation and overstandardization begin to have their effects long before they are recognized as such. The symptom is not any dramatic confrontation among individuals. It is simply an increasing amount of difficulty experienced by everyone in going about his affairs. If too many people decide to take up macramé, each person will find his local store low on the supplies he needs. It is as simple as that. We begin stumbling over each others' feet long before we realize that there are too many of us in the same place trying to do the same things.

CONCLUSIONS

Control theory throws a new light on the subject of conflict, whether it be interpersonal or intrapersonal. When two independent control organizations come into conflict, the result is not simply a vector addition of the efforts. It is an abrupt increase in the efforts, most of the increase of one system's efforts serving only to cancel the increase in the other system's efforts, and producing no useful result for either system (except, perhaps, an increase in the volume of muscle tissue). While it is true that in the process of resolving conflicts the participants may develop new abilities, the glorification of conflict will not tend to develop abilities that are of general usefulness when there is no conflict. It will result only in hypertrophy of some function out of all proportion to the others. The glorification of conflict results in the Muscle Beach syndrome. The abilities developed through prolonged conflict generally go to waste unless another similar conflict can be found. Thus conflict as a goal will not stand up to analysis in terms of a hierarchy of goals.

Once the peculiar disadvantages of conflict between control systems are grasped, it is seen that conflict is the key to understanding many social problems. This conclusion is, of course, quite in line with common sense and experience, and by itself is nothing earth-shaking. But control theory shows us in great detail just *why* conflict has bad effects, and it leads us to see a relationship among conflict, overstandardization, and overpopulation, a relationship that has long been intuitively obvious but which now assumes the proportions of a natural law. Whatever our beliefs about the benefits of conflict in driving people to greater achievements, control theory makes it clear that conflict itself cannot be good for us, any more than breaking a leg is good for us just because it exercises our self-repair machinery. It is true that in the course of trying to resolve conflicts, trying to find a solution to the equations of life, we often come across new skills and knowledge that remain of permanent value when conflict is removed. But we should not forget that there are many ways to accomplish any given purpose, in a rich enough and sparsely-enough populated environment, and our goal is to be able to accomplish our purposes, not deliberately to seek impediments.

Through an understanding of social systems in terms of individual control systems acting independently of each other, I think we can arrive eventually at some clear statements about what is going wrong in the world we share, and perhaps even begin to see a way out of some major problems. A few cherished beliefs may become casualties, but if we can come to understand the real reasons behind the increasing tension and violence of our world, and to see that the main problems arise from attempts to violate immutable laws of nature and logic, we should not find incorrect beliefs too hard to abandon.

REFERENCES

[1] Ashby, W. R., *Design for a Brain.* New York: Wiley, 1952.

[2] Campbell, D. T., "Social attitudes and other acquired behavioral dispositions", in, Koch, S. (Ed.), *Psychology: A Study of a Science,* Volume 6, *Investigations of Man as*

Socius: 74–171, New York: McGraw-Hill, 1963.

[3] Campbell, D. T., "Downward causation in hierarchical selective-retention and/or feedback systems", Paper presented at the Symposium on Cybernetics in Psychology, American Psychological Association Convention, Washington, D.C., September, 1976.

[4] Glaserfeld, E. von, "Piaget and the radical constructivist epistemology", in Smock, C. D. and Glasersfeld, E. von (Eds), *Epistemology and Education:* 1–26, Athens, Ga: Mathemagenic Activities Program – Follow Through; Univ. of Georgia, 1974.

[5] Kuhn, A., *The Logic of Social Systems;* San Francisco: Jossey-Bass, 1974.

[6] Powers, W. T., *Behavior: The Control of Perception,* Chicago: Aldine, 1973.

[7] Powers, W. T., Clark, R. K. and McFarland, R. L., "A general feedback theory of human behavior", Part I, *Perceptual and Motor Skills,* Monograph Supplement, 7, No. 1: 1960; Part II, *Perceptual and Motor Skills,* Monograph Supplement, 7, No. 1: 309–323, 1960.

[8] Wiener, N., *Cybernetics,* New York: Wiley, 1948.

INFORMATION AS A SYSTEM–RELATIVE CONCEPT

ANATOL W. HOLT
Massachusetts Computer Associates Inc.

Nothing is more crucial to modern technology than information processing – and nothing is more crucial to modern society than its technology. Enormous industries producing a vast array of products and services are devoted to the storage, transmission, transformation and retrieval of information – publishing, broadcasting, switched communication nets such as the telephone system, and everything connected with computing, to name some of the outstanding examples.

In spite of the technological significance of information relative to the just mentioned industries and the social, political and legal significance of it to information-oriented societies, there are as yet no powerful theories in which a technical concept of *information* is developed and applied to the solution of practical problems involving information. The sort of theory that we do not possess for information is the sort that we *do* possess – thanks to physics – for the analysis of problems related to energy. Where information is concerned, one and only one technically serious theoretical approach exists, namely, the one of Claude Shannon published in 1949 under the ambitious title "*The Mathematical Theory of Communication*" [2]. While the work was brilliant, the title was premature. To be of help with cryptography problems or with making effective use of a telephone cable is indeed an important contribution, and it should give one faith that a mathematical theory of communication (and hence, of information) can and will be born. But a token from heaven that the Messiah is coming should not be confused with his arrival. We will return to the subject of Shannon later. Let us consider what the proper object of a *mathematical* theory of communication might be.

To begin with, it is more likely to succeed if it restricts its attention to communication in the context of the practical affairs of life – not, in other words, communication between lovers, or communication between artists and their audiences. A further useful restriction is the following: *Communication, where it is clear in what organizational capacities, or roles, the communicators influence one anothers activities by communicating.*

One could express this last restriction in the following shortened form;

communication as an expression of organization since, what an organization does, when once instituted, is to bring about a regimen of communication as between its members. This restriction means that, at least for the beginning, one does not think of the sun as a *communicator* sending *messages* to the scientist who studies its magnetic storms.

The function of organization is to achieve *coordination* as between the activities of its members on behalf of some goal – such as the production of articles for sale, or the enabling of exchange, as in a market – coordination in time and place. The "time" is not, in the first instance, relative to clocks and calendars but relative to the activity cycles of the organizational agents themselves. Communication must insure that so-and-so and so-and-so shall succeed in meeting, or that so-and-so shall have done this before so-and-so has done that. Clocks and calendars are devices (but not the only devices) that are used to make such relations hold. Similarly, places are not in the first instance referred to physical space coordinates, but to the disposition of the organizational agents relative to one another. What matters is who is near whom at some time–"near" in the sense of "now, able to influence".

In organizational settings, one can think of "information" thus: *if one agent exerts influence on the actions of another, then he transmits information to him.* The content of the information is expressed in the particular restrictions it imposes upon its recipient with respect to what he may do after he has received it. Under well-defined circumstances, to say that the information received *restricts* what the recipient may do is equivalent to saying that the information received *enables* what the recipient may do. An agent for whom certain next alternatives for action are organizationally defined cannot do any one of them without eliminating the other alternatives. This elimination will either be the consequence of messages he receives from others or the consequence of his choice – and the exercise of choice may be thought to require information from an organizationally unspecified source.

The actions of an agent that are influenced by others will themselves have the force of communication – i.e. influence the actions of others – and in particular, they must communicate to those others that influenced him if *coordinated* activity between the agents of the organization is to result. This point is far from obvious, but can be clarified by illustration. Suppose John slices bread and Henry slices cheese to make cheese sandwiches. Each slice of bread is to get a slice of cheese and there is never to be more than one slice of bread on hand for which there is not a ready cheese slice, nor more than one slice of cheese for which there is not a ready bread slice. The speed with which slices can be produced is not constant and so, to fulfill the conditions John must always be prepared to wait before cutting the next slice of bread, just as Henry must be prepared to wait before cutting the next slice of cheese. The choice between cutting and waiting on both John and Henry's part is governed by communication from the other and, what is more, the communications are interdependent.

The fact that communication in organizational contexts involves the circulation of influence – i.e. the "flow" of influence over circuits is related to the central importance of feedback in communication – first recognized, isolated, and intensively studied from a mathematical viewpoint by Norbert Wiener.

What classes of problems in our society demand a mathematical communication theory and the associated concept of information as indicated above?

First in line are the technical and legal problems associated with the use of communications and computer technology. It has become apparent in the course of the last five years (and should have been apparent for much longer) that these two technologies are conceptually so intimately related, that it will ultimately be impossible to see them as separate at all.[2] That this point was not grasped during the first twenty years of the computer era is, I think, attributable to the historical accident that the practical intention that drove the invention of modern computers was the intention of solving numeric equations – i.e. evaluating functions – instead of, for example, the intention of controlling the operation of machinery such as looms, printing presses, milling machines, etc., or the intention of text storage, editing, formatting, etc. In the case of function evaluation, the problems of coordinating the activity of what is inside the machine with what is outside of the machine is relatively trivial. At base it amounts to this: the man with the problem submits his argument values via an input device to a machine that is guaranteed to be waiting for him, and the machine produces the function values via an output device physically near the input device where the man (or a delegated substitute) is guaranteed to be waiting for the responses. Because of the extreme simplicity of this scheme, the input/output (i.e. communication) part of the computer remained relatively primitive while the internal logic became ever faster and more complex.

Since, internal to the computer, as everyone knows, "information" is processed – and, indeed, in large quantities – it is, according to our theoretical view of information, *inevitable* that the machine, internally, consist of a large number of parts each of which may be viewed as an agent in an organization. The structure of that organization defines the required patterns of coordination between the activities of these multitudes, and thus the patterns of communication between them. The generally prevalent lack of technical understanding of the relationship of computing to communication has, I believe, resulted in internal computer organizations which are badly matched to the greatly varying purposes that today's machines are supposed to fulfill. Inefficiency is one of many expressions of such a mismatch. Most of the millions of separate parts in a modern conventional computer must, most of the time, be told to wait – a complaint frequently heard against another organization, the Army, and for an organizationally identical reason, namely, centralized control. We must here, however, refrain from a long digression into the problems of computer organization and return to our main theme.

The ever increasing reliance which, in our society, is placed on computer and communication technology for the mediation of influences of human agencies upon one another makes it necessary that we understand how to:

- Specify the organizational functions which we wish a particular installation to fulfill;
- Verify the consistency of such specifications (e.g. absence of deadlocks, endless looping, virtual disappearance of resources, critical races, rule conflict, etc., etc.);
- Verify the correctness of function of implemented installations with respect to specified objectives;
- Calculate the side effects which these installations bring with them – at least those which follow formally from the explicitly stated requirements;
- Determine procedures for modifying an ongoing organization, either in its mechanized parts or in the constitution of its human agencies, with minimal disruption of its function;
- Trace the dynamic distribution of economic burdens incident upon the operation of a system.

Here, it is important to notice that all of these problem classes exist with respec to any organization, whether they include the products of computer and communi cations technology or not. It is only in the presence of these technologies that a powerful mathematical theory to help solve such problems has become an item on the critical list. The reasons for this include the following two important ones.

- In the presence of these technologies, enormous organizational changes are implemented in very short spaces of historical time. Since, in the past, major organizations – such as governments, transportation systems, fiscal systems, etc. – could usually change only a little bit at a time, the processes of gradual adaptation and natural selection (as in biological evolution) could help to safeguard against disaster. These safeguards are no longer with us.
- In the presence of these technologies, there are long and complex chains of organizational influence mediated entirely within electronic communication and computing links – and over these pathways the review of human common sense cannot be exercised (and common sense cannot be automated, no matter how much chess playing can).

Earlier on I referred to legal as well as technical problems. Where legal matters are concerned, I had in mind principally everything that has to do with the regulation of information – its generation, subsequent use, transfer, validation and revalidation, destruction, etc. Here, the most burning issues to my mind are those having to do with the relationships seen by the law of "information" to organization and purpose. Critical to ones functioning in society is the ability to know, when informing or being informed, with *whom* one is dealing – in other words, when participating in information exchanges, to do so in the light of adequately structured expectations as to the consequences of participation. It is this ability which is eroding dangerously in the midst of the rapid development of data banks and switched as well as unswitched long-distance communication nets, including computers. The issue of privacy which has received a lot of public attention in recent years seems to me one aspect of the general problem to which I have referred [1]. The conceptual foundation without which an effective mathematical theory of communication cannot come into existence is also a vital necessity for the success of information regulation in our society.

Let us now look back at Claude Shannon and the mathematical theory of communication, as he understood it. In the second paragraph of this introduction he says:

> The fundamental problems of Communication is that of reproducing at one point either exactly or approximately a message selected at another point. Frequently the messages have *meaning*; that is, they refer to or are correlated according to some system with certain physical or conceptual entities. These semantic aspects of communication are irrelevant to the engineering problem [2; 3].[3]

The first sentence of this quote would have been more accurate had its subject not been "communication" but "certain communication devices (such as telephone, mailing, broadcasting and publishing systems)". The sentence would have been yet more accurate – and much more significant – if it had said of the messages to be reproduced, not only that they should be good enough copies, but also that they should arrive *at the right time.* The question of "the right time" comes to one's attention more forcibly in connection with mailing than it does with the usual views of broadcasting or telephony, because the speed of transmission in the case of the latter is very fast compared to the reaction times of the senders and receivers, when these are human agents (but not, if they are computers). Even then, it depends upon whether one counts the time of establishing the telephone connection between conversation partners as part of the "communications problem" or whether one assumes that communication theory only has something to say when once the partners are hooked up to each other – i.e. appropriately coordinated – or synchronized. From our point of view, this latter is an unfortunate assumption since we believe the function of communication, in general, to be the achievement of coordination as between the activities of otherwise separate agencies.

There is another assumption in the background of Shannon's theory which is related to the issue of coordination between sender and receiver. It is apparent from his definition of channel capacity that he takes the following for granted: the (idealized) sender and receiver each have a clock which they can use to measure time intervals in perfect agreement with each other (i.e. one of them would judge the interval i_1 to be shorter than the interval i_2 if the other one would agree with him). *Thus, by prior agreement, with reference to their respective clocks the sender and receiver can coordinate their activities without communicating* – as indeed they must, for Shannon's channel to work. I mean that, synchronized to the times at which the sender decides whether to change or not change the state of the line, the receiver must detect whether the state of the line at his end did or did not change. It is my belief that an appropriately founded theory of communication must proceed from the premise that the activities of two *separate* agents in an organization can be coordinated *if and only if* they communicate – i.e. exert influence upon one another. The proper defense of this belief must await the possibility of examining its consequences in the development of communication theory.

I would now like to comment on the second and third sentences of my quotation from Shannon's introduction. Shannon sees message meaning as resting in the message's referential relations to entities which lie outside the domain of what channels transmit and, therefore, outside of the domain of the communications engineer and, therefore outside of the domain of the mathematical theory he propounds.

With regard to the domain of concern for communications theory which I described above, one naturally thinks otherwise. In this domain, message meaning – or content – is interpreted as the effect it has on its recipient in respect to his organizationally defined communicative behavior. To put this another way, the "meaning" is understood in terms of the influence which a given message has on subsequent messages. For example: the intended meaning of a message to a message relay station is the generation of a new message which is influenced by the incoming message, *and by it alone*. If, therefore, one wished to call the newly generated message "the same" or equivalent to the received one, one would have to add that its meaning in respect to its *next* recipients may very well be different. The messages these recipients generate by virtue of its receipt may not be a function of that message alone, but a function of other messages which influence them as well. All in all, with respect to the theory of communication which I see emerging (and which I am laboring to help into existence) I believe that the by now classical partition of Charles Morris' semiotic into syntactics, semantics and pragmatics (echoed by Warren Weaver in his part of The Mathematical Theory of Communication) is unfunctional. Messages do not change their *form* unless they are seen as affecting different organizational agents. Such changes are normally accompanied by differences in message effect, and hence differences in their semantic as well as pragmatic content.

I would like to close by saying that, although I seem to have dwelled on criticisms of Claude Shannon and criticisms of computer designers, I regard the contributions of both as of obvious practical value and also, of immense value in providing the soil out of which a genuine mathematical theory of communication can and will grow to maturity.

ENDNOTES

[1] It would appear that if Henry can be counted upon to produce a slice of cheese faster than John can produce a slice of bread then only John would need to communicate to Henry and not Henry to John. This does not change the fact that John's proper organizational functioning is influenced by the assumption about Henry and his cheese suppliers as to speed of cheese slice production relative to bread slice production. And this influence is best expressed in communication terms.

[2] See Miller, A. R., The Cybernetic Revolution, Chapter 1 *in* [1].

[3] In the original, the word "meaning" is in italics.

REFERENCES

[1] Miller, A. R., *The Assault on Privacy,* Ann Arbor, Mich.: The University of Michigan Press, 1971.

[2] Shannon, C. E. and Weaver, W., *The Mathematical Theory of Communication,* Urbana, Ill.: The University of Illinois Press, 1949.

ON LARGE SOCIAL SYSTEMS

INTRODUCTION

"Largeness", and "complexity", are relative terms. We use them to describe attributes of systems that we are unable to comprehend or that force us to admit great difficulty in constructing predictive or control models for their composite behavior. I suggest that the difference between large and small has little to do with size or numerosity but with the simplicity of a system's representation. The history of science is full of examples of objects of study that were once considered too large but then became manageable after irrelevant details were ignored in favor of underlying patterns, after such systems were broken into component parts, or after simple languages were found for symbolic manipulation. The history of science also teaches us to be aware of unjustified simplifications: system properties that appear irrelevant at one point in time may become a dominant force at another; models that are adequate under particular structural conditions may become obsolete once structural changes occur; rational manipulation of one portion of an environment may have unanticipated consequences for another portion which may upon return of their effects, constrain the very effort that caused them.

I believe there are basically three dimensions of largeness or of complexity. The first has to do with the *combinatorial variety* of the elements involved. Take, for example, a moderate size city of fifty thousand inhabitants. This is still a computable number. Assume now that each inhabitant at any moment in time is involved in one out of, say, a hundred different activities (including sleeping, riding a bicycle, playing chess, reading this paper). This is actually much too small to be realistic. At each moment in time that city then occupies one out of $100^{50,000} = 10^{100,000}$ states. This number exceeds anything astronomical and, in fact, exceeds anything computable by a factor of about $10^{99,900}$. Obviously, this Euclidian way of representing the states of a large system leads to combinatorial varieties so extraordinarily large as to prevent comprehension of the whole. And yet, a small municipal administration may well be able to cope with that city (a) by limiting itself to some governable aspect of city life, (b) by employing suitable organizational representations thereof or (c) by a suitable form of aggregation.

The second dimension of largeness and complexity has to do with *dynamics of change.* Emery and Trist [1] once coined the term "turbulent fields", which I understand to denote a portion of the social world in which causal connections and communication links are multiple, changes are rapid and occur at unequal rates, and control cycles of higher orders coexist. In such situations the ability to predict is extremely limited, and social organizations placed in such environments cannot adapt successfully simply through direct interaction. And yet, whatever portion of the social world is identified as a turbulent field one almost inevitably finds included within it social organizations that not only maintain themselves but also seem to thrive and grow.

This is evidence for the contention that large dynamic systems can be coped with but perhaps not from the position of a removed and inactive external observer. What needs to be understood is how social organizations are able to cope with the turbulence in their environment.

The third dimension of largeness has to do with the *complexity in depth* that the human mind exhibits: fringe consciousness and context sensitivity, metaphorical and analogical patterns of thought, the multitude of levels of reference and the logical indeterminancies resulting from their simultaneous presence, etc. Complexity in depth is magnified in conversations such as psychoanalytic discourse and the like, where cognitive interaction produces super-individual, i.e. social realities. Again, it is easy to compute the number of thoughts or behaviors that human beings are capable of from the ten billion neurons in their brain; but this number only tells us that we are on the wrong track towards understanding. And yet people do communicate with each other, empathise with and predict or even control the thoughts and feelings of others, all of which suggests that coping with depth is not entirely out of reach.

A few years ago, in answer to a question on the uses of computers in society, John G. Kemeny flatly denied the availability of adequate mathematical languages to represent such large social systems as cities, governments, and the parties to an international conflict. Statistical techniques are capable of summarizing and identifying patterns in very large volumes of data. However, they assume that the units of enumeration are statistically independent. And, while simulation techniques have greatly advanced in recent years, available languages favor applications to structurally simple aspects of the problems under consideration. Nevertheless work is being done, and progress is being made but perhaps not at the rate at which systems grow and various problems of largeness are created.

The most successful work on large social systems is associated with economics. I suppose this success is largely due to standardized units to measure value, a restriction which allows for simple forms of aggregation. Along with this comes a tradition of sophisticated mathematical theorizing. Neither of these conditions holds for the other social sciences, which are simply less unified. The virtue of aggregation (by addition or multiplication, for example) is that it ignores or avoids references to particular individuals, to specific economic activities or to the unique circumstances in which values are measured, and thus keeps combinational variety within manipulable limits. Indeed, the three contributions in this section address themselves primarily to the problem of combinatorial variety although the advanced economic models mentioned in the first paper do incorporate some aspects of turbulence and explanatory theories for economic behavior such as that described in the last paper have to cope with the problem of depth.

In the first paper of this section, Lawrence R. Klein, who has been associated with the whole group of Wharton economic forecasting models, presents an overview of methods for making economic models grow in size as well as in the fidelity of their predictions. According to Klein, the predominant method is *to link* different *economic models* so as to achieve a more complete accounting system for a given economy. But there are also attempts *to enlarge the scope of economic models* to include demography, sociological processes, engineering, health, politics and the like and thereby to establish a social accounting scheme that includes the economy as one component. Klein also reviews efforts under way *to couple and integrate regional models* into larger ones, efforts to join both models of different sectors of a national economy and models of geographical-political subdivisions of the world economy. Furthermore, Klein discusses the problems of expanding the time horizon of prediction by means of such models and the inclusion of micro level and distributional characteristics.

The second paper in this section, by Frederick Kile, should be considered the first progress report on a simulation study involving ethical presuppositions. The work is motivated on two levels. On one level, it is felt that since social models are primarily constructed by business, industry and various branches of local and national government including the military, they will omit specifically Christian and more generally ethical points of view. To rectify this trend, Aid Association for Lutherans is sponsoring the development of a regionalized world model that includes a large number of economic variables but also several noneconomic ones. In terms of Klein's discussion, the approach is one of expanding the scope (of a set of difference equations) into the axiological domain. On a second level, Kile feels that social models, however specialized they may be, will institutionalize a social reality including the implicit value assumptions underlying model construction once these models are considered valid and taken as a basis for public policy decisions. Here Kile proposes to analyze the explicit and implicit socio-ethical presuppositions by which a model builder or a whole modeling community proceeds. Although the work has only begun, the first task that Kile seems to envision is to describe the national characteristics (profiles) of selected members of the international community. The paper points in new directions and raises issues that are often ignored when large simulation models of social systems are constructed.

The third paper in this section stems from an unusual collaboration between a developmental-behavioral social scientist, Doreen Steg, and an economist, Rosalind Schulman. Concerned with complexity due to depth in human learning, Steg has been interested in the sensory feedback involved in regulating individual behavior relative to an environment. In the context of this work she distinguishes between adaptive and adapting behavior. Accordingly, adaptive behavior is observed when individuals change to accommodate to the requirements of their environment, whereas adapting behavior is recognizable when individuals change their environment to suit their own needs. The paper points out that much psychological experimentation, particularly that governed by the stimulus-response paradigm, may yield knowledge about adaptive behavior but not about adapting behavior. The identification of adapting behavior requires data on circular causal chains. In view of available evidence that individuals behave both adaptively *and* adaptingly, the latter being already identifiable in four-month-old babies, this criticism should be discouraging to traditional behaviorists and may pave the way for cybernetic contributions to learning.

Surprisingly, Schulman comes to much the same conclusion, even through her systemic aggregative analysis of individuals by economic variables. Attempting to explain consumer behavior, she finds that patterns of demand for goods and services become increasingly uninterpretable, given traditional assumptions, as soon as a population reaches a certain level of affluence. The shift from economically deprived to affluent societies seems to be one from adaptive to adapting behavior. In light of this surprising commonality of human learning and aggregated consumer behavior, the authors examine several other areas of human transaction: education, ethics, culture, society, and politics and suggest that learning cannot be ignored in any field of endeavor. In conclusion, the authors postulate that when a material-reward-oriented society has attained a modest but adequate standard of living for a majority of its population, the behavior of individuals in that society becomes adapting rather than adaptive. Advancing theory and research will have to consider this postulate seriously.

REFERENCE

[1] Emery, F. E. and Trist, E. L., "The causal texture of organizational environments", *Human Relations,* 18: 21–32, 1965.

THE NEXT GENERATION OF MACRO MODELS, THE PRESENT AND STEPS IN PROGRESS[1]

LAWRENCE R. KLEIN
University of Pennsylvania, Philadelphia

The history of Macro (econometric) model development, starting with Tinbergen's work in Holland, U.S. and U.K., can be told in terms of two parallel developments in economic analysis 1. The mathematization of Keynes and 2. The provision of national income accounting systems by Kuznets and others. The focus soon turned on building systems that would generate estimates of aggregate income and employment. These were frequently small systems 10, 15, 20 equations, possibly larger but under 50. These systems pre-dated the computer and were necessarily on the small side.

One group kept the systems small, as a matter of principle (Friend, Taubman, Fair, St Louis FED), but generally speaking the thrust of research was towards larger models. Nowadays, the 100 equation model, fully automated for estimation and application is fairly common. The outstanding feature of such systems, however, is not their size; it is their accounting structure. Most of the contemporary models are built around the accounting system of the National Income and Product. I shall call this the NIA system.

The NIA system, for macro analysis, is usually split into four sectors – households, business firms, government and foreign. For some purposes, banks are treated separately from nonfinancial business. The model then fills in the entries in this (double entry) accounting system. This becomes a nondoctrinaire approach to macro model building, and therefore widely acceptable in the profession. That models are now geared to the NIA system can be seen in the fact that results of most models are not now displayed as a solution to a system of simultaneous dynamic equations but as a set of NIA tables. A "table-maker" routinely transforms the mathematical/statistical solution in the standard NIA tables.

Quite independently, the standard input-output analysis has been constructed around the central relationship

$$(1-A)X=F$$

Figure I
RELATIONSHIP BETWEEN INTERINDUSTRY TRANSACTIONS, FINAL DEMAND AND FACTOR PAYMENTS

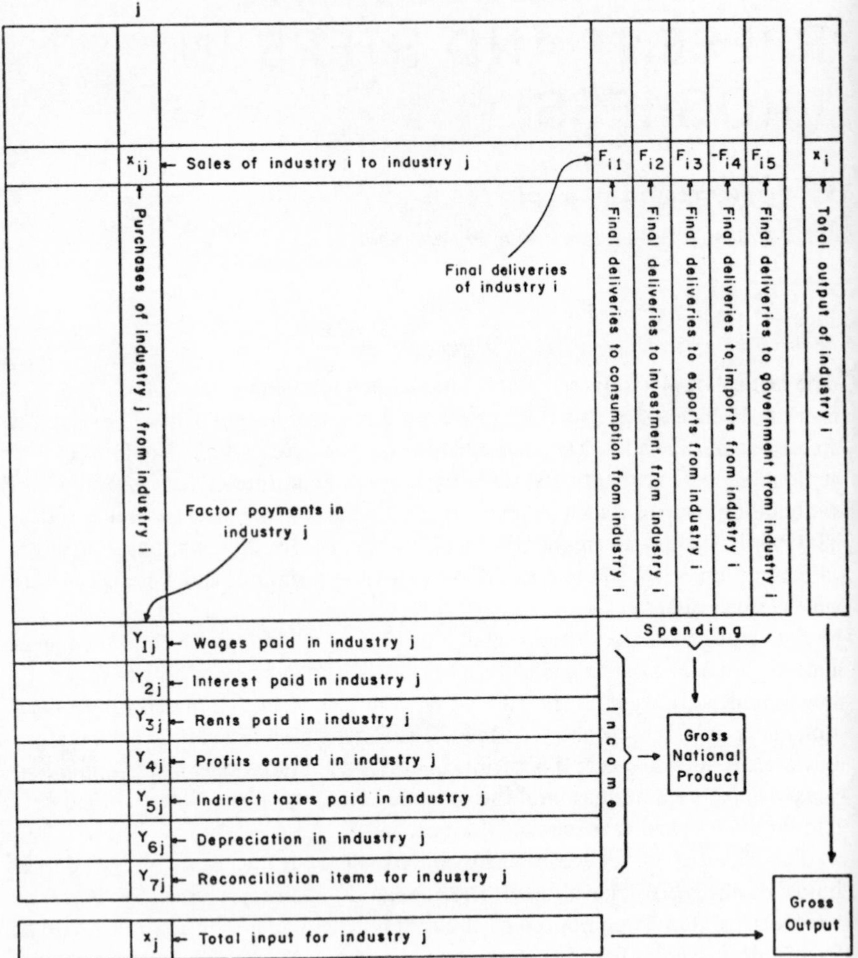

This, too, fits into an accounting system for the flows of intermediate products. The elements of F are, however, final products and directly related to the NIA system. Also, a relationship between NIA and I–O accounts that is not apparent from the above equation is the relationship between national income, gross output, and value added. The diagram in Figure 1, shows the full relationship. A main thrust of the Brookings Model [6] was to link together the I–O and NIA systems. This was independently done by Richard Stone. A more detailed implementation of the process started in the Brookings Model has been carried out by Ross Preston [7]. The Candide Model for Canada also has this combined I–O and NIA structure. Additionally, Candide makes use of rectangular I–O matrices.

The final step in this overall accounting process will be the integration of flow-of-funds accounts with the I–O and NIA system in the framework for one large model. At present, this is being done for a model based on the NIA and F/F systems together. It will be some time before a full linking with the I–O accounts can be made.

Although the term "flow-of-funds" implies the use of a time dimension associated with movement, spending, and consummation of transactions; it is probably more clearly discussed in the framework of balance-sheet-economics. The NIA system makes up social income statements, while the I–O system shows accounts for intermediate operations related to the income statements as in Figure 1. The accounting statements for an enterprise are rounded out by the balance sheet. By analogy, there is a wealth statement for society. This is, in some detail, the "stock" form of the F/F accounts. If we form a time sequence of balance sheets and then their first differences, we obtain a true flow-of-funds statement showing the changes in assets and liabilities for each sector of the economy. Changes in assets and liabilities, properly accounted for, produce savings. These are also viewed as direct estimates of sector savings. They can also be looked upon as sources and uses statements.

Strictly speaking, the F/F accounts deal only with financial assets and liabilities; whereas complete balance sheets would have to include real physical assets as well. Nevertheless, if we have modeled the F/F accounts as they now exist, we shall have practically handled the problem of complete modeling of the whole accounting system.

Accompanying the NIA tables for a national economy, there must be one capital account to close the double entry system. This final account will, in a simplified case, have the form

Capital A/C

Personal savings	Gross investment
Business savings	
Public savings	
Foreign savings	
Capital consumption	

By the savings-investment accounting identity, both sides will balance. The link between the NIA system, which generates all the items in this T–A/C and the F/F system of balance sheets is the net worth identity

$$(NW)_t = (NW)_{t-1} + S_t$$

NW = net worth, *S* = gross saving (including industry capital consumption allowance). This is the composition of the net worth entry in each sector's balance sheet. The net worth entry is also the difference between assets and liabilities; therefore net changes in assets and liabilities equals savings as generated by the NIA system.

In most macro models, savings are generated in the NIA type system as a residual, i.e. as receipts minus expenditures. From the F/F system, savings can be deduced from the portfolio analysis of asset-liability holdings and their changes. The only problem is to achieve consistency between the residual or indirect estimates of savings and the direct estimates. In this respect we pay heed to the warnings of Tobin and Brainard about the pitfalls in financial model building [4].

Among present day macro econometric models, two are outstanding in their treatment of both NIA and F/F systems simultaneously. Much, but not all, the F/F system is linked with the NIA system in the MPS model of Ando and Modigliani (also known as the FED model)[2]. A more recent model by Bosworth and Duesenberry also links the two systems [3].

The combined NIA and I–O systems have never built complete financial systems in the sense of the F/F accounts. They have included only the banking system, mainly the commercial banking system. When the new found methods for completing the F/F system together with the NIA system has been fully worked out and then combined with the whole I–O system, we shall have completed a significant milestone in econometric model construction, and that is likely to occur in the next few years.

The end product will be a large detailed system, still classified as a macro product but with much sector, market, and industry detail. It will have to exceed 1,000 equations by a wide margin.

Why do we want or need such a large system with much fine detail? In recent years, so many of our economy wide problems have been generated from strongly localized causes, that we would lack understanding of the issues if we did not have large systems. The problems of food and fuel escapes us in small macro models. It is mainly through exogenous changes in particular I–O sectors and small categories of final demand that we have been able to monitor, predict, and understand many of the things that happened to our economy in 1973–74. Without large coupled NIA and I–O models, we could have made poor interpretations of the oil crises and subsequent policies associated with Project Independence.

In capital markets, the whole spectrum of interest rates reached new heights and resulted in highly unusual relationships among different rates. Presumably a full F/F system will show supply-demand imbalances or pressures in each of several financial markets and this gives us better insight into interest rate determination throughout the spectrum than we get by simple term structure relationships. Credit rationing and squeezes on particular markets, like housing, should also be more understandable and predictable in such a system.

ENLARGING THE SCOPE

The preceding section outlined a straightforward approach – round out and fill in the equations of a complete economic accounting system. There will undoubtedly be disputes as to whether this is the best way to proceed, but it is underway and, from many points of view, predominant. It is, however, largely pure economics, and another natural way to consider new directions for model building is in branching out to encompass processes that are not strictly economic.

The entry of systems engineers into the general field of model building, exemplified by the *Limits to Growth* volume has goaded econometricians into more active participation in an area where their ideas had been developing but not translated into large scale activity. Many individuals had noted ties between demography or other social processes and economics, but had not undertaken systematic work towards the building of large interrelated models. In some respects, variables have been transferred from the exogenous (or random residual) category to the endogenous category. This is especially true of consumer and producer expectations determined in sampling surveys. But such extensions tend to be marginal in scope. In the *Limits to Growth* there are interrelated models of food, resources, population, economics, and other things. Economists have in my opinion, rightly criticized such attempts at model building as being inadequate and not up to professional standards, but these criticisms can be too sweeping as far as objectives are concerned.

The next generation of models should branch out to several new fields

1. demography
2. other sociological processes
3. engineering
4. health
5. politics

These are not exhaustive groups, but they do pose a considerable challenge and provide a great deal of work for many years to come. They may seem to be a bit arbitrary, but there is nothing as definite as a widely accepted social accounting scheme underlying their generation to serve as a guide in model building.

Demography Several large scale models now include variables for total population, some industrial or occupational decomposition, participation rates, age-sex-race composition of labor force, family composition. It is evident that many of these variables depend on economic factors that are generally developed within large scale models. These factors are different wage rates, income standards, unemployment rates, and productivity rates.

Demography is well suited to endogenous incorporation into macroeconometric models, although few do so. The data base is sound and generally accessible The relationships are as well defined as economic relationships, well documented, close fitting, and dynamic. To bring demography right into our models, we shall have to extend our thinking to social mores, technical progress in birth control, health care facilities, and other seemingly foreign fields; nevertheless the rewards should be great and achievable. This type of extension may prove to be more significant for long run than for short run analysis, but even in the latter case there should be some important contributions to the explanation of short run variations in labor force, which are indeed large and important.

Other sociological processes Besides demography, which appears to be an immediate and obvious candidate for inclusion, there are other ripe areas of sociological interest. These might be criminology, class structure, urbanization, religion, and education. The statistics of crime appear to be ripe for more intensive and extensive analysis. Crime leads to public expenditures, economic costs, economic inefficiency, the existence of unreported economic activities, and many other activities with large economic content. The economic base for crime may, to a great extent, be explained by major macro model variables, and the feed backs on the economy are evident.

Class structure is related to income and wealth distribution but not congruent to those phenomena. Class structure affects spending behavior, residential patterns, and work habits. All, then, are part of the economic system. The rural-urban shift is a major aspect of American economic life and is clearly discernible in the economic processes of other countries. The number of farm workers; farm incomes, residential demand, educational outlays and many other economic variables have important impacts in present models. Further endogenous explanation is needed here.

Religion is economic in the non profit sectors of the national income or flow-of-funds accounts. Income streams in these sectors vary with economic conditions At the same time, religion affects politics and social structure, which, in turn, impact on the economy. Variables associated with education have already been inserted into models as single economic relationships to show employment quality or technical progress in production functions. Also education has been closely related to local government spending decisions. There are many ways that education can influence or be influenced by the economy although the effects may be slow moving.

The demographic and other sociological variables are not only interrelated with economic variables in the usual kind of relationships that we study in econometric models; they also have their own interrelationships. Actually work is underway on the construction of macrosociological models, to a large degree by analogy with macroeconometric models, but these systems are not fully integrated so as to produce simultaneous generation of economic and social variables.

The relationships are specified on intuitive or commonsense grounds without having such powerful engines as optimization, market clearing and dynamic stability that play such important roles in specifying economic systems.

Engineering Pollution generation control, sanitary waste systems, conservation, congestion are all phenomena that are having a lot to do with our present economic lives, yet variables representing these phenomena are not worked into our usual models. Pollution will be produced by the tempo of production and general economic activity. Its control will change prices and costs. With urbanization, there will be traffic congestion, requiring engineering skill for alleviating the bottlenecks. There is much scope for an engineering contribution to the structure of socio-economic models. Some notable attempts at the construction of engineering production functions have shown how important technical information can be in design of economic relationships [1; 5]. These attempts are greatly in need of amplification and extension. There is much to be done by way of tightening economic relationships by the addition of engineering information. Such information is generally *a priori* and not subject to sampling error in the usual sense. The more such information is used in estimation of some of the parameters of econometric models, the less burden is placed on sample data for estimating the remaining parameters and the more is efficiency enhanced. Engineering decisions usually need economic input for analysis of costs and revenues before implementation can be completed; therefore the process is one of dual effects.

Health Like weather and demography, health has direct external effect on the economy. Epidemics may occur randomly, but they are felt on the economy. When they can be directly measured, they should be used as explicit input variables. Other health phenomena are man made or controlled. The construction of hospitals; the training of personnel; the prices of health delivery; the statutory aspects of health insurance are not uncontrolled random events that disturb the economy; they are bound up with economic decisions and have important effects. They shape efficiency of the work force and move the pharmaceutical industry. Medical personnel occupy a sensitive place in the income distribution. There are ample reasons to include health delivery and the general state of health in an expanded model. Research in this field draws upon system analysis as does engineering, but it is also sociological in nature, too.

Politics Party voting, party financing are obvious candidates for modeling in line with economic activity. But the general degrees of political satisfaction on the part of the public at large is important, too. Data will be difficult to find or construct for fruitful work in this area. There will be no ready-made files like those in vital statistics or population records except for voting and party registrations. Subjective political attitudes will have to be ascertained in sample survey collections, and such data will be subject to wide margins of error. They do, however, provide interesting scope for analysis. Consumer attitudes have been found to be useful in modeling spending behavior and it should not be surprising if political attitudes turn out to be useful in modeling other parts of consumer, producer, or government behavior.

Exogenous variables for political decisions of public authorities are widely used in econometric models. These represent monetary and fiscal policy choices, covering such things as tax rates, tariff rates, public expenditures, discount rates, open market operations, reserve rates, and many others. Many of these variables should be made endogenous because politicians respond systematically to economic events. The problem is difficult, because political bodies are small; they are often erratic; they may be covert; and they may change frequently with prevailing moods or election campaigns. The political process poses a severe challenge, but it is a challenge that should be met. Expectations of great precision would not be warranted, but some modest first steps are in order.

THE TIME HORIZON

Most applications of models are now limited to the near term – by months or quarters over a horizon of two years. Of course historical simulation studies within sample limits are often longer run, but extrapolations are limited. Much analysis has gone into the refinement of the short run extrapolation, either for pure forecasting or for hypothetical simulation. The historical data base is limited by availability of materials, and this makes long run extrapolation shaky in the sense that there is little precedent for guidance, but from a purely mechanical point of view the systems are capable of generating long term solutions if enough input information is used.

During the past few years new studies for the longer term have been initiated. The combined NIA and I–O model of Preston has been regularly extrapolated for decade periods [7]. Plans are now underway to make extrapolation studies for two or three decades. This puts econometrics squarely in the midst of analyses of the year 2000. A compelling reason for turning attention to this new long-run period is to contribute to energy economics research. The new technological processes being considered for 2000 and beyond need extrapolations of the expected economic environment, and model builders are trying to fill that need.

The concern with the longer time horizon fits well with demographic modeling. If econometric model builders turn increasing attention to the longer run, their work will blend well with that of demographers because many population processes take several years for completion or even decisive turns.

DISTRIBUTIONS

Macro models built from economic and other time series have concentrated attention on aggregate magnitudes for the most part. The large NIA/I–O models do, in fact, give many results for individual industries and sectors. This degree of micro analysis does not, however, qualify their research as being concerned with underlying distributional materials. The size distributions of income, employees in an establishment, and similar micro-economic information have not found their way into most existing macro models.

In many respects, statistical studies of distributions are interesting in their own right as separate, stand-alone investigations, but in the present context, the issue is whether this material can be used in closed system model building. Individual household or firm data can be used to estimate some of the relationships used in model construction or possibly some of the individual parameters (or functions of parameters). This has always been done to some extent, dating from the work of Tinbergen and later myself, where Engel curve data have been used to help estimate consumption function properties. While the time may not have arrived when full micro-simulation models as proposed by Guy Orcutt can be fully implemented, it is entirely feasible to use micro data more extensively. Time series of parameters or statistics of income distributions to take account of changing inequality of distribution should be among variables of models and also generated by the models, as well. Other distributions should be used in the same way for explanation of both consumer and producer behavior. The concentration of firms may affect price movements, at least that is what the advocates of anti-monopoly policy claim for their proposals as counter inflation measures.

Performance characteristics of economic systems as viewed through the workings of macroeconometric models are inadequate since they show only average or aggregate performance. Such models would be vastly more informative if they were to generate estimates of parameters or nonparametric statistics associated with distributions of the main aggregates. All this involves much research yet to be done, but it is a fruitful and probable direction to be followed.

SUPER AND SUB-NATIONAL MODELS

Macro models were first built for individual nations – Holland, U.S. and U.K. Now this activity has spread to practically every country in the world. Simultaneously, model building is also growing at the level of national geographic-political subdivisions and for the world economy as a whole. The former activity includes regional models, state models, SMSA models, city models. There are examples of all these in action with frequent application to local issues. At a polar extreme, my colleagues have built a model of a single county, Luzerne, in

Pennsylvania, with a population of only 300,000 to study flood damage and relief.

Individual satellite models of national subdivisions, industries, or markets are being built on an increasingly large scale for study of special problems, but it has not usually been possible or recommended that the model of the whole be constructed from the sum of the parts. This is principally due to the fact that data sources are weak or absent for many subnational magnitudes, especially area exports, imports, and profits. There will, nevertheless be a proliferation and extension of subnational models. They already are proving to be useful and attractive.

At the other end of the spectrum, international models are becoming increasingly studied as a consequence of grave changes and disturbances in the world economy. Project LINK is, perhaps, the most ambitious and most developed of such supernational models. LINK has gone far in dealing with international trade flows for goods. For the future it needs extension to services, capital flows, and world commodity trade. The endogenization of exchange rates is a goal, but a fairly remote one, as yet. Of course, new countries are always being added. The greatest challenges are in the fuller incorporation of socialist and developing countries.

THE RATIONALE

Why bother with all the detail, problems of data management, and "noise" of the really large model? Users of models appear to have insatiable appetities. No matter how much additional detail the model builder supplies, users keep returning and asking for finer breakdowns, longer extrapolations, and results of highly localized impacts. Bigger is not better, but if we model builders can retain the overall accuracy of the central aggregates (growth rate, inflation rate, trade balance, unemployment rate, interest rate) and effectively manage the large amount of data, then we should, by all means, stay in pursuit of ever greater amounts of information, transcending disciplinary lines where necessary.

Mainly, such added detail from the next generation models will be useful for planners in the public domain. Given the present ability and increasing aptitude to automate data handling together with complete system application, the planning uses are capable of being implemented. Simulations, automatic control, and search for optimality are all types of application that work best with mathematical type systems run through digital computers. What seemed to be only theorizing some years ago, may become a reality in the coming generation.

ENDNOTE

[1] The paper was presented at the Eastern Economics Association convention, October, 1974. It is included in this volume with the author's permission because it covers substantially similar grounds of the paper presented during the conference a few weeks after the above. (Ed.)

REFERENCES

[1] Anderson, R. J. Jr., "Application of engineering analysis of production to econometric models of the firm", Ph.D. dissertation, Philadelphia: University of Pennsylvania, 1969.

[2] Ando, A. and Modigliani, F., *The MPS Econometric Model: Its Theoretical Foundation and Empirical Findings,* forthcoming.

[3] Bosworth, B. and Duesenberry, J., "A flow of funds model and its implications", *in Issues in Federal Dept. Management,* Boston: Federal Reserve Bank of Boston, Conference, 1973.

[4] Brainard, W. and Tobin, J., "Pitfalls in financial model building", *American Economic Review, Proceedings,* **58**: 99–122, 1968.

[5] Chenery, H. B., "Engineering production functions", *Quarterly Journal of Economics,* **63**: 507–531, 1949.

[6] Duesenberry, J., *et al.* (Eds.), *The Brookings Quarterly Econometric Model of the United States,* Amsterdam: North Holland, 1961.

[7] Preston, R. S., *The Wharton Annual* and *Industry Forecasting Model,* Philadelphia: Economic Research Unit, University of Pennsylvania, 1972.

A REGIONALIZED WORLD MODEL TO DISCLOSE THE NATURE OF IMPLICIT AND EXPLICIT SOCIOETHICAL PRESUPPOSITIONS

FREDERICK KILE
Aid Association for Lutherans
Appleton, Wisconsin

INTENTION

This project was undertaken specifically from a Christian and more generally from an ethical standpoint. Our concern stems from the assumption that computerized social modeling will become a highly sophisticated and widely used planning tool within the next decade. I think especially of dynamic and even planner-interactive models which far surpass simple scheduling models in both breadth and depth.

We foresee use of social models by business and industry, and in the executive and legislative branches of government as well as by a host of quasi-independent public or governmental planning commissions and agencies. Initially, if not in the long run, these models will have a tendency to become self-fulfilling prophecies or in some cases self-negating prophecies. But far more than that: even where models are used with caution to avoid the excesses of positive feedback from model to constituency and back again, these models will be subject to inclusion of unstated assumptions.

Therefore our project is intended to:

1. Address the Christian community regarding the overall impact of social modeling.
2. Address the modeling community regarding the ethical implications of computer-based social planning.
3. Build a credible social model with specifically stated (explicit) socioethical assumptions.

4. Emphasize the social implications of unstated (implicit) socio-ethical modeling assumptions.

We believe these four steps outline the best way to pursue our intention of broadening the conversation about modeling. Personally, I feel thankful that a well informed base for this dialog is beginning to emerge, since the time until social models will be commonly used as planning tools is short.

MODELING APPROACH

We elected to work with world models because they comprise a language of discourse common to all social modelers. There probably are regional or national models at more advanced levels of sophistication but both audience and the supply of constructive critics are restricted for models with a narrower base.

REGIONAL WORLD III

The evolution of Regional World III has progressed through four clearly identifiable stages with many substages. Our first attempt was a checkerboard-like pseudo world with purely hypothetical locations and entirely artificial data. The equations for this pseudo world were largely multiplicative. This approach provided us primarily with a DON'T DO list.

The next model was Regional World I (RW I), an additive model which spanned a limited number of nations. RW I divided the economy of each nation into an agricultural, a resource, and an industrial sector. The model was genuinely limited but showed us how to achieve a dynamic simulation with multiple regions and a sectored economy.

Our third model, Regional World II (RW II), enlarged the scope of RW I to include about 75% of the world's population. This model contained nine industrial production sectors and permitted some verification comparisons with observed world socioeconomic data and trends in data.

The present version of our work, Regional World III (RW III), is subdivided according to nations and selected aggregations of nations. About 98% of the world's population is represented in RW III. A few less populous nations are treated separately when exceptional conditions warrant this: geographical size, great agricultural potential, abundant resources, e.g. Canada, Australia, Iran. Table 1 lists regions presently included in RW III.

TABLE 1

Regions Presently Included in Regional World III

1	U.S.	14	Nigeria
2	Canada	15	Malaysia, Thailand, Burma, Singapore
3	U.K.	16	Arab Nations
4	India	17	Black Africa (except Nigeria)
5	Fed. Rep. of Germany	18	Latin America
6	U.S.S.R.	19	Eastern Europe
7	Brazil	20	Remaining Western Europe
8	China	21	Australia, New Zealand
9	France	22	Vietnam (including Laos), Cambodia
10	Japan	23	Philippines, South Korea, Taiwan
11	Indonesia	24	Turkey
12	Bangladesh	25	Iran
13	Pakistan	26	South Africa, Rhodesia

REGIONAL PROFILES

The profile of each region in RW III consists of a population representation, six production sectors, three resource sectors, an agricultural sector, and a utile (available credit) sector. Representative profile parameters (selected from a list of almost 300) are given in Table 2.

TABLE 2

Representative National Profile Parameters

- Population
- Population in agriculture
- Birth rate (displayed as births per thousand)
- Death rate (displayed as deaths per thousand)
- Emigration
- Industrial production
  - Industrial base
  - Agricultural equipment
  - Advanced technology
  - Other capital goods
  - Overhead (a representation of consumable production)
  - Industrial exports
- Metal Ore reserves
- Petroleum reserves
- Coal reserves
- Calories produced per capita
- Agricultural surplus
- Available utiles (a measure of spendable funds)
- Economic power (a measure of credit)

PERIODIC ADDITIONS TO REGIONAL PROFILES

Model development has traced a pattern similar to the well-known plateau phenomenon in human learning. As a plateau is reached, increments of performance improvement for each new fine-tuning effort are progressively smaller. Typically, a breakout from one of these plateaus follows introduction of carefully chosen new parameters or social/economic sectors into the model structure. Generally, following one of these breakouts the model has achieved new high levels of correspondence with the observed world before a new plateau is reached. At the present stage of development we are ready to undertake a substantial redesign which will include modularization (to facilitate policy testing), an age specific population representation, nuclear power generating capacity, capital invested in nonproductive activities (chiefly military equipment), and capital investment in pollution abatement equipment. Subsequent factors to be introduced will include clearer identification of agricultural production with the inherent natural production potential of the region, waste recycling, and introduction of explicit credit mechanisms.

MODEL FUNCTION

Model function is based on difference equations. Discrete integration is coupled with an approach which compares a decimal fraction of a particular system value with a desired decimal fraction and adjusts the next increment to that value in accordance with prescribed algorithms. While any model of this type owes a definite debt to the work and formulations of J. W. Forrester at MIT, we have made a distinct effort to free our work from the constraints attendant to using an established modeling philosophy. We have also profited from Wassily Leontief' work to the extent that we do a partial accounting of economic imput and outputs. More detailed input/output computations will accompany further economic refinements in the forthcoming modularized model which will be known as RW IV.

FORTRAN IMPLEMENTATION

System equations are coded in FORTRAN IV with model characteristics stored in a series of one- and two-dimensional arrays. Some subroutines are used, but the bulk of the processing is done in a series of loops in the main program. The present program consists of about 1,000 executable statements and is run on an IBM 370/158 computer. Almost all tests are made in the batch mode with printed output since we have effectively outgrown the capacity of our Time

Sharing (TS) system to provide CRT output.

Present FORTRAN implementation requires a substantial number of two-dimensional arrays. True input/output consistency over a multi-region model can only be achieved by use of higher-order arrays. We plan to increase use of higher-order arrays to represent more realistically the interdependence of diverse regions and also to illustrate the potential for social and political intervention in socio-economic processes.

Acknowledgment: The rapid progress of this project is due in large part to the work of Arnold Rabehl, Scientific Systems Programmer, whose creative programming approach has given us a series of models with growth capabilities far beyond our original expectations.

DETAILS OF REGIONAL WORLD III

Metric units are used throughout. The base year for calculations is 1968. The monetary unit is the Utile, equal to one 1968 constant dollar (U.S.). A sketch of the model calculation sequence follows:

1. The data set is read in and other year-end values for the zero-th year are calculated as needed. Regional parameter values for the first year are then calculated as follows:
2. Relative industrial and technological strengths are calculated.
3. Agricultural production is calculated, based on relative technological strength, the number of persons actively engaged in agriculture and the investment in agricultural equipment. Food production is calculated in calories per capita per day and converted to metric tons of grain equivalent. Export surplus or import need is calculated.
4. National economic power is measured according to available credit, industrial production, resource exports, and food production.
5. Nations are separated into economic classes depending on agricultural strength, technological level, exports, investment per capita, and available credit.
6. A world food price is calculated based on world food import need, and world export supplies.
7. Food is imported as needed, if credit and food supplies permit. Food exports are prorated from available supplies.
8. National energy demand is calculated. Energy supplies are calculated using renewable energy production, nonrenewable energy production, energy export availability, and imports. Exports are made according to each nation's share of world exportable supplies and actual world imports.

9. Energy shortfall, if any, is allocated to the various economic sectors according to a unique set of allocation priorities for each nation.
10. Nonrenewable resource use (other than energy) is calculated in the manner described for energy.
11. Industrial production demands are calculated by sector, based on measures of internal need for each region.
12. Industrial production is calculated according to equations describing the industrial production base, percentage of unilization of production capacity during the prior year, and the demand figures computed in 11 above. Production available for export is calculated.
13. Industrial goods are imported according to target, actual value, and the assumed possibility of import. We assume for example that certain nations will experience difficulty importing advanced technology and thus we reduce the coefficient governing their prospects of receiving this type of equipment. As with food and energy, available credit is needed to import industrial goods.
14. Available credit is recalculated based on money spent and received in international trade.
15. New values are calculated for each economic sector according to depreciation, production, and imports.
16. Population is recalculated according to birth rate (dependent on material standard of living, technology, and nutrition level), death rate (dependent on material standard of living, pollution, technology and nutrition level), and migration (dependent on nutrition level). The population in agriculture is recalculated based on nutritional need and industrialization. The nutrition level (in calories per capita per day) is recalculated for each nation based on needs and food supplies.
17. Data are printed out or displayed as called for, intervention is then permitted (although it could easily be permitted at other junctures within a given year), and the values for the next year are computed in the same manner (using year-end values for year n as initial values for year $n + 1$).

A BRIEF DISCUSSION OF IMPLICT, EXPLICIT, AND WHAT LIES BETWEEN

A Definition of the Terms "Explicit" and "Implicit".

The term "explicit" as we use it refers to those presuppositions and intentions of the modeler which he/she:

1. Is consciously aware of, and

2. Designs into the logic and algorithms of the model in such a way as to make them clear to those who study or use the model.

The term "implicit" as we use it refers to those presuppositions and intentions of the modeler which he/she:

1. Is consciously aware of and includes in the model, but unintentionally fails to make clear to those who study or use the model, or
2. Is aware of and includes in the model but seeks to conceal from those who study or use the model.

THE VAST GRAY AREA IN MODELING DECISIONS

Unfortunately, many modeling decisions do not fall neatly into one or the other of our categories so we need to describe this catch-the-remainder category. The modeler may scan alternatives and select a *modus operandi* with only a vaguely defined sense of what he/she would like to do in a particular case. If the modeler acts on a vague "hunch" the action taken becomes a sort of cross-breed which frequently causes trouble to both modeler and model interpreter. This trouble arises because the intuition of the modeler is likely to be poorly represented by an implicit structure which has not been carefully thought out. Explicit choices which are spelled out achieve their intended effects only marginally in this kind of structure.

Recognition of the challenge to every social modeler causes us to restate our own long-range intentions to bring them into sharper focus. Our goals are:

1. To analyze and address the significance of social modeling as an art or discipline.
2. To clarify the significance of implicit and explicit social decisions in the modeling process.

Clearly then we are charged with showing how a modeler can state his/her presuppositions and implementation decisions as explicitly as the potential social impact of the model warrants. We must also seek to establish how critics and users can interpret a model to explicate the modeler's intentions and presuppositions beyond reasonable doubt.

Examples of Implicit and Explicit Modeling Decisions

A primarily implicit modeling decision is: that a quantified measure of human welfare should in the long run be more decisive to model function than a quantified measure of economic welfare.

An explicit modeling decision is: that the model should first have a nation spend its available monetary credits to buy food for its people and only after that use what purchasing power remains to buy hard goods.

Although a policy insertion is an explicit act, subtleties of model structure may produce an implicit change in the model which affects the outcome of a particular test much more (or quite differently) than the explicit change was intended to. Anomalous effects such as this point to the gray area between explicit and implicit modeling decisions. We should note that this same phenomenon occurs in the observed world and thus our basic task is not so much to avoid unwanted side affects in model policy testing, but rather to bring these side effects into congruence with side effects produced by analogous policies introduced into the observed ("real") world.

The vast gray area in modeling can be further characterized by a question a modeler might ask himself early in the design process: "Will this model be more useful as a social model if I devote a greater portion of my effort to explicitly identifying social considerations such as the quality and breadth of the food supply, the possibility of migration from one region to another, and cultural concern for human welfare ("soft" variables), *or* will the model be a more effective social tool in the long run if I devote more attention to implementing more readily quantifiable considerations such as economic production, birth rate, capital investment, and natural resources ('hard' variables)?"

MODEL DESIGN PHILOSOPHY – A TEST-TO-FAILURE APPROACH

Our design philosophy is to block-diagram interactions among selected social phenomena, write equations to represent the interactions indicated by our diagram, code the equations, and run tests until the system fails, i.e. provides totally unacceptable data or causes a computer error. We then look for the reasons for failure, determine the extent of redesign needed and run to failure again. We feel that operation at the limits of the model demonstrates most effectively where to go next. As I indicated, we have felt the need for total redesign twice and for substantial modification on several occasions. However, we also feel that each new failure has been at a point nearer reality than its predecessor. As an example, at one time the model exhibited failure through oscillation of the food supply on a national (but not world) basis. The solution to this failure mode entailed introduction of broadly-based negative feedback which stabilized model performance over a substantial range of operation. This solution introduced a tendency for the model to seek a steady state of operation quite unlike the world which it was supposed to represent. Further changes eliminated much of this arbitrary stability and the food sector of the model was temporarily frozen until other sectors of the model were sufficiently refined to call for additional modifications in the food area. This process is part of the plateau phenomenon and will very likely continue recursively as long as our project is active.

MODEL RESULTS – A CRITIQUE

Because we follow a test-to-failure approach we are usually looking for the weakest areas of model function rather than the strongest. However, this method reveals model strong points as well as weaknesses. Let's examine the strengths and weaknesses of RW III at its present stage of development.

Strengths

Perhaps the greatest strength of the design we use is its ability to handle an increasingly larger number of nations with no perceptible strain on its carrying capacity. Moreover, the model is able to track the observed world fairly closely in the following ways:

1. Differences in regional strengths and weaknesses are clearly represented by appropriately different types of test results.
2. The model demonstrates many of the weaker aspects of the socio-economic behavior of each region thus pinpointing possibilities for social instability and suggesting where appropriate intervention is advisable to avoid social collapse (including war, famine, bankruptcy, production failure, resource depletion, etc.)
3. The model has clearly demonstrated the capacity for expansion to a still more broadly-based socio-economic fabric.
4. The model is amenable to a broad range of policy testing. By policy intervention we are able to simulate specific types of aid: money and food, industrial construction, and technological assistance sent directly from one nation to another. This permits, for example, a test which furnishes a reasonable approximation to the Marshall Plan of the Post-World War II era. The forthcoming revision of the model, RW IV, will incorporate greatly enhanced policy test capabilities.

   The policy intervention capability also enables simulation of war or other catastrophe at a designated point in time. To simulate effects of a war, we need merely adjust values of variables accordingly. If the modeler wishes to simulate a war between nation K and nation L, he simply decides what sectors of the population and economy would be affected and to what portion of their former values they would decline, enters the appropriate changes in values variables for K and L at the end of the year chosen for the simulated war, and allows the test run to continue. We can also simulate economic singularities such as an oil embargo by similar adjustment of appropriate parameter values.

Weaknesses

The most glaring weakness of the model is its strong dependence on subjective selection of constants by the modeler (fine-tuning). I believe we are gradually minimizing the impact of fine-tuning, but the problem will not disappear completely in the foreseeable future.

We cite the following as other examples of model weaknesses:

1. We have experienced substantial difficulty in writing equations to simulate food production over wide ranges of climates, agricultural investment, arable land, etc. Model projections of food production still tend to diverge from experienced levels. This phenomenon may reflect the observed world more nearly than a first glance would indicate when one considers that grain production in the Soviet Union rose 50% from 1975 to 1976, based very largely on relatively unpredictable annual rainfall patterns.
2. We have achieved only a very marginal simulation of the actual workings of international credit. I might interject here that recurrent turmoil on the international monetary scene indicates we are not alone.
3. Prediction of unusual events such as an oil embargo is not possible, although as indicated above, the phenomenon may be simulated after the modeler has decided it will occur.

DATA BASE

The quality of available data, especially for many developing regions, would be laughable if we weren't concerned with a serious topic. Some modelers have suggested that for this reason models must be tailored to available data. I believe that data availability will improve as more advanced model structures are developed and the need for these data becomes apparent.

PROGRESS TOWARD PROJECT GOALS

This presentation is one of a series of efforts to initiate an ongoing dialog on the impact and ethical implications of the widespread use of social modeling. We hope to receive productive criticism and to find friendly adversaries to keep the debate growing and to share insights into modeling. It may be that we will find others willing to run comparison tests on identical data to demonstrate the relative strengths and weaknesses of differing approaches to social modeling.

The RW III model has demonstrated to us that social considerations may be explicitly specified in a computer model of a social situation. We intend to continue testing over an expanding range of explicitly defined social and ethical factors with the hope of compelling general recognition that some computer models of society are "sterile" and essentially unproductive, while others deal more adequately with the range of factors necessary to produce a meaningful social simulation. This thesis is an extension of our doctrine of test-to-failure and then redesign. If we can show that an increasing number of well-chosen and well-implemented social factors leads to increasingly more satisfactory modeling of society, we shall have established by induction the case for this type of modeling as opposed to straightforward quantification of economic factors. To move from this line of reasoning to a general consideration of the implicit as well as the explicit ethical basis for social models will not be difficult. Then perhaps the debate on the ethical aspects and social impact of modeling will grow concurrently with the development and use of social modeling as an administrative tool and we shall have achieved our goal.

HUMAN TRANSACTION AND ADAPTING BEHAVIOR[1]

DOREEN RAY STEG and ROSALIND SCHULMAN
Drexel University, Philadelphia

INTERACTION OR TRANSACTION?

The Newtonian construction – unexcelled for its efficiency within its sphere – viewed the world as a process of "simple forces between unalterable particles" . . . Space and time were treated as the absolute, fixed, or formal framework within which the mechanics proceeded – in other words, they were omitted from the process itself . . . (i.e. interaction). Einstein's treatment, arising from new observations and new problems, brought space and time into the investigation as among the events investigated (i.e. transaction)[6: 111–112].

A transactional approach is seeing together what has been seen separately and held apart.

INTRODUCTION

Since 1953 we have been observing an economic phenomenon for which there was no apparent explanation. That phenomenon was declining price elasticity of demand as measured by income elasticity of consumption. In other words, an ossification of purchasing pattern was spreading through every income class of American, except the poor. *Businesses were going bankrupt without apparent cause in the midst of unprecedented prosperity.*

By 1963 when accurate data became available [24], it became evident that there had been, over the period, a change in elasticity (the ratio between a difference in expenditure for a particular good and an income change). *Elasticity was declining due to increase in income and education* (no other variable proved significant); of the two, *education showed the most significant pattern.*

It has always been postulated by economic theorists that mean elasticity of consumption declines for goods[2] as income rises. But this time the data showed an even greater decline with an increase in education. This was contrary to all predictions and could not be explained on the basis of any previous postulate, (whether Keynesian, Friedman's permanent income hypothesis, or the adaptation of the permanent wealth theory; or Duesenberry's previous standard of living hypothesis). When Steg's papers [17; 18; 19; 20; 21; 22] were finally digested it became evident that they were an explicative formula for this economic behavioral phenomenon. The consumer was "learning" and exhibiting an "adapting behavior" cybernetic mechanism. Only a continuous feedback[3] (hence cybernetic), explains the human transactions in today's consumption. Unfortunately, it is not only the manufacturers who are unaware of either its existence or meaning.

In a series of papers Steg has treated deviation-counteracting feedback in human behavior, i.e. negative feedback, and suggested that at least two distinctive human behaviors become operative in society with possible mixtures of both. These are adaptive behavior, or behavior where a system adapts itself to the requirements of the environment, and adapting behavior where the system changes the environment to suit itself, using a variety of tools: social, economic, psychological, physical, and even political [17; 19; 21; 22].

In the paper on "Communication and Feedback in the Technology of Consumption" [13], Schulman has shown that in a society where discretionary (non-necessity) purchases [11; 24] and consumption are available to the majority of the population, the consumer becomes a "least cost" buyer for necessities, and refuses to follow previously established patterns of authority in purchasing discretionary goods. In a free or semi-free market this plays havoc with fashion's dictates; causes individualistic, fractionalized consumer reaction in the marketplace – to the point of purchase refusal if desires are not met; and expresses itself in overt criticism, lobbying and myriad other group and nongroup activities to influence the market and the manufacturers, economically, politically and socially.

In other words, the consumer is exhibiting an "adapting behavior" pattern, which will increase in intensity as the society becomes more affluent.

BEHAVIOR: ADAPTIVE VERSUS ADAPTING

Adaptive

Automatic activity of man, animal or machine is an *adaptive* control system, by its very nature. It is safe to assume that, as with the laws of physics, the laws governing control systems apply equally to animal, man or machine. In the language of the system engineer, this is a closed-loop control system. The control system pattern consists of (1) an input signal that triggers some action, (2) a feedback signal of the result of this action to compare with the input signal, (3) a closing of the loop and a summation of the two signals and (4) effective action to counteract this summating signal. A persistent residuary signal can be made to affect memory which results in "learning". In a control system, work is triggered as a result of an actual error input.[4] The error is essential to the activity of any control system. These mechanical patterns apply equally to automatic machinery, animal behavior, and man's everyday automatic activity.

Adapting

An important deviation from the automatic pattern occurs when the automaticity of a system is eliminated. Nonautomatic activity will not necessarily be subject to the adaptive nature of the control system and trigger *its* energy to cancel the disturbance.

With the automaticity eliminated the response to a disturbance is chosen after the disturbance has been analyzed as to its source, the energy involved in the disturbance, the possible response and resulting consequences, including analysis and assessment of energy sources and energy balances. In other words, *understanding is replacing automatic response.*

To recapitulate, an adaptive control system is subject to the effect of the environment on its sensing elements and has no freedom to control the effect of the environment on its sensing elements. It can only adapt the system by using its own energy to satisfy the requirement from the environment conveyed through the sensors.

Opposed to this automaticity is the human ability of adapting an environment by means that extend human reach in a specific fashion, including in the process the use of tools, machines, and psychological, socio-political, economic, educational and other instruments. Specifically, the human mechanism directs the signal-triggered action with a view to the adaptation of the environment to eliminate the differential between the fed-back signal resulting from the modified environment and the original input signal. The mechanism involved in the latter system or disturbance is subject to the "filter" of intelligence, thus creating an "art image" of the environment to serve as a blueprint for the adapting process [17; 18; 19; 22]. The system involved in specifically human activity is operable only when an action is triggered to adapt the existing, "given", "objective" environment to an "art" or "dream image".

Adapting Behavior of the Consumer in Economic Life

At a point of economic affluence in any society (where a large majority of the population is living at a level considered by its culture to be "modest but adequate" [11; 13], and has major discretionary purchasing power in terms of whatever costs are being considered within the parameters of the culture), consumers of goods and services begin to exhibit patterns of "adapting behavior". These patterns are quantifiable and measurable in terms of price elasticity of demand $\left(\sum_{Q}^{P}\right)$ and income elasticity of consumption $\left(\sum_{C}^{Y}\right)$, both over time and at specific single periods of time, for specific characteristics of both goods and prices [13].

The data show that consumer demand becomes inelastic as income and education increase, even for discretionary purchases; and that the substitution effect thus becomes a more important part of the change in purchasing patterns than the income effect. The substitution effect causes the buyer to follow a "least-cost" pattern of purchase, whatever the important component of cost may be for the individual – money, time or convenience. Declining elasticity causes a more rigid purchasing pattern on the one hand (I want what I want, *when* I want it), and on the other a more flexible willingness to switch from one good to another (when the two goods have almost identical characteristics) on the basis of price in terms of money, time or convenience. It also shows that the consumer has learned to say "No".

Consumers purchase bundles of "characteristics" [7; 8; 9] and not individual goods and services. Demand loses elasticity for many characteristics with rising income and education (no other variables proving significant). This loss of elasticity and consequent increase in the importance of the least-cost (substitution) effect extends to almost all characteristics and costs with the exception of those falling within the "responsive" and "adapting" behavior patterns where the human system changes the environment to suit itself.

As economic affluence increases, the effects of inelasticity of demand become evident in all material-reward oriented cultures, no matter what their political, economic or social systems. When a dictatorial society decides *not* to make desired consumer goods available – the most profitable industry in such a society will be underground, outside-the-law or smuggling oriented.

It is extremely important to relate the Lancastrian concepts as "bundles of characteristics" with declining price elasticity of demand. First of all, just as there are "bundles of characteristics" there are various types of prices. There are money prices, time prices (the amount of hours spent in any transaction), convenience prices (the amount of effort required for any transaction), and almost any other type of price which a consumer is normally willing to pay. What confuses the average noneconomically oriented individual is that a price may also be a good in itself, as a commodity or characteristic. For example, just as we pay in money, when we borrow money, we pay a price for money. Just as we pay in time, when we use a time saving device, we are paying a price for time. Just as we may pay in convenience or effort, when we use a particular salesperson in a particular store, we may be paying a price for saving convenience or effort.

It is an economic axiom that although a particular consumer demand for a particular commodity is completely satiable, the totality of all consumer demands for all commodities is insatiable. This same axiom is applicable to characteristics and to prices. At the present stage of American society the desire of people to save money may be much less important than the desire of people to save time. This is particularly true as more women with families enter the labor force. It is even more true of the white collar than the blue collar worker, and the preponderance of our labor force today is white collar workers and has been so for over the past twenty years.

It is therefore incumbent upon the economist in any discussion of adapting behavior to understand clearly the differentiation between money elasticity of demand, time elasticity of demand, and convenience elasticity of demand. These may and do differ by income group, by education and by extent of participation of the family within the labor force. The perfect example of this type of differentiation can be obtained from a new commodity appearing on the market within the past three years which was specifically tailored for the working wife and mother and which has had a phenomenal success in a field where the hope of new demand or increasing elasticity of demand was almost forgotten – food. The introduction of the differentiated meal by Birds-Eye in the form of different combinations of vegetables took the market by storm. It is interesting to note that pricewise, compared to the normal frozen vegetable, the Birds-Eye Hawaiian vegetables, the Birds-Eye Italian vegetables, the Birds-Eye French vegetables, etc., started as being almost double or triple in price. Nevertheless these vegetables not only sold but sold out.

What was happening could only be analyzed on the basis of the characteristics which the consumer was purchasing. The consumer's money and time price demand, that is, the consumer's value for money and time per se, was virtually inelastic in a working woman, but the consumer's demand for convenience had almost infinite elasticity and the woman was willing to pay the money price. When actual nutritional surveys were done of the contents of these packages plus the amount of time required to prepare both from scratch and from prepared foods it was found that the time savings approached anyplace between 40 minutes and 1 hour and 8 minutes in the preparation of these particular combinations. Therefore the woman purchasing these vegetables and the family using them was paying for convenience and for time, not for the contents of the package. In figuring the time at the minimal rate for household help of $2.00 per hour it was figured that the price per portion was something like 60c less, including the ingredients, than when these particular combinations had to be prepared at home. For the family where both adults are working the money elasticity is extremely inelastic, time elasticity is very inelastic, but the need for convenience becomes extremely elastic and the greater the amount of convenience given by whatever good it is per portion or item, the higher the price in money and in many cases in time the consumer will be willing to pay.

Today, Birds-Eye is going into the preparation of entire meals in the same fashion: where money price is high, time price is very low and convenience price is practically near zero. Recipes are given out with every package showing how meals may be prepared in less than five minutes. Since the value of the money as money is much lower than the value of the time and effort saved, the money price charged can be higher. Therefore simple projections of elasticity of demand on the basis of percentage of income change spent on a particular good, drops in importance. What becomes all important is what price is being paid for what characteristic.

The only price and characteristic in which consumers will have elastic demand in the future, as we become more and more of an upper middle income society (by the year 2020 it has been estimated that approximately less than 7% of our total population will be living in the poverty or near poverty brackets), will be convenience and variety. Demand in total is becoming inelastic in terms of money, inelastic in price of time, but elastic in convenience and effort price and elastic for variety.

Unfortunately the data are not presently available except in isolated cases (such as the food mentioned above) to enable us to do a true characteristic-differential price analysis of elasticity for all goods. However, new phenomena are arising, reported in the financial news, magazines, and in other public media which gave rise to the speculation that as far as income is concerned, the 1971–72 data will show almost total inelasticity for practically every commodity with the exception of services [1] (included in services are, of course, such items as medical care, recreation, both participatory and nonparticipatory, and travel).

This does not mean that in terms of another kind of price than money, demand will not be elastic. It will, but the difficulty is that we presently do not have the data, although we have the techniques for estimating the particular convenience price elasticity for something other than an item like Birds-Eye Hawaiian vegetables.

The consumer is becoming selective. Part of the equation for determining the elasticity of demand for any product or characteristic is the stock of goods already possessed by any particular consumer. The larger this stock the less elastic the demand, because purchase can always be postponed until complete satisfaction is obtained.

One of the phenomena of the past three years, with increasing inflation, has been the rise of the discount store in the business district such as Wall Street, downtown Philadelphia, etc. These stores are competing on a price basis with stores which otherwise were the only ones to have a low time and low convenience price, the center city large major market merchandisers. When a Silo Discount Appliance Shop opens diagonally across from Wanamaker's Department Store in the city of Philadelphia, it is providing a lower money price with the same time and convenience price as Wanamaker's and therefore is doing very well. We have seen the extension of this into men's and women's apparel, household appliances and recreational goods, personal services such as beauty parlors, etc., book and record stores, and even such items as variety stores for hardware, paper goods, etc. All these stores are now locationally able to give a low convenience price while competing with the money price and the time price of the larger normal previous distributional outlets.

The day when major appliances will probably be sold from catalogs is not far distant. It is already reaching major proportions from catalog stores such as Montgomery Ward and various other catalog outlets. The catalog store is, of course, the ultimate in low convenience price although, as many know, the money price is fairly high and the time price is extremely high because one has to wait for the merchandise to arrive (which may take any place from 2 to 6 weeks). However, there is no time price or convenience price in purchasing the item since this is done at home, or at work, on off minutes or at leisure. Consequently, one can say that they are charging a very high money price for time and for convenience and since the consumer's elasticity of demand for money is low whereas the elasticity of demand for convenience is extremely high, a high money price and a zero convenience price will meet the needs of this particular consumer.

In the same fashion the development of the so-called telephone store combined with catalog for the purchase of food is the ultimate in a zero convenience price, extremely low time price and a rather high money price, as it has been developed in Los Angeles. This is a computerized telephone catalog operation where the order is called in, recorded on the computer, filled by computer, trucks are routed by computer and the order is delivered directly to the door. This type of operation, together with the superstore where everything exists under one roof, is obviously the ultimate answer to a high elasticity of demand for convenience in that it has an extremely low convenience price – the lowest possible in the circumstances.

Another perfect example of a low money elasticity of demand concurrently operating with a high elasticity of demand for variety exists in so-called cable or pay television. In this case the consumer does not mind paying the money because his money elasticity of demand is so low that the money charged can be rather expensive – but the price he is paying for variety is very low, in that he can see what he wants when he wants it without leaving the comfort of his own armchair.

Adapting Behavior: Thinking and Education in Consumption

In an adapting control system, the response to an input signal is not necessarily in a specified relationship to the input, due to differential individual perception and valuation. Perception and understanding are shown by empirical data to be altered by education.

In 1960–61, for education of head of household of eight years or less, $\sum_C^Y$ for "gifts and contributions" reaches an inflection point[5] at income of about \$10,000; for head of household having graduated college, the inflection point is reached at income from \$5,000 to \$6,000. (The relationship of "gifts and contributions" to income is inelastic until the inflection point is reached and elastic thereafter). Similarly, the inflection point for $\sum_C^Y$ for "medical care" is reached at \$5,000, in 1960–61 income, for households having nongrade school-graduate heads; but at \$2,000 for households with college graduate heads [11: 403–411; 12].

Although other consumption patterns are differentiated by education, the tw cited above are important because the first ("gifts and contributions") is an example of a feeling of social responsibility existing at lower income levels for highly educated families – to be specific at roughly half the income level of low-education families; and the second ("medical care") is an example of appreciatio of the necessity for preventive medical advice occurring again at a lower income level for the more highly educated family. When the income elasticity of consum tion$\left(\sum_C^Y\right)$ is greater than 1, or elastic, it specifically means that as family income rises, an increased proportion of such income increase is spent on the good or service. Thus, a college-graduate-headed family, with an increase in income from a mean of \$5,500 to a mean of \$6,750, will increase its charitable contributions by more than double its percentage income rise, and the same is true of medical care expenditures.

The input signal is identical for both groups of families – an increase in incom but the response by each group is not a function of the signal above; it is a function of the signal and the educational level of the family. Incidentally, for both goods and services studied, the response at each income level is *not* a function of race.

As defined by Dewey, art is "to select what is significant and to reject by the very same impulse what is irrelevant and thereby compressing and intensifying the significant" [5: 208]. We should add to the statement that both the "significant" and the "irrelevant" are dynamic concepts that continuously change position. Because machines have only automatic, adaptive responses, and thus have built-in the qualitative aspects, or "significant aspects", "creativity" is impossible.

Education (formal and/or informal) is the phenomenon which initiates a control activity, triggered by the element of relation, association or construction that appears, for example, when an artist produces an image unlike the one achieved by a camera. It also appears in all scientific discovery, as a change from the accepted previous concept. In other words, education centers on the "art" created image and its involvement in control system activity.

Adapting behavior depends on education and not training alone. Training involves learning some specified pattern of behavior, be it prestidigitation or tightrope walking, while education is *new concept formation.* The result of education is creativity, while the result of training is performance involving skill.

If the adapting control process "filters" disturbances, or input signals, in the closed-loop servo-system which controls human action, education is then taking place.

The servo-mechanism of the human control system continuously develops anc grows as thinking develops and grows. Inquiry and correlation of experience are tools used in this process of education; they are elements which trigger the controls. As for experience itself, we can no more know what a particular "experience" will do to education than what a "pencil" will write. Experience, of course, is a pre-requisite, just as one needs a pencil or something to write with.

Any realization of something being wrong is a discovery. It contradicts the previously assumed satisfactory order. Anything that has been "logical" up to this point becomes "illogical", becomes "wrong", becomes an "error", and will make room for the elimination of error – for a new logic – for the "ought" instead of the "is". This realization that something is wrong (which initiates the process) is a prerequisite required for new concept formation. There is a difference between man and animal, or man and machine which is made to simulate man's behavior. The computer essentially accomplishes its function by operating on a multitude of types of problems with techniques for solving them. Thus, a problem fed into the computer in a sense triggers the answer that was originally built into it. But, to reiterate, human problem solving is a matter of education and growth. It creates or formulates problems and at times their solutions.

We have thus a model of thinking which contains *quality* as an essential element and operates pragmatically as a closed self-organizing loop. It accounts in a new way for teleological processes like problem solving, "planning", and mechanistic behavior. It allows for an infinite variety of awareness-cognition-response feedback systems.

Adapting Behavior and Learning

Learning in education is the possibility of going outside of a frame of activity. The difference between man and animal or machine is specifically that a machine that has "automatic" activity has, of course, been programmed to so act. It can automatically perform activities which it was designed to perform. An animal or man can also be programmed, i.e. the responses are limited to programming or designing, just as in behavioral terms, persons automatically respond as experience, reinforcement or "programming" has determined that they shall. The responses are the result of training. Brainwashed man is as programmed as a machine. The learning in this case is programmed, hence automatic. But it is questionable whether one can train all men. The possibility of training may be inversely related to the distance the individual has progressed from the animal state.

Similarly, what is happening in the market place is the development of a responsive environment for consumers with increasingly adapting behavior patterns. Those markets that are not responsive (in its cybernetic sense) environments are going bankrupt.

A system is an organized whole of parts. Hence: $\sum_{i=1}^{N} E_i$ is a system. However, is the system the same if E_2 comes first and E_1 comes second? The answer is no.

Just as in a responsive environment, learning occurs when the individual controls and influences his environment [14; 18; 20; 21; 22; 23], so, the environment, not only a learning environment but the total environment must be made responsive to the individual's actions. It may be that we have now reached the stage wherein, when an individual gets the time to learn in an environment responsive to his desires,[6] he may now begin to think he can influence his environment, and attempt to influence society to give him what he wants. Patterns of consumption acquired in consumer behavior seem to indicate this. Note the meat boycotts, rise of indigenous buying groups, refusal to accept authority, changes in purchases of clothing, furniture, cars, the increased boutiques, and the rise of the consumer veto.

In the United States, choice is present as a result of education and the market place. In the Soviet Union, choice is present as a result of education only. Once choice is present, once options are available, the environment must become responsive to the individual's desires (note: needs are necessities and inelastically demanded, wants and desires are elastically demanded) for the environment to survive. In this case, however, the environment is the usable environment, in the sense of goods and services, and in political, moral and/or social behavior, or education. Furthermore, an individual's gain need not entail another's loss. We are faced here with non-zero sum games. Consumers' surplus, according to John Bates Clark, is a case in point where an increase of one individual's satisfaction does not decrease the satisfaction of another. The principle of insurance is anoth case. Adapting behavior on the part of the individual implies maturity and non-degradation of the environment.

To create a properly functioning social entity that is active in animal husband agricultural pursuits and other activities under the adapting control system (adapting the environment instead of being adapted to it) a system of communic tion is required in which understanding, as an element of an adapting control system, plays a major role. The communication system suitable for an adapting process requires means of communicating elements leading to understanding. A characteristic specific to an adapting system is a type of communication, the nature of which goes beyond transmittal of information. It involves an element of possible reaction to the signal on the part of the receiver that permits understanding of this reaction by the originator of the communication. This particular phenomenon implies a closed loop communication.

In addition, it is an adapting communication system in its own right. Thus the early establishment of the U.S. Agricultural Experiment Station was sited within a radius of a single day's travel there and back for the average farmer in the vicinity. As modes of transportation became capable of encompassing longer distances, the Stations could move further apart; but the necessity for communication was the main determinant of the time-distance.

In the same fashion, industrial and technical research has always tended to cluster about centers permitting constant transaction and communication between individuals. "Science centers" have always existed since the temples of Mesopotamia and Egypt. Knowledge and invention cannot be pursued in a vacuum.

Adapting Behavior and Ethics

There is a further phenomenon that arises when the individual subjugates the environment without developing understanding and control. The meaning of the word "rule" as used herein is to describe a hypothetical relationship of cause and effect as applied to behavior in a control system. It implies a mathematical formula relationship as an end product. Thus, the law or rule of gravity translated into a mathematical equation by Newton used the word gravity in describing a phenomenon, the nature of which was, and still is, a mystery.

When we use the word "ethic" we are discussing an interrelationship. When we use the word "rule", we are describing a causal relationship. Anthropomorphically, a rule was a causal relationship derived from the "Lord". But this is not our concept of ethics.

Communication in an adapting system of man – man stands for a relationship between man and man as described by the term "Ethics", and comprises a special condition that implies *mutual* understanding, awareness, consciousness and reasoning, in contrast with a relationship lacking these ingredients.

We have distinguished between an *adaptive system*, which is a *self-organizing system*, and an *adapting system*, where there is *organizing control*, requiring consciousness. Consciousness is the acquired characteristic of an adapting control. Thus, training is not sufficient nor satisfactory for moral behavior.

As previously noted, communication in an adapting control system relates to a relationship between man and man and comprises a special condition that implies *mutual* understanding, awareness, consciousness and reasoning. This characteristic is acquired by each new generation from the previous one by means of the educational process. Thus, the educational process has a prerequisite: mutual understanding.

The social form of government under an oligarchy, ruling according to rules of slavery or domination of any sort, is the expression of undeveloped understanding by the few in their effort at using the adapting controls to tame the many. (Slavery is defined as external control of the individual's ability to work, think, move and establish familial relationships.) Slavery is the expression of the system of adapting controls characteristic of man, but paradoxically enough, so is freedom from slavery.

To put the above in technical control language would sound something like the following: Moral rules contain the desired quiescent state of a system. This implies that control action does not take place when the moral rules conditions are satisfied. The reaction of the system when the moral rules are not satisfied can thus be considered of two kinds:

1. To satisfy the requirement of a moral rule.
2. To eliminate the requirement for a moral rule.

When control is vested in an oligarchy, the moral rules to satisfy that oligarchy are to be found in category 1. The reluctance of human nature to follow such rules falls in category 2. Control of the adapting type is required in order to eliminate this conflict between the oligarchy and the reluctance of people to follow.

Free social forms are only possible with the overwhelming majority understanding the nature of the adapting control system of man.[7] This makes equal opportunity, voluntary cooperation and competition for all, a satisfactory social environment. Communication in such a system establishes a relationship between man and man as described by the term ethics and consists of decisions concerning continuous choices. Education for ethical behavior is education for choices to be continuously made.

Control Systems versus Reinforcement Control

It has been observed experimentally that providing knowledge of results, rather than reducing or withholding knowledge, does lead to more effective learning. Immediate knowledge is more effective than delayed knowledge, but it will not automatically enhance efficiency of performance and learning. Yet, it is generally assumed that learning can be enhanced if it is followed by reinforcement.

Dynamic sensory feedback provides an intrinsic means of regulating motion in relation to the environment, while knowledge of results, given after a response, is a static after-effect, which may give information about accuracy, but does not give dynamic regulating stimuli. Dynamic feedback indication of "error" would thus be expected to be more effective in performance and learning, than static knowledge of results.

Furthermore, the efficacy of reinforcement assumes an active need or drive state, while feedback theory assumes that the organism is built as an action system and thus energizes itself. Hence, body needs and wants are satisfied by behavior that is structured primarily according to perceptual organizational mechanisms, and require programs that communicate. We can now judge why reinforcement of a child turning his head to the right, being reinforced by a sucrose solution sucked from a bottle, takes hundreds of tries, and Bruner's baby with the $20,000 pacifier takes only a few tries, about five seconds, before he learns to focus a picture of his mother, and he isn't even hungry [4]. The bottle experiment is a stimulus-response model, while the pacifier experiment is a true cybernetic feedback model [2; 3; 4].

Pacifier Experiment (Cybernetic Control Model)	*Bottle Experiment* (Stimulus-Response Model)
• No physiological deprivation	• Physiological deprivation
• No hunger	• Hunger
• Free movement	• Swaddled
• Closed loop	• Open loop
• Internal control	• External control
• Voluntary control	• Stimulus control
• Intrinsic means of regulating motion	• Extrinsic reinforcement schedule
• Means and ends not bonded	• Means and ends are bonded
• Systematic relation to the learned behavior	• No systematic relation to the learned behavior
• Learning requires *no* reinforcement	• If learning occurs, it is transient, requiring reinforcement
• Behavior is the control of input	• Behavior is the control of output
• Dynamic continuous feedback	• Static after effect of knowledge
• Few trials for learning	• Many trials for learning
• Self-determined learning	• Doubtful feasibility of conditioning
• Primitive adapting system	• System adaptive only if and when successfully engaged

To summarize: Use of linear programs (including branching) in all teaching deliberately limits the media of communication, the experience of the student and thus the depth of understanding that he achieves.

Instead the student should be provided with a broad context of experience by resorting to all of the activities and to all of the communicative media at our disposal. This includes verbal and nonverbal material. Thus, the student learns by responding to the perceptual organization of his environment.

Beyond deviation-counteracting feedback or negative feedback, there is also operative a deviation-amplifying parameter, or positive feedback [23].

The world of advertising media in our present "free society" has been geared to the development of reinforcement, stimulus-response models and not cybernetic control, because the media has usually assumed *adaptive* behavior on the part of the consumer. By assuming that the consumer is an adaptive personality and therefore learns what is being taught without wanting to use the learning as a means of further expression, the media in advertising have assumed that constant repetition would cause the consumer to learn, without thinking either of the repetitive method or of the application. Reinforcement control without obtaining consumer reaction (other than in the most general fashion of like or dislike, I or nul-I, percentage listening versus percentage not listening, percentage tuned in versus percentage not tuned in, and the entire world of Neilsen Ratings) is solely for an adaptive behavior society.

Cybernetic control on the other hand, assumes that the response of the media to the needs of the consumer dictates the type of approach to the consumer and that this approach is changeable as the consumer responses are obtained. The chart showing the differentiation between reinforcement and cybernetic control has been given above.

If consumer response to produce differentiation can be looked at as a form of adapting behavior (inelastically demanded product for which according to empirical evidence and theory the substitution effect of least cost is greater than the income effect) no manner of reinforcement control can influence the buying of that good which is cheapest in time, money or convenience. As some of our manufacturers have found to their sorrow (the latest bankruptcy in men's clothing being century-old Botany Industries), no amount of reinforcement control could possibly influence an adapting society, which is exactly what happened.

In the early nineteen-sixties, President John F. Kennedy sent, for the first time, a message to Congress on Consumer Rights:

1. The Right to Safety – To be protected against the marketing of goods which are hazardous to health and life.
2. The Right to be Informed – To be protected against fraudulent, deceitful or grossly misleading information, advertising, labeling or other practices, and to be given the facts needed to make an informed choice.
3. The Right to Choose – To be assured, wherever possible, access to a variety of products and services at competitive prices; and, in those industries where government regulations are substituted, an assurance of satisfactory quality and service at fair prices.
4. The Right to be Heard – To be assured that consumer interests will receive full and sympathetic consideration in the formulation of government policy and fair expeditious treatment in its administrative tribunals [10: 3–4].

It should be noted that of these four Consumer Rights first mentioned in John Kennedy's message, three are solely for an adapting behavior society: the right to be informed, the right to choose, and the right to be heard, which are of course, the basis of cybernetic control or feedback.

Interestingly enough it has been increasingly proven by a number of empirical studies that those advertisements giving accurate information have been and are much more effective than "puffery" as it is known in advertising circles. For the first time in history during the decade of the latter part of the 1960's until the present day, manufacturer after manufacturer has been forced to establish a consumer department or a complaint department which is a direct form of industrial "ombudsman" ready to give information and service to the consumer. Word of mouth campaigns in today's society are having much more influence than the greatest dollar amount of Madison Avenue shellac. Many manufacturers have found that such consumer activities as hiring billboards to express dissatisfaction with product, using sky advertising to advertise defects, increased use of the courts for suing for defective product, and increased bombardment of consumer complaints has forced them into increasing quality control and in some cases to total redesign.

The increased number of subscriptions and memberships in Consumer's Union and other impartial product evaluation organizations within the past five years has caught manufacturers by surprise. Sears Roebuck after its first introduction has never readvertised "Tuffie" jeans, because every time they come into the store they are sold out (after the CU report that they outlasted regular jeans by a factor of three times normal wear). The consumer response to the gasoline and oil crisis is another example of adapting behavior – and the slowness of consumer desire to invest in new automobiles after this crisis has completely confounded the manufacturers – no matter what the advertising says.

The consumer is well able to differentiate between prices, such as the difference between the money price of an automobile, the time price of an automobile, the gasoline price of an automobile, the service and repair price of an automobile and the upkeep price of an automobile. Increasingly he is beginning to demand increased warranties from the dealer even though he cannot obtain them (ostensibly) from the manufacturer.

Where money is important the consumer is beginning to shop in groups. The increased rise of the co-op, particularly where health foods or so called organically grown foods are concerned, is an interesting case of consumers having extremely low elasticities of demand in terms of money but extremely high elasticities of demand for lack of pesticides. Whether or not this matters to the body physiologically is not important, it obviously matters to the consumer psychologically.

Adapting Behavior and Societal Development

While there have been numerous studies that have attempted to integrate the negation of material or cognitive reward by the substantiation of conditioning, *only a material-reward approach can be successfully projected.* If deviation-amplification sets in, cognitive development occurs. What appears to follow is social and affective development [20; 21; 23].

At this time it must be maintained that the individual in a material-reward oriented society that has not attained the "modest but adequate" pattern of living of that society, in an economic sense, is adaptive. As the individual becomes less financially restricted (more and more able to obtain the "normal" accepted level of material life), behavior becomes adapting. In other words, he can use ever greater economic, political or social leeway to change his environment. The precipitating factor is education.

Thus Florence prior to the Renaissance, was the richest country (in terms of time and access) in Italy because it was the one area with sufficient food (which in the Middle Ages was the equivalent of a "modest but adequate" level of material living). Hence, they had discretionary purchase power and from this a positive time correlation with the start of the Renaissance, which was an integrated scientific-cultural-social development within the limitation of the technology of the Middle Ages.

The only two other non-slave based pre-technological societies which attained discretionary purchasing power in terms of their respective cultures were the Inca society in central America and the land of Israel at the time of Solomon. Although the former society is presently undocumented except from tales of the Spanish conquistadores and ruins, the latter has been aptly described in the Old Testament as the 40 years of fulfillment of the promise of the Lord to Solomon at Gibeon (Chronicles II, 12), "I will give thee riches and wealth, and honour, such as none of the kings have had that have been before thee, neither shall there any after thee have the like". And for 20 years Israel, in a land having perhaps less than 3 million inhabitants, built the Temple, with an equivalent man-year labor expenditure of 20 times 153,600 men.

Our present technology is a system of three types of production of which the second has only become technologically feasible in most industries producing discretionary goods within the past 50 years.

1. *Mass Production* – infinite runs of identical goods.
2. *Differentiated Production* – combination of mass produced parts forming an infinite variety of permutations and combinations – (note that an individual hardly ever, if ever, sees two cars alike) possible combinations are myriad. Similarly, clothing parts are mass produced, but assembly is individual. Therefore, a particular production run can be as short as one wants to make it.
3. *Hand Production* – Here there are, of course, infinite combinations possible.

The second and third types of production act to allow for discretionary income. The first type answers the living base. Naturally, combinations of the three types of production run the same gamut as combinations of goods and their characteristics.

As demand becomes inelastic, as items become necessities, or individual reaction to cost changes become inflexible, the substitution effect takes over. The black box makes no distinction between equivalent characteristics except cost to the individual, be it in dollars, time, inconvenience, longevity and so on.

As the United States approaches, in the 1970's and 1980's, the same relative level of discretionary income as Florence or Solomon's time, we see burgeoning the means to grow from this country an indigenous cultural-technological-sociological expression which may well be the beginning within our own technological ability, of a new integration.

It has only become possible with the work of Kelvin Lancaster to distinguish elasticity of demand for characteristics of goods by components of price in such a way that integrated comparisons of different mores become scientifically feasible. Consequently, for the first time it may be possible to generalize from single country experience on the probability that comparable changes for characteristics (even if goods are noncomparable) occurs in materially oriented countries, no matter what their system of ownership, political power, or social structure. However, we cannot compare patterns in a material and a non-materially oriented culture (i.e. Zen Buddhism).

In primitive societies, and in societies where non-material rewards operate, the mass of the people are adapted to the requirements of the few that are leading. Material rewards must be of this life, at this time, and cannot be a credit for the next life or transformation.

As mentioned above, there are material-reward oriented societies where even though scientific development has occurred, we still have a majority adapted to the requirements of a minority, but with consequences of inefficiency, bottlenecks, breakdown and resource waste.[8]

In the U.S. the condition of greater interdependence arose with the age of the rail. Today, almost instantaneous communication creates even greater interdependence. Therefore, with a more educated population, one gets group dynamic adapting behavior. No matter how small a group (2%, 20% or 90% of the population), each can create as much hell as the other. Interdependence carries over into social action. An individual transacts with the closest individual with whom he is already involved in some capacity, thus leading rather easily into cooperative venture (i.e. students getting together to picket a laundry that ruined John's shirt).

This kind of social interdependence occurs over the entire range of income groups except where the individual is absolutely indigent. It extends from the very top to the very bottom. It changes and is amorphous, since one is dealing with a cybernetic flow, a situation having social feedback, dynamic give and take, occurring between people. As understanding replaces automatic response, social transactions develop. The number of possible permutations is infinite! One can never predict people's actions with certainty as people have become accustomed to behaving adaptingly, but the laws of probability permit approximation.

As noted above, art is something no one can teach. One cannot choose what is significant for another. One cannot make a person enhance or distort something in a way that one does not himself know how to distort or enhance. Yet, such oblique or surrealist views and disorderly processes are a necessity for adapting behavior to occur. Furthermore, the selection of the "significant" depends on choice being present, otherwise no selection can occur. All scientific discovery depends on such processes, as a change from the accepted previous concept.

We now postulate that countries such as the U.S.S.R. and China will not grow unless there is enhancement of adapting behavior. And the U.S. as we know it, will probably continue to change. The present state of the United States is that of an adapting society, rich today and tomorrow, given the technological possibility of unlimited nonpolluting power and water, within a few generations.

Statistics and informations coming out of the U.S.S.R. (not given by the official publications) indicate that Russia has been having probably the highest labor turnover in history in the new scientific towns east of the Urals. The new generation graduated from college is assigned to jobs in these areas. In order to make these areas attractive to these workers (mostly white collar scientific and technical personnel) Russia has been allocating to them most of the durable goods available in the country, such as first priority on automobile purchase, television purchase, household appliances such as washers, dryers, etc. The young generation graduating from the schools has signed up for five year contracts in the east Ural technological communities. Once there, they obtain their durable goods and at the end of five years, 60% of them (according to the best available figures) head back to major population areas within the U.S.S.R. One thing the Russian government cannot provide for these areas is, obviously, climate. The second is a lack of cultural interchange which is required for personal growth and development. Consequently, the turnover of personnel in these areas is the largest in the Soviet Union no matter what the government can do.

This is not a directed movement. It is absolutely indigenous and occurring without the concurrence of the government, because unless the labor contracts are made for ten years there is nothing the government can do to stop it. The shortage of scientific and technical personnel is such in Russia that pirating is rife and the young people can at the end of their contract obtain jobs which are more attractive in terms of physical and cultural environment.

Since these young scientific and technical workers are the middle class of Russia they are adopting adapting behavior patterns in the meaning of the term used in this paper. As more of the Russian population approaches this degree of affluence it also will adopt adapting behavior patterns. The Russian population is presently forcing the government into production of consumer goods. If this is not adapting behavior, then it has never before been exhibited in the Soviet Union. There is no doubt of the fact that the indirect influence of consumer demand is making itself felt in the Kremlin. How soon it will make itself felt in Peking is a question of how soon the society can become more affluent.

In the U.S. the governmental system has worked more on intra-party accommodation than on inter-party rivalry. If there is an infinitely possible variety of permutations of feedback effects, then there is going to be inter-party accommodation in the future. The effect of division into parties becomes much less important. What emerges is a form of concensus called the *pragmatic position.* Presently in many questionnaires, checks are asked for the following categories in voting: Democrat, Republican, Leftist, Rightist, Independent, and Pragmatic (*votes on issues*). The Republican "Turks" band together with the Democratic "Turks" on particular actions and do not gang up on one another, thus getting a shift into permanent tailspin, yielding one position on any particular issue. What is occurring is that the two party system is disintegrating into a kaleidescope of issue-dominated, shifting time, accommodative groupings. "A" may agree with "B" on issue 1 and with "C" on issue 2, but this does not prevent "A" from disagreeing with both "B" and "C" on issue 3. As stated above, the system also changes according to order as well as quantity in a non-zero sum game.

Since there is no human life without order or rule or ethic, what we are developing is a continuous-shift pragmatic grouping society.

Feedback, in the cybernetic sense (as opposed to that feedback which means knowledge of results) occurs not only in economics but in politics, in the family, in life. (Note the emergence of communication as a descriptive and explicative framework in psychiatry.) It is affecting the behavior of an entire population. We now have evidence and measurable data in economics that indicate the presence of this phenomenon in consumer behavior. The presence of this adapting behavior does not mean that it is limited to economics only. Why should it be isolated to this field? It is a phenomenon that is pervasive, be it in education, the psychology of learning, social development, ethical behavior, or political development. Adapting behavior is much more than just adapting economic behavior.

In 1910 pre-World War I United States, there was a small upper elite and a large lower class. By 1960, the Newport estates, the yachts and the 27 servants had just about disappeared. The shift occurred in that pre-World War I, the majority was adapted to the requirements of the minority. It led to where the minority became adapted to the majority. The shift is still going on and consolidations continue to take place.

At present there are societies that are technologically and scientifically advanced and are still slave societies. The question arises as to whether in such a society technology and science can be used for or by the slaves. If the slaves are the masters of technology and begin to have discretionary income, time and purchasing power, then agitation will set in. The move towards adapting behavior will occur.

There can also be a technologically advanced society that may not provide a sufficiency of goods for a majority of people. This can also be a master-slave society. But once a group of educated people, be it slave class or not, has discretionary purchasing power in terms of income or time, they become adapting creatures as exhibited in consumption patterns, where slavery cannot co-exist.

In a technological society where the masters use the technology for the slaves and the slaves "receive" the "good life", there are several problems that become associated:

1. The masters will tire out sooner or later if they do all the work.
2. The masters become the working class. Eventually no one will want to be in it.
3. Such society is similar to a bee hive, or an ant heap where the masters become the workers and the queen and the slaves become the drones. Inbreeding would eventually kill off the masters, if the slaves would not be permitted to mix.

There is evidence that learning in the sense of new concept formation develops with conscious individual growth and assertion [17; 18; 21; 23]. If such learning is not enhanced, education is not taking place and only training is allowed, such society will show few discoveries (going beyond the present state of the art). At best it will have innovations (using the present state of the art) but growth will stagnate and the society will be static and deteriorate.

For about 7,000 years of recorded history, since Shub Ad of Ur, the vast majority of people have worried about one thing . . . food. The entire human energy output has been expanded on trying to eat. Today this is still true for the vast majority of the world, for most of the third world, the greater portion of the Middle East, South America and Asia. This is also still the case in some parts of Europe. Only two centuries, $\frac{1}{14}$th of the time span, has given us progress to the point where some people stopped starving in some countries. Basic progress has occurred since the time of mass production, during the second to the eighth decades of the 20th century. In the decades since the 1950's the greatest strides ever have been made. We thus have major development in less than $\frac{1}{10}$th of the time of recorded history. In the U.S. the discretionary income class became a majority after 1964 – less than ten years ago. (We now have evidence that the majority of the blacks have entered the middle class.) [25] The black leadership that led the agitation of the blacks were from the lower upper and upper middle classes. It is thus understandable why this is not the case yet with the Chicanos and the Amerinds (American Indians) and not quite the case with the Puerto Ricans. But that is beginning.

The adapting behavior pattern can best be illustrated by the turnaround in the acceptance of consumer nondurable and durable goods as evidence of the "good life" in American society which occurred with the generation of the 1960 college students. This has had profound effect upon our entire society. We have seen an entire generation conduct a revolution in lifestyle as well as a revolution in dress, in eating habits and patterns, etc., which will probably continue for a long period of time. It is interesting to note that those participating in the lifestyle revolution also have extremely low money elasticities of demand for such goods as they do require, such as hi-fi equipment, automobiles or bicycles, etc. In other words they do not seem to mind what price they pay as long as they obtain what they want. However, they have an extremely high elasticity of demand for variety for those goods which they want, and are willing to experiment in obtaining those items of dress and living which they consider necessary.

We now have the spectacle of kitchen appliance boutiques or kitchen utensil boutiques springing up all over the United States without previous advertising, simply occurring by word of mouth, that the Swedish plastic spoons are better than the U.S. plastic spoons (they don't disintegrate in hotwater). This kind of thing is the perfect example of adapting behavior. Wanamaker's fifth floor now has a boutique for kitchen utensils such as wooden spoons and mixing bowls and pots. We have seen also a rise in the sale of cookbooks, which has almost outdistanced percentagewise the increase in sale of Bibles.

The importance of food and its preparation as well as the importance of home work and home art is a perfect example of adapting behavior. When you want variety you are willing to pay a higher time price to get something different by making it yourself. The increasing sale of the home sewing machine, the springing up of art needlework, painting classes, sculpting classes and all the other appurtenances of the so-called "good" life in the United States is taking place in every small town in the country. People are now beginning to assign to their home production of luxuries more importance than they assign to their ability to purchase them. Hand work is coming to have a higher price than dollars, or the ability to purchase.

The entire movement towards a demand for a good which does not take time for people to use is another example of adapting behavior. People require durability because they do not want to take the time to shop once they have something. They want it to last. They don't want to have to go out and buy it again. For this they are willing to pay a higher dollar price, but the time price is very low and the convenience price of not having to shop twice becomes zero.

CONCLUSION

We have found a pattern of adapting behavior where the environment is changed to suit the requirements of the system, as opposed to the system changing to suit the requirements of the environment. This is a true cybernetic activity in response to economic stimuli. Only education is correlated with this pattern of behavior. *Not* even race is so correlated.

We found adapting behavior or *learned behavior*, in a situation that is not a learning situation, for instance, in a consumption pattern. This means that that kind of learning can extend through life and exhibits a pattern in which the environment is changed to suit individual requirements. Severe problems ensue if this is not recognized. For instance, we have had an enormous increase in bankruptcies amidst the most prosperous economy ever in the past 20 years. In advertising, for instance, not all the repetition in the world has an effect on human consumption. Note the radical change in some of today's advertising in an attempt to influence consumption.

Exhibited adapting behavior is evidenced in a cybernetic situation wherein the individual *takes* choices, relates to information *selectively*, and refuses to be "brainwashed" or influenced in choosing what he desires. This applies in every political environment, such as the new towns in the U.S.S.R. This adapting behavior is exhibited when a particular economic threshold is reached. The individual quickly exhibits this behavior in educational, social and political areas as well. The economic threshold is where the majority of the nation becomes middle class, or to put it in economic terms, when the majority have achieved discretionary income.

It is therefore postulated, that all societies which have attained freedom from abject want will eventually approach the point at which affluence in time, goods and/or money will make it necessary for the societies to respond to adapting behavior, a distinctive human transaction. Of necessity this implies technological development. Such patterns of adapting behavior can be pervasive in all human relationships, in every field of endeavor.

ENDNOTES

[1] The modest empiricism in human transaction herein exemplified is developed from a general theory of adapting behavior [15; 16].

[2] It increases for services.

[3] As opposed to the "Psychological Abstract" definition of feedback as "knowledge of results".

[4] The term "error input" is an engineering term commonly accepted to mean a disturbance

[5] Point of slope change.

[6] Princeton experiment with negative income tax.

[7] Today there are no free social forms in our society. Just because the majority has the power does not mean that they are using it with understanding.

[8] The largest industry in the Soviet Union is smuggling and other forms of illegal or underground activities including reproduction of banned books, records, etc.

REFERENCES

[1] Bergstrom, T. C. and Goodman, R. P., "Private demands for public goods", *American Economic Review,* **63**, No. 3: 280–296, 1973.

[2] Bruner, S., *Studies in Cognitive Growth: Infancy* (Volume III Heinz Werner Lecture Series). Worcester, Mass.: Clark University Press with Barre Publishers, 1968.

[3] Bruner, J. S., "Up from helplessness", *in* DeCecco, J. P. (Ed.), *Readings in Educational Psychology:* 71–75, Del Mar, California: CMR Books, 1970.

[4] Bruner, J. S. and Kalnins, I. V., "The coordination of visual observation and instrumental behavior in early infancy", *Perception,* in press: 1–19.

[5] Dewey, J., *Art as Experience,* New York: Minton Balch, 1934.

[6] Dewey, J., and Bentley, A. F., *Knowing and the Known,* Boston: The Beacon Press, 1949.

[7] Lancaster, K. J., "Change and innovation in the technology of consumption", *American Economic Review,* **56,** No. 2: 14–23, 1966.

[8] Lancaster, K. J., "A new approach to consumer theory", *Journal of Political Economy,* **74:** 132–157, 1966.

[9] Lancaster, K. J., *Consumer Demand,* New York: Columbia University Press, 1971.

[10] Magnuson, Senator W. G., "Consumerism and the emerging goals of a new society", *in* Gaedeke, R. M. and Etcheson, W. W. (Eds.), *Consumerism: Viewpoints from Business, Government, and the Public Interest,* San Francisco: Canfield Press, 1972.

[11] Schulman, R., *The Economics of Consumption for a Changing Society,* Philadelphia: Drexel University Press, 1972.

[12] Schulman, R., "Bibliography for consumer economics", Philadelphia: Drexel University, mimeo, 1969.

[13] Schulman, R., "Communication and feedback in the technology of consumption", *VIIth International Congress on Cybernetics:* 737–744, Namur: Association Internationale de Cybernetique, 1973.

[14] Scriven, M., "Teaching ourselves by learning machines", *Journal of Philosophy,* **67,** No. 21: 898–908, 1970.

[15] Steg, D. R. and Schulman, R., "A general theory of adapting behavior", *VIIth International Congress on Cybernetics:* 831–843, Namur: Association Internationale de Cybernetique, 1973.

[16] Steg, D. R. and Schulman, R., "An interdisciplinary theory of adapting behavior", Office of Naval Research Technical Report, NR 151–356X: 274–289, September 1974.

[17] Steg, D. R., "Some aspects of teaching and learning", *in* Villemain, F. T. (Ed.), *Proceedings of Philosophy of Education Society:* 132–139, Chicago, Illinois: Southern Illinois University, Edwardsville, Illinois, May, 1964.

[18] Steg, D. R., "Programmed teaching and learning", *in* Villemain, F. T. (Ed.), *Proceedings of Philosophy of Education Society.* Chicago, Illinois: Southern Illinois University, Edwardsville, Illinois, May, 1963.

[19] Steg, D. R., "System rules and ethics", *in* Villemain, F. T. (Ed.), *Proceedings of Philosophy of Education Society:* 20–25, Chicago, Illinois: Southern Illinois University, Edwardsville, Illinois, 1966.

[20] Steg, D. R., Mattleman, M. and Hammil, D., *Effects of Individual Programmed Instruction on Initial Reading Skills and Language Behavior in Early Childhood,* International Reading Association, 6 Tyre Avenue, Newark, Delaware, April 1969.

[21] Steg, D. R. and D'Anunzio, A., "Some theoretical and experimental considerations of responsive environments, learning and social development", *in* Rose, J. (Ed.), *Progress of Cybernetics,* Volume 3: 1145–1161, New York: Gordon and Breach Science Publishers, 1970.

[22] Steg, D. R., "A philosophical and cybernetic model of thinking, or a feedback analog to thinking and other consequences", *in* Scandura, J., Durnin, J. and Wulfeck II, W. (Eds.), *Proceedings of the Society for Structural Learning:* Philadelphia, Pennsylvania,

University of Pennsylvania, April, 1973.

[23] Steg, D. R., D'Anunzio, A. and Fox, C., "Deviation-amplifying processes and individual human growth and behavior", *in* Rose, J. (Ed.), *Advances in Cybernetics and Systems:* 1646–1665, New York: Gordon and Breach, 1974.

[24] U.S. Department of Labor, Bureau of Labor Statistics: *Survey of Consumer Expenditures: Consumer Expenditure and Income, Urban U.S., 1960, 1961 and 1960–61,* B.L.S. *Report 237–38,* Washington, D.C.: Government Printing Office, 1963–70.

[25] Wattenberg, B. J. and Scammon, R. M., "Black progress and liberal rhetoric", *Commentary,* **55**, No. 4: 35–44, 1973.

ON SMALL HUMAN SYSTEMS

INTRODUCTION

When a system involves a great many people, theories about such a system can no longer represent human beings as individuals. They are then either reduced to their statistical properties or assessed by the net contribution they make to the fabric of existing organizational forms. Economic modeling of large systems is a way of coping with otherwise unmanageable collectivities by applying a uniform accounting scheme in which the complexity of human personalities become irrelevant. The polar-opposite to this image of man is held in psychology where individuals tend to be considered complex, richly structured and possibly the only worthwhile objects of study. Papers in this section lie in-between these extremes in that they focus on the dynamics of social systems which are not too large for their components to be individually recognizable and structurally describable. The traditional domains of this intermediate focus are social psychology and small group research which attempt to explain respectively, the behavior of individuals in terms of the perception of their environment, and the behavior of a group in terms of the roles assigned to and messages communicated between its members.

The first paper in this section, by Joseph N. Cappella, is an attempt to extend a social psychological model of attitude change and dependence on perceived others into a dynamic theory of dyadic interaction. The psychological model, essentially Newcomb's *ABX*-model, posits that there must be balance between an individual *A*'s liking or disliking another individual *B*, both of whom are oriented to or focus attention upon some object *X*. Newcomb's model is concerned with the perception of one individual only and is static except that it postulates attitude changes to proceed from an imbalanced to a balanced cognitive state. Cappella's contribution is to combine this model with notions of communication and information processing into a mathematical model of mutual influence. The key to this new model is that messages with attitudinal contents are alternately generated, transmitted and received. These induce changes in the attitudinal structure, and lead to the generation of messages in response. In the course of this circular flow of information, attitudes change towards some mutually acceptable stable points, i.e. a balance *across* individuals as well as *within* individuals. The paper analyzes a six equation nonlinear model of this process, presents its approximations and discusses possible extensions in terms of the cybernetic character of the model.

The second paper in this section, by Rolf T. Wigand, is concerned with the dynamics of interaction among social organizations rather than individuals. Wigand's review of the organizational literature, including social psychological findings, leads him to extract four class variables that seem to be basic in accounting for interorganizational behavior. There is (a) an interorganizational communication variable, (b) a perceived organization-set independence variable, (c) a goal attainment variable and (d) an environment variable. In these terms he posits a dynamic mathematical model which leads to a variety of hypotheses about communication between organizations. The multiple equilibrium conditions of the proposed model and some of the measurement problems arising from its application are discussed with reference to solutions published elsewhere. As in Cappella's paper, cybernetic considerations become inevitable when communication is conceptualized interactionally and dynamically.

The paper by Mary E. Lippitt and Kenneth D. Mackenzie takes quite a different route. While Cappella and Wigand develop their models from available literature and published empirical findings, this work stems from years of observation and the testing of numerous explanatory devices on naturally forming groups: committees within universities. The theory derives from theoretical groundwork provided by Mackenzie, who has been engaged in all kinds of small group experimentation. While tested within the university context, the results seem to be generalizable to what the authors call "Authority Task Problems" (ATP). Authority-task problems arise from certain changes within social organizations that occur at a faster rate than do changes in their formal structure. Consequently, the functioning organization becomes increasingly inconsistent with the formal structures. These inconsistencies accumulate and some eventually require solution. Faced with an authority-task problem, the administration seeks a solution which is both feasible and politically acceptable. "A Theory of Committee Formation" predicts which of seven solutions are taken according to a set of eleven contingencies.

Basic to the theory is that group structures are not viewed as independent of the members of the group. Rather, they are seen as a need-satisfying interaction pattern (see Mackenzie's paper in another section of this volume). It is also presupposed that the ATP–solving processes in a group can be represented as a sequence of what the authors call "milestones", that structures are established through some kind of behavioral voting, and that one can use these structures to define the roles assigned to members of the group. The heart of the theory is a mapping function which unfolds in the course of a group's choosing among available strategies to solve a given ATP. When ATPs recur within social organizations, the theory predicts with remarkable accuracy how they will be resolved. The paper concludes with a few suggestions among which is one that the theory provides a mechanism by which Parkinson's law of the Rising Pyramid can be derived.

James R. Taylor builds his paper on a presumed analogy between the human brain and social organization. Both are composed of many components. Both develop structures in time which are not predetermined but stem, at least in part, from interactions with their respective environments. And in both cases structures are behaviorally manifest. Indeed, if Cybernetics is to be identified as the science of organization, as Heinz von Foerster has argued [1], there should be some validity to this analogy.

In his paper, Taylor suggests that a variety of phenomena drawn from the psychological literature on individual choice reaction behavior and the small group literature on problem solving experiments can be explained within a single theoretical framework, the cybernetic theory of self-programming networks. He presents arguments that in both situations the process is decomposable into more elementary processes, phases or stages that are marked by "milestones" and structural transitions: that the process involves information processing and error correction relative to some purpose or goal and is subject to communication channel capacity limitations; and finally, that the process requires some control or coordination of members to maintain or modify structures so that the whole can move toward the completion of some task. The paper, being a progress report, includes a sample of the transcribed communications among group members while solving a problem and illustrates how this will be coded and subsequently analyzed.

The last paper in this section, by Nancy Dworkin, Yehoash Dworkin and Bernard Brown, is neither theoretical nor analytical in intent. Rather, it presents the design of a social control mechanism and reports the success of its implementation in an educational setting. Concerned with the fact that more and more children drop out of school, the authors contend that high failure rates may be due to narrowly defined and unilateral controls and inadequate measures of achievement which do not reflect performance in the classroom.

The system they propose essentially redirects the circular flow of information between teachers and students so as to maximize learning rather than achievement. This is accomplished by interposing and adding to the communications between teacher and child two kinds of information classifiers, according to certain rules. Response classifiers distinguish between appropriate, partial and inappropriate solutions and reinforcement classifiers contain statements identifying the heuristics used by the child, the correct elements and models that lead to appropriate solutions, and statements that encourage the child to continue. The circular flow of information thereby enforced has the effect of making control mutual rather than one sided and authoritarian. Effectiveness is measured by the efficiency with which the two principals interpret information passing through the system.

This control mechanism is supplemented by carefully graded tasks, to teach a variety of cognitive skills, which are called "Focusing/Reasoning", "Programming/Logic", and "Modality Variation/Patterning". The authors report on the remarkable success with which the control mechanism is able to move even learning disabled children at their own pace to appropriate completion of school problems growing out of the standard curriculum. The work exemplifies how cybernetic knowledge about small human systems can be validated by the design of social control mechanism rather than by the analysis of existing patterns of social communication and control.

REFERENCE

[1] Foerster, H. von (Ed.), *Cybernetics of Cybernetics,* Urbana: Biological Computer Laboratory, University of Illinois, 1974.

A DYNAMIC MATHEMATICAL MODEL OF MUTUAL INFLUENCE ACCORDING TO INFORMATION PROCESSING THEORY

JOSEPH N. CAPPELLA[1]
University of Wisconsin, Madison

Theory and research in attitude change has been devoted primarily to the process of attitude alteration in the passive communication situation. That is, the speaker generates a message which affects an audience's attitudes but no influence on the part of the audience directed at the speaker is allowed. However, there is no reason why mutual influence cannot be treated within the framework of standard attitude models. In fact, if we are to begin understanding and explaining *social* processes of the diffusion of knowledge, of information, the alteration of public opinion, and the evolution of cultural mores and norms, then it is imperative to conceive of attitude change processes as processes of mutual influence. This paper seeks a very limited goal: the derivation and analysis of a dynamic mathematical model of mutual (dyadic) influence based upon information processing theory [9]. The work presented here is only a step toward developing and analyzing other attitude change models (dissonance, congruity, balance, social judgement, and stimulus response) within the dyadic framework. The work of Hunter and Cohen [10] has completed the task for the passive communication context.

We can best understand basic dyadic processes by adapting a terminological and structural framework appropriate to two-person interactions. The framework is that supplied by Newcomb's *ABX* model [11; 12], which we shall refer to as the *IJX* model for reasons of uniformity of notation. Newcomb assumes that the components of the interpersonal situation consist of two persons, I and J, and an object of mutual relevance and importance, X. The relations among the components are attraction between I and J and the orientations or attitudes of I to X and of J to X. In addition, each individual is presumed to have a perception of what the other's orientation toward X is. With the addition of these two perception relations, Newcomb can now divide the *IJX* situation into parts on the basis of which of the six relations are relevant. These parts are the two individual (or intrapersonal systems) and the collective (or objective) system. The individual system is constructed from cognitions available to the focal individual. These include *I*'s attraction to *J*, *I*'s attitude toward *X*, and *I*'s perception of *J*'s attitude toward *X*. A similar set of relations constitute *J*'s individual system. The collective system is constructed from two individual systems and, hence, is constructed from information which *at any point in time* is unavailable to either of the individuals in the interpersonal situation. The collective system consists of four relations: *I*'s attraction to *J*, *J*'s attraction to *I*, *I*'s attitude toward *X*, and *J*'s attitude toward *X*.

The beauty of Newcomb's structuring of *IJX* situations is found in the types of actions which an individual may undertake when individual strain is experienced. That is, when an individual acts to alter the *IJX* situation, those actions may be inner-directed toward changes in the individual system, or they may be outer-directed toward inducing changes in the collective system. Changes in the individual system would involve changes in attitude, attraction, or perception of the other. Actions directed toward the collective system would take the form of communicative acts which would presumably have persuasive effect on the other. We shall focus solely on collective system and the interactions which occur in its domain. Newcomb's approach to dyadic situations and dyadic influence isolates eight variables: *I*'s and *J*'s attractions for one another, A_{ij} and A_{ji}, *I*'s and *J*'s attitudes (perceptions) of the object *X*, P_i and P_j, *I*'s and *J*'s perception of the other's attitude, Q_{ij} and Q_{ji}, and the content of the messages which *I* and *J* transmit to one another (as measured on an attitude continuum), M_{ij} and M_{ji}. To begin to analyze the *dynamic* interrelation of these variables necessitates mathematical analysis.

While Newcomb's structuring of dyadic influence situations is useful, Newcomb does not specify in detail how attitudes, perceptions, and attractions should change as a function of the messages sent by the other. In order to do so, we may turn to any of the several attitude change models of the passive paradigm [10] to overlay on the Newcomb paradigm. As a first effort, we take up one of the simplest – information processing theory [9; 10].

THE MATHEMATICS OF INFORMATION PROCESSING THEORY

In overlaying the assumptions of information processing theory on Newcomb's paradigm, we will be forced to consider not only changes in attitudes as a function of messages, but also changes in perception of the other, changes in attraction toward the other, strategies for generating message content, and the transmission of messages. In essence, we are taking the view that each individual system is a subsystem of the collective system. Each subsystem is characterized by three state variables: attitude (P_i or P_j), perception of the other's attitude (Q_{ij} or Q_{ji}), and attraction to the other (A_{ij} or A_{ji}). The two subsystems are linked only when there is transmission from I to J, N_{ij}, or from J to I, N_{ji}, or when both are transmitting. The effect that each subsystem has on the other depends upon the content of the messages sent, M_{ij} or M_{ji}, and their rate of transmission. Each of these aspects is taken up in turn.

Changes in State Variables for Each Individual Speaker

As Hunter and Cohen point out, the fundamental tenets of the information processing models of attitude change are

> . . . that (1) the magnitude of change is proportional to the discrepancy between the receiver's attitude and the position advocated by the message and (2) the change is always in the direction advocated by the message [10: 30].

These changes arise from the internal comparison processes which individuals undergo when an incoming message is compared to their own attitudinal position. The greater the difference between the incoming message and the individual's attitude, the greater the expected change in attitude. That is,

$$\begin{aligned}\text{change in } P_i &= P_i(t) - P_i(t-1)\\ &= \Delta P_i\\ &= a(M_{ji} - P_i)\end{aligned}$$

where M_{ji} is the message sent by J to I and a is a constant of proportionality which is greater than zero but less than one. As the discrepancy or distance between I's attitude and J's message increases, so does the expected amount of attitude change. That is, the amount of attitude change is a linear function of the amount of discrepancy between M_{ji} and P_i. Although the assumption of linear change in attitudes is controversial, even proponents of nonlinear attitude change [19] admit that deviations from linearity occur only under relatively extreme conditions.[2] The above change equation has a simple verbal interpretation. I's attitude toward X at time t is given by I's attitude toward X at the previous time $(t-1)$ plus an increment in the direction of I's perception of J's attitude toward X. The fractional amount of that increment is given by a.

As Hunter and Cohen [10: 34] point out, information processing theorists have given a great deal of attention to the effects of source credibility in inducing the desired amount of attitude change. In purely interpersonal situations where the credibility of the source can be identified almost completely with his or her attractiveness, then the amount of attitude change depends upon the attractiveness of the source. That is, the more attractive the source, the greater the attitude change at least when the attraction is positive. However, when the source is thoroughly disliked so that attractiveness is negative, then two choices arise concerning the form of the credibility function. First, credibility can be assumed to be positive, even when attraction is negative. Or, credibility can be assumed to be positive for positive attraction and negative for negative attraction. Since credibility is always a multiplicative factor for the message-attitude difference, then the latter version of the credibility function will produce changes in attitude *opposite* to that advocated by a message delivered by a disliked (negative credibility) source. This is called a "boomerang" effect. However, the experimental production of boomerang effects is very difficult to achieve and so will not be taken as the mechanism for incorporating credibility effects. Rather, credibility will be assumed positive along the entire attraction continuum. The function should increase from zero for an infinitely unattractive source to one for an infinitely credible source. A function which achieves this is $e^{A_{ij}}/(1 + e^{A_{ij}})$ so that

$$\Delta P_i = a \frac{e^{A_{ij}}}{1 + e^{A_{ij}}} (M_{ji} - P_i).$$

Thus, the greater the attractiveness of the other, the greater the fraction of proposed change that is realized.

The change model for attitudes will be complete when the factor of transmission from J, N_{ji}, is included. In the passive communication context the difference equation above can be applied again and again for each message that is generated. However, in the interactive mode, we desire to have the transmission process included explicitly so that once the process of interaction is begun, it will be terminated *by the interactants* with a termination of transmission. Also, as the rate of transmission increases, the more messages J sends to I and, hence, the faster I should change toward the message. When transmission is zero, then the change in attitudes should also be zero. This suggests that transmission, like credibility, is a multiplicative factor in the change equation for attitudes:

$$\Delta P_i = a \frac{e^{A_{ij}}}{1 + e^{A_{ij}}} (M_{ji} - P_i) N_{ji}. \tag{1}$$

Figure 1 shows the effects of discrepancy $(M_{ji} - P_i)$, attraction, A_{ij}, and transmission, N_{ji}, on the change in attitude. In Figure 1a, the more positive M_{ji} is than P_i, the greater will be the positive change in P_i (that is, in the direction of the message). For the same amount of discrepancy, the greater the transmission from J, the greater the amount of change in I's attitude. In frame b of the same figure, we see that for a fixed amount of discrepancy, the greater the attraction, the greater the attitude change. For a fixed discrepancy and fixed level of attraction to the source, the greater the transmission the greater the change in the direction advocated by the message.

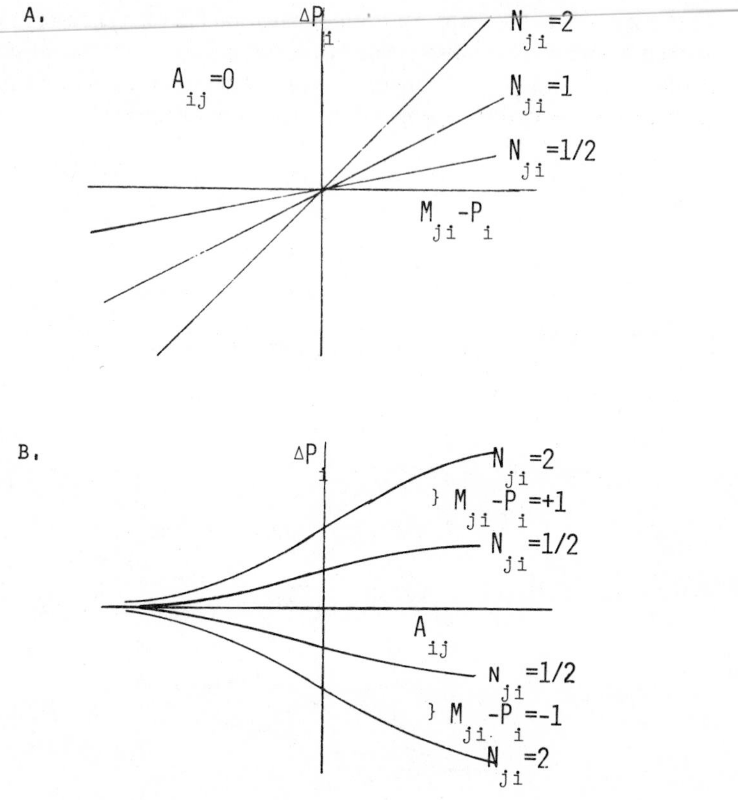

FIGURE 1 Changes in Attitude versus discrepancy between message and Attitude with Attraction equal to 0 (a) and versus Attraction with discrepancy between message and Attitude equal to +1 and −1 (b): Both Show Varying Levels of Transmission.

Having laid the ground work for a change in I's attitudes as a function of J's messages according to information processing theory, developing the change equation for Q_{ij} is an easy matter. The change in I's perception of J's attitude should be exactly analogous to the change in I's attitude as a function of J's message. That is, the more discrepant J's message is from I's position, the more change in Q_{ij} in the direction of the message should be observed. Also the more attractive J is to I, the more change in Q_{ij} that should be realized. And as the transmission from J increases, the rate of change of Q_{ij} in the direction of the message should increase as well. Thus,

$$\Delta Q_{ij} = b \frac{e^{A_{ij}}}{1 + e^{A_{ij}}} (M_{ji} - Q_{ij}) N_{ji}. \qquad (2)$$

Figure 2 presents the graphical form of equation (2) for the cases of constant attraction and constant discrepancy with varying levels of transmission. Comparing Figures 1 and 2 for equations (1) and (2) shows in a striking manner the similarity of the two change equations. The only differences between equations

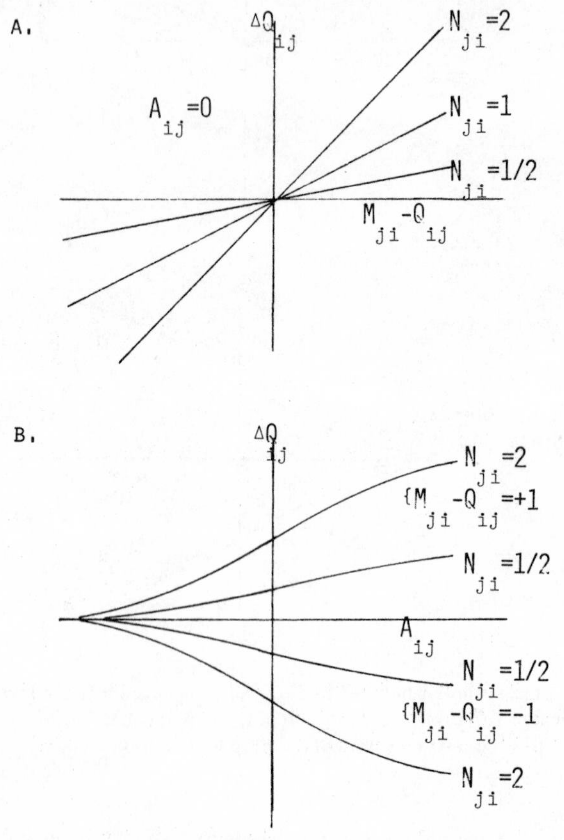

FIGURE 2 Changes in Perception of the Other versus Discrepancy between Message and Perception of the Other with Attraction equal to 0 (a) and versus Attraction with Discrepancy between Message and Perception equal to +1 and −1 (b): Both Show Varying Levels of Transmission.

(1) and (2) are found in the parameters a and b. Although they both have the same purpose, a is not necessarily equal to b and their ratio indicates whether a given message elicits more change in $P_i (a > b)$ or in Q_{ij} $(b > a)$. Based upon a study by Wackman and Beatty [18], cited in Wackman [17], we shall always assume in our examples that *perceptions of the other are less resistant to change than are the attitudes that one holds.* This means that for the same discrepancy,

I's perception of J's attitude will change more in the direction of J's message than will I's own attitude, hence $b > a$.[3] At this point, it is interesting to note that M_{ji} plays a dual role in equations (1) and (2). In equation (1) its effect is that of persuasion and in equation (2) its effect is that of informing I of J's position. This duality is not unreasonable since in attempting to persuade J of his position, I is simultaneously offering J information on the exact nature of his position.

Unfortunately, deriving a change equation for attraction from information processing assumptions is not as easy as it was for perceptions and attitudes. The reason, as Hunter and Cohen [10: 38] note, is that the information processing theorists were not interested in change in the attraction of the source as a function of the message. However, certain requirements for the change in attraction can be stipulated: First, based upon the work of Byrne [3] and his colleagues, we expect that changes in the attractiveness of the other will depend upon the degree of similarity that is perceived by the focal individual. Second, changes in attraction should be both positive and negative so that attraction is capable of either increasing or decreasing. Obviously, if attraction can *only* increase or *only* decrease, then the patterns of attraction which can emerge from such a model will be less than interesting. We believe that any model allowing *only* increases or only decreases in attraction lacks face validity and conflicts with everyday acquaintance processes. Newcomb's famous field study of the acquaintance process [12] observed and measured both increases and decreases in attractiveness between members of a housing unit and related those changes to initial socioeconomic and religious similarities of the subjects at least at the early stages of acquaintance formation. The reason that this point is being emphasized is that dissonance theory of source change (as Hunter and Cohen show) is one of pure source derogation. We feel that such a model of attraction change is too limited to apply to dyadic processes. Similarly, a straightforward extension of Byrne's so-called "law of attraction" would posit

$$\Delta A_{ij} = n\Delta|M_{ji} - P_i|.$$

However, this model has the peculiar characteristic that if I and J initially hate one another but upon interaction find that they agree (that is $M_{ji} = P_i$), then they will remain unattracted to one another despite being in agreement. We find this implausible and at odds with the evidence presented in Newcomb [12].

Rather, we shall posit a model of attraction change which is basically social judgmental in character. That is, we assume that when $M_{ji} = P_i$, the change in attraction is positive and maximum. When M_{ji} and P_i are discrepant, then whether the change in attraction is positive or negative depends upon what amount of discrepancy the focal individual is willing to accept. That is, if person I is willing to accept a certain amount of disagreement but no more before his or her attraction to J begins to decrease, then that amount defines the boundaries of his or her acceptance region. The change equation which will describe the above process is (see Hunter and Cohen, [10])

$$\Delta A_{ij} = c \frac{(t_{ij}^2 - (M_{ji} - P_i)^2)}{1 + t_{ij}^2} N_{ji}. \tag{3}$$

This equation is graphed in Figure 3. $2t_{ij}$ is the width of the acceptance region centered at P_i. When $M_{ji} - P_i > t_{ij}$ or $M_{ji} - P_i < -t_{ij}$, then the change in attraction decreases. When $-t_{ij} < M_{ji} - P_i < t_{ij}$, then the change in attraction is positive and is a maximum for $M_{ji} - P_i = 0$ or perfect perceived agreement. When $M_{ji} - P_i$ falls

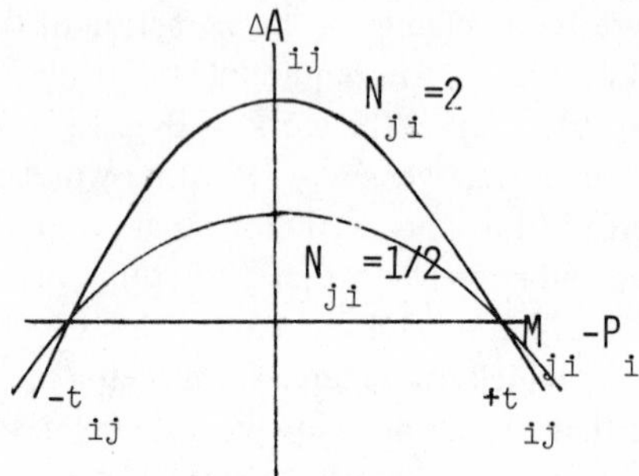

FIGURE 3 Changes in Attraction versus Discrepancy between Message and Attitude for Varying Levels of Transmission.

exactly on the border between the acceptance and rejection regions, then the change in attraction is zero.

Obviously, the behavior of equation (3) depends upon the value of t_{ij}. We shall assume with Hunter and Cohen [10: 49] and Sherif, Sherif, and Nebergall [15: 189] that the width of the acceptance region depends at least in part upon the attractiveness of the other. The more attractive the other, the wider the acceptance region. Specifically, it is assumed that $t_{ij} = e^{A_{ij}}$. In this way, the more positive the attraction, the more likely that discrepancies will fall within the acceptance region and produce even greater attraction. Also, the more negative the attraction, the more likely that discrepancies will fall outside the acceptance region and produce further decreases in attraction.

The equations (1) through (3) with the counterpart equations for person J constitute the state equations for the two subsystems I and J. We next consider how each individual system generates its output in the form of messages and their rate of transmission.

The generation of message content We shall consider two models of the generation of message content. Each of these models will be highly speculative since the question of *what* is said in interaction has not been well researched.

First, suppose the subject always speaks his mind. Then the message will just be his attitude or

$$M_{ij} = P_i. \tag{4}$$

This prediction constitutes the first model of message generation and has been the one most commonly adopted in interactive models of attitude change [1; 16]. It will be called the "veridical" model.

In the veridical model, the speaker says the same thing regardless of who the listener might be. But suppose that he seeks to ingratiate himself with J by shifting his message in the direction of his perception of J's attitude. That is,

$$M_{ij} = pP_i + (1 - p)Q_{ij}$$

where p is a weighting factor between 0 and 1. If we presume that individuals are more ingratiating for more attractive others and less ingratiating for less attractive others, then the weighting factor p would be a function of attraction. Furthermore, the more attractive the other, the closer p should be to unity. This implies that p could be chosen to be $p = 1/(1 + e^{A_{ij}})$. Rewriting the above equation with this new expression for p, we have

$$M_{ij} = P_i + \frac{e^{A_{ij}}}{1 - e^{A_{ij}}}(Q_{ij} - P_i). \tag{5}$$

When I thoroughly dislikes J, then he speaks his mind (that is, is veridical) and does not seek interpersonal rewards from J by ingratiating him. When I likes J a great deal, then he seeks to further the favors and good graces from J by saying what he thinks J wishes to hear. We shall call this the "shift" model because of the cynical connotations associated with an "ingratiation" model.

While several other content generator models could have been specified, the veridical and shift models are both simple and yet differ starkly in their assumptions about speakers. These conditions are important for early stages of model building.

The transmission of messages Recall that Newcomb's discussion of message transmission was as a possible response to individual system strain. He presumed that the reaction to individual system strain through communication to the other actually took the form of attempts to influence the other's point of view concerning X. Consequently, influence attempts directed toward the other should arise from forces created by perceived discrepancies on X. That is, we assume that transmission is intended to alter the other's attitudes toward X and not to alter I's attraction to J. If N_{ij} is the number of influence attempts generated by I toward J, then we assume that

$$N_{ij} = d \frac{|Q_{ij} - P_i|}{\sqrt{1 + (Q_{ij} - P_i)^2}} \left[1 + \frac{e^{A_{ij}}}{1 + e^{A_{ij}}}\right] \tag{6}$$

where d is a positive constant.

Equation (6) is the product of a discrepancy term and an attraction term. It was chosen to yield the following specifications: (a) For constant attraction, the greater the discrepancy perceived by I, the greater the transmission from I. (b) For constant perceived discrepancy, the greater the attraction which I has for J, the greater the transmission from I. (c) For large negative attraction, transmission is still positive and depends upon the amount of perceived disagreement. The first two characteristics have been well researched and validated by Festinger and his colleagues [6; 2; 8; 14] and the third represents the fact that when individuals are constrained to remain in a dyadic situation (e.g. a work group), the social norm of reciprocal interaction outweighs the (negative) relationship between attraction and transmission.

Summary

Newcomb's structuring of *IJX* situations has led us to view the process of mutual influence as the interaction between two subsystems, each described by three state variables (i.e. attitude, perception of the other, and attraction to the other) along with the corresponding change equations. The interaction between subsystems is described by the message transmission rates and two (possible) message content generation schemes. The equations are summarized in Table 1 and constitute the model to be analyzed.

ANALYZING THE DYNAMICS OF CHANGE IN *IJX* SITUATIONS

Attempting to analyze the long-term and short-term behavior of the mathematical system described in Table 1 either for the case of veridical or shift message generation is no easy matter. Without simplifying assumptions we are faced with a system of six highly nonlinear equations. The strategy we have taken to build understanding of the general model is to solve successive approximations of the complete model. The following sections will present results for the cases of (I) constant attraction and constant transmission for shift and veridical messages, (II) constant attraction, varying transmission for shift and veridical messages, (III) variable attraction, constant transmission for shift and veridical messages and, finally, (IV) varying attraction, varying transmission for both shift and veridical messages.

TABLE 1 Change Equations and Input–Output Relations (Message Transmission and Message Content) for *IJX* Dyads.

STATE EQUATIONS FOR I AND J
$\Delta P_i = a\,C_{ij}\,(M_{ji} - P_i)\,N_{ji}$
$\Delta Q_{ij} = b\,C_{ij}\,(M_{ji} - Q_{ij})\,N_{ji}$
$\Delta A = c\,\dfrac{t_{ij}^2 - (M_{ji} - P_i)^2}{1 + t_{ij}^2}\,N_{ji}$
INPUT-OUTPUT RELATIONS
TRANSMISSION
$N_{ij} = d\,\dfrac{\lvert Q_{ij} - P_i \rvert}{\sqrt{1 + (Q_{ij} - P_i)^2}}\,(1 + C_{ij})$
CONTENT
1. VERIDICAL STYLE
$M_{ij} = P_i$
2. SHIFT STYLE
$M_{ij} = P_i + C_{ij}\,(Q_{ij} - P_i)$
$t_{ij} = e^{A_{ij}}$ AND $C_{ij} = t_{ij}/(1 + t_{ij})$

Limitations of space will not allow a detailed presentation of the analysis procedures[4] but allows only a summary of the model and its central long-term and short-term characteristics. In general, standard techniques for linear dynamic systems were used for the linear approximations and computer-generated numerical solutions using a fourth-order Runge-Kutta procedure[5] assisted in analyzing the nonlinear models.

I. *Constant attraction and transmission for shift and veridical messages* Under assumptions of constant attraction and transmission, the equations describing the system are linear:

$$\frac{dP_i}{dt} = ak_{ij}(M_{ji} - P_i)N_{ji} \tag{7}$$

$$\frac{dQ_{ij}}{dt} = bk_{ij}(M_{ji} - Q_{ij})N_{ji}$$

$$\frac{dP_j}{dt} = ak_{ji}(M_{ij} - P_j)N_{ij}$$

$$\frac{dQ_{ji}}{dt} = bk_{ji}(M_{ij} - Q_{ji})N_{ij}$$

where k_{ij}, k_{ji} are constants and equal $t_{ij}/(1 + t_{ij})$ and $t_{ji}/(1 + t_{ji})$ respectively. The messages are either veridical, $M_{ij} = P_i$ and $M_{ji} = P_j$, or shift, $M_{ij} = P_i + k_{ij}(Q_{ij} - P_i)$ and $M_{ji} = P_j + k_{ji}(Q_{ji} - P_j)$.

When N_{ij} and N_{ji} are nonzero and attractions are not infinitely negative, it is easy to show [4; 1] that (a) the system of equations (7) has an infinity of critical points such that attitudes and perceptions of the other all converge to the same point for both I and J, and (b) the dyad will always converge to one of these critical points. These results hold for both the shift and veridical models. They mean that under constant communication and attraction both individuals will end up with the same attitude and will be accurate in their perception of the other's attitude *regardless of the initial attitudes or perceptions.*

On the other hand, the *point* of final convergence depends crucially on the transmission and attraction constants and whether messages are veridical or shift (ingratiating). With veridical messages, the convergence point is $(k_{ji}N_{ij}P_i(0) + k_{ij}N_{ji}P_j(0))/(k_{ji}N_{ij} + k_{ij}N_{ji})$. The person who shifts most is the one whose "attraction-transmission" coefficient (that is, $k_{ji}N_{ij}$ or $k_{ij}N_{ji}$) is smaller. With shift messages, the convergence point is complicated[6] because it depends on initial perceptions of the other as well as initial attitudes. However, the "attraction-transmission" coefficient plays a similar role. Thus while convergence to *some* point of complete equality is independent of initial attitudes and perceptions, the particular point of convergence depends upon where the individuals begin.

A final observation is useful. In the extreme case in which one of the individuals does not transmit information but the other does, the shift model converges to a point such that the silent individual *perceives* no discrepancy in attitude between himself and the other but actual discrepancy in attitude exists at the collective level. The veridical message model does not exhibit this misperception characteristic.

II. *Constant attraction and varying transmission with veridical and shift messages*
The equations describing this approximation are exactly those of the system for

equations (7) except that N_{ij} and N_{ji} are no longer constant. For varying transmission $N_{ij} = d(1 + k_{ij})|Q_{ij} - P_i| / \sqrt{1 + (Q_{ij} - P_i)^2}$ with a corresponding equation for N_{ji} with i and j permuted.

For both the veridical and shift cases, this model has two infinite sets of equilibria. When $P_i = P_j = Q_{ij} = Q_{ji}$, then there is attitudinal agreement, accuracy, and the absence of transmission for both I and J. The system has stopped changing at a point qualitatively similar to that of the completely linear model. However, when $P_i = Q_{ij} = M_{ji}$, then system change ceases as well. But in this case I perceives no disagreement and sends no further messages to J while J continues to perceive disagreement (P_j and Q_{ji} are not equal) and continues to send messages to I.

Interestingly we were able to establish [4] that the latter set of equilibria are much more likely to occur than the former and the latter set are "stable" whereas the former are unstable. The reason is that the IJX situation converges to a common equilibrium point $P_i = P_j = Q_{ij} = Q_{ji}$ only under the very restrictive initial conditions that (a) both attractions are equal, (b) both initial perceived discrepancies are equal, (c) both message-attitude discrepancies are equal, and (d) both message-perception discrepancies are equal. Even the slightest deviation from equality in any of these initial conditions will cause (for example) I's attitudes and perceptions to converge on the J's message and, hence, cease transmitting to J before J's attitudes and perceptions can converge on I's message.

Thus, under the assumptions of this model, the most probable final state has one person perceiving agreement and silent, the other person perceiving disagreement and transmitting, and the collective system actually showing disagreement.

III. *Varying attraction and constant transmission with veridical and shift messages*
There are six rather than four nonlinear change equations for the varying attraction case:

$$\begin{aligned} \frac{dP_i}{dt} &= aN_{ji} \frac{t_{ij}}{1 + t_{ij}} (M_{ji} - P_i) \\ \frac{dQ_{ij}}{dt} &= bN_{ji} \frac{t_{ij}}{1 + t_{ij}} (M_{ji} - Q_{ij}) \\ \frac{dA_{ij}}{dt} &= cN_{ji} \frac{t_{ij}^2 - (M_{ji} - P_i)^2}{1 + t_{ij}^2} \end{aligned} \tag{8}$$

with comparable equations for P_j, Q_{ji}, and A_{ji}. We revert to assuming that N_{ij} and N_{ji} are constant and introduce the assumption that $t_{ij} = e^{A_{ij}}$. Now shift messages have the form $M_{ij} = P_i + t_{ij}/(1 + t_{ij})(Q_{ij} - P_i)$.

The most important mathematical property of the system of equations (8) is that it has no equilibrium points. Without equilibria it is impossible to indicate final states for the system but the direction in which the state variables are moving can be indicated. In fact, we found the system tended in two qualitatively different directions: (a) toward convergence of attitudes and perceptions on a common limit with *both* attractions tending toward positive infinity and (b) toward both attractions approaching negative infinity with attitudes and perceptions tending toward different asymptotic values. Whether the system headed toward the former or latter configuration depends upon the initial values. Whenever I's and J's initial messages are within the other's acceptance region (width given by $2t_{ij}$), or when both initial messages are slightly outside the region, or when one message is within the other message outside the other's acceptance, then the IJX situation will change toward increasing mutual positive attraction and toward agreement and accuracy in the collective system and perceived agreement in both individual systems. On the other hand, whenever both I's and J's initial messages are well outside the other's acceptance region, then the IJX situation will change toward increasing mutual dislike and will exhibit discrepancies in agreement and in perceived agreement at the individual level, collective level or at both levels.

The above conclusions hold for both the veridical and for the shift models of message generation. However, their implications differ in significant ways. Suppose first that I and J disagree so that $P_i \neq P_j$ and both are well outside each other's acceptance region. If even one of the two persons in the IJX situation is accurate, let us say $Q_{ij} = P_j$, then under the shift model I's initial message will be shifted in the direction of J's *actual position* and the likelihood that I's initial message will fall into J's acceptance region is increased. If the shift message does fall into J's acceptance region, then the IJX situation will tend toward a qualitatively different state (mutual positive attraction and zero discrepancy) than the veridical model would predict (mutual dislike with nonzero discrepancy). On the other hand, if I and J initially agreed but were both vastly *inaccurate*, then the veridical model would predict change toward mutual positive attraction and zero discrepancy and the shift model would predict mutual hostility and finite discrepancy. Of course, this does not cover all possibilities. But the two cases cited do serve to point out that the choice of message model is crucial in predicting where the IJX situation is tending with varying attraction and constant transmission assumptions.

IV. *Varying attraction and transmission for shift and veridical messages* This is the model of Table 1 in its full generality and combines the results of the two previous approximations – varying attraction with constant transmission and constant attraction with varying transmission. That is, the direction in which the IJX situation is moving depends both upon the symmetry-asymmetry of initial conditions as with the case II approximation and the location of initial messages relative to the other's initial region of acceptance as with the case III approximation. The interaction of these two sets of initial conditions gives rise to eight possible combinations of initial conditions given in Table 2. However, cell IV–S

TABLE 2 Possible Combined Initial Conditions from the Case II Approximation (Symmetric versus Asymmetric Initial Values) and the Case III Approximation (Initial Message Location Relative to Other's Acceptance Region).

	Symmetric	Asymmetric
Both Messages Within	I-S	I-A
Both Messages Well Outside	II-S	II-A
Both Messages Just Outside	III-S	III-A
One Message Within & One Message Outside	IV-S	IV-A

is empty since it is logically contradictory to require all initial conditions to be symmetric with respect to each person but to have one initial message outside and the other message inside the person's acceptance region.

While the varying transmission and varying attraction model is the most complex and least tractable of all the models considered thusfar, it is also the most interesting. The final states as a function of initial value configurations are summarized in Table 3. The results are tentative because they are based only upon numerical analyses of what is a complex mathematical system. Nonetheless, even if the results are approximately correct (as they certainly are from the numerical trajectories generated) the varying transmission and varying attraction case allows the *IJX* dyad to achieve final states which were unavailable in the varying attraction case alone. In particular, the "both inside-asymmetric" case shows that it is possible for *I* and *J* to be mutually attracted but for actual discrepancy to exist in the collective system.

With varying attraction and transmission we have discussed the veridical and shift models together. The results of Table 3 are directly applicable to the veridical case with the minor observation than when initial conditions are asymmetric $M_{ji} = P_j$ and only Q_{ji} is arbitrary. The other major difference between the shift and veridical cases is to be found in the location of initial messages for the two cases relative to the other's acceptance region. As we noted in the constant transmission-variable attraction case, if *I*'s and *J*'s initial attitudes were within the other's acceptance region but both were very inaccurate, then veridical messages would be within and shift messages outside the acceptance regions. Each case would produce quite different final states. On the other hand, if *I*'s

TABLE 3 Final States for Veridical and Shift Messages for Case IV as a Function of Position of Initial Message and Symmetry of Initial Attitudes, Perceptions, and Attractions.

I-S:	I-A:
$A_{ij}, A_{ji} \rightarrow$ CONSTANT>0	$A_{ij} \rightarrow +\infty$, $A_{ji} \rightarrow$ CONSTANT>0
$P_i = Q_{ij} = P_j = Q_{ji} = P^*_i$	$P_i = Q_{ij} = M_{ji} = M^*_{ji}$ $P_j = P^*_j$, $Q_{ji} = Q^*_{ji}$
II-S:	II-A:
$A_{ij}, A_{ji} \rightarrow -\infty$	$A_{ij}, A_{ji} \rightarrow -\infty$
$P^*_i \neq Q^*_{ij} \neq P^*_j \neq Q^*_{ji}$	$P^*_i \neq Q^*_{ij} \neq P^*_j \neq Q^*_{ji}$
III-S:	III-A:
$A_{ij}, A_{ji} \rightarrow$ CONSTANT<OR>0	$A_{ij} \rightarrow +\infty$, $A_{ji} \rightarrow$ CONSTANT
$P_i = Q_{ij} = P_j = Q_{ji} = P^*_i$	$P_i = Q_{ij} = M_{ji} = M^*_{ji}$ $P_j = P^*_j$, $Q_{ji} = Q^*_{ji}$
IV-S:	IV-A:

$A_{ij} \rightarrow$ CONSTANT>0, $A_{ji} \rightarrow$ CONSTANT<OR>0
$P_i = Q_{ij} = M_{ji} = M^*_{ji}$
$P_j = P^*_j$, $Q_{ji} = Q^*_{ji}$

IN THIS TABLE "*" INDICATES A CONSTANT, ASYMPTOTIC VALUE, AND "→" INDICATES APPROACH IN THE LIMIT AS TIME GOES TO INFINITY.

and *J*'s attitudes were well outside the other's acceptance region but both were quite accurate, then shift messages would be likely to be within the other's acceptance regions while veridical messages would not. Once again, we see that the veridical and shift message model give similar analyses but involve very different implications.

Discussion and Extensions

Because of space limitations, we will not take time here to evaluate the present model in light of previous research or to propose research strategies for the future evaluation of the model. This has been done elsewhere [4]. Rather the long-term significance of the work presented here will be found not in the testability or validity of any *single* model but in comparisons among *alternative* models of the same process.

As was noted at the outset, Newcomb's structuring of *IJX* situations provides no "content" for describing the change relationships among variables but only identifies the variables. Any of several attitude change models (social judgment, dissonance, congruity, etc.) could be superimposed on the Newcomb framework to give it content. However, from the viewpoint of cybernetics Newcomb's choices are particularly important.

First the distinction between the individual and collective systems makes possible the system-subsystem distinction with each individual constituting a subsystem. Our discussion has treated both subsystems as identical and, hence, greatly reduced the mathematical complexity. Each subsystem is described by structurally similar equations. A future goal of this research is to investigate system behavior when the subsystems are communicating with different "styles". For example, *I* could be veridical and *J* ingratiating. Second, each subsystem is "reading" the behavior of the other subsystem through Q_{ij} and Q_{ji}. Thus each subsystem gathers information of the behavior of the system. Further, each subsystem acts on this information by comparing where he thinks the other subsystem is (Q_{ij} for person *I*) to where he is with regard to *X* (that is, P_i for person *I*). When there is the perception of discrepancy, difference, or error, the subsystem acts to reduce that perceived error by transmitting information to the other subsystem.

The form or content of the transmitted information depends upon the particular message content function adopted. In more concrete terms, what is said depends upon the style of the communicator (e.g. veridical or ingratiating). Finally, and most interestingly, the message received from the other is the goal against which current subsystem behavior is judged. That is, person *I*'s attitudes, perceptions, and attraction are evaluated relative to the incoming message and changed so as to be more in accord with that goal. The goal is a more or less important determinant of changes in attitudes, perceptions, and attraction depending upon the credibility or attractiveness of the message source. Furthermore, the goal (that is, the incoming message) is a standard of comparison which itself is changing as the source's state variables are changing. With message content a function of the sender's internal states, and incoming messages the standard against which each subsystem compares its internal states, then we have a cybernetic system whose behavior is changing relative to goals or standards which are themselves changing.

What Newcomb's analysis of *IJX* situations does is identify a "system-read" variable for each subsystem, identify a set of state variables for each subsystem which are identical for each subsystem, and posit incoming messages as the standard against which state variables are evaluated. We have added (a) the association of message content with the internal state of each subsystem, and (b) the transmission of messages as a function of subsystem perceived discrepancy. In addition, we have overlayed what is a general cybernetic paradigm with the assumptions of information processing theory. The goal of the work only begun in this paper is to overlay that same general cybernetic model of *IJX* situations with other standard attitude change theories such as S–R, social judgment, dissonance, and congruity. As this work nears completion our hope is that the several cybernetic models of mutual influence in dyadic situations will yield not only alternative but competing predictions as to changes in attitudes, perceptions, and attractions. In this way we might be capable of at least falsifying certain models of the mutual influence process by rejecting the predictions made by the mathematical realizations of a general cybernetic framework.

ENDNOTES

[1] The author wishes to acknowledge the assistance which John E. Hunter provided at every stage in this work. Without his early influence this work would never have been undertaken and without his painstaking editing and critical scrutiny would never have been completed.

[2] The nonlinear models of attitude change are associated primarily with Social Judgment theory [15]. Work by the author in overlaying the Newcomb paradigm with Social Judgment assumptions is well-along.

[3] In the numerical calculations for the trajectories of the differential equations the ratio of b to a was always taken to be 3 to 1.

[4] For complete details of the analysis see Cappella [4] especially Chapter 3 and the Appendices.

[5] The program employed was a packaged routine adapted by the author but developed at Northwestern University's Vogelback Computing Center. Up to fifty first-order differential equations can be efficiently and cheaply solved using this routine. An interactive version of this program in BASIC and suitable for use on a PDP–12 is available from the author on request.

[6] Its value is

$$\frac{N_{ij}k_{ji}(b(1-k_{ij})P_i(0)+ak_{ij}Q_{ij}(0))+N_{ji}k_{ij}(b(1-k_{ji})P_j(0)+ak_{ji}Q_{ji}(0)}{ak_{ij}k_{ji}(N_{ij}+N_{ji})+b(N_{ij}k_{ji}(1-k_{ij})+(1-k_{ji})N_{ji}k_{ij})}$$

REFERENCES

[1] Abelson, R. P., "Mathematical models of the distribution of attitudes under controversy", *in* Fredericksen, N. and Gulliksen, H. (Eds.), *Contributions to Mathematical Psychology,* New York: Holt-Rinehart, 1964.

[2] Back, K. W., "Influence through social communication", *J. of Abnormal and Social Psychology,* **46**: 9–23, 1951.

[3] Byrne, D., "Attitudes and attraction", *in* Berkowitz, L. (Ed.), *Advances in Experimental Social Psychology,* New York: Academic Press, 1964.

[4] Cappella, J. N., "A dynamic mathematical model of dyadic interaction based upon information processing theory", Unpublished Doctoral Dissertation, Dept. of Communication, Michigan State University, 1974.

[5] Cohen, A. R., "A dissonance analysis of the boomerang effect", *J. of Personality,* **30:** 75–88, 1962.

[6] Festinger, L., "Informal social communication", *Psychological Review,* **57:** 271–282, 1950.

[7] Festinger, L., *A Theory of Cognitive Dissonance,* Evanston, Illinois: Row, Peterson, 1957.

[8] Festinger, L. and Thibaut, J., "Interpersonal communication in small groups", *J. of Abnormal Social Psychology,* **46:** 9–23, 1951.

[9] Hovland, C. I., Janis, I. L. and Kelley, H. H., *Communication and Persuasion,* New Haven: Yale University Press, 1953.

[10] Hunter, J. E. and Cohen, S. H., "Mathematical models of attitude change in the passive communication context, Unpublished Manuscript, Dept. of Psychology, Michigan State University, 1974.

[11] Newcomb, T. M., "An approach to the study of communicative acts", *Psychological Review,* **60:** 393–404, 1953.

[12] Newcomb, T. M., *The Acquaintance Process,* New York: Holt, Rinehart, and Winston, 1961.

[13] Osgood, C. E. and Tannenbaum, P. H., "The principle of congruity in the prediction of attitude change", *Psychological Review,* **62:** 42–55, 1955.

[14] Schachter, S., "Deviation, rejection, and communication", *J. of Abnormal and Social Psychology,* **46:** 190–208, 1951.

[15] Sherif, C. W., Sherif, M. and Nebergall, R. E., *Attitude and Attitude Change: The Social Judgment-Involvement Approach,* Philadelphia: Saunders, W. B., 1965.

[16] Taylor, M., "Towards a mathematical theory of influence and attitude change", *Human Relations,* **21:** 121–140, 1968.

[17] Wackman, D. B., "Interpersonal communication and coorientation", *American Behavioral Scientist,* **16:** 537–550, 1973.

[18] Wackman, D. B. and Beatty, D. J. F., "A comparison of balance and consensus theories for explaining changes in *ABX* systems", Unpublished Manuscript, Presented at the International Communication Association Convention, Phoenix, Arizona, 1971.

[19] Whittaker, J. O., "Resolution of the communication discrepancy issue in attitude change", *in* Sherif, C. W. and Sherif, M. (Eds.), *Attitude, Ego-Involvement and Change,* New York: John Wiley, 1967.

A MODEL OF INTERORGANIZATIONAL COMMUNICATION AMONG COMPLEX ORGANIZATIONS[1]

ROLF T. WIGAND
Arizona State University, Tempe

The importance of autonomous and competing organizations for viable democratic processes is emphasized by scholars in sociology, economics, political science, and other fields. Models resulting from such theoretical positions assume that the processes of exchange, competition, cooperation, coordination and communication are inherent in social reality. Some scholars, however, have criticized that such theoretical models emphasize a static view of social processes [Cf., 20; 70; 71; 85]. This paper argues that the study of interorganizational relationships is one area that can appropriately incorporate these social processes. The notion of interorganizational relationships becomes an important analytical tool for comprehending these processes.

If researchers are to gain a significant insight into certain aspects of organizational behavior, there is a need to consider interorganizational communication [together with 25]. Most organizational communication research deals with individuals within the organization and not with the organization *per se*. Whatever type of organization, the researcher should understand the environment within which the organization operates. Through understanding the pressures acting on a focal organization and its individual decision makers, the researcher can comprehend sets of objectives, goals and criteria relevant to the organization. Considerable emphasis is directed to the idea of organizational change or components thereof. Various approaches in organizational development, group dynamics, etc., stress the importance of the concept of change. Most researchers in the past became aware of a need for organizational change because such symptoms as performance measures, disturbances or breakdowns in communication, etc., suggested it. Few studies, however, attempt to identify and measure a set of variables that are causative of change and/or whose recognition necessitates a specific, desired change. This author attempts to conceptually differentiate between endogenous and exogenous variables that may affect a change in the behavior of organizations.

Initially, the types of variables that are largely instrumental for influencing an existing communication behavior among organizations will be identified. There are two main sets of class variables that are discussed in this context:

(a) It is attempted to focus on the endogenous *information processing* of the organization. This is understood as the flow of and the behavior in certain communicative acts occurring within the organization that is structurally represented in the form of a communication network and subsets thereof;

(b) there is one exogenous *environmental variable* that exerts influence on the focal organization as well as on a set of organizations.

This paper also focuses on a set of phenomena, specifically relationships, that encompass the individuals within the organizations, the organization *per se*, as

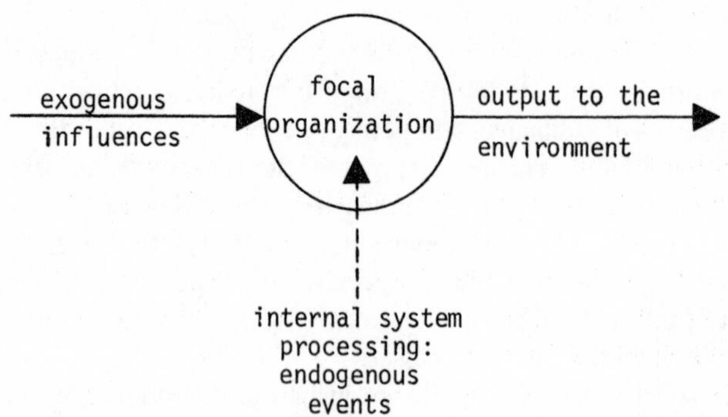

FIGURE 1 The organization viewed as the focus of analysis in an environmental context.

well as the immediate organizational environment. These relationships may be represented in a model that reflects the real situation and formulates in a hypothetical form the logical patterns of causes and effects which link together the specified phenomena. Furthermore, the author constructs a set of relationships that will permit a model of the total system to behave in a manner which compares favorably with the system in reality.

The description of these phenomena, allows for the generation of differential equations for processes that will produce these phenomena. Under the consideration of several interacting variables, a differential equations model is presented and its utilization allows for replication of the basic characteristics of the inter-organizational relationships. It is emphasized that this behavior is not static, but dynamic in nature.

INTERORGANIZATIONAL COMMUNICATION RELATIONSHIPS

Organizations are social systems, i.e. systematic ensembles of interdependent, interhuman activities intended to achieve joint objectives by coordinating joint efforts of a group of people following a predetermined program of conduct [Cf., together with 2]. A complex of roles is formed by such a social system and is constituted by individuals and groups that are linked together by their mutual recognition and realization of certain values and norms. In this process, organizations are evolutionary formations, which emerge, exist and change for the realization of basic human values. A set of organizations operating in a given, joint environment is to some degree interdependent and may be viewed as a system. *An interorganizational relationship is defined as the interaction between two or more organizations for the realization of their respective goals which is affected by the nature of the interaction pattern and the conditions under which such interactions occur.*

Most organization scholars are concerned with *intra*organizational phenomena and only a few have studied *inter*organizational phenomena. A large number of studies have investigated communication that passes through channels that are highly structured or diffused, open or closed. It is known that the shape of communication networks decisively affects the quality and role of communication as well as the behavior of the network participants. A few studies, however, have looked critically at certain formations in natural, complex organizations while considering the influences of the environment. The concept of the environment, including its components and relevant dimensions, is not well explicated and specified in the literature [21; 23; 43; 46; 57]. Emery and Trist [23] emphasize the processes occurring in various subsets of the organization and the environment in which it operates. The scheme of these authors still seems to emphasize system-internal and intra-system processes, although it allows for "processes through which parts of the environment become related to each other – i.e. its causal texture – the area of interdependencies that belong within the environment itself".

It is this latter environmental sphere, described as the causal texture of the environment, that is the primary area of discussion for the purposes of this paper. In part this area has been further described by Evan [27] as the "organization-set". In Evan's conceptualization – developed from Merton's "role-set" – the unit of analysis is an individual organization or a class of organizations and its interactions that are mapped with the relevant network of organizations in its environment.

All such interorganizational relationships occur in some sort of communicative form: they may be formal, social, using various channels for the transmission of messages (telephones, letters, etc.). They may flow between and among organizations, groups, individuals and combinations thereof. A number of writers are concerned with such variables as the size of the organization, propinquity, interdependency, informal interactions, etc. A sizeable number of studies have emphasized the importance of interorganizational relationships in the light of rehabilitation and mental health [10], delinquency prevention and control [53; 60], politics [58], education [16], economic networks [4; 5; 29], medical care [49], services for the elderly [54], community action [80], urban structure [78; 79] and community disaster situations [28; 34].

The nature of organizational environments was explored with regard to the idea of turbulence [23; 72]. A few studies focused on the impact of the environment on organization-internal processes. Thompson and McEwen [75] and Dill [21] demonstrated that the condition of the organizational environment may alter the goal setting behavior of organizations. Yuchtman and Seashore [86] specified organizational effectiveness with respect to the organization's success in obtaining resources from the environment, which in essence was hypothesized by Terreberry [72] that organizational change is largely influenced by environmental factors. Thompson [74] and Lawrence and Lorsch [45; 46] also suggest certain ways in which environmental forces affect organizations. Simpson and Gulley [69] studied voluntary organizations with diffuse environmental pressures. Variations in cultural values and norms were found to affect the internal structure of organizations [18].

Increasing attention is focused on the idea of *exchange* and transactional interdependencies [22; 38; 40; 49; 50; 60; 61; 74]. Levine and White [48] propose an *exchange model* of interorganizational relationships in which organizations that share domain consensus are able to unilaterally, reciprocally, or jointly allocate scarce resources of clients, labor services, and other resources. Analogous to such an *exchange model* Homans' [41] model envisages human behavior as a function of its payoff: in amount and kind, an organization's responses depend on the amount and quality of reward and punishment that its actions elicit. Reid [60; 61] proposes a thesis of relations among autonomous organizations and suggests that there are three basic modes of behavior in interorganizational relationships: independence, interdependence, and conflict.

Additional difficulties are encountered in measuring and describing the condition of the environment through which interorganizational communication flows and is influenced by. The environmental condition may be ascertained by describing the characteristics of the larger social and industrial units in which the organization is located – community, industry, region, etc. Weick [81] emphasizes the *enacted environment* which identifies the information space outside the organization and is understood as a composite of the various viewpoints of the organization's members. Emery and Trist [23] identify four main types of environments, each of which is based on a significantly different conception of the information space of a given organization:

(a) the *placid, randomized environment* is a state in which the organizational goals and the pertinent noxiants are considered to be relatively stable and are distributed randomly;

(b) the *placid, clustered environment* describes a condition in which the goals of the organization and the noxiants are nonrandomly distributed, i.e. they have developed a pattern and are clustered;

(c) the *disturbed, reactive environment* is characterized by the fact that there are a number of similar organizations operating competitively in the same general environment; and

(d) the *turbulent environment* is recognized by the organization because of the unstable, unpredictive, complex condition that is generally difficult to cope with.

Each of these four descriptive states of the environment may significantly influence the communication behavior and the interorganizational relationships of organizations. These relationships are viewed as they are reflected in the nature, perception and flow of interactions between and among organizations. The set of organizations to be selected for analysis purposes will depend on the degree to which they operate in the same, relevant environment.

This author suggests that a minimum set of variables can be identified that are characteristic of the most salient aspects of interorganizational relationships. The variables may be viewed as the state variables whose values and variances define the state of the communication characteristics existing within a given set of organizations. These characteristics are reflected in various communication networks and in the relationships to be detected within such a network. The entire process – modified by the environmental conditions – that influences the communication patterns of a focal organization can be represented in the form of a graphic model (Cf., Figure 2). A discussion of the selection and the relationships among such variables, together with the development of a dynamic model is presented in the following sections.

THE DEVELOPMENT OF A DYNAMIC MODEL

This section concentrates on the selection and identification of important variables which describe a proposed interorganizational model and the mathematical specifications of the existing relationships among these variables. Part two of this section describes the qualitative and dynamic properties of the developed model. After these properties are explored, the following section will present a number of hypotheses based on the theoretic discussion and the development of the dynamic model. Finally, some suggestions toward the measurement of the class variables are made.

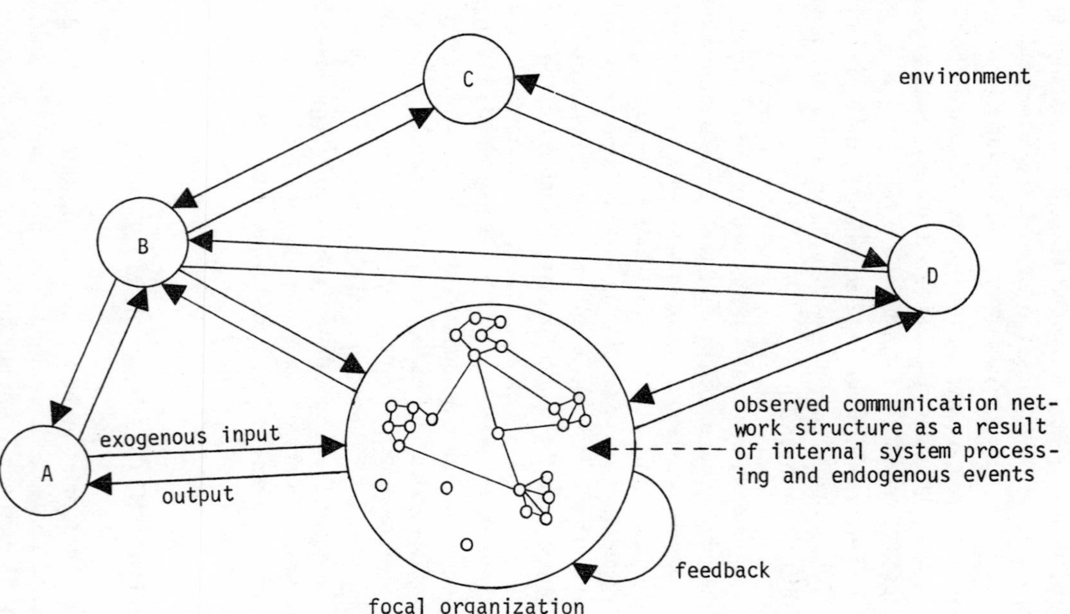

FIGURE 2 A graphic representation of interorganizational communication.
[*NOTE:* Organizations A, B, C, D and the *focal organization* are members of the organization-set operating in the same environment.]

1. *Interorganizational System Variables*

During the construction of any model, it is important to identify the essential and characteristic set of state variables that describes and suggests the critical properties of the system [6]. In regard to the critical properties of the system, the appropriate literature was reviewed, and the author compiled a list of what he considered to be relevant variables for the development of the proposed model. The construction of a matrix, with one row and one column for each variable, aided in the schematic representation of variables from the literature which indicate a causal effect on another variable. From the compilation of this matrix, these variables appear to fall into three basic *classes of endogenous variables* and *one exogenous class variable.* In considering all four variables and while preserving the interrelationships among them, the model allows for the examination of these relationships and variables as an integrated unit. The functional relationships among the class variable determine the model's state transitions. The functional relationships were expressed graphically in the form of curves from which a set of algebraic functions are constructed. These functions express each variable in terms of other variables.

For the purpose of the model, the following endogenous class variables are selected:

(a) an interorganizational *communication* variable;

(b) a perceived organization-set *interdependence* variable; and

(c) a *goal attainment* variable.

To this list of variables, a fourth, exogenous class variable is considered that reflects the influences and conditions of the environment:

(d) an *environment* variable.

The interorganizational *communication* variable is a measure of the communication transfer among a set of organizations operating within the same relevant environment. In all organizations, the occupants of some positions perform a liaison function with other organizations and form, for example, official, professional, social, and political organizational linkages or ties. The divergence between the organizational linkage systems is the source of the specific dynamics of the interorganizational system as well as the source for distintegrative tendencies. With regard to the situational context, communication may be measured as the frequency, amount, importance, intensity, content or the lack thereof. As suggested earlier, communication is essential for interorganizational activities. In the proposed model, communication is considered to be both an influence and is influenced by the interdependence variable.

The *interdependence* variable is a measure of the degree to which the individual organization perceives a need to behave in unison as a member of its relevant organization-set. This need is a measure of the perceived forces impelling the organization to coordinate, cooperate, merge or compete with elements of its organization-set. Although the need for interdependence is assumed to be determined within each organization individually, the organization-set's contextually defined state of need is considered as the result of forces that are exogenous to any focal organization. Some measures of the interdependence variable may be the degree of adherence to collective goals, joint profit maximization, etc.

The *goal attainment* variable describes a long-term state of affairs [1] and is a measure or an index of performance. Goal attainment of organizations is understood to be one preferred and observable state or several substates which are not identical with the state of the elements of the organization. Other terms for goal attainment are achievement, effectiveness, performance, objective, profit realization, coordination efforts toward a joint goal, etc. Some of these goals may be unobtainable, but nevertheless they exist as the ultimate goals towards which the organization is proceeding and against which certain actions can be measured. Obviously the goal attainment variable is, in part, dependent upon the operating conditions existing within the environment of the organization-set.

As previously suggested, these endogenous class variables (communication, interdependence and goal attainment) have to be seen in the light of the prevailing conditions of the *environment* that may influence the organization-set. The distinction between the world as perceived and the world as acted upon defines the basic condition of survival of organizations [Cf., 67; 68]. Environmental pressures acting upon organizations may function as constraints on the performance of the system and are reflected as constraints in the model. The compelling conditions and influences of the environment are then added as a fourth, exogenous variable to the list of class variables that comprise the state of the model.

These four class variables constitute the set of critical state variables upon which the model is built. Following, the description of and the relationships among these variables will be explored with regard to organization theory and research in the areas of organizations and small groups. This elaboration will place the model in perspective to current theory and research.

2. *The Relation Among the System Variables: Three Propositions*

In the proposed model, the relationships among the class variables are stated as propositions:

(1) the interorganizational *communication* variable has a direct relationship with the interdependence variable;

(2) the *interdependence* variables varies directly with the interorganizational communication variable and with the goal attainment variable;

(3) the *goal attainment* variable is directly related to the interdependence variable and the environmental variable.

ad loc. (1) Although there is some variation in the findings, the relationship expressed in the first proposition between *communication* and interdependence has been widely supported in the literature. In the area of small group research, it is a well-established fact that groups exert pressures on their members toward uniformity and thus interdependence is beyond dispute [14; 30; 31; 36; 42; 47]. Other studies have attempted to designate transactional interdependencies among organization-sets [22; 38; 48; 49; 50; 60; 74]. The concept of interdependence allows the researcher to focus on the problem of interorganizational exchanges and thus interdependence becomes a critical tool for the analysis of this process. The majority of studies concerned with organizational interdependence views the organization as an entity that requires inputs and outputs for its functioning, thus linking together a number of organizations via the process of exchanges and transactions. Aiken and Hage [3] studied organizational interdependence for certain social service organizations by operationalizing organizational interdependence as a measure of the joint programs that a focal organization has with other organizations. Similarly with Guetzkow [38], these authors found that the greater the number of joint programs, the more organizational decision-making is constrained through obligations, commitments, or contracts with other organizations, and the greater the degree of organizational interdependence. The fact that communication enhances interdependence has been reported also in studies by Barnard [8], March and Simon [51], Thompson [76], and Terreberry [72].

ad loc. (2) The first part of proposition (2), namely that *interdependence* varies directly with communication, follows partially from proposition (1) and is understood to be reciprocal. Homans [42] hypothesizes that the relationship between interdependence and communication is reciprocated [Cf. also 51] and is considered as one of the key concepts in viewing group activities. From the second part of proposition (2), i.e. the interdependence variable varies directly with the goal attainment variable, it follows that a high level of goal attainment may constitute an increasing relationship with the degree of organizational interdependence. March and Simon [51], however, state that a decrease (deterioration) in goal attainment or performance necessitates the organization to reassess its relationship to other members of the organization-set such that operating conditions are perceived as more competitive rather than cooperative.

ad loc. (3) The third relationship among the model's state variables expresses that the *goal attainment* variable is directly related to the interdependence variable and the environmental variable. In the field of economics one can observe that the level of collective goal attainment existing among members of the organization-set increases as the members adhere to a group goal such as joint profit maximization, market dominance, attempts to create an oligopolistic market or to form cartels. Furthermore, the improvement of environmental operating conditions produces an increase in the level of goal attainment (e.g. the effects of the well-publicized energy shortage of 1973/74 on the oil and related industries). Phillips [59] developed a theory of interfirm behavior positing that firms are members of groups and that the explanation of group behavior requires assumptions beyond those relating to the motivation of the individuals in the group. He continues to state that assumptions with respect to individual motives are necessary but not at all sufficient to explain the group behavior of firms. This theory of interfirm organization is based on the premise that it is incorrect to assume that individual firms attempt unilaterally to maximize anything at all, whether it is profits, sales or even a "general-preference function" if all the dimensions of the function are variables *internal* to the firm [59].

A number of researchers have viewed "goal attainment" in the light of the existing conditions in the organization-relevant environment. Tolman and Brunswik [77], Emery and Trist [23] analyzed the causal texture of organizational environments arguing that the main problem in studying organizational change is that the environmental contexts in which organizations operate are themselves changing. Thus, changes occurring in the environment are said to have such an impact that they demand consideration for their own sake when viewing one focal organization, several organizations or the entire organization-set.

The postulate that behavior is a function of the interaction of an organism with the environment is widely accepted and the theoretical as well as practical implications are investigated [9; 12; 17; 33; 55]. Furthermore, Thompson and McEwen [75] state that the setting of goals is essentially a desired relationship between an organization and its environment. Change in the organization or in the environment requires review and maybe alteration of goals. These authors and others [11; 35] suggest also that the setting of goals is not to be viewed as a static but as a dynamic element.

The earlier stated and above described three verbal propositions can be rewritten in mathematical form. Thus, the following corresponding equations can be developed:

$$C = C(I) \tag{1}$$
$$I = I(C, G) \tag{2}$$
$$G = G(I, E) \tag{3}$$

where C represents the interorganizational *communication* variable; I stands for the *interdependence* variable, G equals the *goal* attainment variable and E symbolizes the *environmental* variable. From these relationships, a differential equations model will be developed that allows for demonstration of certain dynamic aspects. The equilibrium points will be investigated and some observations are made with regard to changes among the variables.

3. *A Dynamic Model*

Previously, it was stated that environmental conditions and contexts change in addition to changes in organizations. The impact of technological [23] and societal change causes these environmental conditions to change rapidly at an increasing rate. This, in turn, elicits changes in the level of goal attainment as well as the perceived interdependence of organizations operating in the same environment. Change in the interorganizational communication behavior and in the degree of organizational interdependence occur at a relatively slower speed. When considering (t) for a specific time period, the stated relationships in equations (1), (2) and (3) can be rewritten as follows:

$$\frac{\mathrm{d}\,C(t)}{\mathrm{d}\,t} = \psi[I(t), C(t)], \tag{1}$$

where the partial derivative of $\frac{\mathrm{d}\,C(t)}{\mathrm{d}\,t}$ with respect to I is >0 when I is large, and <0 when I is small, and where the partial derivative of $\frac{\mathrm{d}\,C(t)}{\mathrm{d}\,t}$ with respect to C is <0;

$$\frac{\mathrm{d}\,I(t)}{\mathrm{d}\,t} = \omega[C(t), G(t), I(t)], \tag{2}$$

where all constants are assumed to be positive, except for the partial derivative of $\frac{\mathrm{d}\,I(t)}{\mathrm{d}\,t}$ with respect to I is <0;

$$G(t) = G[I(t), E(t)]. \tag{3}$$

The first rewritten equation suggests that the influence of interdependence upon the interorganizational communication variable can fluctuate with the level of interdependence. One can infer from this conditional relationship that a low level of interdependence, interorganizational communication flows smoother and in a less distorted fashion to a focal organization. In response, the focal organization will reciprocate this communication at an increased amount. When interdependence is very high, this interdependence may be perceived such that communication is considered as no longer necessary.

Given equation (2) above, this equation can be rewritten through the use of equation (3):

$$\frac{\mathrm{d}\,I(t)}{\mathrm{d}\,t} = \omega\{G[I(t), E(t)], C(t), I(t)\},$$

which can be rewritten as:

$$\frac{\mathrm{d}\,I(t)}{\mathrm{d}\,t} = \phi[C(t), I(t), E(t)]\,. \tag{2}$$

The dynamic model has been reduced to two basic differential equations:

$$\frac{\mathrm{d}\,C(t)}{\mathrm{d}\,t} = \psi\,[I(t), C(t)]\,, \tag{1}$$

and

$$\frac{\mathrm{d}\,I(t)}{\mathrm{d}\,t} = \phi[C(t), I(t), E(t)] \tag{2}$$

that describe the behavior of the interorganizational communication variable with respect to t as well as the behavior of the interdependence variable with respect to t.

The dynamic system as expressed in equations (1) and (2) and the characteristics thereof can be demonstrated graphically in form of a plane [32; 66; 84]. This plane is constructed through an interdependence axis (I) and an interorganizational communication axis (C) and should be merely regarded as a typical phase plane for certain specific sets of equations. Within this plane the dynamic interactions between the communication and interdependence variable can be demonstrated. This interaction is specified by the trajectory's equation which regulates direction and adjustment:

$$\frac{\mathrm{d}\,I}{\mathrm{d}\,C} = \frac{\mathrm{d}\,I/\mathrm{d}\,t}{\mathrm{d}\,C/\mathrm{d}\,t} = \frac{\phi[C, I, E]}{\psi[I, C]} \tag{4}$$

Equation (4) allows for the specification of the rate of change of the interdependence variable with respect to the communication variable for all possible values of I and C. If all possible pairs of values are plotted in this plane, only one specific response is possible and the resulting curve is the integral curve. The aggregate of all these curves represents the direction field within the model [32; 66]. In Figure 3 the direction field is represented by arrows.

The direction field is determined by the equilibrium relationships of the interdependence variable as well as the communication variable. To specify this relationship, one must find the equilibrium points for both variables, I and C, i.e. where

$$\frac{\mathrm{d}\,I}{\mathrm{d}\,t} = \phi[C, I, E] = 0, \tag{5}$$

and where

$$\frac{\mathrm{d}\,C}{\mathrm{d}\,t} = \psi\,[I, C] = 0. \tag{6}$$

Consequently, the system's trajectory, $\frac{dI}{dC}$, is horizontal along this curve. The trajectory will move to the right or left depending on whether ψ is either a positive or negative number. Furthermore, the magnitude of this shift is determined by the size of the partial derivative with respect to I, i.e. if $\frac{dC(t)}{dt}$ with respect to I is > 0 then I is large, when < 0 then I is small. Analogously, the points of equation (6)

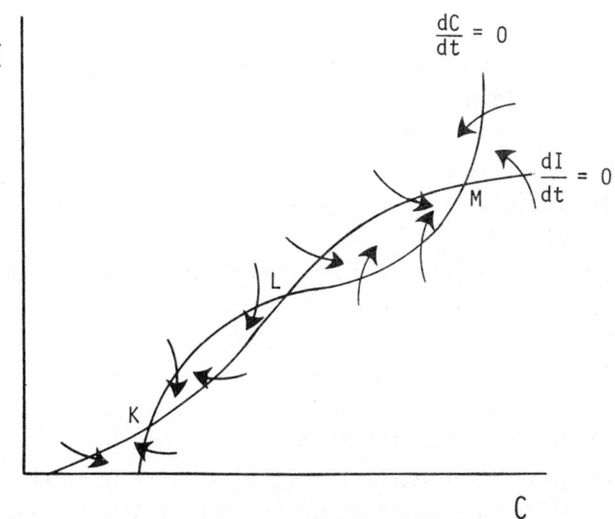

FIGURE 3 The Interdependence–Communication (IC) plane with its direction field.

will generate a trajectory along which there is no tendency for the communication level to change. The resulting trajectory along this curve has to be vertical and will alter up or down according to whether ϕ is a positive or a negative number.

When equations (5) and (6) meet at intersections, the equilibrium points are reached implying that at the same point in time both interactions have been satisfied. As demonstrated in Figure 3, points K, L, and M are such equilibrium points. These points as well as the direction field can be explored further with regard to various other influences, self-recovery mechanisms, stability, the earlier established relationships among the state variables as well as with regard to research and findings of the literature which, however, would go beyond the limits of this paper.

The following section presents several hypotheses that are generated from the above described relationships among the state variables.

A SET OF HYPOTHESES ABOUT INTERORGANIZATIONAL COMMUNICATION

The theoretical framework of organizational behavior relevant to interorganizational communication was discussed initially. Four class variables have been presented and their simultaneous interaction patterns constitute the dynamic model of interorganizational communication. From these explored relationships and discussion, the following hypotheses are presented and are considered as testable within the framework of the model and the following discussion on measurement:

(a) The greater the pressure toward interorganizational interdependence and the greater the perception of homogeneous goal attainment, the greater is the tendency to communicate toward these positions.

(b) The amount of communication between organizations is a function of the magnitude of the discrepancy between interdependence and goal attainment.

(c) A high degree of centralized communication varies inversely with a high level of organizational interdependence among the organization-set.

(d) The greater the degree of similarity of goal attainment between the organization-set and the focal organization, the greater is the amount of interdependence between them.

(e) The greater the perceived level of the environment, specifically, the size of the organization-set, the smaller is the degree of interdependence.

(f) The amount of communication depends upon, and increases with, the level of organizational interdependence.

(g) Greater interdependence among the organization-set produces more highly interconnected, interorganizational communication networks.

In part, this set of hypotheses has found support in the area of small group research and to some extent at the organizational level. How the variables suggested in the model and in these hypotheses can actually be measured is described in the following section.

SOME SUGGESTIONS TOWARD MEASUREMENT

The interorganizational *communication* variable may be measured according to three basic techniques. Some methodological difficulties may intrude this process in the attempt to delineate precisely between the transfer of information and general, social communication among organizations. On the one hand, one may be concerned with the assessment of information in its natural unit, the bit, which measures the change in uncertainty resulting from the receipt of the message [52]. On the other hand, one may realize that measuring information *per se* is not an adequate representation of interorganizational communication since general, social communication may be rather instrumental in establishing interdependent relationships. Obviously, some tradeoff needs to be made between precision and quality of "communication". In addition, information-theoretic measurement presents considerable operational difficulties for the social scientist. There is some question to what extent such measurement can be utilized in macro-social analysis at the present time.

As a second technique, Cadwallader [13] suggests the assessment of the volume of mail, telegrams, telephone calls, and memoranda which could be sampled at input terminals, output terminals, and other crucial points in the network.

The most suitable approach toward the measurement of communication behavior in large, social systems seems to be the network analytic approach in which the basic unit of analysis is a relationship or a communication link. In the past, an important drawback constituted the storage of sociometric information in the form of sociomatrices, i.e. as the network becomes large in size, meaningful and manageable analysis becomes increasingly difficult. Even the use of computers in storing sociometric data in matrix form is inefficient and prohibitively expensive as the network becomes large. An algorithm was developed by Richards [62] that overcame this problem. In the meantime, this approach has been computerized in a complex program that allows for the efficient and inexpensive analysis of social systems of up to 5,000 individuals.

Communication networks are generated when analyzing the communication relationships or links among members of an organization along a predetermined dimension. Each individual's communication relationship can be measured with regard to content, frequency, duration, and strength as well as directionality and reciprocity. The recognition of various patterns in existing relationships among network members allows individuals to be hierarchically (Cf., 67] classified into various roles: group and bridge members, liaisons, isolates, etc. Once a communication network is categorized, the structural properties of particular network patterns become important and can be described and measured with various graph-theoretic and information-theoretic approaches. Some of these appear in the form of indices such as connectivity, integrativeness, flexibility, assessibility, and others. Network analysis is described in some detail by Richards [63] and Wigand [83]; for an application of the method see Wigand [82].

As suggested earlier, the *interdependence* variable is an organization-internal measure. It may be the aggregate of forces, observed or perceived, that impell the organization to coordinate, cooperate, merge or to compete with other organizations in unison as a member of its relevant organization-set. Sometimes, cooperation may occur in the joint acquisition of resources, in the situation where organizations face greater rivals than themselves, in the form of joint production, marketing and advertising efforts. The precise measurement of this variable may be more specifically an indication of the intensity of interdependence.

The third endogenous class variable, *goal attainment,* is in its essential form a measure of survival, growth and/or well-being [72]. A goal *for* the organization becomes a goal *of* the organization, if that goal has been authorized. This implies: the formulation of the goal must be decided upon by those persons or that group who have been in power to do so and when decided upon, declared as binding for the organization. Consequently, the goals of the organization are the goal formulations which have been authorized by the core group [Cf., 19; 44; 65]. The need to survive may be the most essential goal and objective of any organization. Nearly every organization demonstrates some desire to grow, no matter how one defines growth. Growth, it seems, is an almost built-in objective of every Western socio-economic system that resembles analogous ideologies such as manifest destiny and providence. Well-being [37] incorporates such notions as enjoying the daily activities in organizations, social responsibility, social meaningfulness and other social goals that constitute tradeoffs with regard to profitability and efficiency.

A statement about goal attainment is a statement about the nature and character of the organization and as a variable should be measured in the context of the particular research situation. For example, if the emphasis is on the economic characteristics of an interrelationships, then a measure of goal attainment may be the organization's profit realization with regard to a set goal. For organizations where there is no profit maximization, goal attainment may be a measure of clients served, problems solved, etc., with regard to a prespecified goal as in the case of social service organizations. Organizational goal attainment tends to be stable in nature since it represents the sum total of the members' aspirations and motivations in measured form with regard to survival, growth and well-being. In deriving goal attainment criteria, one ought to note the criteria which are applicable to individual parts of the organization should be equally applicable to the totality. This implies, if a criterion for goal attainment, when applied to organizational subsets, is q_i for each of the n parts and is q when seen in the light of the totality, then it is not a necessary condition that if $\frac{\mathrm{d}\, q_i}{\mathrm{d}\, t} > 0$ for all i, then $\frac{\mathrm{d}\, q}{\mathrm{d}\, t} > 0$. Considering the individual responses to the goal attainment variable, various methods such as ranking, preference scales, the application of utility theory [Cf., 15; 39; 56; 64] can be employed to ascertain precise and representative measurement.

The exogenous class variable, the condition and influences of the *environment*, may in its measured form constitute the aggregate of all organization-set members' responses with regard to competitiveness, cooperativeness, legal constraints, difficulties with the labor market or to enter a market as well as possible unobtrusive measures and observations external to the individual organization. This variable may be the aggregate of direct responses, an index, or may be also a measure of economic, social and cultural indicators.

SUMMARY

It was pointed out that there is a necessity for interorganizational research if social scientists are to gain significant, additional insight in the study of organizations. From the literature review four class variables were identified that appear to subsume most other variables studied by scholars in the field of interorganizational research and constitute the state variables for a dynamic model. Three endogenous class variables are an interorganizational communication variable, an interdependence variable and a goal attainment variable. A fourth, exogenous class variable was added which reflected the condition and influences of the environment. From the relationships among these class variables a set of propositions was developed as well as the dynamic aspects of the model were explored. The interactions among the variables, multiple equilibria points, and the direction fields were described. A set of hypotheses concerned with interorganizational communication were presented. Lastly, some suggestions toward the measurement of all four class variables were offered.

ENDNOTE

[1] The research for this paper was in part supported by a research grant from the National Science Foundation. Computer time and other assistance were rendered by the Department of Communication and the Computer Institute for Social Science Research, both Michigan State University, as well as the Departmento de Comunicación, Universidad Iberoamericana, Mexico City, Mexico. The author wishes to acknowledge the helpful comments of Joseph N. Cappella, Donald P. Cushman, Vincent R. Farace, Ralph L. Levine, and Floreda D. Lux-Wigand.

REFERENCES

[1] Ackoff, R. L., *A Concept of Corporate Planning,* New York: Wiley, 1970.

[2] Ackoff, R. L., "Systems organizations, and interdisciplinary research", *General Systems,* **5**: 2–3, 1960.

[3] Aiken, M. and Hage, J., "Organizational interdependence and intra-organizational structure", *American Sociological Review,* **33**: 912–930, 1968.

[4] Anderson, R. C., "A sociometric approach to the analysis of interorganizational relationships", Technical Bulletin, Institute for Community Development and

Services, Continuing Education Service, Michigan State University, East Lansing, Michigan, B-60: 1–25, 1974.
[5] Anderson, R. C., "The perceived organized structure of Michigan's Upper Peninsula – a sociometric analysis", Paper presented to the Rural Sociological Society, 1965.
[6] Ashby, W. R., *An Introduction to Cybernetics,* London, England: Chapman & Hall, 1956.
[7] Ashby, W. R., "The effect of experience on a determinate dynamic system", *Behavioral Science,* **1:** 35–42, 1956.
[8] Barnard, C. I., *The Functions of the Executive,* Cambridge, Mass: Harvard University Press, 1962.
[9] Barton, A. H., *Organizational Measurement and its Bearing on College Environments,* New York: College Entrance Examination Board, 1961.
[10] Black, B. J. and Kase, H. M., "Inter-agency cooperation in rehabilitation and mental health", *Social Service Review,* **37:** 26–32, 1963.
[11] Boulding, K. E., *The Organizational Revolution: a Study in the Ethics of Economic Organizations,* Chicago: Quadrangle Books, 1953.
[12] Brunswik, E., *Perception and the Representative Design of Psychological Experiments,* Berkeley: University of California Press, 1956.
[13] Cadwallader, M., "The cybernetic analysis of change in complex social organizations", *American Journal of Sociology,* **65:** 154–157, 1959.
[14] Cartwright, D. and Zander, A., *Group Dynamics;* 80, Evanston, Illinois: Row, Peterson, 1960.
[15] Churchman, C. W., Ackoff, R. L. and Arnoff, E. L., *Introduction to Operational Research,* New York: Wiley, 1957.
[16] Clark, B. R., "Interorganizational patterns in education", *Administrative Science Quarterly,* **10:** 224–237, 1965.
[17] Cronbach, L. J., "The two disciplines of scientific psychology", *American Psychologist,* **12:** 671–684, 1957.
[18] Crozier, M., *The Bureaucratic Phenomenon,* Chicago: The University of Chicago Press, 1964.
[19] Cyert, R. M., Feigenbaum, E. A. and March, J. G., "Models in a behavioral theory of the firm", *Behavioral Science,* **4:** 82–83, 1959.
[20] Dahrendorf, R., "Out of utopia: toward a reorientation of sociological analysis", *American Journal of Sociology,* **64:** 115–127, 1958.
[21] Dill, W. R., "Environment as an influence on managerial autonomy", *Administrative Science Quarterly,* **2:** 409–443, 1958.
[22] Dill, W. R., "The impact of environment on organizational development", *in* Mailick, S., and van Ness, E. H. (Eds.), *Concepts and Issues in Administrative Behavior:* 94–109, Englewood Cliffs, N.J.: Prentice Hall, 1962.
[23] Emergy, F. E. and Trist, E. L., "The causal texture of organizational environments", *Human Relations,* **18:** 21–31, 1965.
[24] Etzioni, A. (Ed.), *Complex Organizations,* New York: Holt, Rinehart & Winston, 1962.
[25] Etzioni, A., "New directions in the study of organizations and society", *Social Research,* **27:** 223–228, 1960.
[26] Evan, W. M., "The organization-set: Toward a theory of interorganizational relations", *in* Thompson, J. D. (Ed.), *Approaches to Organizational Design:* 173–191, Pittsburgh, Pa.: University of Pittsburgh Press, 1966.
[27] Evan, W. M., "Toward a theory of interorganizational relations", *Management Science,* **10:** B217–B230, 1965.

[28] Farace, R. V. and Wigand, R. T., "Crisis relocation planning information strategy", Technical Report, Michigan State University, Department of Communication, East Lansing, Michigan, 1974.
[29] Farace, R. V. and Wigand, R. T., "The communication industry in economic integration: the case of West Germany", Paper presented to the International Communication Association convention, Chicago, 1975.
[30] Festinger, L., Schachter, S. and Back, K., *Social Pressures in Informal Groups,* New York: Dryden Press, 1950.
[31] Festinger, L. and Thibaut, J., "Interpersonal communication in small groups", *Journal of Abnormal and Social Psychology,* **46:** 92–99, 1951.
[32] Ford, L. R., *Differential Equations,* New York: McGraw-Hill, 1957.
[33] Forehand, G. A. and Gilmer, H., "Environmental variation in studies of organizational behavior", *Psychological Bulletin,* **62:** 361–382, 1964.
[34] Form, W. H. and Nosow, S., *Community in Disaster,* New York: Harper & Row, 1958.
[35] Galbraith, J. K., *The Affluent Society,* Boston, Mass.: Houghton & Mifflin, 1958.
[36] Glanzer, H. and Glaser, R., "Techniques for the study of group structure and behavior: II. empirical studies of the effects of structure in small groups", *Psychological Bulletin,* **58:** 1–27, 1961.
[37] Goodeve, C. F., "Science and social organization", *Nature,* **188:** 4746, 1960.
[38] Guetzkow, H., "Relations among organizations", *in* Bowers, R. V. (Ed.), *Studies on Behavior in Organizations:* 13–44, Athens, Georgia: University of Georgia Press, 1966.
[39] Gupta, S., "Choosing between multiple objectives", Unpublished Paper, Management Science Center, University of Pennsylvania, Philadelphia, Pa., 1969.
[40] Homans, G. C., "Social behavior as exchange", *American Journal of Sociology,* **63:** 597–606, 1958.
[41] Homans, G. C., *Social Behavior: its Elementary Forms,* New York: Hartcourt, Brace & World, 1961.
[42] Homans, G. C., *The Human Group,* New York: Hartcourt, Brace & World, 1950.
[43] Jirasek, J., *Das Unternehmen – ein kybernetisches System?* Hamburg, Germany: GFM–Verlag für Markt- und Unternehmensforschung mbH, 1968.
[44] Kirsch, W., "Unternehmungsziele in organisations-theoretischer Sicht", *ZfbF:* 670 ff, 1969, cited in Klein, H. K. and Wahl, A., Zur Logik der Koordination interdependenter Entscheidungen in komplexen Organisationen, (Part II), *Kommunikation,* **6:** 142, 1970.
[45] Lawrence, P. R. and Lorsch, J. W., "Differentiation and integration in complex organizations", *Administrative Science Quarterly,* **12:** 1–47, 1967.
[46] Lawrence, P. R. and Lorsch, J. W., *Organization and Environment: Managing Differentiation and Integration,* Boston, Mass.: Harvard University Press, 1967.
[47] Leavitt, H. J., "Some effects of certain communication patterns on group performance", *Journal of Abnormal and Social Psychology,* **46:** 38–50, 1951.
[48] Levine, S. and White, P. E., "Exchange as a conceptual framework for the study of interorganizational relationships", *Administrative Science Quarterly,* **5:** 583–601, 1961.
[49] Levine, S., White, P. E. and Paul, B. D., "Community interorganizational problems in providing medical care and social service", *American Journal of Public Health,* **53:** 1183–1195, 1963.
[50] Litwak, E. and Hylton, L. F., "Interorganizational analysis: a hypothesis on coordinating agencies", *Administrative Science Quarterly,* **6:** 395–420, 1962.
[51] March, J. G. and Simon, H. A., *Organizations,* New York: Wiley, 1958.

[52] Miller, G. A., "What is information?", *The American Psychologist,* **8**: 3–11, 1953.

[53] Miller, W. B., "Inter-institutional conflict as a major impediment to delinquency prevention", *Human Organization,* **17**: 20–23, 1958.

[54] Morris, R. and Randall, O. A., "Planning and organization of community services for the elderly", *Social Work,* **10**: 96–102, 1965.

[55] Murray, H. A., *Explorations in Personality,* New York: Oxford University Press, 1938.

[56] Neumann, J. von, and Morgenstern, O., *Theory of Games and Economic Behavior,* Princeton, N.J.: Princeton University Press, 1953.

[57] Perrow, C., "A framework for the comparative analysis of organizations", *American Sociological Review,* **32**: 194–208, 1967.

[58] Perrucci, R. and Pilisuk, M., "Leaders and ruling elites: the interorganizational bases of community power", *American Sociological Review,* **35**: 1040–1057, 1970.

[59] Phillips, A., "A theory of interfirm organization", *Quarterly Journal of Economics,* **74**: 602–613, 1960.

[60] Reid, W., "Interagency coordination in delinquency prevention and control", *Social Service Review,* **38**: 418–428, 1964.

[61] Reid, W., "Interorganizational coordination in social welfare: a theoretical approach to analysis and intervention", *in* Kramer, R. and Specht, H. (Eds.), *Readings in Community Organization Practice,* Englewood Cliffs, N.J.: Prentice-Hall, 1967.

[62] Richards, W. D., Jr., "An improved conceptually-based method for analysis of communication network structures of large complex organizations", Paper presented at the International Communication Association convention, Phoenix, Arizona, 1971.

[63] Richards, W. D., Jr., "Social network analysis: an overview of recent developments", in this volume.

[64] Rivett, P., *Principles of Model Building,* London, England: Wiley, 1972.

[65] Simon, H. A., "A behavioral model of rational choice", *Quarterly Journal of Economics,* **69**: 99–118, 1955.

[66] Simon, H. A., *Models of Man,* New York: Wiley, 1957.

[67] Simon, H. A., "The architecture of complexity", *Proceedings of the American Philosophical Society,* **106**: 477–481, 1962.

[68] Simon, H. A. and Newell, A., "Simulation of human thinking", *in* Greenberger, M. (Ed.), *Management and the Computer of the Future:* 95–114, New York: Wiley, 1962.

[69] Simpson, R. L. and Gulley, W. H., "Goals, environmental pressures, and organizational characteristics", *American Sociological Review,* **27**: 344–351, 1962.

[70] Smith, D. H., "Communication research and the idea of process", *Speech Monographs,* **39**: 174–182, 1972.

[71] Steinbuch, K., *Automat und Mensch, kybernetische Tatsachen und Hypothesen,* Berlin–Heidelberg–New York: Springer Verlag, 1965.

[72] Terreberry, S., "The evolution of organizational environments", *Administrative Science Quarterly,* **12**: 590–613, 1968.

[73] Thompson, J. D. (Ed.), *Approaches to Organizational Design,* Pittsburgh, Pa.: University of Pittsburgh Press, 1966.

[74] Thompson, J. D., "Organizations and output transactions", *American Journal of Sociology,* **68**: 309–324, 1962.

[75] Thompson, J. D. and McEwen, W. J., "Organizational goals and environment: Goal-setting as an interaction process", *American Sociological Review,* **23**: 23–31, 1958.

[76] Thompson, V. A., *Modern Organizations,* New York: Knopf, 1961.

[77] Tolman, E. C. and Brunswik, E., "The organism and the causal texture of the

environment", *Psychological Review,* **43:** 43–72, 1935.

[78] Turk, H., "Comparative urban structure from an interorganizational perspective", *Administrative Science Quarterly,* **18:** 37–55, 1973.

[79] Turk, H., "Comparative urban studies in interorganizational relations", *Sociological Inquiry,* **38:** 108–110, 1969.

[80] Warren, R. L., "The interorganizational field as a focus for investigation", *Administrative Science Quarterly,* **12:** 396–419, 1967.

[81] Weick, K. E., *The Social Psychology of Organizing,* Reading, Mass.: Addison-Wesley, 1969.

[82] Wigand, R. T., "Communication, integration and satisfaction in a complex organization", Paper presented to the International Communication Association convention, New Orleans, Louisiana, 1974.

[83] Wigand, R. T., "Communication network analysis: a computerized approach toward the engineering of systems and organizations", Paper presented to the II Inter-American Conference on Systems and Informatics, Mexico City, Mexico, 1974.

[84] Williamson, O. E., "A dynamic theory of interfirm behavior", *Quarterly Journal of Economics,* **79:** 579–607, 1965.

[85] Wrong, D., "The oversocialized conception of man in modern society", *American Sociological Review,* **26:** 183–193, 1961.

[86] Yuchtman, E. and Seashore, S. E., "A system resource approach to organizational effectiveness", *American Sociological Review,* **32:** 891–903, 1967.

A THEORY OF COMMITTEE FORMATION

MARY E. LIPPITT
University of Minnesota, Minneapolis

and

KENNETH D. MACKENZIE
University of Kansas, Lawrence

INTRODUCTION

A preceding paper entitled "Where is Mr Structure?" introduces several ideas of a recent theory of group structures [2, in this volume]. This theory, like any new theory, requires more thought, more testing, and ultimately major revision. There are many problems currently being developed that are not in the first two volumes and there are many new problems that have only recently become perceived by us. This paper describes the early stages in the process of formulating one extension for the formation of committees. We have become fascinated by the processes of committee proliferation and we think that it is possible to extend our ideas about group structures to this problem. As will become evident, the focus of our attention is on how they come to be created rather than on how they behave. We are not opening up the normative questions of whether or not it is optimal to form a committee or who should be on a new committee, how one should decide the charge to give the committee, or under what conditions a committee should be encouraged to function. Our goals are very modest, here. We seek to understand the type of problem solving that leads to the formation of a new committee. We also seek to illustrate some of the ideas presented in "Where is Mr Structure?"

We see organizations as having many structures. We can watch these structures change. We can, at least in little laboratory organizations, study the processes of structural changes. We assume that structures are need satisfying patterns of interactions. We can define and empirically demonstrate the close relationships among the structures, group problem solving processes, and group role matrices. We can measure these entities and derive models and measures to study efficiency, degree of hierarchy and other interesting problems. We see the dynamics and active processes by which structures, problem solving tasks, and incentives mutually adjust to one another. This way of viewing group structures emphasizes these active processes. We believe that our view more faithfully and fruitfully captures the behaviors of groups than the process passive "Mr Structure" views.

There is a class of problems, we call authority-task problems, that result from inconsistencies between the "Mr Structure" concepts of structure embodied in the establishment of pyramids of authority and the reality of the active, dynamic processes of structural change and multiple structures. The desire to establish a "Mr Structure" pyramid of authority results in grouping together similar activities to form positions or offices. Then the process of assigning authority begins, wherein a position is given authority over other positions, and a pyramid begins to form. However, the assignment of authority often precedes any real understanding of the task processes and also, once established, tends to lag, and often seriously, the structural adjustments that are taking place in performing group tasks. In short, the authority system is often out of touch with the realities of the problem solving systems. Thus we perceive numerous gaps between the authority role system and the role system based upon actual behavior.

This gap, called the authority-task gap, generates a type of problem that we call an authority-task problem. The actual task process dynamics can create new authority-task gaps. When one becomes aware of an authority-task gap, this creates an authority-task problem, ATP. We believe that most issues involving administrators in organizations such as state supported universities, regulatory bodies, and government agencies have their source in an authority task gap. Such a gap may arise from personnel change, changing task processes, outside intervention and other exogenous shocks, and inappropriate authority assignment. The gap's existence may set off a variety of processes and responses which the organization will desire to monitor and control. Our interest is focused on the group's behavior when it is confronted by an authority-task problem.

We posit two stages in the solution of the ATP. First, the manager must find a technically feasible set of activities and allocation of their performance which would remove the problem, and second, the manager must work out acceptance of the allocation of activities. In *A Theory of Group Structures* [4; 5], this second stage is called the *consummation* of the role matrix. Two stages are necessary because a solution removes the problem only if it is implemented. The theory of committee formation presented here suggests mechanisms by which ATP's are handled in an organization. It is also suggested that committee formation to solve ATP's can be a significant control device on group processes. However, unlike some forms of problem solving, the problem must be solved and be capable of implementation in order to be acceptable. We shall examine the solution of an ATP as seen by a manager.

STATEMENT OF THE BASIC PROBLEM

Confronted with an authority-task problem, the manager responsible for solving it appears to recognize two stages in his task, namely, to see a technically feasible solution derived and to see it implemented. The solution translates to changes in the group's task role matrix, R_{T*},[1] where implementation is consummation of the new group task role matrix. With this task in mind, the manager looks across his organization and notes the existence of a number of conditions which could affect how he arrives at a feasible solution and then how he implements it by gaining consummation of a new task role matrix. He also notes a number of conditions on the ATP. In observing these conditions he acts as if he faces a list $\langle \mathbf{X} \rangle$ of variables.

Observing the consistency in the processes by which groups solve ATP's we suggest that the manager responsible for solving any particular problem acts as if there exists a function by which he takes the values of the various X's in $\langle \mathbf{X} \rangle$ and determines a strategy for performing his task of removing the problem. Let us view this formulation simply as $\mathbf{y} = f\langle \mathbf{X} \rangle$, where $\mathbf{y}$ is the strategy, $\langle \mathbf{X} \rangle$ is the list of considerations representing the conditions on the problem and the group, and $f\langle \mathbf{X} \rangle$ is the *mapping function* that describes how he picks a strategy given the list $\langle \mathbf{X} \rangle$. The mapping function transforming observations of what exist into decisions on what to do may be remarkably similar across managers and across organizations where ATP's are encountered.

A central portion of this theory of committee formation is the mapping function. The description of the list of variables $\langle \mathbf{X} \rangle$ and the mapping function $f\langle \mathbf{X} \rangle$ are presented in Figures 1 and 2. The core of an authority-task problem solving process, leading in some cases to committee formation, is described by the mapping function. We think that these processes can be tracked using this device. We assume that the ATP comes to the attention of the appropriate authority. First, he must evaluate the conditions on the problem and on the organization, where the conditions are described by the list $\langle \mathbf{X} \rangle$ of binary variables. The determination of values of the X's may be described by some subroutines, where again we observe that the manager acts as if there exist subordinated mapping functions for the subproblems of evaluating each variable in $\langle \mathbf{X} \rangle$. Having estimated the values in $\langle \mathbf{X} \rangle$, we assume that he makes his choice by acting as if he consults the mapping function, $f\langle \mathbf{X} \rangle$.

We have depicted seven possible actions, or strategies, available to the manager. They are: (1) do nothing, (2) appoint a committee to provide a recommendation, (3) send the problem to a standing committee for recommendation, (4) form an operating group, a task force, to solve the problem, (5) direct implementation of your solution, (6) give the problem to your superior to solve, or (7) hire a consultant. After an action has been chosen and implemented, a response from the group is observed. If the response is favorable, consummation of the new task role matrix may occur and the ATP may be solved. However, the group's response may not always be favorable. The group may balk and refuse to implement the solution, and then they may spread the issue. A committee formed to derive a

FIGURE 1

A MAPPING FUNCTION AND ITS VARIABLES FOR COMMITTEE FORMATION

X_1	Does an ATP exist or will one arise in this issue?	$\begin{cases} X_1 = 1 \text{ if yes} \\ X_1 = 0 \text{ if no} \end{cases}$
X_2	Does the issue need to be resolved?	$\begin{cases} X_2 = 1 \text{ if yes} \\ X_2 = 0 \text{ if no} \end{cases}$
X_3	Does LCA know a technically feasible solution?	$\begin{cases} X_3 = 1 \text{ if yes} \\ X_3 = 0 \text{ if no} \end{cases}$
X_4	Does LCA have capacity to solve ATP?	$\begin{cases} X_4 = 1 \text{ if yes} \\ X_4 = 0 \text{ if no} \end{cases}$
X_5	Does LCA have authority to preempt solution activities?	$\begin{cases} X_5 = 1 \text{ if yes} \\ X_5 = 0 \text{ if no} \end{cases}$
X_6	Is the problem recurring?	$\begin{cases} X_6 = 1 \text{ if yes} \\ X_6 = 0 \text{ if no} \end{cases}$
X_7	Would the solution be accepted, or could an acceptable solution be negotiated by the LCA?	$\begin{cases} X_7 = 1 \text{ if yes} \\ X_7 = 0 \text{ if no} \end{cases}$
X_8	Can LCA pass the buck?	$\begin{cases} X_8 = 1 \text{ if yes} \\ X_8 = 0 \text{ if no} \end{cases}$
X_9	Does an appropriate Standing Committee exist?	$\begin{cases} X_9 = 1 \text{ if yes} \\ X_9 = 0 \text{ if no} \end{cases}$
X_{10}	Is it feasible to hire a consultant?	$\begin{cases} X_{10} = 1 \text{ if ye} \\ X_{10} = 0 \text{ if no} \end{cases}$
X_{11}	Is there a severe time constraint or a deadlock?	$\begin{cases} X_{11} = 1 \text{ if ye} \\ X_{11} = 0 \text{ if no} \end{cases}$

FIGURE 2

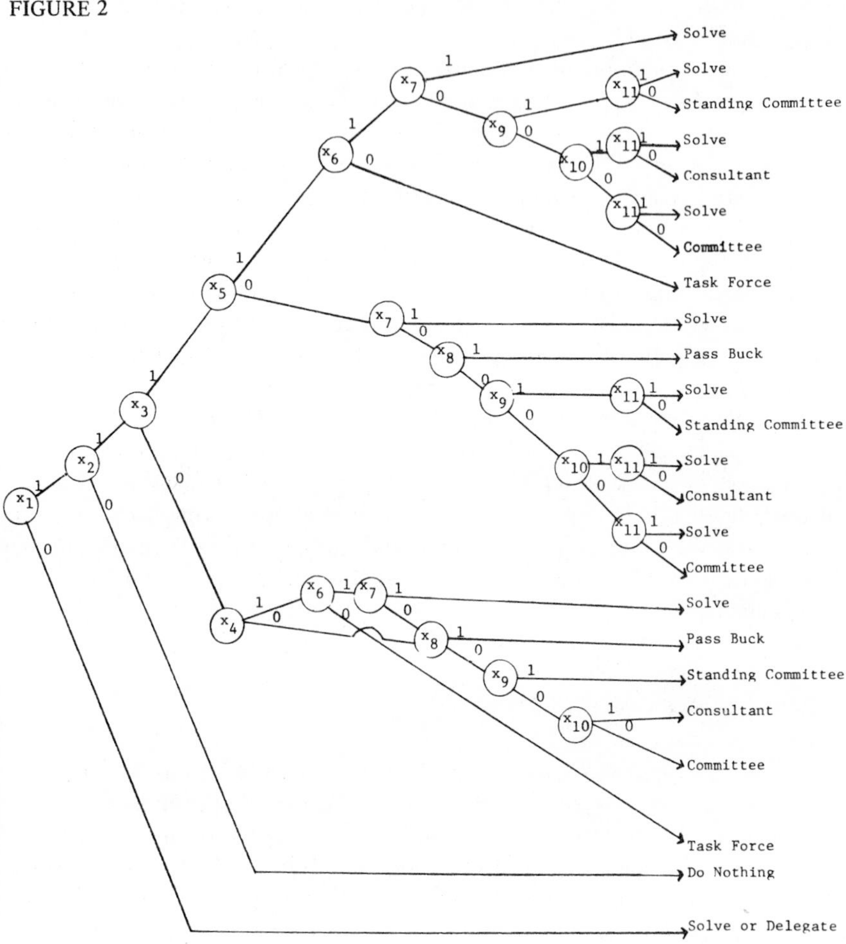

The nodes of this mapping function are the eleven variables in <X>. The path from one node to the next is determined by the value, 1 or 0, of the corresponding variable.

feasible solution may fail and/or come up with at best a split recommendation, and so on through a range of responses. Based upon the response observed, the manager obtains new information on the actual conditions and reevaluates some of the X's. For example, the group's response may create a new ATP, and the manager returns to X_1 in the mapping function, or it may point back to a variable within the function where he misestimated an X, so he must start over from that point to find a new strategy. In cycling through the mapping function a new y is determined, based upon new knowledge of the X's, and the process of observing the response repeats itself until the ATP is solved. Clearly, one ATP can create a succession of new ATP's and committees.

The mapping function operationalizes our belief that the manner in which resolution of issues proceeds in a group has a direct bearing on the form of the resolution and the stability of it once adopted. In the organization which we view daily, the University of Kansas, there are a host of problems which are regularly routed directly to a new or standing committee, and there are instances where failure to do so creates new ATP's for the administration. We believe the answer to the question of why committees are prevalent in this institution and other similar organizations lies in an exploration of the incongruities between task and authority structures. Many of the incongruities arise for reasons argued in "Where is Mr Structure?" [in this volume], namely, that we construct, manage, and study organizations using a deficient concept of structure.

The following new theory has been derived out of *A Theory of Group Structures* and years of observation of committee formation within universities. The ideas on group structures have been extensively tested by Mackenzie, and the mapping function ultimately derived from new ideas on committee formation has been tested by Lippitt [1] and Lippitt and Mackenzie [2]. There are portions of the new theory which have yet to be tested, but the results thus far are encouraging.

DEFINITIONS AND CONCEPTS

With a brief sketch of the theory completed, let us begin to lay down a base of definitions and concepts upon which to construct a theory of committee formation. Some definitions and concepts come from *A Theory of Group Structures* [4; 5], and some come as a logical extension of it. This theory views structures as a need satisfying interaction pattern, and is based on the presuppositions that the problem solving processes of a group can be represented by a sequence of milestones, that structures are established through behavioral voting, and that one can use these structures to define a group role matrix.

In order to differentiate the authority role system from the task role system we shall call Mackenzie's group role matrix, the *group task role matrix.* Activities, displayed in a group task role matrix, R_T*, constitute the basic element of the theory of structures and the theory of committee formation. The group's task role matrix is defined by the set of task activities performed by the group and the set of participants. The *ij*-th entry in the matrix is one if the *i*-th participant performed activity a_j, and zero if he does not.

There is also a group authority-role matrix, R_A. The group authority-role matrix is defined by the set of activities performed by the group and the set of participants, where the entries are one or zero, depending upon whether or not a participant has authority to preempt a particular activity.

At this point, let us be specific in our use of the term authority. *Authority* is that right which is delegated, ultimately from the State, to preempt activities in organizations. Authority can be delegated through the organization, while power and influence cannot. To have authority over an activity means one has the ability to dictate or preempt the state of that activity, either 1 or 0, performed or not performed, in a group's task role matrix. Location of an entry in a group's task role matrix requires naming both an activity and a participant. One may have authority to determine that a particular activity will be performed in the organization, but may not have the authority to make a particular person perform it. Because of the problems inherent in this case, we observe that authority tends to segment the task role matrix, grouping activities and people who perform them into positions. If a superior has authority over most of the activities a subordinate performs, then the group begins to act as if the superior has authority over the subordinate. The organization chart is often employed to depict authority relationships as they existed or were thought to exist when the organization chart was specified. The difference between it and the actual task-role system may be considerable. The organization chart can be represented by a third role matrix, R_C, whose ij-th entry is unity if person i is the immediate superior of person j and zero otherwise.

Thus, we see three role systems: (1) the group task role system as represented by $R_T{}^*$; (2) the authority role system as represented by R_A; and (3) the organization chart represented by R_C. R_C may be inconsistent or unclear in its relationship with R_A. And both R_C and R_A may be inconsistent or uncertain with respect to $R_T{}^*$. It is $R_T{}^*$ that describes what is actually going on. Even though the organization may tend to see itself in terms of R_C and some members see it in terms of R_A, the existence of uncertain relationships with $R_T{}^*$ gives rise to authority-task gaps and when these are noticed, to authority-task problems.

An issue is said to arise when a set of activities in $R_T{}^*$ are placed in "recall", that is, their 0 or 1 value becomes uncertain for some reason. The set of activities recalled define the issue. The person responsible for resolving the issue will be the lowest official in the organization who has authority in R_C over all of the persons involved in the issue. We shall call him the *lowest common ancestor*, LCA. Even though the LCA has authority over the people involved according to the organization chart, R_C, he may not have the right to preempt each of the activities involved in the issue, as we can discover by checking R_A. Thus, inconsistencies between $R_T{}^*$, R_A, and R_C can result in an LCA's having the responsibility for resolving an issue involving activities which he does not have the right to preempt. Such an issue is called an authority-task problem, ATP. Our theory begins at the point where an ATP comes to the attention of the LCA.

Solutions to ATP's must be viewed in terms of both the set, A, of activities to be added to or deleted from $R_T{}^*$ and the actual entries for each of the participants. A *technically feasible solution* is one that, if implemented, removes the ATP. Thus we view a feasible solution to an ATP as a pair $(A, R_T{}^*)$.

An important concept used in developing our theory of committee formation is the *behavioral constitutional* approach to role changes [4]. Changes in a group's task role matrix, $R_T{}^*$, occur through an exchange of behavioral votes. The behavioral constitutional approach to studying this voting process is actually a calculus of information-processing, where the researcher attempts to infer the rules by which group members process the votes cast within the group. In the laboratory, votes are coded from the written messages passed between group members. Once the rules are inferred, the process of the group can be tracked and in some cases, predicted.

Two other concepts are required before introducing a central concept of behavioral treaties and then launching the theory. We often observe responses to attempts to resolve an issue to include what we call *issue spreading*. When an ATP arises, the LCA generally attempts to localize the problem, we think in order to avoid involving activities over which he has no authority and which would thus threaten the rise of new ATP's. An issue is said to have spread when more activities are recalled. There are a variety of strategies for issue spreading by subordinates. They may, for example, spread to activities over which the LCA is more vulnerable, that is, where an authority-task gap is perceived to exist. They may also choose to spread the issue to activities for which the LCA they had been dealing with is no longer the appropriate one for solving the problem. The LCA may himself engage in issue spreading for purposes of passing the buck to one of his superiors or by choosing to become an advocate of subordinates as an issue goes to a higher official for resolution. Buck passing may not require issue spreading, however. If an LCA notices a problem which may involve one of his superiors, one for which a superior would preempt him, or a problem which he perceives as "too hot to handle", he will attempt to get rid of it to the next LCA. Buck passing requires acceptance by both parties before the ATP becomes the responsibility of the next LCA.

A CONCEPT OF BEHAVIORAL TREATIES

Because structures are need satisfying interaction patterns, a stable structure or allocation of activities represents a form of unanimity among the participants. Whether the structure represents a regular task process milestone or a solution to an ATP, it must reflect unanimity among a minimum number of controllers. We assume for any issue there exists a *minimum number of controllers*, among whom agreement on the value of entries in $R_T{}^*$ constitutes a preemption and thereby resolves the issue, as if actual unanimity had been achieved. By axiom, a *preemption* dictates the state of a role set element. Unanimity among the minimum number of controllers is reached through a process of bargaining, side payments, threats, offers and counter offers all of which are used as means of influence, with the intent being to influence perceived costs and benefits of various activities. All of the forms of influence and the manner in which they were exercised are taken into account when one decides to accept a particular allocation of activities. Thus, there is an underlying notion of equilibrium between benefits and costs associated with a stable allocation. We shall describe that equilibrium by the terms of a *behavioral treaty*, where the treaty contains the conditions and contingencies by which a stable allocation will remain stable.

In voting to resolve issues involving the allocation of activities in $R_T{}^*$, we observe the group to act as if each member has a preference function for the performance of each activity $a_i \in \{A\}$ where $\{A\}$ is the set of activities in $R_T{}^*$ which is recalled. The preference function has the general form of equation (1), where p_{ka_i} is the preference of member k for activity a_i.

$$p_{ka_i} = e_{ka_i} B_{ka_i}, \tag{1}$$

where

$$B_{ka_i} = \begin{cases} 1 & \text{if } a_i \text{ is elected to be added or maintained in the role set of } x_k. \\ -1 & \text{if } a_i \text{ is elected to be deleted or maintained out of the role set of } x_k. \end{cases}$$

and

$$e_{ka_i} = \begin{cases} 1 & \text{if the elected state of } a_i \text{ is desirable to } x_k. \\ 0 & \text{if } x_k \text{ is indifferent.} \\ -1 & \text{if the elected state of } a_i \text{ is undesirable to } x_k. \end{cases}$$

The role set of person x_k is said to be *enhanced* when activities which are positively evaluated ($p_k > 0$) are added or when activities which are negatively evaluated ($p_k < 0$) are deleted. *Detraction* from x_k's role set occurs when either activities which are positively evaluated are deleted or when activities which are negatively evaluated are added. In resolving an issue, which includes implementing a solution, the group acts as if each member desires to enhance his role set as much as possible and detract from it as little as possible. We often observe participants in organizations to not be sufficiently interested in the outcome of the reallocation to get involved in it. If a member is indifferent about each of the potential activity changes, or if there are few opportunities to enhance his role set and few to detract from it, we might expect him to sit out the issue, and let others resolve it. To allow for this behavior, we need some notion of the stakes involved in each issue. The stakes for x_k on an issue are defined by equation (2).

$$S_k = \sum_{a_i \in A} |p_{ka_i}| \tag{2}$$

The higher the stakes for all participants involved, the greater the likelihood that one person's gain will be another's loss. A *preclusion* is said to occur when the enhancement of one member's role set results in detraction from another member's role set. For each pair of participants (say k and ℓ in an issue, preclusions are computed by equation (3).

$$p_{k\ell} = \sum_{a_i \in A} |e_{ka_i} B_{ka_i} - e_{\ell a_i} B_{\ell a_i}| \tag{3}$$

If a large number of preclusions are possible in an issue, we can expect a very lively influence process aimed at resolving it. There may be threats cast regarding the performance of certain activities or promises of benefits for performing certain others. When resolution of the issue finally occurs and a new need satisfying structure is adopted, it will have been forged from a dynamic exchange of influence votes. An intricate web of expectations will have been created to stabilize the new structure and prevent continuing preclusions. We suggest that groups act as if there are *behavioral treaties* whose terms contain the minimum conditions and contingencies by which further preclusion attempts are avoided so the participants can continue to prefer the currently adopted structure to a different one.

We suspect that treaties may play a very dominant role in problem solving behavior where authority-task gaps occur. When a gap exists, it is not clear whether an LCA has the ability to preempt or not, and it is only made clear when one is attempted. There seem to be a set of rules by which an LCA decides to attempt a preemption when it is not apparent that he has the authority to do so. We believe those rules arise out of previous interactions with approximately the same group of subordinates. The traditions of their past relationships, favors traded, bargains struck, and so forth, shape expectations of future behavior. These expectations may be viewed as embodied in a behavioral treaty, and if a departure from expectations is serious enough (e.g. perhaps preclusions occur) then there is a perceived treaty violation, following which they will shape new expectations and create a new treaty. When a treaty is disturbed, either party runs the risk of getting less favorable terms in the new treaty, since we expect participants to seize the opportunity for change by attempting to enhance their role sets, while avoiding detractions.

The notion of a treaty violation is a key to the subroutine for evaluating x_7, the acceptability of a particular solution. If the LCA has a solution, (A, R_T^*), we believe he asks whether his attempt to implement the solution would be perceived as a treaty violation. If it is not expected to be a violation, then he assumes $X_7 = 1$, and proceeds to implement. If he expects to be viewed in violation of a treaty, he asks whether the treaty change might be viewed as desirable. For example, in some instances, subordinates are frustrated by a superior's inability (or unwillingness) to make a decision. Thus, a treaty change under which the superior makes certain decisions may be very desirable in the eyes of subordinates. However, if the LCA perceives that subordinates will not view the change favorably, we believe that he attempts to infer subordinates' perceptions of their own blocking power. If they think they can make a block stick, they are likely to try one if the LCA goes ahead with his treaty violation. Presumably, the higher the stakes and the more preclusions seen to occur, the more likely a treaty violation will be met with an attempted block. The result of these exchanges could very well be one or more new ATP's. Based upon this type of analysis, the LCA decides whether he will assume the solution to be acceptable or unacceptable, and proceeds to check other aspects of X_7, including whether he can negotiate an acceptable solution.

A NEW THEORY OF COMMITTEE FORMATION

Let us assume that an authority-task problem occurs in an organization, and that it comes to the attention of the lowest common ancestor for solution. His task is to select the best problem solving process to follow, where "best" includes deriving a solution which is technically feasible and at the same time, one which can be consummated. A "best" process also is chosen with an eye toward the type of responses which could be expected, not only to the solution but to the very process by which it was derived and consummation attempted. A problem solving process that causes a great many more ATP's to arise is not much of a candidate for "best".

The primary model in this theory of committee formation is a mapping function, where problem solving processes are chosen by the LCA after first evaluating eleven binary variables representing conditions on the ATP and on the organization. If a process has previously been attempted, then the function provides a model by which to analyze the response to a prior strategy and to choose a new one.

The mapping function used by our LCA in choosing a problem solving strategy requires evaluation of eleven binary variables in $\langle \mathbf{X} \rangle$. The first variable, X_1, simply asks whether an ATP exists or will arise in the issue; X_2 asks whether the ATP needs to be solved which, according to the subroutine for X_2, depends upon the costs of not solving it. If failure to solve the ATP will cause more ATP's, and/or if the ATP itself will not go away if ignored or delayed, then he decides the problem needs to be solved. Next the LCA must determine whether he knows a technically feasible solution, X_3. This requires change in terms of both the activities and who performs them in the organization (A, R_T^*). Does the LCA have the capacity to solve the ATP is the decision required at X_4. Here capacity is viewed in terms of time, with a notion of opportunity cost crudely represented in the subroutine for this variable. Time as the determinant of capacity is used in X_4 because the other reasonable determinant of capacity – ability to solve the problem – is captured in X_3. Mackenzie [3; 4; 5] has demonstrated the effects of such time constraints on capacity.

A critical variable in the mapping function and thus in choosing problem solving strategies is X_5, which asks whether or not the LCA has the authority to preempt solution activities. It is at this stage that the authority-task gap becomes a telling factor. Since the problem is already assumed to be an ATP, the LCA must identify all the activities in the solution, and consult R_A to decide whether he has the authority to preempt those activities. This admits the possibility that an ATP can be resolved by a solution which consists only of activities which the LCA does have the authority to preempt. If the LCA does not have authority, he may ask whether a committee has previously been formed to find (A, R_T^*) for the ATP at hand. The idea here is that a committee report tends to bestow *tentative* authority upon the LCA to implement the solution the committee recommended. We believe that this tentative authority derives from the fact that when a committee of peers sanctions a solution, it becomes a treaty violation to block the implementation of that solution. We suspect that a split committee decision does not lend sanction to tentative authority to the degree that a unanimous recommendation does. For this reason, when committees split we may see more committees formed or perhaps a consultant hired.

The means by which an ATP is solved should be affected by X_6, whether the problem is expected to recur or not. Recurring problems may require a new standard operating procedure for a solution. Furthermore, one expects that whenever it is feasible, nonrecurring ATP's might be solved in temporary organizations, such as task forces or committees, outside the regular organization in order not to disrupt : R. L. Swinth [7] has described how such groups could operate.

The next variable, X_7, is also critical. Variable X_7 asks whether the solution chosen by the LCA or an *ad hoc* decision maker such as a committee or consultant would be accepted by the group or whether the LCA could negotiate an acceptable solution. The idea here is that if an (A, R_T^*) is unacceptable to the consummators, the LCA has a political problem on his hands, regardless of whether he has the authority to ramrod the solution through. Treaties enter into this previously described subroutine as the important factor in estimating acceptability In the absence of a known acceptable solution, the LCA determines whether he can negotiate an acceptable one. Research suggests that key factors in evaluating this aspect of X_7 include the ability to clearly identify the group or factions with whom to negotiate, the size of the group, ability to clearly define the issue, and the availability of bargainable subissues.

In X_8, the LCA asks whether he can pass the buck to the next LCA. We mentioned previously the conditions under which the LCA is expected to notify the next LCA of an ATP for him. Again we note that buck passing requires the consent of both the LCA and the next LCA. At X_9, the LCA asks simply whether an appropriate standing committee exists to solve the problem. Here we note that failure to recognize a standing committee may be perceived as a treaty violation, rendering a solution by the LCA unacceptable in X_7. For X_{10} there is a subroutine which examines the feasibility of hiring a consultant. Here there are obvious economic costs to be considered. We suspect that many committees are often formed in lieu of hiring consultants because committees are viewed as zero-cost work groups especially if they are not profit centers. If salaries are viewed as a fixed cost, and a major cost to the organization, then committee formation can be viewed as a means of spreading fixed costs. If a consultant is hired, it often appears to be for the purpose of sanctioning tentative authority of the LCA as discussed in X_5. One probably, then, chooses a consultant who will affirm the LCA's favored solution, presuming he has one.

Finally, at X_{11} it is asked whether a severe time constraint or a deadlock exists. This variable appears only in the upper branch of the mapping function, where a technically feasible solution is known, because its evaluation as 1 results in the LCA's imposing a solution himself rather than sending the problem to a committee, standing committee, or consultant. A severe time constraint, as long as it is perceived by subordinates as well, gives the LCA the tentative right to bypass an otherwise appropriate, and expected, solution process. When a deadlock is perceived by the LCA, it may be because other strategies have been tried and failed, and parties are so inflexible in their positions that no further bargaining is expected to be successful. The LCA chooses to impose a solution himself then, often not expecting the solution to work, but hoping to at least get the issue out of deadlock by his action.

We shall not track all of the branches of the mapping function, but shall discuss those which lead to a prediction of committee formation.

Strategy: form a committee We obtain a prediction of committee formation along the following paths through the mapping function:

$$[X_1 = 1, X_2 = 1, X_3 = 1, X_5 = 1, X_6 = 1, X_7 = 0, X_9 = 0, X_{10} = 0, X_{11} = 0] \quad (1)$$

$$[X_1 = 1, X_2 = 1, X_3 = 1, X_5 = 0, X_7 = 0, X_8 = 0, X_9 = 0, X_{10} = 0, X_{11} = 0] \quad (2)$$

$$[X_1 = 1, X_2 = 1, X_3 = 0, X_4 = 1, X_6 = 1, X_7 = 0, X_8 = 0, X_9 = 0, X_{10} = 0] \quad (3)$$

$$[X_1 = 1, X_2 = 1, X_3 = 0, X_4 = 0, X_8 = 0, X_9 = 0, X_{10} = 0]. \quad (4)$$

(Variables with no listed values are considered uncritical with respect to choice of paths through the mapping function in each of the four cases.)

Under the conditions of (1), a problem exists which must be solved, and furthermore, the LCA knows a technically feasible solution. He has the authority to impose it and the problem is a recurring one, but the crucial variable is his estimation that the solution, though technically feasible, is unacceptable to the group and he does not perceive himself as able to negotiate an acceptable one. Finding no standing committee, determining that it is infeasible to hire a consultant, and perceiving no severe time constraint or deadlock, the LCA appoints a committee. Even though he had the authority to preempt, the use of it would apparently have been perceived as a treaty violation at X_7, so the unacceptability of the solution caused a committee to be formed.

In (2), the initial three conditions are the same, including a known technically feasible solution, but the LCA does not have the authority to impose the solution. Determining that he cannot successfully negotiate an acceptable solution, he tries instead to pass the buck, find a standing committee, or hire a consultant. Failing all of these, he determines that no severe time constraint or deadlock exists, and he forms a committee. Trying to impose a solution without the authority to do so may violate a treaty, but more importantly, leaves him vulnerable to a successful block (and perhaps preclusion) by subordinates, or a preclusion in the form of an overruling from his superior. The response of being overruled by a superior is important to the LCA because it limits his right to manage. An overruling by a superior may unwittingly, or perhaps quite intentionally, create an authority-task gap by demonstrating to the LCA and the members of the organization that the lower LCA did not actually have the authority he was thought to have. Presumably new treaties will then be made between the lower LCA and his subordinates regarding his use of the preemption over those activities now in the gap.

In (3), a situation is depicted in which an ATP exists which must be resolved, but unlike the previous two cases, no technically feasible solution is known, although the LCA has the capacity to find one. The problem is viewed as a recurring one. At X_7 the LCA determines that he cannot negotiate a technically feasible and acceptable solution, so he checks the feasibility of passing the buck, looks for an appropriate standing committee, and determines the feasibility of hiring a consultant to solve the problem. Failing these, he forms a committee to recommend a solution.

Finally, the conditions in (4) describe a situation in which an ATP exists which must be solved, but for which the LCA does not know a technically feasible solution nor has the capacity to find one. Determining that he cannot pass the buck to his superior, that there is no appropriate standing committee for the problem, and it is infeasible to hire a consultant, the LCA chooses to appoint a committee.

Using the mapping function to process group responses to LCA's choice Let us consider now the range of responses to the LCA's choice of a strategy for removing an ATP. Some responses will indicate to the LCA that he misevaluated one or more of the variables in the mapping function, and will point him back to that variable for reevaluation of it and of succeeding variables in the function. Some responses, notably issue spreading, raise new ATP's.

The first response to note is consummation of the solution, which removes the ATP. Another response is a block on the part of consummators, which points the LCA back to X_7. He must consider whether he can negotiate an acceptable solution and if not must change $X_7 = 1$ to $X_7 = 0$ and proceed. An unacceptable solution from a consultant, committee or task force can occur, returning the evaluation to X_5. Unsuccessful buck passing points back to a reevaluation of X_8 in the function. If the ATP did not go away as expected, the LCA must return to X_2, changing X_2 to 1 and proceeding. A nonrecurring problem which recurs is a response which creates a new ATP, and sends the LCA back to X_1. When a known solution ($X_3 = 1$) is consummated but turns out to be incapable of removing the problem, a new ATP arises. Issue spreading likewise causes a new ATP. The last response to consider is the possibility that subordinates form a committee. This action is a signal that the solution promoted by the LCA was unacceptable, and points him back to a reevaluation of X_7. If X_7 was originally zero, then the LCA has a new ATP on his hands.

In developing the subordinated mapping functions to describe the search for values of the X's, one must build into them contingencies representing prior strategies which might have been taken. For example, in determining the value of X_5, the authority variable, the LCA would presumably ask whether a committee had previously been formed and sanctioned tentative authority for him. In the subroutine for X_7, acceptability of the solution, there must be at least one path, or set of contingencies, under which the LCA polls his subordinates through a questionnaire. There may be other special contingencies which must be identified in the subroutines or perhaps be added to the main mapping function. Since we do not expect the main mapping function to be universal, it may even be that for some organizations the contingencies in the subroutines belong in the main mapping function.

Using the models just presented, namely the mapping function and the subroutines, we can readily see the effect of an authority-task gap on problem solving processes. Committees, we believe, arise to resolve issues which cannot be resolved through the exercise of authority because of the existence of an authority-task gap. When we open up and explore the concept of authority, we see managers who are vaguely aware of R_A, but have mainly compressed it into an R_C, and who are looking both upward and downward to protect their domain of authority, all the while they are trying to solve problems in R_T*.

CONCLUSIONS AND SUMMARY

If our analysis of the authority-task gap and its relation to committee formation is reasonable, we should be able to test the following kinds of conclusions. When a small number of gaps, or no gaps, exist in an organization we would expect few if any committees to be formed. For example, we might contrast an industrial organization, where the task process is more stable and predictable, with a university, where the task process is extremely diverse and unstable. The authority-task gap is absent for the most part, we suspect, in the smaller industrial enterprise because those who are in authority, controlling the task processes, know those processes. Clearly, that is not the case in a public university, so we expect frequent gaps and a proliferation of committees.

Related to the analysis above, we predict that where superiors in R_C do not have knowledge of the organization's task processes they will not create an appropriate R_A. An inappropriate R_A leads to many authority-task gaps, causes many ATP's, and thus results in frequent committee formations.

When an organization undergoes rapid change in its task process we expect there to be more gaps, more ATP's, and therefore more committees. As a corollary when complexity of the task process is increased we again expect more gaps, more ATP's, and should observe more committees. An example of this latter case is the effect of new HEW Affirmative Action regulations on universities. Because of the changes associated with search activities, tenure and promotion decisions, and curriculum, we should see an increase in the numbers of committees formed to deal with these issues. Casual observation at the University of Kansas tends to affirm our suspicion.

This theory of committee formation is based upon the existence of authority-task gaps which in turn arise out of authority and position relationships that are inconsistent with the actual group processes. The main variables used to predict the formation of new committees are contained in a mapping function. These variables are estimated using subroutines. Given the variables' values and the mapping function, we can predict which of seven options will be taken by the LCA; (1) do nothing, (2) form a committee, (3) give the problem to a standing committee, (4) appoint a task force, (5) impose a solution, (6) pass the buck, and (7) hire a consultant. This decision will be based upon (a) what solution to the ATP is required and (b) the possible reactions by his group to the solution. The group's reactions are presumed to be based upon the stakes involved in the issue, preclusions arising from it, and judgements about possible treaty violations. The response by the LCA to their reactions is to redefine certain variables in the mapping function and to cycle through it again. Thus, an unacceptable solution can create a succession of new ATP's and in some cases additional committees.

We should like to end this paper with another speculation. We think that ATP's, which are based upon inconsistent authority relationships, create a need for more administrators. We have shown that an ATP can result in the formation of a new committee. Although we have not gone into how the LCA selects a committee, it seems reasonable to expect him to place a trusted subordinate on a committee in order to provide some direction and control. This trusted subordinate will usually be someone who is not a principal in the dispute. He will often be drawn from the staff of the LCA. However, the time he spends on the committee reduces the time that he has available to spend on other duties. If he spends enough time on these committees he will need an assistant to help him do his regular work. Thus, we have another administrator. However, the newly appointed administrator will want to have his duties and authority defined. This appointment can result in new ATP's and new committees and eventually in new appointments. Thus we are in the position to describe mechanisms by which the Parkinson Law of the Rising Pyramid can be derived [6]. We are intrigued that Parkinson's amusing analysis did not even consider committees!

ENDNOTES

[1] Technically, $R_T{}^*$ is the timely second pass group role matrix [cf. 4, chapter 5] but is referred to as the group task role matrix here for brevity. Operations defining various types of role matrices and their analytical manipulation are described in the cited work. Use of the word "task" is for clarity.

REFERENCES

[1] Lippitt, Mary E., Development of a Theory of Committee Formation Ph.D. dissertation. University of Kansas, Lawrence, Kansas,1975.

[2] Lippitt, Mary E., and K. D. Mackenzie, "Authority-Task Problems." *Administrative Science Quarterly*, 21: 643–660, 1976.

[3] Mackenzie, K. D., "Measuring a Person's Capacity for Interaction in a Problem Solving Group" *Organizational Behavior and Human Performance*, 12, No. 2: 149–169, 1974.

[4] Mackenzie, K. D., *A Theory of Group Structures, Volume I: Basic Theory*. New York, N.Y.: Gordon & Breach Science Publishers, 1976a.

[5] Mackenzie, K. D., *A Theory of Group Structures, Volume II: Empirical Tests.* New York. N.Y.: Gordon & Breach Science Publishers, 1976b.

[6] Parkinson, C. N., *Parkinson's Law*, New York: N.Y.: Ballantine Books, Inc., 1959.

[7] Swinth, R. L., "Organizational Joint Problem Solving." *Management Science*, 18, No. 2: B68–B79, 1971.

MODELING THE TASK GROUP AS A PARTIALLY SELF-PROGRAMMING COMMUNICATION NET: A CYBERNETIC APPROACH TO THE STUDY OF SOCIAL PROCESSES AT THE SMALL GROUP LEVEL[1]

JAMES R. TAYLOR
Université de Montréal, Montreal, Canada

GENERAL OBJECTIVES OF THE PAPER

Over the past several years there has emerged a field of studies, broadly inspired by cybernetic models of information processing, which provides the general lines of a powerful theory for the study of organismic functioning at the level of the individual in a variety of contexts. This field has come to be known as cognitive psychology. It is the thesis of the present paper that the principles which have been developed in this individual-oriented domain are equally applicable to the study of group problem-solving behavior. A first purpose of the paper is to determine what modifications to the theory must be envisioned when we turn to examine group processes. This attempt to extend the range of an existing theory is motivated by the belief that to the extent that a system of explanation can be shown to hold for phenomena at quite different levels of complexity, its power and its plausibility are enhanced, and new directions for research are suggested.

The second objective of the paper is methodological. Recording and data analysis, where the object is the behavior of groups, are based on complex sequences of behavior, generated by the groups in response to the experimental task variations. In order to test hypotheses which derive from the overall theoretical position which has been adopted, it is necessary to develop appropriate data-simplification procedures. In the second part of the paper, the lines of a method for data recording are briefly outlined, and preliminary steps of data-analysis are described.

A. THE THEORETICAL BACKGROUND

The Analysis of Individual Choice Reaction Data

Shannon's landmark monograph on the theory of communication inspired a number of studies on individual information processing limitations during the early 1950's. Shannon's theory was at first interpreted as an exemplification of the S–R paradigm current in behavioristic schools of psychology: the organism was regarded as a throughput device, a channel, or transducer; the stimulus display was interpreted as Shannon's *source*, and subjects' responses as his *destination.* The entropy of stimulus and response was measured, information gain or transmission was computed, under a variety of experimental conditions, and from these empirically-determined values, an attempt was made to measure the capacity of the individual as a *channel* for the communication of information.

While this literal-minded application of Shannon's theory led to initially-promising findings, it rapidly ran into difficulty. Serious anomalies in the data began to crop up with increasing frequency, leading to the introduction of *ad hoc* modifications of the theory. In turn, the application of the Shannon paradigm was shown to be based on a misreading of the original theory [14: 6–7].

A different approach was proposed in two articles by Craik [4; 5], and it is the second approach which underlies much of the current investigation of cognitive phenomena and is the starting point for our present discussion. Craik, borrowing from the language of contemporary control theory, advanced a view of the organism as an *intermittent correction servo-mechanism.* Craik had noted, in his research using tracking tasks, that as the difficulty of the task increased (i.e. as the variety of the stimulus ensemble or rate of presentation increased), subjects could no longer track variations in the path of the stimulus object in a continuous manner: they resorted to behavior consisting of discrete adjustments. Craik reasoned from this observation that the nervous system does not function like a telephone switchboard which processes input until, as input rates are increased, it jams; rather the organism must operate more like a computer, which accepts inputs, effects transformations (logical and computational, using stored and new material), produces outputs, and finally monitors feedback.

Craik's position (and my own) may be considered to exemplify three main interlocking assumptions:

the assumption of *decomposable processes*,
the assumption of *error correction*,
the assumption of *central integration (or regulation) of behavior.*

Decomposable Processes

The reaction time of a subject in a choice reaction experiment reflects the operation of distinct underlying subprocesses which concatenate to make up a complete reaction pattern. These subprocesses include at least stimulus recognition, response selection and response execution, and it is possible that these stages may in turn be decomposed into more elementary units.

The brain is made up of a collection of cells linked to form a communication net the structure of which can vary over time. The structural properties of the network can be thought of as a set of relationships defined on the set of cells, where measures of relationship derive from the set of communication activities (messages) of the system. Thus from observation of process, structure can be inferred (by "structure" we refer to functional rather than physical properties of the system).

An important corollary of the decomposability assumption is that structures and the role of components of the network may change at different phases of the reaction sequence.

Error Correction

The organism is assumed to be coupled to an environment: the output of one serves as input to the second. The notion of error-correction, or purposive regulation, or "reflex action" has been described as follows [18]:

> The general pattern of reflex action, therefore, is to test the input energies against some criteria established in the organism, to respond if the result of the test is to show an incongruity, and to continue to respond until the incongruity vanishes, at which time the reflex is terminated . . . . Stimulus and response must be seen as phases of the organized, coordinated act . . . Because stimulus and response are correlative and contemporaneous, the stimulus processes must be thought of not as preceding the response but rather guiding it to a successful elimination of the incongruity. That is to say, stimulus and response must be considered as aspects of a feedback loop.

The theory of decomposable processes specifies that an organism has available to it a repertoire of subprocesses – of different ways it can behave – which can be combined into different reaction patterns. The larger the repertoire, the richer the combinatorial possibilities, the greater the potential adaptability of the organism.

The theory of error correction assumes that, for any given state of the environment, and for any choice of behavior by the organism, there is some set of outcomes which is assigned a positive or negative evaluation by the organism. If the outcomes are "unfavorable", that is to say if the organism perceives itself to have made a mistake, it will change its behavior; if the outcome is favorable, it will not.

The ability of the organism to maintain the set of outcomes within acceptable limits depends both on the variability and complexity of the environment and on the variety of responses which is available to the organism and its understanding of constraints in the environment. As the entropy of the environment increases, the information available to the organism must also increase. (It is for this reason that Ashby has termed the organism a *correction channel*, in the sense defined by Shannon) [1: 211]. *Information load* on the organism is defined as the ratio of the variety of states taken by the environment to the variety of behaviors available to the organism. Extremely high and low values produce conditions which are termed, respectively, *overload* and *underload*. In general, it is assumed that load increases produce behavioral change.

Central Integration of Behavior

The assumption of centralized regulation of brain operations has been stated as follows:

> The RAS [Reticular Activating System] acts as a kind of traffic control system, facilitating or inhibiting the flow of signals in the nervous system . . . The astonishing generality of the RAS gives us a new outlook on the nervous system. Neurologists have tended to think of the nervous system as a collection of more or less separate circuits, each doing a particular job. It now appears that the system is much more closely integrated than had been thought. This should hardly surprise us. A simple organism such as the amoeba reacts with totality toward stimuli; the whole cell is occupied in the act of finding, engulfing and digesting food. Man, even with his 10 billion nerve cells, is not radically different. He must focus his sensory and motor systems on the problem at hand, and for this he obviously must be equipped with some integrating machine . . . The RAS seems to be such a machine. It awakens the brain to consciousness and keeps it alert; it directs the traffic of messages in the nervous system; it monitors the myriads of stimuli that beat upon our senses, accepting what we need to perceive and rejecting what is irrelevant; it tempers and refines our muscular activity and bodily movements. We can go further and say that it contributes in an important way to the highest mental processes – the focusing of attention, introspection and doubtless all forms of reasoning [8: 8].

In similar vein, Moray [19], positing a limited capacity central processor "whose organization can be flexibly altered by internal self-programming", argues that the total capacity of the brain can be allocated in different ways, according to the task, or the phase of the task involved.

The central regulation of behavior is itself an activity which requires capacity.

The analysis of group problem-solving reaction data

It is the thesis of this paper that the assumption of decomposable processes, error correction and central integration of behavior provide a useful tool for the investigation of group problem-solving behavior.

Let us first state our assumptions more formally.

A *multi-stage process* can be thought of as a vector of states of a system, having the following form:

$$\mathbf{P} = [\mathbf{P}_0, \mathbf{P}_1, \mathbf{P}_2, \ldots \mathbf{P}_t, \ldots \mathbf{P}_T] \tag{1}$$

where $\mathbf{p}$ is interpreted as a vector $\mathbf{x}_{ij}(t)$ which qualifies a set of entities of the system with respect to their attributes or properties at time "t". $\mathbf{P}_0$ is taken to be the initial state of the system, and each succeeding value of $\mathbf{p}(\mathbf{p}_1, \mathbf{p}_2,$ etc.) can be viewed as the state of the system one time unit later. A *transformation* is a function $\mathscr{T}(\mathbf{p})$, $t = 1, 2, \ldots, t, \ldots, T$, having the following property: a transformed point, $\mathbf{P}_t = \mathscr{T}(\mathbf{P}_{t-1})$, is a member of the set $\mathbf{P}$ for all $\mathbf{p}$ in $\mathbf{P}$. A multi-stage process can be represented canonically as follows:

$$\mathbf{P} = [\mathbf{p}, \mathscr{T}(\mathbf{p})]. \tag{2}$$

A nonstationary, or decomposable, process can be defined as follows: $\mathbf{P}_t = \mathscr{T}_t(\mathbf{p}_{t-1})$, $t = 1, 2, \ldots, t, \ldots, T$, is a member of the set $\mathbf{P}$ for all $\mathbf{p}$ in $\mathbf{P}$.

A *multi-stage decision*, or regulated, process has the following form:

$$\mathbf{P} = [\mathbf{p}_0, \mathbf{p}_1, \mathbf{p}_2, \ldots, \mathbf{p}_t, \ldots, \mathbf{p}_T; \mathbf{q}_0, \mathbf{q}_1, \mathbf{q}_2, \ldots, \mathbf{q}_t, \ldots, \mathbf{q}_T]. \tag{3}$$

where $\mathbf{p}_t = \mathscr{T}(\mathbf{p}_{t-1}, \mathbf{q}_t)$, for all $\mathbf{p}$ in $\mathbf{P}$. The "q" vector describes choices which may be thought of as incorporating information concerning previous errors.

Measures of structure, including measures of centrality, may be obtained for both the "p" and "q" vectors, and such measures may be multi-dimensional, where the process is time-dependent (thus reflecting changes of structure over different phases of the task, or learning processes, where process is measured over successive runs of a task having similar characteristics).

Let us now see how these assumptions, which derive from the cybernetic approach, can serve to guide research at the small-group level.

Decomposable Processes

Actual research in the small-group field exemplifying the approach which has been outlined proves to be disappointingly meager. Flament [6; 7] has provided a useful analysis of a multi-stage process in group problem-solving activities, but fails to consider nonstationary or decomposable processes. The serious neglect of the study of processes, as the term is defined here, is nowhere better illustrated than in the review article of Kelley and Thibaut [12] where the authors distinguish between phases of *information assembly and response distribution* but do not indicate in the literature they reviewed any attempts to define or measure such phases. Rosenberg, in his review of models of small-group process, concludes: "A model of the process by which a group hierarchy develops and stabilizes over time, that is, a dynamic model of changes in group participation, would be an important extension of the static models discussed above. This problem has as yet received scant attention" [20: 235].

Recent work by Mackenzie [16] goes some distance towards rectifying this situation. Mackenzie shows that task processes can be broken down logically, depending on the nature of the task, into phases (which he terms milestones). He finds in his research that process can vary not only from task to task but also from group to group, and, with respect to a single group, from one problem to the next. The Mackenzie description of task process is dynamic in that it is derived from observation of messages which reflect changes of state of the group's information processing variables, and sequences of transitions between states.

In the same work Mackenzie shows that several measures of group structure can be computed, based on observation of group process, and that such measures can be multi-dimensional, and can take into account changes of structure over time.

This recent research indicates that the assumption of time-dependent decomposable processes is valid for problem-solving groups. In a later section of this paper, a possible discovery procedure for the determination of phase boundaries will be briefly discussed.

Error correction

The assumption of error correction or purposive regulation would suggest that groups, presented with a problem, would vary their behavior until they had "learned" an appropriate response, and would then stabilize around that pattern; that presentation of a new and more difficult problem would produce deviations from the previously-learned patterns, and hence changes in group structure; that group perception of error (as for example when one group member fails to respect learned time constraints) will lead to changes of group process and structure; and that changes in process and structure will follow increases in information load.

Mackenzie's research indicates clear support for several of these suppositions. Group processes and structures do vary; where problems are uniform in difficulty and similar in type, there is evidence of learning; particularly for simple tasks, groups tend to choose structures which are correlated with efficiency; group restructuring follows a change in the type of task where tasks vary markedly in difficulty; failure of a key individual to complete his assignment within an acceptable time-frame produces a capacity for change of group structure.

These results strongly suggest that further investigation of the role of dynamics of error correction processes will prove useful.

Central integration of behavior

It is obvious that a problem-solving group, particularly one which consists of undergraduate subjects working together in a laboratory situation where they have met for the first time, is not an organism in the same sense as an individual human being [13]. We should not therefore expect to find a mechanism similar to the RAS. Nevertheless, without some form of centralized regulation of its activities, groups would be highly inefficient, and considerably more unstable in their behavior and subject to uncontrolled oscillations, than observation suggests they in fact are.

The problem is complex. Ashby [1: 240–243] has noted the isomorphism between the case of multiple controllers and N-person game theory. Unfortunately, attempts to model problem-solving situations using a principle of "local rationality" [2; 3; 17] have failed to predict accurately observed group behavior. In fact, the validity and generality of the assumption so common in game-theoretical approaches that each individual first rationally assesses his own interest and then enters a most profitable coalition as the result of negotiation may well be questioned. (See [15] for further discussion.) Such a position tends to assume that social interaction is an incidental by-product of individual optimization needs, an assumption that is difficult to square even with ordinary everyday experience.

The opposite point of view can be maintained. Grice [9], for example, posits the existence of a "cooperative principle" which states that, whatever the *ultimate* purpose of a talk exchange, it has *a prior* purpose of achieving a "maximally effective exchange of information" [11: 447]. This assumes that participants in a conversation follow mutually known and accepted (and hence learned) rules of social interaction, which are acquired as part of each person's period of socialization, much as he learns rules of language. The rules may vary from one type of social interaction situation to another. Thus we hypothesize that groups behave as integrated collectivities unless and until the situation makes such collaboration distinctly undesirable for one or more members of the group, in which case he has always the option of exercising his veto power [1; 16]. We assume in fact the existence of a general norm of collaboration, and of acceptance of an integrated control system as long as this produces outcomes which are seen to be efficient. This assumes further that centralization of regulatory functions in one or more individuals is generally considered legitimate (although what constitutes "legitimacy" remains somewhat mysterious).

This approach, in its main lines, has been adopted by Mackenzie in his theory of behavioral constitutions [15]. He has shown that structures are adopted as the result of "voting", and that once a procedure has been adopted, the group expects it to be consummated. In other words, constitutions have a legitimizing function, with respect to the emergence of defined social roles, and of centralized control.

One difficulty with the Mackenzie research stems from the choice of experimental conditions in that restriction to written communications as a means of interaction makes organizational activities rather expensive in time and effort. Thus in Mackenzie's work regulatory activities, or voting, occur relatively infrequently. Possibly for this reason the centrality of information-processing and regulatory structures was not measured separately. In our work, we have introduced more rapid and flexible electronically-mediated communication media, in order to produce a richer source of data and to permit a more detailed analysis of regulatory processes.

In Summary

I have tried to show in the preceding discussion that a variety of phenomena drawn from the psychological literature on choice reaction experiments and the small-group literature on problem-solving experiments can be explained within a single theoretical framework, that is, the cybernetic theory of self-programming communication networks. Unfortunately, evidence at the small-group level is sparse, but sufficient, I think, to indicate the potential value of the approach. An even more challenging task, but one I am convinced is not beyond our means, is the application of the theory to the study of more complex social systems.

B. METHODOLOGICAL PROBLEMS

Objective of the Recording Procedure

The goal of this section is to show that from protocols of group behavior, data can be recorded which provide useful information concerning the group's information-processing, error-correction and self-programming activities as it organizes itself into a communication net. This information will in turn allow us to make statements concerning the nature of the group's processes and structures.

The methodology which is proposed derives from an approach suggested by Harris [10], but differs in that it attempts to incorporate recent advances in the field of transformational linguistics. The overall intention of the proposed method is to re-code surface linguistic and nonlinguistic behavior of members of the group into a standardized form, where underlying patterns can be isolated. Modern transformational grammar provides an ideal tool for this end, in that transformationalists both distinguish between surface and deep structures, and also have explicated noninformation-losing transformation rules which can be used to infer the underlying structure.[2]

Since the length of this paper precludes anything but a quite superficial treatment of the methodology, I will show briefly how one excerpt taken from one of the groups protocols recorded in our laboratory can be analyzed. The group, incidentally, was asked to complete a common-symbol task typical of the Bavelas communication-net experiments: each subject received two symbols at the beginning, and the task of the group was to produce a comprehensive, non-redundant list. The excerpt given here covers only the data-sharing or symbol-distribution phase (the experiment used French-speaking subjects: a translation is shown at the right-hand side).

Table 1 shows the same text in recoded form. Column 1 of that table indicates the number of the intervention. Column 2 gives the name of the originator of the message, coded as follows:

G_1 = Mireille
G_2 = Alain
G_3 = Gaétan
G_4 = Réal
G_5 = Yoland
G_g = Everyone

Column 3 indicates the intended receiver of the message, where this is clearly manifested, either by a direct interpellation ("Oui, Mireille", "Gaétan", etc.), by the use of the second person singular form of the verb where the context shows without ambiguity to whom the message is addressed, or by other evident contextual clues. Column 3 gives my interpretation of the function of the message, within the framework of analysis developed in this paper. I define four functions: *control, information, error-signaling,* and *confirmation*. Control functions can be further analyzed into three sub-categories (not shown here); *programming* (or making procedural suggestions, illustrated by interventions 3, 8, and 18, for example); *routing* (or determination of an order of presentation), and *scheduling* (or specifying of the time at which an intervention is to occur). Since this is an initial run for the group, the routing procedure has not yet been fixed, and routing and scheduling are confounded. Interventions 1, 6, 10, 11, and 19 fall into this category. Interventions which report data are classified as informative. Error messages may report noncomprehension (intervention 2), comprehension (intervention 4), failure to identify a speaker (intervention 12), varying degrees of certainty (interventions, 13, 16 and 17). They may also report mistakes (not illustrated in this passage). Confirmatory messages simply indicate acceptance of a suggested procedure (intervention 19), or of an information (not illustrated in this passage).

Column 5 gives the type of communication; this is equivalent to what would be termed by transformational grammarians a "pre-sentential modifier" in the deep structure of the sentence. In general, the type of the sentence is correlated with its function: thus, for example, a scheduling message is usually couched as a simple question, an error message often takes the form of a negative, an informational message is always assertive. Additional clues to the function may be obtained from the use of a modal auxiliary and the tense (columns 6 and 7). Error messages are frequently phrased in the past tense. A control message, where the function is procedural, may be stated in the imperative (intervention 8), but is more often stated as a request (interventions 3, 17 and 18), in which case the

ORIGINAL TEXT SAMPLE

	Text Excerpt	
	Original [French]	Translation [English]
1. Yoland	Oui Mireille, Quel numéro as-tu?	Yes Mireille, What number do you ha
2. Gaétan	J'ai pas compris	I didn't understand.
3. Alain	Mireille, Voudrais-tu parler un peu?	Mireille, Would you like to talk a bit?
4. Mireille	Moi, je reçois bien; j'ai deux numéros. [Les montrer à l'écran]	Me, I'm receiving well; I have two nu [Shows them on the TV screen]
5. Alain	Les tiens, c'étaient 64, 63?	Yours were 64, 63?
6. Mireille	64, 63. Les tiens?	64, 63. Yours?
7. Alain	Les miens, c'est 14, 21.	Mine are 14, 21.
8. Mireille	Montre donc tes numéros.	Show your numbers.
9. Alain	[Les montrer à l'écran]	[Shows them on the screen]
10. Gaétan	21, 11.	21, 11.
11. Réal	32, 08.	32, 08.
12. Gaétan	Qui vient de parler?	Who spoke just then?
13. Mireille	C'est Yoland qui vient de parler? Ah! c'est Réal, je pense. Réal, Quels sont tes numéros?	It was Yoland who just spoke? Oh! It was Réal, I think. Réal, What are your numbers?
14. Réal	32, 08.	32, 08.
15. Mireille	32 et 08?	32 and 08?
16. Alain	Mireille, Je t'entends mieux que Réal.	Mireille, I hear you better than Réal.
17. Mireille	Mais moi, je ne suis pas certaine d'avoir compris Réal. Gaétan, est-ce que tu veux vérifier les numéros de Réal?	But me, I'm not certain that I unders Réal. Gaétan, do you want to verify Réal's numbers?
18. Alain	Eh! Regardez! Si on mettait nos numéros juste en face comme ça, quelqu'un là . . . . [Les montrer à l'écran]	Hey! Look! If we put our numbers ju in front like that, someone . . . . [Shows them on the TV screen]
19. Mireille	En haut, O.K. Yoland, tes numéros?	Up, O.K. Yoland, your numbers?
20. Yoland	21, 11.	21. 11.
21. Réal	Par qui on commence, là?	Who do we start with, then?

TABLE 1

CODED TEXT SAMPLE

No.	S.	R.	Function	Type	Aux Modal	Tense	Channel-related	Date-related
1.	G_5	G_1	Control	Q				G_1 * N, N = wh-
2.	G_3		Error	Neg		Past	G_3 understand S, S undefined	
3.	G_2	G_1	Control	Req	Wish	Cond	G_1 speak S, S undefined	
4.	G_1		Error	Ass			G_1 understand S, S undefined	
	G_1		Inform	Ass				G_1 * 2(N), N = 64, 63 (Visual)
5.	G_2	G_1	Control	Q		Past		G_1 * Nn, Nn = 64, 63
6.	G_1	G_2	Inform	Ass				G_1 * Nn, Nn = 64, 63
	G_1	G_2	Control	Q				G_2 * Nn, Nn = wh-
7.	G_2	G_1	Inform	Ass				G_2 * Nn, Nn = 14, 21
8.	G_1	G_2	Control	Imp			G_2 display Nn	
9.	G_2		Inform					G_2 * Nn, Nn = 14, 21 (Visual)

10.	G_3		Inform	Ass				G_3 * Nn, Nn = 21, 11
11.	G_4		Inform	Ass				G_4 * Nn, Nn = 32, 08
12.	G_3		Error	Q		Past	G = wh-, G speak S	
13.	G_1		Error	Q		Past	G_5 speak S	
	G_1		Error	Ass	Think	Past	G_4 speak S	
	G_1	G_4	Control	Q				G_4 * Nn, Nn = wh-
14.	G_4		Inform	Ass				G_4 * Nn, Nn = 32, 08
15.	G_1	G_4	Control	Q				G_4 * Nn, Nn = 32, 08
16.	G_2	G_1	Error	Ass			G_2 understand S, G_1 speak S >	
							G_2 understand S, G_4 speak S <	
17.	G_1	G_2	Error	Neg.	Be certain	Past	G_1 understand S, G_4 speak S	
	G_1	G_3	Control	Reg.	Wish		G_3 asks G_4 S, S =	G_4 * Nn, Nn = wh-
18.	G_2	G_g	Control	Req.		Cond.	G_g display Nn	
19.	G_1	G_2	Confirm	Ass			G_g display Nn (OK)	
	G_1	G_5	Control	Q				G_5 * Nn, Nn = wh-
20.	G_5		Inform	Ass				G_5 * Nn, Nn = 21, 11

auxiliary "wish" is used, or, as in intervention 18, the conditional. In either case, the effect is to "soften" the force of the command.

Columns 8 and 9 indicate the content of the interventions. The original interventions have been re-shaped to show underlying patterns. The following symbols are used:

* = "have"
N = "number"
Nn = "numbers"
2N = "two numbers"
wh- = "which, what, who, etc."
S = "sentence"
Ss = "sentences"
= = "are"
> = "better than, more than, greater than, etc."
< = "worse than, less than, etc."

Other transformations are straightforward: "understand" includes "understand", "receive" and "hear"; "talk" is coded as "speak Ss", "display" includes "show", etc.

A useful by-product of the proposed method is that it permits us to identify milestones. Thus, under the heading "data-related communications" all statements are of the form "G_i * N_j". In this particular experiment, i = 5, and j = 2. When "i" and "j" have taken on all values within the indicated range, the data-sharing milestone condition is attained. This occurs at intervention 20. The intervention by Réal at line 21 indicates that the group itself is conscious that it has completed the initial data-sharing phase. This indicates that processes can be decomposed into their phases by the use of a quite simple discovery procedure.

Analytic Procedures

In general, the analytic procedures which aim at isolating emerging and changing group structure adopted are similar to those proposed by Mackenzie [19]. The latter records interactions in a *group milestone role matrix*, where each entry indicates the number of messages sent by one group member G_i to another, G_j, during the time period about activities related to the m-th milestone. However, the method proposed here permits a more differentiated analysis of control activities than that suggested by Mackenzie. To take only one example, we can analyze scheduling activities of the group during this initial milestone. Here we find that Yoland controls Mireille, Mireille controls Alain, etc. The overall control structure for scheduling of data-transmission is shown as follows:

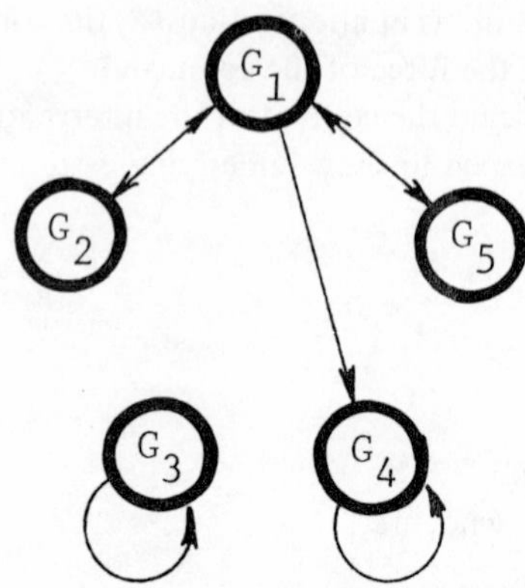

FIGURE 1 Overall control structure for scheduling of data-transmission

Similar analysis may be carried out for other control activities, such as adoption of operating procedures (programming activities), and error-correction activities. In addition, the method proposed allows us to make inferences concerning such facets of the group's activity as the relationship between error-correction and behavior change, and the relationship between degree of hierarchy of a group and its relative efficiency.

ENDNOTES

[1] The work reported in this paper has been undertaken with the assistance of contracts from the Social and Economic Planning Branch of the Department of Communications, Ottawa. A full report of earlier work will be found in Taylor [21]. This report furnishes an extended discussion of choice reaction experiments within the theoretical framework presented in this paper.

[2] It may be well to note that there is far from unanimous agreement concerning the appropriate set of transformations for the English (or any other) language. This does not affect the utility of the transformational approach for the limited objectives of this study.

REFERENCES

[1] Ashby, W. R., *An Introduction to Cybernetics,* London: Chapman and Hall, 1956.

[2] Christie, L. S., "Organization and information handling in task groups", *Journal of the Operations Research Society of America,* **2**: 188–196, 1954.

[3] Christie, L. S., Luce, R. D. and Macy, J., "Information handling in organized Groups", *in* McCloskey, J. F. and Coppinger, J. M. (Eds.), *Operations Research for Management,* Baltimore: Hopkins, 1956.

[4] Craik, K. J. W., "Theory of the human operator in control systems, I, The operator as an engineering systems", *British Journal of Psychology,* **38**: 56–61, 1947.

[5] Craik, K. J. W., "Theory of the human operator in control systems, II, Man as an element in a control systems", *British Journal of Psychology,* **38:** 142–148, 1948.

[6] Flament, C., *Réseaux de Communication et Structures de Groupe,* Paris: Dunod, 1965.

[7] Flament, C., *Théorie des Graphes et Structures Sociales,* Paris: Mouton, 1968.

[8] French, J. D., "The reticular formation", *Scientific American,* **66:** 54–72, 1957.

[9] Grice, H. P., *Logic and Conversation,* to appear.

[10] Harris, Z. S., *Discourse Analysis Reprints,* The Hague: Mouton, 1963.

[11] Katz, J. J., *Semantic Theory,* New York: Harper and Row, 1972.

[12] Kelley, H. H. and Thibaut, J. W., "Group problem solving", *in* Lindzey, G. and Aronson, E. (Eds.), *The Handbook of Social Psychology,* Reading, Mass.: Addison-Wesley, 1969.

[13] Krippendorff, K., "Values, codes and domains of inquiry into communication", *The Journal of Communication,* **19:** 105–133, 1969.

[14] Laming, D. R. J., *Information Theory of Choice-reaction Times,* New York: Academic Press, 1968.

[15] Mackenzie, K. D., Behavioral Constitutions, Working Paper No. 80, School of Business, The University of Kansas, Lawrence, 1974.

[16] Mackenzie, K. D., *A Theory of Group Structures,* Vols. I and II. New York, N.Y.: Gordon and Breach, Science Publishers, 1976.

[17] McWhinney, W. H., "Simulating the communication network experiments", *Behavioral Science,* **9:** 80–84, 1964.

[18] Miller, G. A., Galanter, E. and Pribram, K. H., *Plans and the Structure of Behavior,* New York: Holt, Rinehart and Winston, 1960.

[19] Moray, N., "Where is capacity limited? A survey and a model", *in* Sanders, A. F. (Ed.), *Attention and Performance I:* Proceedings of a Symposium on Attention and Performance: 84–91, Amsterdam: North Holland Publishing Company, 1970.

[20] Rosenberg, S., "Mathematical models of social behavior", *in* Lindzey, G. and Aronson, E. (Eds.), *The Handbook of Social Psychology:* 179–244. Reading, Mass.: Addison-Wesley, 1968.

[21] Taylor, J. R., The Overload of Communication Systems. Mimeograph. Département de Communication, Université de Montréal, 1972.

INFORMATION FLOW IN THE EDUCATIONAL PROCESS

NANCY DWORKIN
Center for Unique Learners, Rockville, MD.

YEHOASH DWORKIN
Center for Unique Learners, Rockville, MD.

and

BERNARD BROWN
Office of Child Development, Washington, DC

INTRODUCTION

The paradigm traditionally associated with the classroom educational process posits separate control and regulatory functions for teacher and child. Essentially, all of the internal control functions are established by the teacher, who selects tasks, states objectives and administers the measurement instruments by which the system's efficiency will be assessed. Information feedback from child to teacher constitutes the regulatory function, with the primary control decisions remaining in the teacher's domain. Within the closed information loop represented

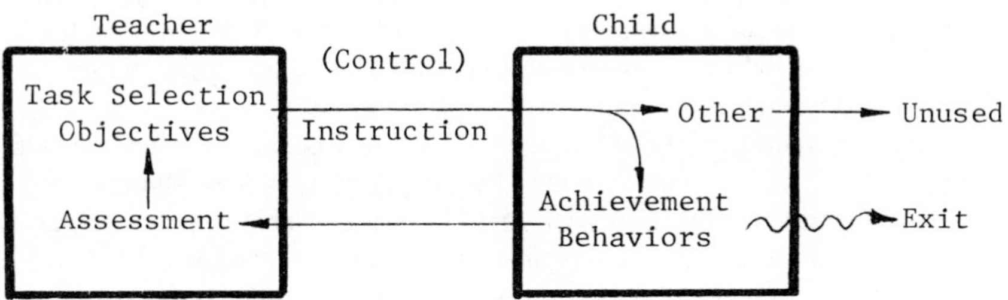

FIGURE 1 Traditional Paradigm for Information Flow between Teacher and Child

by this paradigm (see Figure 1) the measure of the effectiveness of the process is the child's [output] achievement.

The simplicity of the paradigm confronts the analyst with an inherent paradox between predetermined measures of achievement success, on the one hand, and the observation of operating classroom systems in which achievement data represent only a small portion of learning and teaching behaviors. A vast body of theoretical literature also indicates that reinforcement, learning feedback, vigilance, etc., all represent information loops from teacher to child which are not included in the original paradigm. Certainly educational theory has paid a great deal of attention to the function of reinforcement in effecting the speed and profundity of learning. There is widespread agreement that "knowledge of results (KR) is essential for improvement to occur in most activities. This feedback is often supplied by a direct comparison of actual with required performance" [9: 109]. Aside from agreement on reinforcement in the abstract, the issues of frequency and immediacy have also been examined. "At present, there seems to be no contrary evidence to the general conclusion that learning is facilitated by frequent, immediate and positive reinforcement" [7]. Indeed, if the reinforcement "... is delayed, the trainee's motivation may lag, and also, the reinforcement fails to provide information which he may need in order to learn anything" [7]. Reinforcement feedback, while recognized as essential in providing the learning organism with KR has not been entered into the classroom teaching process in any systematic way. Thus, the options for reinforcement are controlled by the teacher with no automatic reference to the child's behavior. While the information loop from teacher to child, relative to task selection and objectives, and from child to teacher, relative to behavior and achievement, are absolute aspects of the control and regulation demanded by the current system, reinforcement feedback operates (if at all) in an extra-system fashion without any of the constraints which would subject it to effectiveness measures. In short, the very information which is most critical in the child's learning process is least subject to control and regulation. Growing out of this lack of control/regulation balance is the more serious amputation of feedback information from child to teacher in any learning context other than achievement standards imposed by the external system (i.e., school norms, district norms, county norms, etc.). "Decisions, goals, and priorities are too frequently merely responses to the limited and narrow pressures of closed information systems that exclude inputs from the very people and environments to which policymakers should respond" [8: 197]. Clearly, school failure rates, now climbing well past the 25 percent mark [3: xi] lead us to two conclusions: that narrowly defined control and regulatory functions have thrown the system hopelessly out of balance, and that achievement measures alone are inadequate for the assessment of the real information exchange which takes place in the classroom.

In addition to the limitations of the information loop as regards reinforcement, there is the complete isolation of behavior from any assessment considerations. Although the child's response to instruction is clearly exhibited in the form of complex behaviors, the control and regulation system only deal with a relatively narrow range of that behavior, namely, achievement. Were we to assume that the only measure of teaching and learning effectiveness was relationship to abstract academic standards (i.e. reading scores, arithmetic scores, etc.), the achievement measure and the simple information loop would be eminently acceptable. It is clear, however, that "A theory of instruction . . . is essentially about what a teacher can do to change the learner's behavior" [1: 77]. Further, "The really

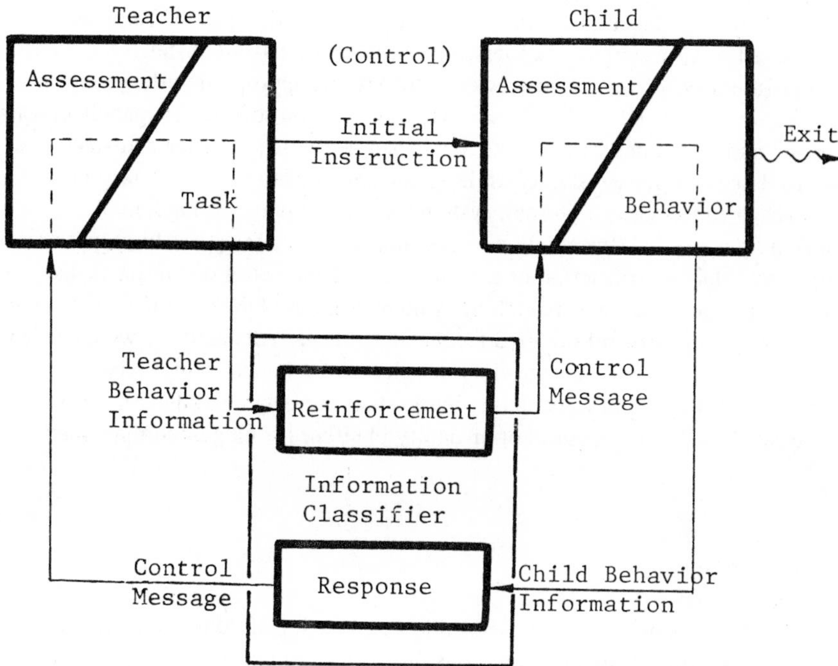

FIGURE 2 Reconstituted Paradigm for Information Flow between Teacher and Child

important thing is not so much the details of behavior, as what its overall purpose is" [1: 78]. Thus, the objective of behavior, and its modification, should be as open to measurement as achievement. Without inclusion in the information loop such assessment is operationally impossible. Our task, therefore, is to effect a change in the learning paradigm through a redefinition of control and regulatory functions, and a reassignment of these functions to teacher and child which will bring the system into balance.

This paper is addressed to an information paradigm which focuses on the interaction between teacher and child relative to the carrying out of specified tasks selected from the larger educational system. Critical are two elements which stand in contrast to the system described above. First, both control and regulatory feedback are bi-directional (i.e. from teacher to child/from child to teacher). Second, both are equally responsible for control and regulation. Along with the changed character of the feedback loop is critical restatement of system effectiveness. Rather than achievement, effectiveness is a function of the efficiency with which the two principals interpret information passing through the feedback loop.

In order to guarantee the possibility of appropriate interpretation of information two sets of logic rules have been developed, one applicable to the teacher's, and the other to the child's, control and regulation of the system. These are designated as information classifiers, and information passing through them changes from regulatory to control feedback on the basis of appropriate interpretation (see Figure 2). The educational technology growing out of the new paradigm was applied and tested in school settings described below [5]. The issue at stake was not to determine the possibility of designing appropriate feedback loops in abstract or for application to new systems. Rather, could existing school machinery accept a reformulation of teaching and planning procedures, without requiring total reconstruction of the classroom? This matter was of particular importance were there where children who had already fallen victim to the tyranny of one-way, achievement-oriented feedback, to the extent that they were labeled as nonlearning. To this end, the children selected for inclusion in the initial "testing" of the system were chosen by teachers and school administrators on the basis of their past histories of difficulty in either learning or communicative realms.

INFORMATION FLOW

The components of the system are strategy and cueing options which utilize learning and reinforcement feedback as the critical elements in managing teacher and child decision-making. Within the paradigm all information feedback is interpreted through the use of logic rules identified as system classifiers. Initial instruction is based on a combination of curricular and administrative demands. Selections are made by the teachers in response to current class standing or the needs of the individual child. Thus, the closed information loop relates to any single instruction or linked instructions relating to specific, unique classroom objectives. Exit occurs when teacher and child have satisfied task demands via appropriate interpretation of feedback through the classifiers. Instruction flows from teacher to child. The child's function is to exhibit a behavior relative to the instruction. The behavior constitutes a bit of feedback and follows an information path to the response classifier. It is then interpreted in terms of: 1) appropriate solution (+); 2) partial solution (±); 3) inappropriate solution (−). From the response classifier, the feedback path returns to the teacher, who now exhibits a behavior (verbal or nonverbal) based on the information received. The information continues on the path to the reinforcement classifier where it is designated as: 1) appropriate solution to be stored for future use (and exit from the system); 2) partial solution to be used as a model for completion of the remainder of the problem; or 3) inappropriate solution followed by system support which will allow risk-taking on the part of the child. The entire process becomes self-regulating until an appropriate solution has been achieved and exit occurs. The classifiers serve to identify the appropriateness of the child's feedback and the desired outcome of the teacher's feedback. Since appropriateness and outcome are tied to each other, both teacher and child are senders and receivers of information, with neither having hegemony over regulation and control. This element makes the paradigm critically different from the traditional process in which information flows in a single direction under the exclusive control of the teacher.

The interpretation of learning feedback in terms of appropriateness represents a vast departure from the use of pure achievement measures. Where the system requires appropriate solution before exit can occur the issue of achievement is recast in temporal terms – namely, how long does it take for child and teacher to arrive at exit, rather than how great is the distance between assumed and observed achievement. More importantly, evaluation of the system's effectiveness does not rely exclusively on achievement since the function of the classifiers is to guarantee a continuous feedback flow, even in the case of inappropriate or partial solutions. Quite evidently, an inaccurate interpretation, on the part of teacher or child, of the feedback flowing out of the classifiers would seriously impair the effectiveness of the system. Thus, the efficiency of the two principals in producing feedback, and in interpreting it, is the internal evaluative measure of the system.

Of equal importance [to measurement] is the interpretation of appropriate reinforcement. Since the information system rests on a balance between control and regulation, with teacher and child equally responsible for both, reinforcement is no longer an optional aspect of teacher choice. Rather, it is intrinsically tied to the logical classification of the child's initial learning feedback. Just as there is a balance between teacher and child assessment, so is there a balance between the logic rules which constitute the response classifier and those constituting the reinforcement classifier.

SIGNAL SYSTEM

Linked with the measure of system efficacy is the clarity of interpretation, on the part of both organisms, of feedback signals. For the adult, a measure of clarity is the fit between expected, or desired results and the child's problem-solving behavior. So long as that behavior is not "bizarre" (relative to the stated problem) the adult is able to interpret appropriateness and issue control instructions. Behavior unrelated to problem-solving can be viewed as system noise for which the teacher can develop buffers or filters related to the reduction or extinction [of those behaviors]. Further, this sensitivity to system noise underlines the operational difference between inappropriate solutions and inappropriate behavior. In the former case, the system's feedback channels redirect the efforts of the child towards models of heuristic precision, allowing for further risk-taking, leading to eventual solution and exit. In the latter case, however, buffers, tied to control classifiers, eliminate the static from the system in order to allow for more space in the feedback channels through which information is to be carried and processed.

For the child, clarity of interpretation represents a more complex problem. The control messages from adult to child are frequently tied not only to the problem-solving behaviors, but to prior actions unrelated to the problems under consideration. Further, the adults wider concerns for extinguishing undesirable behaviors and encouraging risk-taking may, in themselves, constitute a form of static where the child's ability to assess system signals is concerned. Affirmative statements on the part of the adult may indicate appropriate solution and exit, or support of the process by which the child is working toward solution, or a form of distinguishing between current acceptable behavior and prior, illegitimate actions. While the learning organism should be devoting maximum energy to the problem at hand, it is frequently difficult to determine how much effort must be devoted [by the child] to partialing-out the exact implications of feedback statements. While negative signals are usually clear enough, redirecting the problem-solver's efforts away from interfering behaviors and towards exit, positive signals can remain undifferentiated since they have been used in a variety of prior contexts unrelated to the problem.

The system, therefore, must posit clear control and regulatory signals which match specific expectations. "In the case of perception the meaningful is related to the continuous fulfillment of expectation" [4: 100]. For the purpose of matching perception of output signal and input expectation it is necessary to establish three conditions. First, all signals must relate directly to the problem and behaviors under consideration; second, all signals must be mutually established and understood; third, all signals must distinguish between intermediate and exit solutions.

In order to accomplish these ends the feedback signal system must be established prior to the presentation of the problem statement. Teaching and learning organisms can arrive at agreement concerning signal words or symbols prior to the first problem-solving behavior, to be used only in relation to those behaviors *and for no other purpose.* Finally, and perhaps most critical, there can be no similarity (real or perceived) between the signals for successful completion and inappropriate solution. In the latter case the signal must be, further, couched in neutral terms, conveying information pertinent to continued solution behavior. In short, the adult cannot assume that feedback signals are clear to the child without direct appeal to him/her. Without the initiating act of signal establishment all subsequent system processes are open to question.

THE RESPONSE CLASSIFIER: LOGIC RULES

The behavior feedback of the child is of a regulatory nature until it is filtered through the logic rules of the response classifier. They are:

1) an appropriate solution satisfies the initial objectives established by the teacher, allowing the child to exit from the system (+);
2) a partial solution contains at least one element satisfying the objectives established by the teacher, and requires the child to continue in the system (±);
3) an inappropriate solution does not satisfy any of the initial objectives established by the teacher, and requires the child to continue in the system (−).

Once behavior feedback has been filtered through the logic rules, a control message is issued, guaranteeing the responsive behavior of the teacher as a function of the interpretation imposed by the classifier.

THE REINFORCEMENT CLASSIFIER: LOGIC RULES

The reinforcement feedback of the teacher is of a regulatory nature until it is filtered through the logic rules of the reinforcement classifier. They are:

1) when an appropriate solution is achieved, reinforcement must indicate process, and the reasons for appropriateness, setting up conditions for long-range storage, in order that similar problems in the future will elicit comparably appropriate solutions;

2) when a partial solution is achieved, reinforcement must indicate the part [or parts] of the problem which have received appropriate solution, thus indicating those elements of the problem and solution which may be used as models for future problem-solving behavior [short-range];

3) when an inappropriate solution is present, reinforcement must identify the information paths in the system through which the solution process may continue until exist has been achieved.

Once reinforcement feedback has been filtered through the logic rules a control message is issued guaranteeing the problem-solving behavior of the child as a function of the interpretation imposed by the classifier.

RESPONSE/REINFORCEMENT SYNCHRONY

The logic rules make explicit the harmony between learning and reinforcement feedback leading to optimum system regulation:

1) an appropriate solution (+) [child] leads to a reinforcement statement [teacher] → identifying the heuristic and leading to long-range storage [child] → and exit [teacher and child];
2) a partial solution (±) [child] leads to a reinforcement statement [teacher] → identifying the correct elements for modeling future, short-range problem-solving behavior [child] → leading to exit; OR/partial solution/OR;
3) an inappropriate solution (−) [child] leads to a reinforcement statement [teacher] → indicating payoff for further risk taking [child] → leading to exit;

OR/partial solution/OR inappropriate solution.

It is clear that no educational process can afford to minimize the importance of achievement, nor does the system do so, since exit can occur only upon the completion of an appropriate solution. What is unique, however, is that all other forms of feedback (i.e. behavior) on the part of both principals are used as the regulatory mechanisms feeding into control. What is further unique is the use of classifiers which filter feedback and direct information flow between teacher and child, tying response and reinforcement together in a single information loop.

SYSTEM HEURISTICS

The system heuristics encompass two major organizations, the first dealing with problem-solving strategy frameworks and the second with specific cueing sequence. Each of the strategy framework: *Focusing, Programming* and *Patterning*, includes a series of tasks of graduated difficulty, with each task utilizing the problem-solving information developed through its predecessor task.

Cutting across all of the strategies (and tasks) is a cueing system which is operative for each individual task. This system, or sequence, allows for the manipulation of information *vis-à-vis* the initial instruction, in order to guarantee successful solution and exit. It addresses itself to the range of possible solutions and involves:

1) *Identification*
   a reformulation or restatement of the initial instruction, maintaining the complexity of the original problem statement and desired solution;
2) *Isolation*
   a focusing on the problem presentation, with shift in emphasis or presentation order. The objective is to reduce the probability of student error by highlighting specific presentation characteristics. Although the form of presentation may be changed, however, the complexity of the desired solution is maintained;
3) *Redefinition*
   a change in both presentation order and desired solution. The child may be brought into proximity with the problem through the use of kinesthetic, as well as auditory and visual problem-solving devices. Any change may be made in presentation and/or solution so long as a single original element of each is retained. Following this step, however, the problem is returned to its initial state and the complexity of the desired solution.

THE STRATEGIES

The *focusing* strategy concentrates on the manipulation of teaching materials with the objective of reducing competing stimuli, or system noise. Although the distractibility of children is a major concern of educators, the reduction of noise has been most commonly viewed in terms of controlling elements in the external environment which impinge negatively on the learning organism. Although theoretically sound, we have, thus far, been unable to construct learning environments in a vacuum. Indeed, it is questionable whether isolation of the child can have long-range beneficial effects. One might even posit an inverse ratio between the control of competing stimuli [in order to maximize learner attention] and removal of the child, or protection of the child, from the environment in which the learner must ultimately function. In terms of the system model, appropriate solution leading to exit (into the external environment) would, at the same time, imply inappropriate movement where the learner and the external environment had been kept apart from each other.

In order to bring the feedback flow of the system into consonance with noise reduction, a series of techniques are used in which the highlighting and manipulation of material is inherent in the presentation of that material, rather than through the readjustment of the external environment. Selection of optimal presentation techniques is at the discretion of the child who, thus, learns to diminish competing stimuli through the logic of material selection. The strategy has the additional advantage of allowing the adult to identify the "best" way for the child to locate randomly distributed information.

The *programming* strategy concentrates on providing the learner with a series of logic tasks requiring binary decision-making. Each task includes one element of a previously-identified symbol. The child selects the unknown elements of the problem [which are named, but not identified] on the basis of rejecting the "known". Efficiency in traveling through a series of binary decision paths is the measure of the child's success. The task is to select an unknown **X**, which is possible only through the elimination of a known Y.

The *patterning* strategy concentrates on the manipulation of input modalities in the development of problem formats. Mathematical and language arts problems are presented in exclusively visual or auditory modes. Where additional cues are necessary, they are presented in the opposite mode. For the arithmetic problems, order, repetitive placement, and color coding constitute the visual displays while sound families and phonemic combinations are used in the auditory patterns through which language arts tasks are presented.

DIFFERENTIAL PRODUCTION

In addition to information delivery and problem solution each of the strategies illustrates an element of differential production. Focusing presents the simplest production problem. All of the necessary information for problem solving is presented in a controlled series of varying format. The child's task is either to identify specified items or to manipulate specific items without creating new information and without changing the original elements of the problem.

Programming presents an intermediate production problem, providing part of the information necessary to appropriate solution. Although the learner must move from the known to the unknown, thereby developing new information to be used in subsequent tasks, none of the initial elements of the problem are subject to change. Thus, despite the necessity of producing new information, the learner can always return to a stable point of departure.

Patterning presents the most complex production demands. Both the language arts and mathematical displays require total reconstitution of the initial elements in the problem in order to produce completely independent, new information.

Critical to the development of an ordered approach to production is the understanding that the complexity of a problem, while certainly of concern to the teacher, is not nearly as important as is the complexity of the production method It is common for teachers to select problems at levels of complexity quite appropriate to specific children while demanding levels of production which are totally outside the children's range. Since a measure of the system's efficacy is the accuracy with which teacher and child interpret the response reinforcement feedback, it is necessary to guarantee levels of production which will not produce false signals *vis-à-vis* the ability of the child to move towards appropriate solution.

EFFORT REDUCTION

Just as the ordering of the strategies illustrates the principle of differential production, so does the sequencing of the cues illustrate the principle of effort reduction. Each of the three cues demands a different amount of effort by the teacher in order to help the child arrive at appropriate solution. In any learning framework it is desirable to transfer maximum effort from teacher to child. Viewing the expenditure of effort in terms of available energy it is quite evident that the expenditure of a single energy unit from child to teacher, or teacher to child has a value of 1, whereas the expenditure of energy from teacher to class multiplies the initial value unit by the number of transactions necessary to deliver information to every member of the class. Thus, each time any portion of the class experiences difficulty with problem solution, the teacher must be concerned with the selection of cues which, in effect, diminish the available supply of teacher energy. The cueing sequence, as a result, is developed around a geometric progression of energy expenditure. Each succeeding cue doubles the amount of teacher manipulation necessary for exit, likewise doubling the number of energy units (EU) expended, as follows:

Cue I – Restatement of problem – no change in problem display or solution complexity; (EU = 1)
Cue II – Manipulation of problem display – no change in solution complexity; (EU = 2)
Cue III – Manipulation of problem display – manipulation of solution complexity. (EU = 4)

In order to illustrate the critical difference occasioned by the ordering of cues, the following problem is posited.

Teacher A			*Teacher B*		
Children	*Cue Value*	*EU =*	*Children*	*Cue Value*	*EU =*
40	I × 1	40	40	III × 4	160
20	II × 2	40	20	II × 2	40
10	III × 4	40	10	I × 1	10
		120			210

The problem is based on the reduction of uncertainty by 50%, following the initial problem statement and each succeeding cue. What is illustrated in this hypothetical problem is the dramatic impact of cue sequence on the expenditure of teacher energy. In the simplest terms, the A ordering provides the teacher with 90 additional EU's, either for working with children most in need of individual teacher time, or for the presentation of new material.

FIELD USE

The introduction of the system into the Washington, D.C. and Maryland schools served the purpose of testing the information paradigm under actual classroom conditions. Approximately 35 teachers and 100 learning disabled children were involved with both teachers and children equally distributed across lower, and middle-upper socio-economic groups; Black and White; and urban and suburban classifications. There was some question as to whether an information feedback framework would be sufficient to allow teachers and children to experience appropriate completion of tasks, especially when the population involved children with a past history of learning difficulties. The system facilitated interaction between teacher and child to the point where every youngster involved was able to achieve exit in all the strategy areas. Further, negative teacher expectation levels prior to the use of the system were changed to positive expectation at .001 significance [5: chapter 4]. In effect, the expectation results reflected the efficacy of the system's regulatory mechanisms in moving children to appropriate completion of school problems growing out of the standard curriculum, despite disability labels carried [by these children]. With success established as the control mechanism of the system, the ability to handle information flow truly became a measure of the system's effectiveness.

PLANNING IMPLICATIONS

Although not directly germaine to this paper, the system points the way to diagnostic planning as a function of information exchange between adult and child, rather than the assessment of abstract, externally designated symptoms. There are three specific contexts in which the paradigm lends itself to direct planning. First, the initial establishment of signals is an indicator of the communication strengths or deficits of the child; second, the use of specific strategies allows for the identification of problem-solving strengths on the part of the child; third, the selection of strategies and cues by the adult suggests preferred operating procedures on the part of the adult as well as the child.

The most exciting element of the paradigm is the equality granted each of the participants in control and regulatory functions. At any given point the individual [producing response behavior] observes the power of the system in routing regulatory information directly to control. "The process of feedback, by which individuals make adjustments in their interaction rhythms, requires that their rhythms are within reaching distance of each other" [2: 78]. Interpreting the information feedback, each principal engages in "a cybernetic hunting for the most stable rhythm at which the other adjusts" [2: 78].

REFERENCES

[1] Annett, J., "Learning in practice", *in* Peter, B. (Ed.), *Psychology at Work:* 76–96, Warr.: Penguin Books, 1971.

[2] Chapple, E. D., *Culture and Biological Man: Explorations in Behavioral Anthropology,* New York: Holt, Rinehart & Winston, 1970.

[3] DeHirsch, K., Jansky, J. J. and Langford, W. S., *Predicting Reading Failure,* New York: Harper & Row, 1966.

[4] Crosson, F. J., "Information theory and phenomonology", *in* Crosson, F. J. and Sayre, K. M. (Eds.), *Philosophy and Cybernetics:* 99–136, Clarion Books, Simon and Schuster, 1967.

[5] Dworkin, N. E., *Changing Teachers' Negative Expectation Toward Educationally Vulnerable Children Through the Use of an Interactive Process,* Ph.D. dissertation, Hofstra University, 1974.

[6] Dworkin, Y. S. and Dworkin, N. E., "Teacher planning: a function of management", *in* Scandura, J. M., Durnin, J. H. and Wulfeck II, W. H. (Eds.), *1974 Proceedings: 5th Annual Interdisciplinary Conference on Structural Learning:* 135–139, 1974.

[7] Gagne, R. M. and Bolles, R. C., "A review of factors in learning efficiency", *in* Galanter, E. (Ed.), *Automatic Teaching: The State of the Art:* 21–48, New York: Wiley, 1959.

[8] Land, G. T., *Grow or Die: The Principle of Transformation,* New York: Random House, 1973.

[9] Mackworth, J. F., *Vigilance and Attention,* Warr: Penguin Books, 1970.

KNOWLEDGE STRUCTURES IN SOCIETY

INTRODUCTION

Information has always been the most important concept in cybernetics [4] where it is conceptualized as a constraint on entropy, and in C. E. Shannon's work [3], it is a constraint that is transmitted across time and space. Since control always involves constraining or guiding an otherwise free process, the concept of control and that of information transmission is closely related.

Knowledge too is related to information. Living things possess it, utilize it, become linked through it with past and future events, and, by distinguishing between internal and environmental events, establish their identity on it. Thus, knowledge always involves semantical interpretations, which information need not have, and by referring in addition to given problems, values, purposes and goals, knowledge involves pragmatical interpretations as well. Knowledge also constrains a knower's freedom to act, is a form of control, and to the extent it is so related to behavior, it may serve as a powerful construct to account for the multitude of constraints and direction that history, context and purpose impose on the behavior of living things. Knowledge in this sense is always both passive when stored and active when in control, paradigmatically structured and structuring the knower-reality relationship. Knowledge is also distributed in space requiring efforts to locate and use it, involving processes of communication and information processing.

In society, knowledge is distributed over many people, including the files, libraries, and data banks that they operate and the social structures that have evolved in the course of regular interaction with some environment. Information storage and retrieval in society uses many forms of representation, not all of which are well understood and reflected in the design of technical devices [2]. Knowledge structures in society are, then, to a significant degree influenced by if not homomorphs of those social-organizational arrangements through which information is retrieved, processed and relocated. They are inextricably linked to the network of communication through which information can travel and relate different bodies of knowledge. They grow and are selectively maintained in accordance with societal conditions and purposes.

To look at knowledge structures rather than social structures is perhaps nothing but turning a coin to the other side. However, it might just be that this other side is more productive in linking the constraints of history and society with theories of communication and control. The study of knowledge structures in society has neither a disciplinary origin nor a long conceptual tradition to build on. It is therefore not surprising that the contributions to this section come from political science, anthropology, management science, and English, with the last two bending towards computer science.

In the first paper, L. Vaughn Blankenship considers knowledge structures both from an institutional and from a public policy perspective. He identifies six societal information systems consisting of institutions and roles which provide – knowingly or not – information to public policy makers and thereby influence the efficacy of particular courses of action. He then analyses the kind of questions public policy makers typically raise: questions regarding the boundaries and suitable representation of the problem area; questions regarding the seriousness of the problem; questions regarding the efficiency, effectiveness and feasibility of the contemplated actions; questions regarding the group affected by and the long-range consequences of the decisions they might make; and finally questions of retrospective evaluation (how well existing policies actually worked, which groups actually benefitted or lost through their implementation, etc.). With these questions providing criteria, Blankenship suggests that the existing societal information systems be examined as to their capability for providing relevant information. To explain why they provide the information they do and why they don't provide the information public policy decision making would require, Blankenship looks at the internal structure, the norms and functioning of such information systems.

Barbara Frankel considers the kind of knowledge structures that exist within a social organization. Specifically, she attempts to explain how information about ongoing activities, about successes and failures, about planned changes, etc., is exchanged within a relatively isolated social organization. Her fascinating "tale" of how a somewhat utopian, community-oriented and democratically administered rehabilitation hospital for drug addicts and alcoholics struggled with the idea of extending its services to a seven-day/week of operation suggests that knowledge in social organization is not a single, shared, hierarchically structured storehouse of information. Rather, in the day-to-day operation of the hospital and specifically during this "struggle", several bodies of information were found to interact. Frankel suggests that there is, first, a relatively down-to-earth kind of knowledge about how the organization functions, the kind of people that belong, the prestige and status associated with various activities the kind of communication they exchange horizontally and vertically. She calls this the sociological knowledge structure. There is, second, the set of beliefs and propositions about what the organization stands for, the philosophy or ideology which is employed to justify and integrate particular actions, especially in non-routine and emerging situations. She calls this the ideological knowledge structure. Frankel suggests that ideally the two knowledge structures are consistent. However, the innovation that emerged within the hospital was such that it introduced (cognitive) dissonance between the two structures. Each in turn tried to remove this dissonance in its own way, which led to amazing misperceptions of reality and to social conflicts. If Frankel's contentions are correct and generalizable to other situations, this may well lead to a new understanding of social-structural changes based on the way information is organized, stored, generated and exchanged within a social institution.

The existence of knowledge structures in society precedes of course the recognition of their social significance. The first to recognize their own involvement with it were early journalists, whose conflict with the church is well documented. And yet, while it was these professional communicators who raised the flag for the free flow of information and linked the suppression of information to religious or political oppression, very little solid theory emerged that would shed light on the intentional and involuntary guidance function of today's mass media. Richard F. Carter's paper on "a journalistic cybernetic" is a first beginning in this direction.

After discussing various purposes that the mass media may serve, Carter distinguishes between knowledge structures, the hierarchical organization of knowledge about the world as conveyed, for example, by education, and knowing structures, the organization of knowledge resulting from both the surveillance by societal organs of the threats and opportunities in the environment and the attempts to correlate, to coordinate, or to resolve conflicts between the societal components. Journalists are involved in both activities. Naturally, it is important to consider their role within the circular flow of information and, given structural deficiencies, how well they can function. By analogy with the manipulation of a computer by an operator, through whom environmental information may enter a system, Carter distinguishes between the kind of instructions journalists would have to convey and the kind of computational capabilities a society would have to possess in order to provide for adaptation and/or control. Carter concludes his contribution by discussing preliminary work on a pictorial language which is designed to analyze information contents and to complement the communication of ideas by natural language.

John M. Dutton and William H. Starbuck have explored the diffusion of intellectual technology within society over time and present here a case study of the origins, spread, acceptance and application of highway-related simulation studies. The paper's contribution to knowledge structures in society is two-fold. On the one hand, the ideas that govern technological development are in part institutionalized knowledge. The development of such knowledge is supported by organizations, disseminated through various channels, brought in contact with other domains of knowledge and maintained by practical applications. Individuals are involved, no doubt, but as members of specialized institutions. On the other hand, computers are, as the authors point out, an intellectual technology which has an impact on knowledge itself, in the sense that this equipment can direct intellectual pursuits and speed up the testing of knowledge in practical situations. Because the adoption of such a highly specialized technology might proceed differently than the majority of cases previously studied, for example, the adoption of weed killers by farmers or of birth control devices in India, Dutton and Starbuck give separate accounts for the diffusion of the concept of traffic simulation and for the diffusion of actual applications thereof. The distinction proves important in as much as the diffusion of the concept turns out to proceed quite differently from the diffusion of applications. In addition, familiarity with the concept may facilitate or inhibit actual applications. Dutton and Starbuck offer numerous interesting observations of the diffusion process: of the multiple and independent emergence of the ideas, of the role of geographic space and substantive barriers, of the role of funding institutions, the kind of individuals and groups contributing to the knowledge process, etc. Their evidence also supports the proposition that "computer hardware and software were not only a means for but also a stimulant of highway-related simulations". This self-directing process is, I believe, fundamental to both technological development and to knowledge-generating processes in society.

In the final paper of this section, Christopher R. Longyear points out that information always takes the form of representations; and he finds this true in electronic computers, in man and in human society alike. Circular flows of information, which are the essential ingredients of control, assign specific functions to those representations. By examining different forms of adaptation, Longyear comes to the conclusion that adaptation is not only limited by the quantity of information that a system can process but also by the kind of representation it can handle. He suggests at least three levels of information: information about the outside world – these first-level representations are already evident in the way a simple theromstat "views" its world-; information about the nature and limitations of the first-level representations within themselves and within others; and self-generated information (as in dreams, for example) which refers among other things to second-level representations. It is the third-level representation of information that provides structures rich enough to respond to unforseen circumstances and to adapt the environment to individually held goals.

For Longyear, societal knowledge structures seem to involve a device for information storage and retrieval among members of the same informational community. Individuals are members of an informational community to the extent they share the same knowledge. And such informational communities are living in the sense that their knowledge structures grow, adapt, evolve, reproduce copies of themselves and die. While the capacity to store information increases with the size of an informational community, Longyear suggests that the adaptability of knowledge structures to emergent situations and their effectiveness in providing relevant information may be decreasing with the size of an informational community. If true, this would again point to the significance of the organization of knowledge in society.

REFERENCES

[1] Ashby, W. R., *Introduction to Cybernetics,* London: Chapman & Hall, 1956.
[2] Krippendorff, K., "Some principles of information storage and retrieval in society", *General Systems,* **20:** 15–35, 1975.
[3] Shannon, C. E. and Weaver, W., *The Mathematical Theory of Communication,* Urbana: University of Illinois Press, 1949.
[4] Wiener, N., *Cybernetics; or Control and Communication in the Animal and the Machine,* 2nd ed., New York: MIT Press, 1961.

PUBLIC POLICY ISSUES AND SOCIETAL INFORMATION STRUCTURES

L. VAUGHN BLANKENSHIP
National Science Foundation

How can we begin to think about and assess the potential relevance and effectiveness of societal information structures as they relate to public policy issues and choices? This is the broad question which I have set for myself in my talk this morning. It is a practical question in that it is continually being asked about these structures. It is an important question in that it goes to the heart of public policy formulation in the United States.

By "societal information structures", I mean an identifiable and semi-bounded set of social institutions and roles which have, as one of their purposes, the provision of information – symbols in various formats or contexts – to public policy makers in order to persuade them of the efficacy of a particular course of action. It is not necessary that those whom I assign to one of these structures consciously think of themselves in this way. By "public policy makers" I mean individuals whose roles give them some official responsibility for determining the direction and flow of future events and decisions which affect the lives of sizeable numbers of citizens. The class of "public policy makers" is a variable rather than a given depending on the particular policy or problem at issue.

With these definitions in mind, I have identified six societal structures which generate, aggregate, interpret and present different types of information which is directly relevant to the making and implementation of public policy in the United States: (i) the political system which includes the Congress, political parties and, to the extent that they are intentionally organized into interest groups and lobbies, various voluntary associations like corporations, unions, environmentalists, agricultural associations and the like; (ii) the market which, in various ways, summarizes the millions of economic transactions taking place among individuals, groups and nations; (iii) specialized public bureaucracies which deal with reasonably discrete areas of policy interest – energy, science, natural resources, justice, agriculture and the like; (iv) the legal system including both the adversial process by which it proceeds as well as the legal profession, laws and court decisions through which it works; (v) the mass media; and (vi) the specialized knowledge industry including universities, research centers, and scientific and professional bodies.

In the best of all possible worlds, we would like to be able to assess the relevance and effectiveness of a given information structure by answering two questions. First, what is the value of the information it produces for defining or helping to resolve a *range* of policy issues? Among other things, this would require (a) demonstrating how policy resolutions were marginally improved, or denigrated, by virtue of this information; (b) showing that these marginal effects would *not* have occurred in the absence of the information; and (c) attaching some value to these marginal effects. Secondly, how much did it cost to produce the information and are there alternative information structures which could have done equally well, or badly, at less cost?

In the real world, of course, meaningful answers to such questions are unachievable, especially when we are dealing with something as broad as societal information structures. Thus our aim must be more modest, more open-minded, more qualitative and descriptive. It is for this reason that it seems more useful and tractable to talk about the *potential* relevance and effectiveness of such structures for policy making. This *potential* depends on three things: (a) the characteristics of the policies at issue including both the types of questions being asked and the set of relevant policy makers; (b) the characteristics of the information structures themselves including their internal dynamics and distinctive norms as well as the resources which they have at their disposal; and (c) the nature of the links between the two.

A complete answer to the question posed at the beginning of my talk would require a thorough analysis of all three factors. This is too substantial an undertaking for a single presentation. I intend to begin the task, however, by discussing the first point at length and by indicating, in my conclusion, the next steps to be taken.

CHARACTERISTICS OF POLICY ISSUES

Public policy issues are "squishy" problems.[1] Furthermore, decision-structures, i.e. those seeking information which will bear on these "squishy" problems, will vary from issue to issue. These are the two contexts within which information is developed and presented.

"Squishy" Problems

Consider four broad issues which have received a good deal of attention in the past year or so:

- inflation vs. recession in the U.S. and world economies

- the "energy crisis"
- the world food or commodity "crisis"
- environmental degradation and excessive pollution

They have a number of things in common.

They are diffuse in the sense that their boundaries are vague and it is difficult to tell the problem from its environment – everything is connected with everything else and it is almost impossible to know what can be taken as a given and what as a variable. They summarize a multitude of diverse trends, events, feelings and perceptions and we either have no very good models or theories which enable us to identify, extract and act, with confidence, upon their relevant parameters or, as in the case of the inflation vs. recession issue, we seem to have many partial models which lead to seemingly contradictory conclusions.

What kinds of information are required on issues such as these or, to put the matter in more operational terms, what types of questions do policy makers typically ask? The first set deals with *boundary-defining and representational problems.* They want to know whether or not there *is* a "problem" how it can best be represented so as to indicate what its limits are, who the key actors seem to be, and what cause-effect relationships seem to hold.

We all saw pictures of Detroit, Watts and Newark burning; many of us experienced violence on university campuses or saw TV replays of police-demonstrator confrontations at the 1968 Democratic convention; those of us who commuted to work, sat in gas lines, reduced our speed, and turned out thermostats down in the Spring of 1974; and we know we have been paying higher prices for things at the grocery store. Do these discrete events add up to a "problem" requiring changes in public policy or, even more fundamentally, changes in the "democratic-capitalist" structure of society? Are they merely the flow of natural events which, if left alone, will work themselves out? Or are they creations of a mass media seeking sensationalism and higher sales or politicians seeking some issue around which to organize distinctive policy initiatives which will bear their name and indicate that they are providing energetic leadership to "get the country moving again"?

Argumentation over the *existence* of a "real" problem is inextricably linked with discussion about how best to represent it. What is the appropriate model to apply to an understanding of current economic conditions or the "energy crisis"? Is the "energy crisis" the result of actions by OPEC nations or the consequence of years of profligate misuse or overuse of scarce resources by the United States? For that matter, are the resources really scarce or is it that they were underpriced? If the latter, then it is merely a matter of removing restrictions and allowing prices to rise to a point where it becomes profitable to discover new sources of energy or to develope more expensive substitutes. Have the oil companies created the "crisis" by working in collusion with OPEC nations in order to increase their profits at the expense of the consumer and poorer countries? Clearly the way the problem is represented has a good deal to do with indicating who has a responsibility for doing something and what the range of their choices might be.

This points to the second set of questions which policy makers confront: those which deal with *efficiency*, *effectiveness*, and *feasibility* considerations. Given a representation of and set of boundaries to an issue, what set of strategies are available for dealing with it and which set is best? Here policy makers need information on the relative efficiencies of the acts they might take as well as estimates of their probable effectiveness – the likelihood that they will accomplish their intended purpose without incurring so many negative consequences as to cancel out their value.

Public policy is made in an environment involving many actors whose cooperation, or at least benign tolerance, is required to make it work. The process is best described as one involving floating or unstable coalitions. While some actors have a "one shot" or specialized interest in a *single* issue area, say agriculture, farm price supports, the oil depletion allowance, or environmental regulation, others have interests which span numerous areas over time. Thus questions of the administrative and political feasibility of different strategies have a long as well as short run dimension. Selection of the most politically efficacious alternative in the short run may, by affecting the future coalitions which can be built, limit a policy makers ability to make choices in other issue areas of interest to him at a later date. Consequently, calculations of efficiency, effectiveness and feasibility require information on short- as well as long-term consequences.

A third area of concern deals with *distributional questions*: what groups will be affected in what ways by the strategies adopted? Who will bear the burdens of a solution, who will benefit from it? And, most difficult of all, what is the most equitable position to adopt for the short and long run? As we all know, much of the recent discussion of the energy and food "crisis" as well as argumentation over environmental policies have centered around distributional issues. How are the costs of food production distributed among farmers and various middlemen and who is benefiting most from recent price increases? Should prices for farm goods be held down to benefit those living on low or fixed incomes and can you hold down these prices without controlling factor input prices as well? What impact does construction of a nuclear power plant in a semi-remote area have on the natural and social environment? Should farmers, campers, fisherman and hunters put up with such a plant so that people in New York City can have cheaper or more reliable power? What will the development of a shale oil industry in the West do to the local environment? If, as some suspect, it will require large amounts of water, what will happen to other water uses including farmers and ranchers who themselves are producing equally needed commodities?

It is tempting to believe that "squishy" problems of this sort can be turned into more tightly structured ones with the answers being provided by some large econometric model and a well-done cost-benefit analysis. However, as Robert Dorfman has recently observed, "... The critical shortcoming of this procedure is that the overall sum of benefits less costs is important to virtually no one except academic investigators. The people really concerned are concerned about cost and benefits to themselves and their constituents, and only on Sundays about global totals" [1]. This underlines the fact that questions of distribution and equity ultimately blend into the earlier problem of political feasibility and the role of shifting political coalitions in the selection of policy strategies.

A final set of questions are those which are cast in an *evaluation mode*: How well have policies and programs of the past worked? How effective have they been in accomplishing their intended purpose? Who actually benefited, and lost. through their implementation? Questions such as these are typically posed so that the information contained in their answers can be used as the basis for fine-tuning current policies and generating strategies for dealing with new issues. Seemingly they are the least "squishy" of the lot, the most amenable to careful definitions and unambiguous answers. Certainly within segments of the societal information structure I have called the knowledge industry, there are said to exist elaborate and rigorous "evaluation methodologies" for providing reliable and valid information on such matters. However, to paraphrase a current advertising jingle, "the more rigorous they get, the more inconclusive the information they generate looks". The reason for this is straightforward enough: rigor requires structure, simplification and an ability to isolate a problem from its real context. Evaluation issues are inextricably tied to representational, effectiveness and distributional questions and one can deal only very imperfectly with the former without confronting the latter.

A case in point is the recent debate over general revenue sharing. How effective has this program been? To answer this, one must first determine how it was represented and what it was supposed to accomplish. This places one in the midst of arguments among proponents and opponents, each arguing the validity of his partial model and set of intentions. Was this a problem of the financially hard pressed cities needing relief which only the Federal Government with its larger and less regressive tax base could provide? Was it a problem of Federal bureaucracy and "red tape" frustrating the "real" needs and spontaneous, innovative ideas of local government through existing grant programs. Was it a problem of continuing Federal deficits and a need to cut expenses for social services? What one requires is either a meta-theory for choosing among these alternative models or sufficient time and resources to explore them all.

If one gives up in frustration and moves to a more mundane level to ask, simply, "How did communities *spend* this money? Who received what for what purposes with what effects"?, matters don't get perceptibly better. I have spent several hours in rooms filled with economists, political scientists and representatives of the Federal and local governments listening to them explain why these questions can't be answered, at least not very well. Not only is it extremely difficult to trace the flow of such monies through the local fiscal system, but, since it is hard to say what communities would have done in their absence, it is virtually impossible to assess their marginal impact and value. There is one thing we *can* say with considerable confidence, however: Congress will be evaluating the general revenue sharing program in 1975!

CONCLUSION

My brief remarks this morning represent a first step on the way to answering my opening question: How can we begin to think about and assess the potential relevance and effectiveness of societal information structures as they relate to public policy issues and choices? We must begin I have suggested, by examining (a) the types of information required to deal successfully with such issues and (b) the decision-making structure which characterizes each issue area. The first tells something about the different *purposes* for which information is needed, the second something about the social and political context in which it is received and the range of constraints which govern its use.

The next step in the process is to assess each of the six societal information structures according to these criteria. Which structure is most likely to be most effective at defining the boundaries of problems or determining how they are to be represented? Many would argue, with good cause, that what I have called the "specialized bureaucracies" enjoy the best *opportunity* to define the problem. They are, after all, near where the problems are and they generally work on such matters on a full-time basis. Many would be equally distressed at this fact, if true, and, for a number of reasons, argue that problem definitions *should* come from the political system or even the "knowledge industry".

Those who know much about the latter two structures, however, quickly despair. They tell us that the purpose of political parties is to get people elected to office, *not* to argue, fundamentally, about major policy issues and how best to represent them. Likewise, others tell us that a major function of universities is to give Professors a relatively safe home where they can work on their own pet projects, often of interest to only half a dozen or so other researchers in the United States. We could ask similar questions about what I have called *distributional* and *efficiency* issues. Many, for example, would argue that the "invisible hand of the market", if left alone and untrammeled by government intervention, is the best source of information on these subjects.

If we begin to examine each information structure in this fashion, it is apparent that we are quickly driven to begin analyzing their internal structures, norms and functioning. Why do they produce or not produce the types of information they do? Why, for example, do universities or scientific groups seem to be doing much that is irrelevant to policy issues? Or, to put the matter more positively, why do they do much that they do and not other things? Why do political parties appear to be poor institutions for developing new *Weltanschauungen*, new ways of conceptualizing about problems? What changes would need to be made to improve matters?

Ultimately, and in conclusion, this is why I said that my opening question was a very practical one, because, ultimately we are talking about very practical ways of reforming the way in which information for policy making purposes is generated.

ENDNOTE

[1] This is a term used in an insightful paper by Ralph E. Strauch [2].

REFERENCES

[1] Dorfman, R., "Modelling through", *Interfaces,* 3: 1–8, 1973.
[2] Strauch, E. E., "A critical assessment of quantitative methodology as a policy analysis tool", *The Rand Paper Series,* August, 1974.

THE CAUTIONARY TALE[1] OF THE SEVEN-DAY HOSPITAL: IDEOLOGICAL MESSAGES AND SOCIOLOGICAL MUDDLES IN A THERAPEUTIC COMMUNITY

BARBARA FRANKEL
Lehigh University, Bethlehem, Pennsylvania

INTRODUCTION

This paper is, essentially, about two kinds of "knowing" – those types of culturally-shared knowledge which anthropologists, among others, often call the "is" and the "ought". Other ways of phrasing the contrast include such related pairs of terms as real and ideal, statistical norms and jural norms, existential facts and moral prescriptions. The implication contained in the use of such terms is that the universe of possible things one "knows about" subdivides into two mutually exclusive categories, or even that these categories exhaust the universe of possible things one *can* know about. I shall argue, at the end of this paper, that the nature of human communication and learning makes such simple dichotomies less than satisfactory – at least in the case I shall be talking about, which I am sure is not unique. There is a quality of "is-ness" about the ought, and of "oughtness" about the is, which my tale illustrates.

The usual way to organize a paper of this kind is to begin by discussing theories of *X*, to outline the theoretical problem raised by one's own particular case of *X*, to describe one's own case of *X* in terms that make it clear how it provoked consideration of the problem, and finally to draw theoretical inferences and conclusions from the model case.

I propose, instead, to begin by plunging almost directly into my narrative and only afterward considering the theoretical problem it raises and the inferences one can appropriately draw from the events recounted, when these are set in the cultural context of their occurrence. I proceed in this fashion in order to make this a kind of "meta-paper" on the subject of knowledge – an essay the form of which in itself constitutes a commentary on the process whereby we humans obtain our perceptions and cognitions concerning events around us. I hope to demonstrate how the problem was raised in the first place – how I came to "know" it was there, and how I subsequently framed it in theoretical terms and resolved it to my own satisfaction, so that I felt I "knew" the answer to the riddle posed by the seven-day hospital affair.

Concretely put, the experience of doing field work at Eagleville and witnessing the brouha over the seven-day hospital innovation came first. Analysis of it – what Geertz [7: Chap. I] calls "constructing a reading" of the account of events, using it as a kind of text – came only later. The *knowledge* of the event, in other words, came after receipt of a long series of communications, and the whole thing had to be decoded.[2]

THE SETTING

The tale cannot be intelligible without a description of the social and cultural setting in which the events occurred.

Eagleville Hospital and Rehabilitation Center is classified by the State of Pennsylvania as a "special hospital" for the drug-free rehabilitation of persons who have come to be labeled as "addicted" to various illegal or harmful chemical substances, from heroin to alcohol.[3] The core inpatient division houses 126 patients, or residents as they are generally called, of both sexes, all ages and races, mostly working-class or lower class. The real estate belongs to a private charitable organization, and was originally the site of a TB sanitorium, now defunct. It is old and rather shabby, but it is rescued from grimness by its pleasant grounds and rural location.

Between the time Eagleville in its present incarnation first opened its doors in 1966, and the time of my study (1971–72) the staff had grown from its original 25 to about 300. This explosive growth came from the addition of a number of ancillary facilities and services during the previous five years. Besides this, other changes had been rapid, as treatment methods were experimented with, and as both drug addicts and women were introduced into what had originally been an all-male alcoholic client population. At the time of my study Eagleville was unique in that it treated alcoholics and addicts within a joint program, the only segregation being by sex.

The modality of treatment which Eagleville employs is somewhat ambiguously labelled a "therapeutic community". TC is a term covering more than one sort of rehabilitation program, and it is convenient to divide the many varieties of TC's into at least two major types: self-help and professional.

One type of TC emerged from the earliest and most successful of secular Western self-help movements for deviants, Alcoholics Anonymous. Synanon, the original prototype of the self-help residential community for addicts [cf. 2; 14;

21] has, since the early 1960's, given rise to many second- and third-generation offspring. Such communities oppose the notion of professional leadership, though they may hire professionals to render special services, such as medical care. They operate on much the same principle as the well-known "curing societies" of many traditional cultures: people who have recovered from an affliction are believed to be those best qualified to help others suffering from it currently. Self-help communities usually require people to be self-referred, highly motivated to change, and to remain in the TC for long periods of time – even for life. They are frequently rather authoritarian in ideology, and insist that individuals must *earn* the right to a voice in communal affairs by their recognizable progress toward rehabilitation. There is no "staff" vs. "patient" dichotomy – all are persons in different stages of recovery.

The second major type of therapeutic community emerged out of efforts by professional psychiatrists to democratize mental hospital treatment [cf. 11; 12; 13]. The intention was to prevent the chronic back ward cases of "institutional neurosis" which were believed to result not from any inherent psychotic disease process, but from the authoritarian and dehumanizing social milieu of the traditional mental hospital. In the professionally organized therapeutic community all human relationships within the treatment setting are seen as potentially therapeutic (or the reverse). All staff members, therefore, from the kitchen help to the attending physicians, are ideologically defined as members of a treatment team whose interactions with patients are important, and must be regulated accordingly. TCs of this type stress efforts to help patients develop autonomy and self-respect through encouraging them to make decisions and to accept the consequences of those decisions. The emphasis is upon growth through allowing patients maximum freedom to order their own lives. This type of TC generally stresses the symbolic levelling of hierarchies and encourages the forms of "participatory democracy" in its procedures, both for staff and patients [cf. 17; 18].

Eagleville belongs, essentially, to the latter type of TC, even though it employs large numbers of recovered alcoholics and addicts, some of whom hold very responsible positions. Like the Maxwell Jones model after which it was fashioned, Eagleville's philosophy declares that both professionals and nonprofessionals, nonaddicted people and recovered people are essential to realizing the goal of the community, which is the transformation of addict identities into respectable identities. The expertise and sophistication of trained professionals is considered essential, but so is the dedication and loyalty of paraprofessionals whose orientation is to the helping task and to Eagleville, rather than to a discipline or an external reference group. The empathy, role-modeling and special insight provided by recovered persons is also seen as a vital ingredient in the optimum staff mix required to rehabilitate addicted people.

Ideologically speaking, Eagleville's entire population is considered to be "part of the community", though it is only the residents who actually live on the premises twenty-four hours a day. Cardinal values in the ideology are phrased as "honesty, openness and responsible concern" of all members of the TC toward one another, toward themselves, and toward the community at large [15; 16]. Heavy stress is placed upon free expression of feeling as a remedy for the social isolation and constriction of affect thought to be part of the etiology of addiction. Human beings are, in Eagleville's philosophy, almost infinitely perfectible – all have an essential self which is valuable and worth saving, just as all have faults which need correction. All persons have both "sick" and "healthy" parts of these essential selves, all are in need of help from others, and all are capable of giving help to others. In a very essential sense, then, staff and residents (all human beings, in fact) are equals.

Since all people are equal in some sense, it follows that all members of the community are accorded the right to participate in communal decisions, insofar as this is consistent with an ethic of professional responsibility for the welfare of patients. (Obviously this egalitarian ethic can, on occasion, conflict with the ethic of responsibility – and often does.)

The centerpiece of the treatment program is an eclectic version of "encounter group" or confrontation therapy.[4] Four or five mornings a week groups of a dozen or so residents meet with their counselor-therapists for this rather intensely emotional curative ritual. The objective of therapy sessions is to introduce people to their "real selves" by various means which include forcing them to see themselves through the eyes of others. The free expression of strong emotion – love, hate, rage, despair, and fear – is encouraged and rewarded in the therapy session context, and this expressiveness tends to overflow into other contexts of communication at Eagleville as well. This produces a rather highly charged and volatile emotional atmosphere in which crises are frequent occurrences. There are chronic problems in maintaining the fine line that divides freedom and autonomy from license and anarchy within the community. This, in fact, is the essence of problems in social control which exhaust staff members and upset residents at regular intervals, in a kind of perpetual oscillation from crisis to calm and back again [cf. 17].

At the time my fieldwork was conducted the organizational structure of the hospital was something officially called the "Unit System". The official form of this system is diagrammed in the Table of Organization below.

The Unit System has a number of interesting features which space will not allow me to discuss here [cf. 5: chapter 1]. The chart is included here primarily to clarify the relationship between two major segments of staff that are most relevant for present purposes: "program staff" and "nonprogram staff". Many other salient lines of cleavage exist, such as sex, race, seniority of rank, training and previous addictive status. These aspects of staff identity contribute to the complex and ever-shifting set of political alignments which are an enduring feature of Eagleville life, but these will emerge to some degree as I tell my

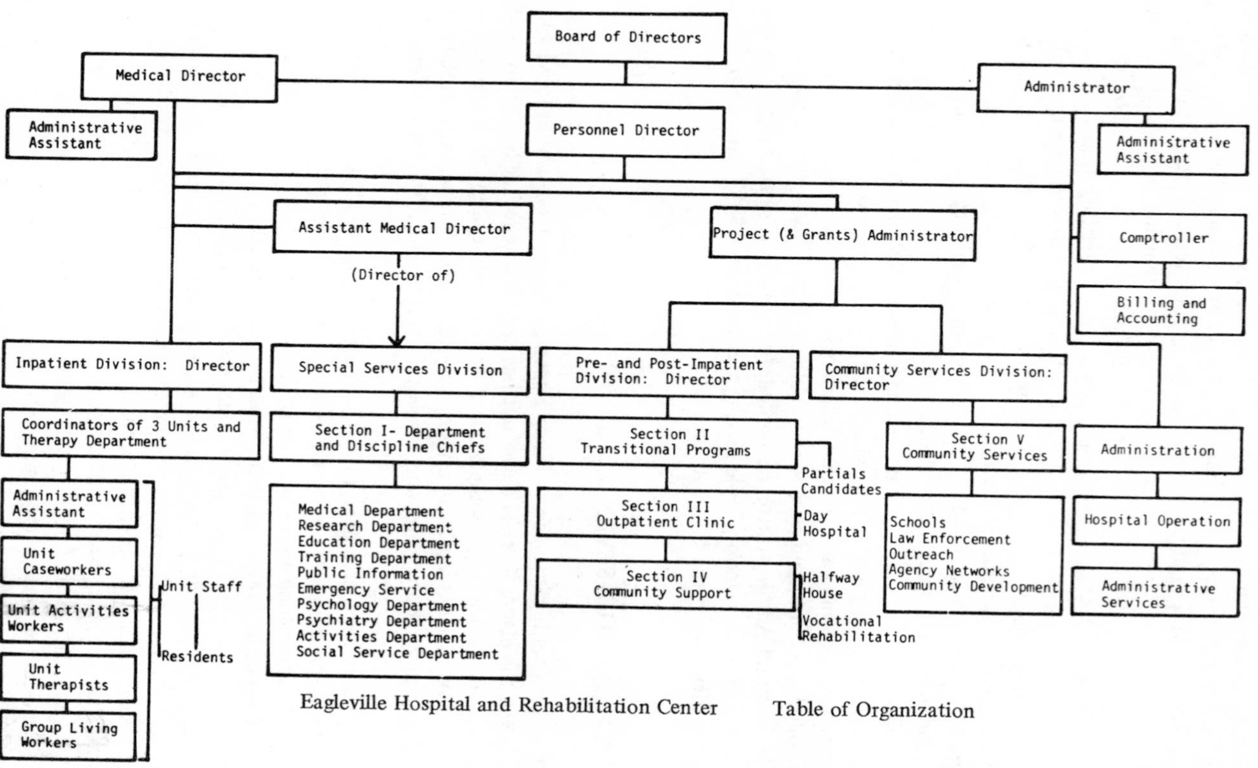

Eagleville Hospital and Rehabilitation Center Table of Organization

(Adapted from EHRC Table of Organization, May 1971)

cautionary tale, and I shall not dwell on them here.

For the present let me say only that, despite the egalitarian ethos which is so important at Eagleville, some people are rather more equal than others – even as in the Real World. Program staff are distinctly an internal elite. These are personnel hired to deal directly with residents, or to train and supervise those who do so. Their daily decisions are crucial to the success of the rehabilitation enterprise. It is program staff who set therapeutic (as opposed to organizational) policy, and who are charged with transmitting Eagleville's therapeutic cultural values to the newer residents. There were, during the fieldwork period, about 90 staff members who could be unambiguously categorized as program staff, plus about 40 others whose status as "part of the program" depended somewhat on their individual interpretations of their roles.

Nonprogram staff, by contrast, included about 120 people whose jobs were primarily concerned with the services needed to run any institution: bookkeeping, records, housekeeping and the like. The largest single bloc of such personnel can be seen on the right-hand side of the Table of Organization, where subordinates of the Hospital Administrator are ranked. In addition to this group, there were a number of nonprogram people serving as secretaries, clerks, and administrators in other divisions of the Eagleville complex.[5]

THE TALE OF THE SEVEN-DAY HOSPITAL[6]

Now to the events that occurred over a span of two years or more, to which I have given the above title. The account has been put together from two kinds of information: data which I collected as a witness on the scene between September 1971 and April 1972, and accounts by several reliable informants for the periods before and after those dates, and for certain events which I did not witness directly.

To begin with, I have been told that sometime late in 1970 the Eagleville residents' representative body, the Delegates' Council, proposed that the community do something about weekends. At that time Saturdays and Sundays were barren days of boredom and depression for many residents. It was felt that this fact was responsible for numerous incidents involving serious rule infractions and "splitting" (sudden departures against staff advice or without staff knowledge). Residents complained that there were too few staff members present to talk to on weekends, that activities were practically nil, and that this created a situation in which people who were "sick seven days a week" were receiving help only five days a week. Why could not the hospital be run on a seven-day basis?

The initial response of program staff took place at two levels, really. At the level at which residents were posing a problem in social control (recurrent misbehavior and attendant crises) there was a practical problem to be met. The first innovation was to hire a professionally trained Activities Director, who soon instituted a fuller program of weekend recreational activities. Somewhat later a plan was devised for augmented staff coverage on weekends, providing for additional pay for participating staff members. Still later a "unit coverage" plan was initiated whereby unit staff members rotated work on weekends, receiving compensatory time off rather than extra pay for these services.

Social control improved a bit, but many Eagleville people were still uncomfortable, for the residents' complaint was also perceived as an ideological reproach. At this level Eagleville's claim that it was a "therapeutic community" was somehow being challenged. A "real community" had to be more than just a "program" for addict rehabilitation; it had to be an authentic way of life, as well. As any fool knows, a total way of life is not suspended for two days a week and resumed on Mondays. Following this line of thought, some of the staff, mainly at the upper levels of the Inpatient Division, began to favor the idea of what came to be called a "seven-day hospital". This, though its precise operations were left vague, would do away with the abandoned feeling created by empty offices and much-reduced staff presence on the weekends. It was suggested that an atmosphere of full operation could be created on Saturdays and Sundays which would not only reduce the number of crises on weekends, but create a closer approximation to a *real* therapeutic community – one more like the self-help type. In self-help TC's, as viewed by Eagleville staff, a genuine full-time lifestyle is shared by all. Eagleville staff members frequently held up this type of enterprise as an ideal – although they simultaneously rejected the harsh "hard-nosed" approach to rehabilitation which self-help communities usually exemplify.

Discussions of the practical and ideological advantages of a "seven-day hospital" took place first among members of the Eagleville Executive Committee. Eventually the Director of the Inpatient Division (himself a recovered alcoholic) was asked to broach the idea to the staff of the three newly formed inpatient units. This move was in accord with Eagleville's ideological stress on "grass roots participation in decision-making". It also demonstrated the keen and constant awareness of top staff that personnel who were to be affected by any major innovation must be persuaded of its value, and willing to cooperate.

One limitation governing such a plan was clear: additional staff could not be hired to provide seven-day operation of the program. It would have to be managed by spreading the existing staff thinner, since money was not available for staff expansion. A further corollary of the plan would be that additional pay for "moonlighting" weekend workers would no longer be forthcoming, because "weekend coverage" was to be discontinued if a seven-day plan took effect. All weekend work would be done by means of staggering schedules and providing compensatory time off on weekdays, in a fashion similar to the current "unit coverage" system. Exactly how this would be done, who would be included, how many weekends people would need to work; what would be done with weekend time and a number of other questions were not yet considered. At this point it was merely asked that program staff discuss the pros and cons of a seven-day operation, to decide whether it would answer a need.

The Director of Inpatient Services was a strong supporter of such an innovation. He also planned to leave soon for another job, and there was later speculation that he had wished to leave a "monument" behind him – a major contribution to the community's well-being, for which he would be remembered in years to come. What better monument, staff members reflected in thoughtful moments of later days, than to bring Eagleville closer to its ideal of becoming a "true community" in which problems of social control would become much more manageable?

No one seems certain of what really happened next. Much later, many lower and middle level staff members claimed they had never been consulted, but had been presented with the seven-day idea as a *fait accompli* which they assumed had been approved by others while they were on vacation during the summer of 1971. Most certainly there was no concrete plan for a seven-day hospital, merely a broad principle for redistributing staff time through some system of staggered workdays to insure a close-to-normal staff presence on Saturdays and Sundays. What is certain, however, is that someone informed the Executive Committee that program staff in the units had approved development of a plan to implement a seven-day hospital operation.[7]

Based on this information, the Executive Committee issued a directive toward the end of summer, 1971, in which they set a mid-December date for implementation of a seven-day hospital plan. In the interval between the directive and the implementation date a concrete plan was to be worked out by all staff working cooperatively, with leadership to be supplied by the program staff's "middle management", the unit coordinators. The planners were free to include all Eagleville personnel in their proposed arrangements (with the exception of kitchen help and nurses, who already worked on a seven-day schedule). The goal was to give Eagleville the "feeling of being a going concern seven days a week rather than only five " – a notion which implied that not only therapeutic services but offices, record room, cashier's window, library and similar facilities should be manned on weekends as well.

All these events pre-dated my arrival at Eagleville. I was present, however, in early November 1971 when signs of a program staff revolt began to surface. A letter was circulated to all staff, along with an announcement of a general staff meeting convened by an *ad hoc* group calling itself the Committee for Concerned Citizens of Eagleville. The six members of the committee represented a rather good cross-section of staff in terms of the aspects of identity which are relevant to the internal politics of the institution. There were four women and two men. Four of these individuals were white, two black. Three had professional status, the others did not. Two of the six were recovered alcoholics, one of whom was a graduate of the Eagleville program; and all but one (a physician) were line level personnel.

The letter which accompanied the announcement was signed by a female therapist who has an M.S. in psychology.[8] In symbolic terms, it was an extraordinarily interesting document – unfortunately far too long to include here. The burden of the lady's complaint, however, was that the way in which the seven-day hospital innovation had been put forward had deprived the lower orders of Eagleville staff of their dignity, their integrity, and their right to make autonomous decisions. They were being victimized by senior staff hypocrisy in the guise of a "myth about dedication". Line staff, especially blacks and paraprofessionals, had been coerced into compliance by a combination of fear for their jobs, and guilt over appearing to be insufficiently "dedicated" if they should object. A long diatribe – considerably less coherent than my summary – was followed by the proposal that a genuinely democratic expression of staff sentiments could only be obtained by holding a secret ballot on the seven-day hospital issue. If staff favored going ahead with the experiment, line program staff should be in charge of making all arrangements, since they would have to live with them. If they voted against the seven-day hospital, the problem of dealing with weekend crises would have to be met by hiring additional staff for seven-day coverage.[9]

A large number of staff members turned out for the *ad hoc* committee's meeting. It was clear, both here and at subsequent meetings, that there were strong feelings – and a great deal of confusion, as well – over the way things had been done concerning the seven-day hospital proposal. It was not at all clear, however, that most line staff and paraprofessionals felt intimidated or feared losing their jobs if they should reject the innovation; in fact, the high level of outspokenness throughout the affair seemed to signify quite the reverse. Nevertheless, the cry of "oppression" was sent up, serving as ideological justification for an unprecedented secret ballot. Such a procedure was quite contrary to the usual Eagleville practice, whereby people took positions on issues in the open. In essence a choice was posed between two cardinal values, "honesty" and "openness", in such a way that the first could be served only if the second were subordinated. The motion for such a ballot carried.

As finally formulated, it offered three alternatives for staff: to continue as per the original directive of the Executive Committee, to reverse the decision to have a seven-day hospital, or to delay a final decision indefinitely in order to give the entire community time to discuss and clarify all the issues. When the ballots were counted, the third alternative – a "stop the clock vote" – carried by a slim majority. My discussions with numerous staff members who had voted for this alternative indicated that they were not so much opposed to a seven-day hospital as they were reluctant to "buy a pig in a poke". They wanted to know details before voting for *or* against a seven-day plan.

The first major phase of symbolic warfare, then, had ended with a moratorium, with victory for neither side. In the months that followed, innumerable meetings and discussions took place in an attempt to formulate what a seven-day hospital actually meant. In the process, residents were consulted to obtain their "input". Program staff struggled to face their cultural inhibitions against reorganizing the "normal" work week, and to define Eagleville for themselves. Was Eagleville really a "hospital" or a "community" – and what were the implications of either view? If a seven-day hospital should be approved, what sort of plan would be both equitable and effective? What would happen to things like visiting hours and weekend passes for residents? How would such changes affect the functioning of staff whose work involved dealing with outside agencies that were closed weekends? Would there be therapy sessions on weekends? Would there be problems of communication created by staggered work schedules? How many successive days would people have to work? All sorts of questions, large and small, were discussed during two intense months in which it seemed that all that could be said had been repeated many times, and that everyone had been consulted – almost.

By mid-January, when it was judged that discussion was sufficiently complete, the seven-day innovation had gained momentum. It had come to be cast in ideologically irrefutable terms, really, for the aim was to solve problems of social control by means of becoming a "more genuine community" in which to live and work. Earlier accusations of "oppression" had lost credibility, for the plan which slowly emerged exempted no one. High and low, program and nonprogram staff alike were to be involved, for only in this way could a "real community" with a seven-day-way of life be created.

In practical terms the plan meant that everyone would work two weekend days per month, with two weekdays of compensatory time off. For executives and nonprogram personnel one of these days would be spent in their usual duties, to provide against empty offices and the psychological effect that buildings were abandoned. The second day would be spent "relating to the residents" in what was billed as a "chance for people to use other parts of themselves" that, in theory at least, were more interesting humanly and more valuable therapeutically than those parts usually in evidence. The benefit to residents would come from the opportunity to learn that nonprogram people were human beings, too, who had something to offer them. The benefit to nonprogram people would come from giving them, at last, the status of full members of the community. The benefit to program people would come from providing them with additional "warm bodies" and a range of nonprogram talents and skills to call upon on weekends, when unit staff would be thinned out.

In mid-January a general staff meeting was called to consider and vote upon a plan designed along the lines sketched above. Most program staff attended, and most nonprogram staff stayed away, as usual. After lengthy discussion, a large majority of those present approved the seven-day hospital plan. With a few organizational details still to be worked out, a mid-March date was set for implementation. It was agreed – in order to pacify the remaining doubters – that the experiment would be evaluated after three months of operation, with a vote on whether it should be continued.

Those who still disagreed were, at this point, trapped in their own symbol system and unable to protest on a basis of principle. No longer "oppressed" by anyone's standards, democratically consulted *ad nauseum*, they were in no position to voice objections. What looked like an ideologically impeccable solution to an acknowledged set of problems had been arrived at. Participatory democracy had functioned in its inimitably chaotic way to produce, finally, a broad consensus.

Just as everything seemed settled, however, a second shock came. When staff members were polled several weeks after the seven-day plan's adoption, so that they could express preferences on a scheduling matter, nearly all nonprogram staff – led by administrative employees – staged a mass revolt. These individuals rarely attended general staff meetings, and had hardly been aware of what was planned for them up to that time. When they were asked what sort of arrangement they preferred for weekend work, they put large red X's across their questionnaires and returned them with memoranda of protest attached.

Some confessed, quite frankly, that they were afraid of the residents and wished no contact with them apart from that entailed by the kind of work they had been hired to do. Others said that they lacked both training and aptitude for forming therapeutic relationships with residents; moreover they were sure that the therapists would not appreciate having them transmit their personal values to the patients. Many people said that they already had trouble keeping up with their work, and could not spare an hour much less a day each month for "fun and games with the residents". Several simply refused to work on *any* weekends, with one woman citing the names of her six small children as six good reasons why she would not.

The burden of working one weekend out of four was not, however, a chief focus of resentment. The real sources of nonprogram staff's malaise were clearly that (1) they did not feel that they were "part of the community" – and many implied that they had no wish to be, and (2) they felt that the allotment of time to "relating to the residents" would interfere with the work they were really hired to do and felt responsible for.

Another series of meetings was begun by the long-suffering Planning Committee which had, by now, emerged among upper level program staff and administrators. It was ideologically impossible, given the Eagleville ethos, simply to order people to cooperate on pain of dismissal. What was required was to hear all grievances and use the arts of persuasion and compromise to integrate nonprogram personnel into the seven-day plan. The meeting sessions permitted ventilation of anger and hurt feelings, and allowed the Planning Committee to reassure nonprogram staff as to their value to the community. In the end, after a number of concessions to individual hardship cases with genuinely pressing work or family requirements, most nonprogram personnel were integrated into the seven-day hospital. In most cases the weekend work arrangements made them "part of the community" at least a few hours a month, when they interacted with residents in the units. At long last, in the early Fall of 1972, the plan went into effect. By this time I had left Eagleville, and my reports come once more from informants.

The promised evaluation of the experiment actually took place after about four months of operation. For the occasion a member of the research staff had compiled statistical comparisons of the periods before and after the seven-day operation. The figures showed that the "split rate" (sudden departures) as well as serious rule infractions had dropped significantly during the four-month experimental period. Since the research staff person was not able to attend the early part of the crucial staff meeting, another staff member was given the data to present during the opening moments of the discussion. By what may not have been pure chance, however, the staff member who had the statistical summary in hand never arrived.

In the absence of hard data, evaluation of the seven-day hospital's success took place through the expression of staff impressions. Most members of line staff concurred in their opinions that there had been no significant improvement in social control within the Inpatient Division during the experimental period. Though some people argued that they liked the feeling that Eagleville was "more of a real community" during the period of seven-day operation, these were greatly outnumbered by those who insisted that the practical benefits had been negligible or nonexistent. By a substantial majority vote, it was decided that results did not warrant the continuance of seven-day hospital arrangements. Proponents of the innovation – mainly, but not exclusively upper level staff – gave two weak cheers for democracy and bowed to the inevitable. Defeat had been snatched, somehow or other, from the jaws of victory. But how?

ANALYSIS

My tale presents several intriguing problems, but here I shall concentrate upon only one. How may we account for what appears to have been a massive misperception of the objective situation by a large majority of staff members at Eagleville?

It is possible to argue that many staff members simply lied to themselves and one another out of selfishness – and this is exactly what embittered proponents of the plan did say. It is true that weekend work, even once in four weeks, *was* inconvenient for many people, and some staff members *had* lost small amounts of income when extra pay for "weekend coverage" was abandoned. Still, the stake all staff members have in the successful operation of the program is rather great; everyone's jobs depend on good results and the clientele they attract, and this should outweigh such petty considerations as those above. Moreover, after several years of contact with Eagleville, I am convinced that there is, in fact, a high sense of mission and genuine dedication among the great majority of staff members. I think, therefore, that an analysis of the seven-day hospital's demise must seek deeper reasons than the short-term, self-interested conscious motives of individuals.

When we witness misperception not just among scattered individuals, but in a rather large group of people who show considerable unanimity about it, I suggest that something is occurring at the sociocultural level that strongly resembles the phenomenon we call cognitive dissonance [4] when it occurs at the psychological level. In brief, I shall contend that Eagleville staff's mass failure to perceive actual improvements in social control during the seven-day hospital period had its source in their shared knowledge structures – the system people at Eagleville use as the basis for appropriate behavior within their small society. More specifically, I think it had to do with a dissonance between *two* such systems, one of which I shall call the "ideological system" and the other the "sociological system", for want of better terms.

By the "ideological system" I mean the interrelated set of cultural beliefs, prescriptions and propositions which serves both to program and to justify action – in this case, action toward the Eagleville goal of rehabilitating addicted people. An ideology is not, in this definition, a propositional system of the same type as a scientific theory – for its organizing principle lies not in its facticity but in its function. The issues of truth or verifiability are unimportant when we evaluate an ideological system; any such system is likely to contain some propositions that are verifiable and some that aren't [6]. The important thing is the relationship between what an ideology asserts to be true and the social goals toward which a group adhering to it directs its activities.

By the "sociological system" I mean the system of cultural categories people employ when placing themselves and others within a particular social structure. This, too, is a knowledge structure under the present definition; social categories must be "believed in" or "known" in order to become part of people's socially constructed reality [1]. I am not, therefore, drawing the usual distinction between "culture" as a system of symbols and meanings and "society" as a system of concrete social relationships. Both ideological and sociological systems, in the sense I use the terms here, are *parts* of the set of meaningful symbols belonging to the culture of Eagleville. The chief difference between the two is that they are conceived, by those sharing the knowledge structures of the community, as dealing with different aspects of reality. These are often called the "ought" and the "is", the "ideal" and the "real" – or any of the other paired terms mentioned at the beginning of this essay. To Eagleville people, as to most of us, the ideology represents the ideal to which all should aspire, while the system of social categories represents the imperfect reality that must be lived with in daily existence.

It is somewhat misleading, however, to think of the "is" and the "ought" as a clear dichotomy. As I indicated rather cryptically at the beginning, these categories are not so separate as they seem. Consider the following, for example:

If you recall, there were two sorts of rationales, really, for initiating a seven-day hospital plan. One rationale had to do with the need to solve very real and practical problems of social control; there were too many residents coming to grief during weekends, and these untoward incidents were detrimental to the success of the program. The second rationale had to do with a desire to implement the Eagleville ideal as nearly as possible – to become more of a "real community" and thereby to achieve a state of being which the anthropologist calls *communitas.* These two purposes, however, are actually inseparable from one another, given the ideology of treatment in a therapeutic community. In a place like Eagleville, the problem of social control is seen not merely as a symptom of residents' "sickness" but as a sign of staff failure to create the "real community" that will cure that sickness. Put another way, institutional failure is a sign of moral failure, and moral failure leads to institutional failure. The relationship between the "is" of chronic social control problems and the "ought" of having the proper sort of community is perfectly circular. At the level of the ideological system, then, staff members know that when there are crises it is because people have failed to live up to its prescriptions for building the sort of community in which crises should not occur – a community that is genuinely "therapeutic" for all its members.

I would not like to suggest that anyone at Eagleville is so lost to reality that they do not realize that all human institutions are bound to be imperfect. Nevertheless, as with all utopian experiments, efforts to approximate perfection must never cease; like Zeno's runner, one may never reach the goal – but one can keep halving the distance to it. When things go wrong, then, people must try harder – or abandon the ideology and the ideal entirely.[11]

In sum, there is a quality of "oughtness" to the "is" and vice versa. What actually happens is always what one deserves, given the ideology. Eagleville is, at this level, an overpoweringly *moral* universe (and not easy to live in, I might add).

Something rather similar also obtains at the level of the sociological system. This, if you recall, is the level at which people "know" that there are different categories of people within the Eagleville community. The ones we have dealt with here are those applying to staff, who may be roughly sorted out into those who are "part of the program" and those who are not.

As I have mentioned earlier, these two broad categories of people not only play different functional roles within the institution, but have different moral standings as well. Program staff have jobs that are viewed as inherently more demanding. They are defined as requiring not only special skills and intuitive gifts for dealing with difficult people such as addicts, but the fortitude to carry on in the face of many disappointments, since frequent therapeutic failures are the fate of all addict rehabilitation programs. Program people are, therefore, an internal elite commanding considerable respect from the rest of the community.

The prestige they have within the community assumes an especially intense importance for the line level of program staff, most of whom are paraprofessionals. Line staff generally have few means for commanding either responsible jobs or deference in the larger society. Many are recovered persons whose experience fits them for few kinds of work outside a rehabilitation program context. Many others are relatively unconventional social types who would find it difficult to fit into most ordinary jobs. Unlike the professionals on the staff they possess no important outside reference group among academically trained peers, and they tend therefore to identify almost exclusively with life inside the community, and their own prestigious place in that life. Their elite status is something they feel – not without justice – that they have richly earned. Clearly, any threat to that status would meet with considerable resistance, not only because of the social reality – the "is-ness" – of that eliteness, but because of its "oughtness" as well. Here again, what exists is also what is morally deserved.

A vital feature of the seven-day hospital innovation was, you may remember, the inclusion of nonprogram people in weekend program activities involving "relating to the residents". In terms of the ideological system, this was entirely appropriate; indeed, it was a step toward the "real community" in which all persons are equal and *communitas* reigns. In terms of the sociological system, however, what it amounted to was a message stating that anyone at all was fit to do program work. If that were the case, though, where was the special prestige and moral value that had always attached to the program staff category?

Obviously, here is where the dissonance in the knowledge structure was located: the ideology said things *should* get better when all men are created equal, but the sociology said things *should* get worse when people without special talent, skill or strength are permitted to assist in the rehabilitation of residents. Put another way, ideology tends to view all staff as potentially interchangeable parts; sociology resists this notion and differentiates staff members from one another.

In the actual event, it turned out that the ideological prediction was the correct one. The "split rate" and evidence of rule violations dropped significantly with the advent of seven-day hospital operation. In the absence of hard data, however, the majority of staff – proponderantly line staff, who greatly outnumbered the upper-level proponents and their scattered allies elsewhere – saw reality differently, and perceived the innovation as a failure.[12]

The cultural dissonance theory accounts for the reaction of program line staff and paraprofessionals, who voted the plan down. It also accounts, by implication, for the quite different perceptions of that minority of program people who favored continuance of seven-day operations. Predominantly upper level, academically credentialed mental health personnel, these persons are far less dependent on the internal prestige and status system of Eagleville for recognition and satisfactions. Even though they, too, are strongly committed to the community, their commitment is less to its sociological system than to its ideological system, which they identify with institutional success in treating addicts. Their prestige is assured within the internal social system in any case, by their positions of power. Prestige in the external society is what they are less secure about, and that can only be gained by demonstrating that the therapeutic community ideology works – that Eagleville's treatment modality is the best one for addict rehabilitation. It is therefore natural that they should perceive any innovation bringing Eagleville closer to the *communitas* ideal as being an improvement in actual institutional effectiveness as well.[13]

Thus far, I haven't dealt with the reactions of nonprogram staff – who, remember, had staged a revolt of their own before the seven-day program was initiated. It is, I think, fair to say that these people, too, were threatened by the seven-day hospital innovation – though their votes were probably not crucial in bringing the experiment's demise, since few of them ever attend the monthly general staff meetings and it was at one of these that the final evaluation took place.[14]

Nonprogram staff members are well aware of their inferior moral status within the sociological system of Eagleville. Those who are concerned about this fact may exhibit one of two reactions: they may aspire to "cross over" and obtain program jobs (which happens with some regularity, especially among the nonprogram people who have histories of recovery from addiction) or they may reject pressures to orient their lives to the internal moral system of the community. People in the first group would obviously tend to react as though they were line-level program staff; they admire and covet that status. People in the second group, as I observed in many talks with them, tend to criticize the insularity of Eagleville. They note that program staff seem to have a lot of trouble negotiating the transition between the societies inside and outside the hospital – what I have elsewhere called the "two worlds problem" [5]. They observe what they perceive as the rather eccentric or unconventional lives of program personnel and fear too close identification with residents as a cause of this, hence a potential disturbing factor in their own private lives. Again and again their message came through loud and clear: Eagleville is their place of employment, not their community.

These people, moreover, tend to be task-oriented rather than feeling-oriented individuals. Though they are aware that within Eagleville their jobs lack prestige, they also know that in the Real World it is important to do a visible, tangible day's work and to receive a day's pay for it. If the larger society is the primary locus of reference for nonprogram staff, its work ethic included, how could they feel comfortable with the notion that they could be spared from "real work" for a whole day every four weeks to indulge in activities labeled "relating to the residents"?

This set of attitudes had been clearly expressed in the written protests of more than forty nonprogram staff members, as well as in my interviews with many of them. Few of them really focused upon the inconvenience of occasional weekend work, since work requirements were something they could readily accept if they were seen as fair and necessary. What they could not accept was being asked to give up weekend time for something they did *not* consider "real" work – for "fun and games with the residents" (whom some viewed with distaste, in any case). And what many truly feared was that they would become *like* program staff if they became involved as "part of the program" and thus "part of the community". This did not appear to these conventional working people as a desirable state of affairs at all.

As in all institutional settings, staff satisfaction turned out to be at least as important as client service in the long run. For their different reasons, both program and nonprogram staff agreed that the sociological system which set them apart from one another was the way things not only *were* but the way they *ought* to be. The sociological muddles created by the seven-day hospital, with its ideological messages about community, were rejected as intolerable by a majority of staff.

EPILOGUE

A late bulletin from Eagleville informs me that the seven-day hospital is being tried again. Even more interesting, the way it is being tried seems to confirm my analysis as to why it didn't take hold the first time. Someone, by either intellect or instinct, has done the following things:

1. The plan was made to appear as the idea, primarily, of middle-echelon staff. They were under some pressure to prevent further weekend crises – which had reached a truly alarming state – and were seeking remedies. Cries of "oppression" could not be raised as easily when middle level staff were the instigators.
2. Widespread staff discussions were initiated for many months before any decision was even requested of line and nonprogram staff. These discussions were problem-oriented "what can we do about social control" sessions, rather than "here is the correct solution" directives from ideological leaders. Presence of a clear problem demanding practical measures seems to have reduced the level of symbolic in-fighting.

3. When a plan emerged, "exempt categories" of personnel were created to avoid both practical problems and the ideological implication that "just anyone" could help the residents. This helped to protect the elite status of program staff.
4. Nonprogram staff may under this plan, *volunteer* for weekend work, and after demonstrating their goodwill toward the community may receive extra pay for doing so. They need not "relate to the residents" unless they wish to.
5. If they elect to do this they must be screened for suitability and briefed for program work by program staff members. This system supports program staff prestige, and again avoids the "interchangeable parts" ideological notion that was so hard for everyone to take under the first seven-day plan.

It is always dangerous for social scientists to attempt predictions, but I suspect that this time the seven-day hospital will finally prosper, bringing my cautionary tale to a happy ending after all.

ENDNOTES

[1] A cautionary tale is one in which someone comes to grief through lack of wisdom, knowledge or forethought. It is intended to serve as an object lesson.

The research on which this paper is based was carried out under a fellowship provided by the Educational Foundation of the American Association of University Women, whose generous support is gratefully acknowledged. Thanks is also due to the Danforth Foundation, whose program of Graduate Fellowships for Women supported the author throughout her graduate training.

[2] It is obvious that one cannot decode messages unless they are received as ordered information rather than random noises, and that this calls for prior knowledge at a lower level, where both words and syntax are meaningful. Discussion of the process at this level is, however, beyond the scope of this paper. It is simply assumed that all persons involved in the events to be recounted share a symbolic system that enables them to decode one another's verbal and gestural messages most of the time.

[3] What an addict "really is" is a subject of much debate, and not essential to define here. It is sufficient for present purposes that Eagleville residents have acquired, one way or another, the social label that has gotten them admitted there as patients. On the definition problem [cf. 3; 9; 10; 20].

[4] The many threads of theory and method which are bundled together in what has come to be called the "encounter movement", the "human potential movement", or the "humanistic psychology movement" are, once again, matters beyond the scope of this discussion. A brief summary is contained in [8]. For greater depth consult [22] and bibliographi in both works.

[5] The fifty-odd people who have not been counted were staff members working off-campus in ancillary facilities belonging to Eagleville, and most of these people were involved only peripherally in the seven-day hospital affair.

[6] The label "seven-day hospital" itself has an interesting ambivalence built into it. A "way of life" such as is practiced in a "true community" is, as understood by most people, something that functions all the time. A "hospital" ordinarily doesn't. Weekend skeleton staff in such institutions generally confine their activities to minimal routine patient care, and except for emergencies little active treatment takes place. The title "seven-day hospital" used at Eagleville may, then, be a symbol of chronic difficulties experienced in bridging the gap between an ethic of professionalism and an ideology of communitarian mutual help, in which professional identity is irrelevant. In my own view the communitarian ideal is, at Eagleville, a medical prescription – thus the medical model of treatment is meta- to the notion of community. Members of the staff, however, tend to see the communitarian ideal as *supplanting* a traditional medical model of treatment.

[7] Whether a deliberate lie was told, or whether there was simply some wishful thinking involved is unclear from the accounts I have been able to obtain. If there was some purposeful deception, the likeliest culprit had left Eagleville before things began to erupt, and an unspoken agreement developed to avoid public discussion of his part in things, since he was not present to defend himself.

[8] This person also had a private clinical practice to which she devoted her weekends. Some staff members nastily suggested that this fact explained her "ringleader" role in the staff revolt, but that suggestion does not account for the fact that her own discontent resonated with that of so many other staff members for at least a brief period of time. Personal charisma is not an explanation either. She was not an outstandingly popular person among line staff, due to a tendency to flaunt her professional credentials before a dominantly nonprofessional peer group.

[9] One notes that the writer was not prepared to live with the current cycle of weekend difficulties, in any event – regardless of the outcome of a ballot. It was someone else who suggested the third alternative for staff to vote upon, which was the one finally adopted by a slim margin.

[10] The term "communitas" is taken from Victor Turner [19: 96–97] where he uses it to refer to "an essential and generic human bond, without which there could be *no* society". Communitas is not a concrete social group so much as a state of being that partakes of both lowliness and sacredness, in which an unstructured social homogeneity exists.

[11] Thus, when there are crises of the sort common to addict rehabilitation programs, Eagleville staff generally responds with strong feelings of responsibility and guilt. They expect a similar response from residents, and to the extent that individual residents are successfully socialized to the ideology this is, indeed, what happens. Emergency meetings around major incidents are filled with breast-beating self-accusations, and everyone – but especially staff – is cautioned to be more alert to others' needs, better exemplars of the community's moral code, and more willing to assume the blame for such things happening.

[12] One can only speculate as to what might have happened if the statistical data *had* been presented early in the crucial staff meeting. I am informed that the compiler of the statistics did arrive before the meeting ended, and was appalled to find that her colleague had failed to show up with the data sheets. She tried to stem the tide, but it was too late. The chairman – a man with a good grasp of political reality – did not attempt to delay the vote though he favored the innovation. Possibly he foresaw what later turned out to be true; that another day would come in which the seven-day experiment could be tried again.

[13] I am somewhat oversimplifying my case here, for the sake of clarity. There were line staff who supported the seven-day hospital, and there were top staff who didn't, though

these were few in number. "Middle management" people were more equally divided, as might have been expected.

It would require a much longer paper to deal with these "deviant" cases adequately. In a general way, however, they might be explained by looking at the very complex web of patron-client type relations that exist between upper level and line level individuals. Recovered Eagleville ex-residents who are now program staff members feel deep allegiance to their former therapists and mentors who are now upper or middle level executives. On the other hand, recently-hired professionals at the upper levels of the program sometimes turn out to be less than fully persuaded by the ideology of the therapeutic community. In short, another set of categories which might be called "ideologues" and "nonideologues" cuts across the sociological categories to some extent – though not decisively in the present instance.

[14] Here I go out on a limb somewhat, since I was not present at the crucial meeting. Informants are uncertain as to how many nonprogram staff members actually attended, since it didn't occur to them to take a count. It was a larger-than-usual general staff meeting, however, and I would expect that those nonprogram people who may have attended would have voted with the majority for the reasons explicated below.

REFERENCES

[1] Berger, P. L. and Luckman, T., *The Social Construction of Reality,* Garden City, N.Y.: Doubleday, 1967.

[2] Casriel, D., *So Fair a House: The Story of Synanon,* Englewood Cliffs, N.J.: Prentice-Hall, 1963.

[3] Diethelm, O., *Etiology of Chronic Alcoholism,* Springfield, Ill.: Thomas, 1955.

[4] Festinger, L., *A Theory of Cognitive Dissonance,* Evanston, Ill.: Row Peterson, 1957.

[5] Frankel, Barbara, *Context, Power and Ideology in a Therapeutic Community*, Doctoral Dissertation, Princeton University, Department of Anthropology, 1973.

[6] Geertz, C., "Ideology as a cultural system", *in* Apter, D. (Ed.), *Ideology and Discontent.* New York: Free Press of Glencoe, 1964.

[7] Geertz, C., *The Interpretation of Cultures,* New York: Basic Books, 1973.

[8] Holloman, Regina, "Ritual opening and individual transformation: rites of passage at Esalen", *American Anthropologist,* **76**: 265–280, 1974.

[9] Jellinek, E. M., *The Disease Concept of Alcoholism,* New Haven, Hillhouse, 1960.

[10] Joint Committee of the American Bar Association and the American Medical Assn. *Drug Addiction: Crime or Disease?* Bloomington, Ind.: Indiana University Press, 1961.

[11] Jones, M., *The Therapeutic Community,* New York: Basic Books, 1953.

[12] Jones, M., "The Concept of a therapeutic community", *American Journal of Psychiatry,* **1**: 647, 1956.

[13] Jones, M., *Beyond The Therapeutic Community.* New Haven: Yale University Press, 1968.

[14] Kanter, Rosabeth Moss, *Commitment and Community: Communes and Utopias in Sociological Perspective.* Cambridge: Harvard University Press, 1972.

[15] Ottenberg, D. and Rosen, A., "Merging the treatment of addicts into an existing program for alcoholics at Eagleville Hospital and Rehabilitation Center", Paper delivered at the International Institute on the Prevention and Treatment of Drug Dependence, Lausanne, Switzerland, 1970.

[16] Ottenberg, D. and Rosen, A., "Treatment of alcoholics and addicts in a single program at Eagleville Hospital and Rehabilitation Center", Paper delivered at the Twenty-first

Annual Meeting of the North American Association of Alcoholism Programs, San Antonio, Texas, 1970.
[17] Rapoport, R. D., *Community as Doctor,* London: Tavistock, 1967.
[18] Rapoport, R. D. and Rapoport, Rhona S., "Democratization and authority in a therapeutic community", *Behavioral Science,* **2**: 128–133, 1957.
[19] Turner, V., *The Ritual Process,* London: Routledge & Kegan Paul, 1969.
[20] Wikler, A., *The Addictive States.* Baltimore, Md.: Wilkins and Wilkins, 1966.
[21] Yablonsky, L., *The Tunnel Back – Synanon.* New York: Macmillan, 1965.
[22] Yalom, I., *The Theory and Practice of Group Psychotherapy,* New York: Basic Books, 1970.

A JOURNALISTIC CYBERNETIC

RICHARD F. CARTER
University of Washington, Seattle

Consider this impressive prospect: A hierarchy of knowledges constituting a structural isomorph of a lawfully ordered universe; a mass communication capability linking all the members of a society; and, a cybernetic mechanism by which those linked individuals can pursue the implications of knowledge for adaptation within the universe – a mechanism which signals a fit or a lack of fit between knowledge's instruction and observed adaptive consequence, a mechanism which, when multiplied, can be placed at any point in the total system where successful attempt to adapt should be reinforced, or where an attempt could fall short because of malfunction, a mechanism which inexorably moves society in adaptive progress because it will even reveal flaws in the theoretical isomorph in a self-correcting operation.

This cybernetic mechanism is powerful, indeed, to have that potential for the society wherein mass communication has so developed and knowledge is so arrayed, in a universe so lawfully structured.

Fortunately, its power is not restricted to that case. For that case does not now obtain. And, with luck, it never will.

That such a case does not now obtain is obvious enough. We are not yet interconnected sufficiently. The theoretical isomorph has not been found; and without its hierarchic contingencies our recourse, given negative feedback, is to trial and error of whatever knowledges seem pertinent to the current adaptation imperative (e.g. today's headlines).

That such a case should not obtain is not so obvious. There are many who view their efforts as directed toward producing the prospect envisioned above. They will admit we are not there yet, but say we are learning and developing steadily. It is pointed out that the cybernetic mechanism tells us something of how this has occurred, by describing the selective impact on repeatable behaviors and reproducible products of observed consequences.[1]

But any basis for optimism regarding the above prospect is the conviction that there is a lawfully ordered universe. (Then the cybernetic mechanism *could* even help produce the structural isomorph needed for progressively adaptive behavior.) If this conviction is mistaken, and I shall argue that it is, then it would be equally mistaken to use the cybernetic mechanism only as an adjustive device for adapting entities, for it is just as useful as a control device when there is no *a priori* instruction – merely prior implication.[2] That is, it will work just as well for trial and more trial as it will for trial and error.

This is what the *journalistic cybernetic* is all about. If there is no ordered universe to be given hierarchic representation in theoretical knowledge, societies are thereby not necessarily just adaptive systems – any more than people are. They cannot be just adaptive unless, by some deception, an order is stipulated. Thus a society, like an individual, needs a capability to produce order beyond discovering and confirming it. Informally, we speak of a "mind". What distinguishes journalism from mass communication is the presumption that the former serves the societal mind. Journalism says that there is more to mass communication than social linkages. There are, for instance, activities of surveillance and correlation [3].

Surveillance and correlation supplement, and complement, education as societal uses of mass communication. Where education pertains to the knowledge structure of society, surveillance and correlation pertain to the knowing structure of society. Surveillance suggests a cybernetic implementing of given instruction, but it is clear in the mass media's attention to controversy *per se* that it also implements, cybernetically, a lack of singular instruction. Correlation, the bringing together of people and their thoughts, is obviously not operating with respect to a given instruction; it is supposed to produce an instruction.

If we do not accept the postulate of a lawfully ordered universe, there is no redundancy (given or emergent) between knowledge and knowing. There is, of course, a functional relation between them. They are complementary, as suggested above; society, like any other system, in subject to two different structural-functional relations. One of these is the well-known redundancy of structure and function within the boundary of a system. The journalistic cybernetic represents an attempt to bring the other, neglected relation to light, and to show how it can be used in an invention to assist societal mental capability.[3]

Not everyone is convinced of a lawful, ordered universe. But, interestingly, this lack of conviction is biased toward the likelihood of there being such a condition. Scientists have moved tentatively toward a conception of a lawful, partially ordered universe [11]. The bias obtains because the newer conception implies that nonorder does not amount to anything, except in relation (quantitatively, as in entropy) to order. That is, order *or* lack of order pertain to the state of any given system; they are not both considered to pertain to the condition that there is more than one system – the qualitative condition which is necessarily implied in partial order, and which is empirically confirmed by the cybernetic mechanism itself when observed as the functional relationship between *two* behavioral entities.

The rationalizations surrounding the unity structure postulate are many.[4] The postulate of a lawfully ordered universe persists, even if slightly encumbered by the insertion of "partial". I shall give a brief argument here to suggest its fallacy – which is a lack of comprehensiveness. The postulate omits a structural condition. In doing so, it leaves functional analysis of behavioral systems incomplete. If we add the omitted structural condition, we can see better the functioning of behavioral systems.[5] Particularly, we can see better how collectivities must operate, and the opportunities for useful societal inventions.

How does a postulate of a lawfully ordered universe – a unity structure postulate – lack comprehension? (It certainly seems comprehensive. . . .) The difficulty occurs in equating order with structure. The pursuit of order by scientists has led to seeking connections (e.g. the missing links). But structure includes discontinuity as well as continuity, if there is partial order.

This distinction between structure and order is critical. For the empirically obvious case of more than one system, such that there is both structural continuity and structural discontinuity, there are different functional consequences of the *two* qualitatively different structural conditions for any behavioral system. There are two kinds of structural-functional relations and both obtain.[6]

The neglected structural-functional relation has to do with structural discontinuity. Here the functions are not redundant to given structure; rather, they are specified only as requisites. They apply as much to invented behavioral systems as to given behavioral systems.

Because these functions are requisite, implementation need not be uniform. That which is demanded but not furnished must be acceptable as furnished, in so far as meeting the requisite function is concerned. Pragmatically, different implementations have different consequences as structured products, even though functionally equivalent.

More formally, we are talking about two structural-functional relations:

$$S_C = F_C \qquad (1)$$

$$S_D \ldots F_R \ldots S_C \qquad (2)$$

In (1), structural continuity is redundant to functional continuity; in (2), structural discontinuity establishes functional requisites whose implementation is a structural continuity.

For any behavioral system, the functional requisites imposed by structural discontinuity pertain to the absence of complete instruction – and thus the need for an observational capability to get *and* use what was not furnished. These functional requisites are given as a *behavioral molecule*, because their implementation must take a particular sequential structure. Given the ubiquity of structural discontinuity (the continuing presence of more than one system), this molecular structure is an important constraint on societal invention. The functional requisites implemented sequentially in a behavioral molecule are:[7]

3.1. Cessation of movement;

3.2. Directing of movement;

3.3. Definition of movement;

3.4. Initiation of movement; and,

3.5. Movement.

Living systems, not being completely informed, are susceptible to frequent changes in direction of movement. The assumption here is that each change observed represents this sequence, in which observation must serve to help stop a previous movement, give direction to a new one, perhaps help provide a defining mechanism between the direction and its being carried out, and help initiate a movement.

The cybernetic mechanism, which is properly viewed in this context, operates primarily with respect to cessation of movement (3.1). A given movement stops or continues according to feedback received. However, by now, we are quite adept at utilizing contributions from one system in the service of another. So we also have such developments as the successive use of cybernetic "checks" in a countdown implementing the initiation of a movement (3.4). And we are very familiar with the use of cybernetics to operate a computer which will, by program, provide us with a "decision" to direct our movement (3.2).

Societies, in meeting these functional requisites, have tried such implementations as:

(3.1) Posting sentinels, holding periodic elections, votes of confidence in parliaments;

(3.2) Decision making, expert planning, playing "follow the leader" – with numerous variations;

(3.3) Providing procedures – such as roles, producing machines whose operation represents given instruction;

(3.4) Last minute checks, analysis of secondary consequences – such as in environmental impact statements, risk calculations – with regard to cost and/or uncertainty; and,

(3.5) Fighting, not fighting, and countless other activities.

There is a surfeit of prior implementations, with few surviving societies to advance the pragmatic significance of their respective attempts. Although not possessed of any hierarchic summary of knowledges, we already have an information overload. Structural discontinuity was responsible, historically, for the varied use of cybernetic mechanisms to meet functional requisites.[8] Because functional equivalence of implementations did not assure their pragmatic equivalence, feedback has continued to promote new implementing attempts. As discontinuity continues, the need to implement continues; and feedback provides an impetus to implement in new ways. So, by now, we are troubled by the vast amount of structural continuities accumulated from meeting the functional requisites.

Faced with a behavioral problem like operating a collectivity, we easily get bogged down trying to use accumulated structural continuities – our "information". Fortunately, we have the option of trying to implement the behavioral molecule anew. We can invent, and reuse old structures in new ways, rather than trying to retrieve or discover a presumedly appropriate given structural-functional redundancy.

A general consequence of structural discontinuity as a persisting condition is that any behavioral system must engage in control behaviors as well as in adaptive behaviors.[9] Failing total instruction, behavioral systems must give less emphasis to finding their appropriate path and more emphasis to avoiding the paths of others – to stopping (3.1, 3.4) [6].[10] And, it becomes very important for societies to have such control capabilities as:

1. An ability to formulate new instruction. If a society stops one movement only to adopt its opposite (or some related alternative), the endless recycling is an arbitrary – and dubious – orbiting.
2. An ability to take advantage of the options involved in the lack of total instruction, so as to be creative and not just "lost".
3. An ability to operate via a "voluntary citizenry". Members of a society must implement the functional requisites for themselves in addition to whatever they do for society. Because of the control imperative in the consequences of structural discontinuity, there is no way an individual can avoid being a personality, whatever allegiances he may proclaim or however he be proscripted. While he lives and observes, he is an elector.
4. An ability to think together. "Understanding" becomes more important when it must exceed (comprehension of) given instruction and the consequence of heeding it. It also becomes more difficult.

That prospective society which was to be fully adaptive had one very attractive feature. It would have exerted only homeostatic controls, once the knowledge structure was complete. It would be a controlled, not a controlling (as well), system.[11] That is, it would be controlling only in the sense that its functions were redundant to its structure, and thus controlling on other systems but itself controlled.

Persisting structural discontinuity prevents such a possibility. The difficulties of forever trying things out only to find that they do not work all that well, to find that there are unforeseen consequences, to find that other possible tactics now look better – these difficulties are not about to disappear. They occur because of the need for control capability as well as for adaptive capability. We are both within and outside systems. We may be able to alleviate our difficulties, but we cannot end them without surrendering our freedoms.

Our best chance is to improve our use of cybernetic mechanisms in the service of knowing structures. We cannot restrict knowing structures just to the duty of confirming a given ordering of nature. We have not; we should not.

Improved feedback has become increasingly important in today's world with respect to control via stopping – surveillance:

Is public opinion less favorable?

Is this program working?

Is the GNP rising at the same rate?

Is he/she/it doing the job?

Is this consistent with our society's norms?

The mass media find much to be self-congratulatory about in their pursuit of such questions.

But what about correlation, about thinking together? What is the improved role for cybernetic mechanisms to be here? What are mass media – and everyone else – doing about improving our ability to exercise control in the directing of movement, to supplement our ability to stop previous moves? What about the role of cybernetic mechanisms when "understanding" is something more than a comprehension of nature's ordering?

For society's members to be able to think together, the concept of understanding must apply to their comprehension of each other's orderings and not just those of nature. This takes us past the individual's comprehension of nature's ordering in two respects:

1. What ordering does each individual have, which would be used to imply the direction to be taken by society; and,
2. There is a mutuality involved – there is understanding only if everyone comprehends everyone else, or could if such an effort is elected.

What this comes to is that society must cope with a much more difficult referencing problem than it has yet managed. For there are ideas to be shared which are not governable by the mediating consequences of natural order as "umpire". The language to provide this capability cannot be merely a straightforward isomorph of natural ordering; one-to-one correspondence is without significance if there is no oneness to be represented.

What we have in a natural language like English is a structure largely representative of spatial and temporal relationships, spatial entities, and functional relationships of the structural-functional redundancy type. But it is a "living" language; feedback in social usage has permitted anything to remain for which referential agreement could be obtained. For instance, useful fictions have thrived (despite the implicit conflict with natural order as umpire).

It is likely that some regard fictions as something which will fade away as knowledges accumulate to their postulated instructive whole. However, that too is a fiction which we are able to entertain via referential agreement.

The same kind of cybernetic capability has rescued us in the sharing of ideas. When reference could not be governed by natural ordering, we have managed with agreements. This time, though, the agreement is with regard to probable consequences – such as direction to be taken (e.g. "Are you with me or against me?"). The usage is quite evident in such societal mechanisms as voting between alternatives, where it is generally conceded that the agreement among those voting a particular alternative does not necessarily imply they see things exactly alike.

As long as the illusion is maintained that society is just a means for adapting to natural ordering, there is no great problem (albeit some difficulty) involved in equating understanding with an agreement on the consequences of various ideas. Consequences will iron out the differences as knowledges accumulate! People will understand better! But this is illusory, given structural discontinuity. A means for sharing ideas, so that understanding pertains directly to those ideas – not their consequences, is needed to take advantage of the freedom to invent and create, in addition to whatever adaptation seems useful.

Although the course is not one to be urged, it should be possible for a society to martyr itself on behalf of its creation, rather than dying of misadventure or maladaptation.

What society needs is a referential capability by which ideas can be shared as such, so that agreement on reference is not indirect and ambiguous because of a reliance on allusions to consequences and natural orderings.[12]

What society has enough of already is political parallel to the individual's difficulty. People cannot usually get together to formulate policies, so a division into initiative and review has emerged to handle the using of ideas. Elections are typically a mode for feedback on the previous ideas as implemented – i.e. seen in their apparent consequences – and we may lose useful ideas by inept reference as well as through exclusion by those responsible for initiation. There is also the occasional despair of leadership whose ideas are assessed with regard to subsequent, not just consequent, events. Mass media, too, have been better able to handle ideas as indicated in consequences – directly observed as events, or as visualized outcomes by those considered informed and/or interested observers. For example, reporters ask questions of scientists and politicians as to the consequences of whatever it is they are talking about.[13]

If we can provide a useful approach to this capability, the cybernetic mechanism will soon tell us of its acceptance, as people find it much easier to get positive feedback in their referential usages – on the expression, if not the acceptance, of ideas. Otherwise, anti-intellectualism is to be expected if expression and comprehension are difficult, and if like consequences seem to mark ideas as pragmatically equal and only aesthetically different.

I would like to present now some preliminary work on the invention of a language adjunct, which is designed to help us communicate ideas. It is supposed to help implement the directing of movement (3.2) for society – whether we view this as a correlation function, a planning mode, or as increased citizen participation. It takes explicit note of the absence of total instruction for behavioral systems. Thus it is responsive to options, rather than to alternative, provisional (necessarily, failing completeness) statements of knowledge in respect to given instruction of a lawfully ordered universe. It is especially designed to make evident options regarding creativity and invention. It constitutes a knowing, not a knowledge, structure.

The general approach is to make more referentially specific the various possible sources of that which, as elected observational option, directs movement in the behavioral system. These sources I shall call *implications* (rather than instructions or information, obviously). Implications are the material of which ideas are constructed; their scope and detail need a commensurate referential capability.

This adjunct language runs along with, and supplements, other languages. It is not intended as a replacement, nor as an additional special language for handling new referents. It has to do with what people have been talking about, but have had trouble saying, for a long time. We are currently using it in research on communication behavior. The uses of elements from this adjunct are regarded as tactics, for which there are applicable strategies and principles as well as relevant situational conditions.[14]

The language adjunct is called PIX.[15] It provides an idiographic, two-dimensional representation – usually a rough drawing – of those implicatory relations which have been found significant in giving direction to movements. There is no chance to delve into the origins in this paper, fascinating though they be, nor can I dwell on the differences in pragmatic consequence from using one or another in a given situation – with which part of our research is concerned. Here what concerns us is their functional equivalence, and thus their previous and possible future use as optional societal thinking modes.

I shall begin with an example of the tough referential problem with which we must contend. Someone says brightly to another person. "If X then Y!" *Some* of the possibilities he may have had in mind are:

X is an attribute which is subsumed by the attribute of Y, so if you have X you'll have Y.

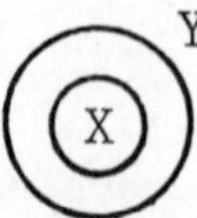

X is an attribute possessed by the object Y; so if you want X, find Y; or, X may help you find Y.

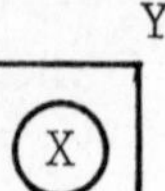

X is an object condition which, if functionally related to Y, an object condition, has a sufficiency capability.

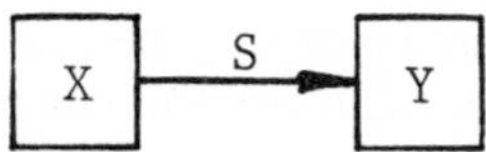

X is an object condition which, if functionally related to Y, an object condition, has Y as a prerequisite.

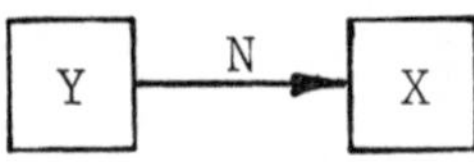

X is a relationship between two objects which occurs only because of their inclusion as a part of the system Y, an object.

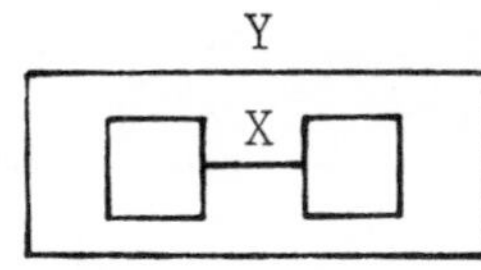

The definitional requisite (3.3) of the behavioral molecule requires that any such implications have a linkage to the time/space dimensions of movement. So, historically, it has been possible "to get the point" (in the most literal sense) without having comprehended which particular implication was to have been conveyed. Thus, for instance, politicians concerned with sufficiency conditions for attaining objectives have managed to converse with scientists concerned with the necessary conditions of a lawfully ordered universe. The cybernetic mechanism of agreement on implicatory consequence has permitted this, so we have managed in spite of rather than because of the referential capability of natural language.

The first example (though not as lengthy as it might have been) did not permit natural language to do as much as it could. Some options could have been ruled out with a few more words and a choice of some different words. But, granting that leeway, what we are finding in our research is that referential agreement on word usage is so poor that comprehension of the particular idea to be conveyed cannot be expected to occur except under the constraint that the conveyor insists (or that the researcher insists).

Without something like the adjunct, it is socially dysfunctional – and sometimes dangerous – to pursue a point. (One is considered to be either trying to prove something to himself or to prove the condition of discontinuity, although the latter would not be expressed directly – more like: "What are you trying to do: confuse me?") Some cybernetic check on understanding is critical in the social undertaking, given the options available among sources of implication. It is enough of a burden that the cybernetic mechanism give a check on whether the intended implication is comprehended, using the adjunct. PIX is useable as this journalistic cybernetic.

So far, what we have developed and used in our research is an alphabet of elements and some PIX to represent commonly used ideas and idea fragments. On being introduced to PIX, there seems to be more affinity for the PIX of ideas than for the alphabet elements – a condition analogous to the functional distinction of words versus letters or sounds. The distinction is clear, for example, between these PIX:

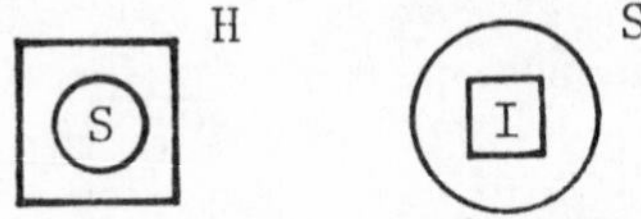

where the situation is that of the "soldier" (S) as seen by the general. Is he seen as a human, H, with soldierly attribute, S, or as an instance, I, of that attribute, S (with compassion, or, as "cannon fodder")?

There are three major types of PIX:

1. The inclusion (and/or exclusion) type, involving views of objects and/or attributes;
2. The connection type, involving some link in time and/or space between objects; and,
3. The system type, involving both inclusion (a boundary condition) of objects and/or attributes and links in time and/or space of objects.[16]

In the first example used, all three are represented in the listing. The first two are inclusion types; the second two are connection types; the last is a system type.

These "types" of implication pertain to a consistently applied extension of the elements as defined. In early usages there are inconsistencies of application analogous to spelling and grammatical mistakes.

The elements of the PIX alphabet currently in use are:

Object	□
Attribute	○
Temporal connection	......
Functional connection	⟶
Functional sufficiency connection	$\xrightarrow{S}$
Functional necessity connection	$\xrightarrow{N}$
Self as object	☒
Spatial gap*	□ □

* Unlike the others, the gap condition cannot be shown independently of (at least) two objects.

The gap is a very important relationship, especially because of discontinuity having implications (i.e. that some objects are being attended to and not others, even though not otherwise connected, that exclusion is to be differentiated from irrelevance).

Elements taken by themselves are, of course, a very cumbersome and quite restricted mode of representation. It is the inclusive and connective relations constructed with these elements which produce the implication needed for directed movement. In a world where things are not fixed in positions, relationships become critical. And, a referential capability responsive to relationships becomes necessary. We are enabled to point, quite literally, by inclusion and/or connection.

The implication (in PIX) is the structural characteristic of ideas. Thus, ideas employing views of things as included or including, of connected or not, are useful for creating new things as well as for describing them. We are not restricted to the more common language situation of reaching an agreement on reference after the fact.

Potentially, societies can invent and create their futures – if ideas are available. Members of a society can help to construct their futures – if ideas can be shared. PIX or something like it provides a structural framework for communicating ideas. Add a cybernetic mechanism and sharing can be pursued.

PIX can be useful as an analytic tool for those who are charged with political initiative, as a way of showing what options society has. In complementary ways, citizens reviewing policy will be better able to sharpen their questions about the views taken, or not taken, by their representatives.[17] The kind of hardware envisioned by societal interconnection will sorely need a software to make use of such opportunities as are thereby afforded – to avert reliance on even more simplistic modes of societal self-determination than we now know, a reliance engendered by the dual frustrations of incomplete, nonsingular instruction from the knowledges available and of views increasingly expressed, but not shared among citizens.

ENDNOTES

[1] Although it is useful to know about how things happen in this sense, there are two other "how" questions which cannot be overlooked: How-historically and how-purposefully. See the discussion by George Gaylord Simpson in [10].

[2] By "implication" there is only the suggestion of that which furnishes direction to behavioral movements. This is discussed later in the paper.

[3] John Platt urges social invention upon us [9] yet a total reliance on the efficacy of knowledges tends to minimize the perceived need for societal inventions regarding knowing.

[4] For instance: There are missing links which will clear up the connection; observers make

errors in describing connections; if science keeps building with blocks of knowledge the edifice will become clear; between *levels* there is hierarchic order; etc.

[5] A fuller discussion of the unity structure postulate, and of this proposed alternative formulation is found in [4].

[6] It is possible to take a restricted focus on behavioral entities such that one relation seems to be a within-entity condition and the other a between-entity condition, but this is incomplete and quite undescriptive of mankind.

[7] A more complete discussion of the behavioral molecule and its components is found in [3].

[8] The course of history viewed from the perspective of discontinuity theory is given in [3].

[9] Ibid.

[10] See [6]. The article discusses control possibilities – and control malfunctions such as stopping too often or not often enough.

[11] See, for example [7].

[12] The pragmatic theory of meaning finds no difficulty in allowing referential usage to be determinable by behavioral consequences. But that works only after the fact, and if used beforehand (to consider optional behaviors) would still be indirect, and questionable too as knowledge (as in criticisms of private meanings).

[13] In the case of social scientists, this can be poetic justice, if they are given to ask of social behavior how it is naturally determined.

[14] Generally we have set aside the bivariate analysis as proper only to testing hypotheses. For improving our knowing of communication, as behavior implementing the functional requisites, we must build with larger bricks than those designed for an edifice of structural-functional redundancies.

[15] Because of its genesis in picture theory, as introduced in [1; 2].

[16] It is the point of another paper [5], that systems approaches to description are helpful because of this dual capability.

[17] The same kind of questions can be asked of social theorists, who not only are not saying the same things but are often not talking about the same things – though using common terms (e.g. "system").

REFERENCES

[1] Carter, R. F., "A journalistic view of communication", a paper presented at the convention of the Association for Education in Journalism, mimeo, 1972.

[2] Carter, R. F., "A general system characteristic of systems in general", a paper presented at the Society for General Systems Research meeting, AAAS, Washington, D.C., mimeo, 1972.

[3] Carter, R. F., "Communication as behavior", a paper prepared for the convention of the Association for Education in Journalism, mimeo, 1973.

[4] Carter, R. F., "Toward more unity in science", mimeo, 1974.

[5] Carter, R. F., "Elementary ideas of systems applied to problem-solution strategies",

in Proceedings, 6th Annual Conference, Far West Region, Society for General Systems Research, 1975.

[6] Carter, R. F., Ruggels, W. L. Jackson, K. M. and Heffner, M. B. "Application of signaled stopping technique to communication research", *in* Clarke, P. (Ed.), *New Models for Mass Communication Research:* 15–43, Beverly Hills: Sage Publications, 1973.

[7] Kuhn, A., *The Logic of Social Systems,* San Francisco: Jossey-Bass, 1974.

[8] Lasswell, H. D., "The structure and function of communication in society", *in* Bryson, L. (Ed.), *The Communication of Ideas:* 37–51, New York: Harper & Bros., 1948.

[9] Platt, J., Book review in *Science*, **180,** No. 4086: 580–582, 1973.

[10] Simpson, G. G., "Biology and the nature of science", *Science,* **139,** No. 3550: 81–88, 1963.

[11] Wiener, N., *The Human Use of Human Beings,* Boston: Houghton Mifflin, 1950.

DIFFUSION OF AN INTELLECTUAL TECHNOLOGY

JOHN M. DUTTON
New York University

and

WILLIAM H. STARBUCK
University of Wisconsin, Milwaukee

INTRODUCTION

This paper deals with the question of how intellectual technologies are diffused in society. We discuss this question in terms of a specific case of intellectual technology, computer simulation. The data are drawn from field interviews and an earlier study of the history of a total of 1921 computer simulations of human behavior [27]. We focus here on a subgroup of 290 studies concentrated in one field of effort; highway-related simulations published between 1950 and 1968.

We define technology as any tool, technique, process, product, physical equipment, or method by which human-capability is extended [23; 26]. Technology thus defined has many different dimensions. It falls along a range of complexity, a range of abstraction, and a range of hardware intensiveness. When compared to product and process technology, intellectual technology tends to be complex, ambiguous, and abstract, but widely varied in its requirements for physical apparatus. Computer simulation, itself, is a hardware-intensive technology. And, because computers are complex-hardware, added difficulties are imposed upon computer simulation users.

Diffusion of technology can mean different things. It can mean the cumulative adoption by different Iowa farmers of weed spray over a ten-year period [24: 109]. It can mean that textile mill craftsmen teach their skills to others or move to a new location to perform [19: 72–77]. It can mean that an individual transfers a finding of his own to a new use, as with Land's development of self-developing photographic film. Diffusion of technology can also mean new applications of a development. Thus Trevethick used the stationary, high-pressure steam engine developed by Murray not to work a movable hammer mounted on a carriage but instead to power a carriage placed on a railway [32].

As with these more familiar product and process technologies, intellectual technologies like Newtonian calculus, methods used to investigate fossil sites, economic input-out analysis, and architectural design techniques can diffuse geographically or be put to new applications or shift across topics.

But intellectual technology transfer does not travel all these paths equally nor does each path yield equal results. Our data suggest there are differences between diffusion of concepts and diffusion of applications. Factors which enhance the diffusion of concepts may not affect or may even inhibit the diffusion of applications. Conversely, factors that quicken transfer of applications may not affect spread of concepts. Finally, factors that translate concepts into applications are different than those that cause new concepts to spring up.

A TYPOLOGY FOR DIFFUSION OF INTELLECTUAL TECHNOLOGY

A number of studies have explored factors that affect the adoption of technical innovations. Sociologists have concentrated on social characteristics of early versus late adopters, surrounding culture, characteristics of the innovation itself, and the role of change agents [24]. Mansfield and others have examined the effects of economic and social variables like present numbers of adopters, expected profitability, and investment required upon the rate of technological transfer [18: 25]. Studies of conditions affecting research productivity such as Ben-David and Zloczower [1] and Gordon and Marquis [17] are also closely related to the diffusion of intellectual technology. Finally, studies of the role of publication in science and technology transfer such as by Cole and Cole [3] and Price [23] have an obvious bearing on the spread of intellectual technologies like computer simulation.

We employ a typology for examining the diffusion of highway-related simulation studies that selects from these existing studies those factors most likely to influence the transfer of intellectual technology. Our typology emphasizes the way these factors were involved in actual instances of the emergence and spread of highway related simulations. We chose to treat instances of concept diffusion and instances of spread of applications as distinct categories of events each influenced by different factors. The factors discussed include: (1) availability of fund support, (2) visibility of study consequences, (3) social support, (4) work site characteristics, (5) task group composition, (6) role of promotional agents in government and academia, (7) mechanisms of communication of concepts of applications, and (8) the role of computing machine hardware and software. These factors are organized, as shown in Table 1, according to their role in diffusion in geographic space, in diffusion across simulation topics, or across different types of computing devices.

Table 1. Selected Instances of Diffusion of Highway Related Simulations

	Concepts	Applications
Diffusion in geographic space	The use of simulation for the study of automobile traffic. Early 1950's sites were University of Michigan, University of California at Los Angeles, and the Road Research Laboratory at Harmondsworth, England.	Actual highway plans developed using simulations for specific locations. A group formed in 1953 to develop a comprehensive highway study for Detroit provided essential personnel that applied computer simulation to later highway studies carried out in Chicago, Buffalo, Rochester, Syracuse, Utica, Schenectady, and Albany.
Diffusion across substantive topics	The transfer of use of computer simulation as a means of studying the behavior of electrical networks to studying the behavior of highway traffic networks. This transfer took place at the Institute of Transportation and Traffic Engineering at U. C. L. A.	The transfer of simulation study capability employed in a study of Pittsburgh highways to the use of simulation to study other city problems including Pittsburgh regional development forecasts, urban renewal, and then to Northeast corridor transportation plans.
Diffusion across computing machines and peripheral devices	A shift from use of special-purpose analogue machines to the use of general-purpose digital computers. Such shifts occured at U. C. L. A. and at Harmondsworth, England.	A shift to a more powerful computer. Such shifts occurred at Detroit, Chicago, and at Buffalo. Shifts in input devices and in output display devices occurred in Chicago.

The Diffusion of Concepts in Geographic Space

The concept of computer simulation of traffic was discussed by Harry Goode in 1951. Goode was located at the Willow Run Research Center of the University of Michigan. At nearly the same point in time researchers at two other locations were independently exploring the same concept. These other locations were the Institute of Transportation and Traffic Engineering at U.C.L.A. and the British Road Research Laboratory located at Harmondsworth, England. The existence of these three multiple locations independently exploring the use of computer simulation to model highway traffic behavior suggests, as other writers have, that when conditions are right an idea will take concrete form, in much the same way that change occurs in other evolutionary settings. In the case of new technological concepts this perhaps means, as with semiconductors, a combination of current need and pre-existing information upon which to build further knowledge [26: 8–10]. It also suggests that many ideas lie latent in the intellectual environment, perhaps especially among active intellectual groups, and that when particular ideas receive support they spring up into greater prominence.

However it was that these three traffic simulation sites came to spring up, each soon learned of the other's existence and communications began by means of correspondence, 'phone calls and site visits. Goode was an activist with contacts in the U.S. Bureau of Public Roads. The BPR had developed a strong interest in systematic highway planning, particularly in urban areas, as a result of traffic studies completed during the 1930's and 1940's under their guidance. These studies revealed that the bulk of highway mileage in the U.S. resulted from travel within cities and their surrounding suburbs, rather than from intercity and rural travel. The Bureau showed an interest in Goode's ideas and encouraged their discussion. But no highway simulations were done at Willow Run.

The group at U.C.L.A. was composed of five young Ph.D's interested in computers and applied mathematics who took positions with the Transportation Institute, largely because they needed fund support. They discovered a mutual interest in simulations applied to highway traffic problems. They were considerably influenced by L. M. K. Boelter, Dean of Engineering, who pressed the group to undertake a project to simulate traffic in some part of Los Angeles. With this encouragement they pursued both analogue and digital computer simulations of highway intersections [10; 11; 12; 13; 21; 29].

Sometime after 1952 a traffic simulation project was begun by a group at the Road Research Laboratory in Harmondsworth, England [20]. This group built a digital device called the Automatic Delay Computer which used telephone relays to simulate driver delays at traffic signals. The device worked; its parameters were matched to observations on driver behaviors and its predictions compared to actual driver results. The slow speed of the device, however, led the group to begin working with the Automatic Computing Engine at the National Physical

Laboratory.

The computer expert who helped the Harmondsworth group was a man named Davies. Davies later visited U.C.L.A. and there met Gerlough of the U.C.L.A. traffic simulation group. Gerlough had already gone to Michigan to see Harry Goode, who spent a whole Saturday discussing traffic simulation with him. Thus before results of any actual traffic simulation efforts were published members of all these three early sites were in face-to-face contact with one another.

We draw the conclusion that concepts initially diffuse in geographic space because multiple locations using the same concepts spring up at about the same time. These locations become aware of each other and provide mutual support for ideas. Face-to-face meetings are preferred to extensive written correspondence because the individuals involved need to develop a common language through which they can communicate their thoughts. Spoken language forces out this commonality more rapidly than does written language. Moreover, communication through academic journals and even conference proceedings is a method that contains major time lags. Face-to-face meetings and especially on-site, laboratory visits reduce this lag by months and often by years. Reliance on high-status academic journals also imposes a fair probability of never communicating truly *avant guarde* concepts across fields because high status journals are often not read across fields.

The Diffusion of Applications in Geographic Space

The spread of highway traffic simulation from the Detroit Metropolitan Area Traffic Study site is an instance of how an intellectual technology applied to a real-world problem diffuses in geographic space. The 1953–1955 Detroit Traffic Study was jointly supported by the City of Detroit, the State of Michigan and the U.S. Bureau of Public Roads as an original systematic highway development effort for a major urban area.

Before the Detroit study, urban transportation planning had not emphasized relationships between traffic, highway location, and land use. Urban highway development often meant widening existing roadways or building one road at a time with no analysis of the impact of alternative road locations on highway usage. This lack of analysis stemmed not only from traditions in highway planning but also from lack of data on existing travel patterns and from a limited ability to analyze large masses of actual travel data.

Following World War II the U.S. Bureau of Public Roads (BRP) encouraged the separate states to develop comprehensive plans for their major metropolitan areas and worked to increase federal funds for urban highway planning and construction. Traffic congestion had become a problem in U.S. cities well before World War II. The consequences of increased auto usage in the U.S. were not fully anticipated. A 1938 travel survey sponsored by the Bureau of Public Roads and the Bureau of the Census revealed that over half of the total mileage travelled in the U.S. took place within cities and their surrounding urban areas. This finding led the BPR to pursue further detailed studies of urban travel. In 1943 a study of auto trip origin and destination was made in Little Rock, Arkansas. In the early 1950's land use studies were combined with auto travel surveys by the BPR. But as yet these data had not been systematically employed in metropolitan area highway planning.

The Detroit approach was to take a new and more comprehensive approach to traffic analysis and highway planning. The governing board formed to guide the study included traffic engineers from Detroit, Michigan highway department engineers, city mass transit representatives, Detroit urban planners, and personnel of the Bureau of Public Roads. J. Douglas Carroll, Jr., was selected to direct the study. In early 1953 Carroll was heading a Flint, Michigan, study of shopping and other travel patterns for the Michigan University Institute of Social Research. Carroll came to the University of Michigan to pursue quantitative research in city planning, having earned in 1950 one of the first doctorates in city planning awarded by Harvard University with a Ph.D. thesis on the journey to work.

Based upon his background in urban planning, highway traffic study and survey research Carroll recommended making a large-scale survey of Detroit metropolitan area travel and land usage patterns upon which to base future Detroit highway planning. His idea was approved and in 1955 the study was completed [5]. During the course of the study Carroll gathered a core group of personnel, including: E. Wilson Campbell, highway engineer; and John Hamburg, systems analyst. At one point in time 300 persons were employed gathering data for the survey. Once the data were collected in Carroll's words,

"Then we had to decide what to do with the data. We had decided the critical issue was land usage and we wanted to look at highways as a system that served land usage needs, rather than just as a set of roadways some of which were over their capacity. The thread that ran through all of our thinking was an emphasis on moving people from place to place, not just the movement of vehicles."

"The BPR had encouraged this viewpoint but people who could use it were scarce. We hired Ladislas Segoe as a consultant. He was an old-time planner with imagination. He had the notion that highways were like a drainage system that took care of traffic flows instead of water flows."

"A major problem for us was to develop a way using our survey data to 'flood' a highway network, either an existing one or a possible future one. Our data processing equipment was primitive. In Detroit we were limited to card punches, sorters, accounting machines, readers, and an old French computer. We were forced to do much of the work manually and could undertake only a few lines of analysis because of this."[1]

Despite these limitations in two years the Detroit group completed a comprehensive land use and traffic study for the city of Detroit. As Carroll said of the Detroit study,

"It was successful. We wanted to do something better than the old-type highway studies. We were able to accomplish a lot in a short time. In the early 1950's you could get good people who worked hard. The spirit was there. Money was available. Costs were low. You could afford to take risks. I was able to get good people I knew from schools and from the literature."

"My preoccupation has been with doing research in real-life situations. In Detroit and in later studies I was able to use funds for risks as well as for straight, traditional delivery. The Detroit site brought out ideas constantly. The BPR was interested and contributed. Others came in to look at what we were doing. They both took ideas from us and gave ideas to us."

In 1955 Carroll was invited to undertake an even more ambitious urban study for the city of Chicago, a project that would stress systematic planning for the future of all transportation in the Chicago metropolitan area, for both private auto travel and mass transportation. The new study again had the support of the Bureau of Public Roads, who hoped highway plans would be developed for every major U.S. city. Influential State of Illinois and City of Chicago persons also supported the study, including leaders in the Illinois Highway Department and the Cook County Traffic division. Carroll accepted. As he put it,

"There was a can-do spirit in Chicago, the tradition that the city had 'broad shoulders' and could tackle major problems."

Thus from 1955 to 1961 a major effort was made to develop comprehensive transportation plans for Chicago. Wilson and Hamburg moved with Carroll from Detroit to Chicago. Key new personnel were added to this core including: Roger L. Creighton, holder of a Harvard masters degree in city planning, who had spent three years after graduation as planning director in Portland, Maine; Peter Caswell, another planner; L. E. Keefer, traffic engineer; Morton Schneider, holder of a graduate degree in physics and applied mathematics, as a systems analyst; and John Howe, a data processing specialist. Professions represented in the Chicago staff or in its linkages to other research-oriented staffs included: traffic engineering, city planning, transit engineering, census taking, data processing, demography, sociology, economics, and geography. Relationships were developed with the Cowles Commission for transportation economics, the geography department and the transportation group at Northwestern University, and the Illinois Institute of Technology [4; 133: 250]. Relations also developed with Carnegie Institute of Technology, and the Harmondsworth group in England, in addition to maintaining a standing relationship with the Bureau of Public Roads.

The picture that emerges from the Chicago Transportation Study is one of bringing to bear resources from a variety of locations and fields on a specific applied problem in a concentrated fashion. High energies were present. Resourceful efforts were mounted. Obstacles were overcome. Support was present in the form of funds and continued encouragement from sponsoring agencies.

During the six years from 1955 to 1961 the project team tackled and found solutions for a number of difficult data-handling and planning problems connected with the use of travel survey data in highway design. In the process computer simulation was developed at Chicago as a major applied tool for this purpose. For instance, a major stumbling block for urban highway planning work was the lack of an efficient method of generating the shortest path between two points in a complex network of city roads. Finding a solution to this problem was crucial to the generation of large numbers of realistic individual auto trips. The solution came through the Armour Research Foundation of the Illinois Institute of Technology. Armour was asked to work on the problem of finding, as an actual city driver would quickly do from experience, the shortest path between a regularly-used pair of origin and destination points in a city highway maze. In the course of their work the Armour people reported on a 1957 Harvard symposium paper by E. F. Moore which described an algorithm for locating the shortest path through a switching network [4: 250]. This algorithm proved usable. Then, however, the problem became one of finding a feasible way to make use of this trip generating capability. Manual methods were out of the question, and even the mechanical calculating approach used in Detroit would be overwhelmed by the magnitude of the computational and information-storage problems involved. If the method were to be used to real advantage, a fast and low-cost method had to be developed by which major alternative designs as to costs versus benefits could be compared, each of which would require a massive data-processing effort. General purpose digital computers had reached a stage of development where they could conceivably cope with the problem. But in 1957 no computer program existed to apply the Moore algorithm to the enormously complex Chicago network of 2,800 miles of streets and millions of daily vehicle trips via thousands of links. Moreover, computers of the period were not yet equipped with operating systems convenient for the user and thus were difficult to program. The Chicago programming problem was finally solved by an intensive and creative effort on the part of staff member Morton Schneider. The computer then being used in the Chicago study lacked the size necessary to accept the program. Therefore Schneider took the program to Cincinnati where he successfully operated the program on an IBM 700 series computer owned by General Electric. Subsequently, the Chicago project acquired a computer for their own use that was capable of generating the simulated trip data necessary for extending the study.

In addition to the trip generating problem the Chicago staff attacked and developed solutions for a number of other large-scale simulation problems, including: better visual output display devices, dividing trips into city (internal) and out-of-city (external) categories, forecasting land use, comparison of the relative advantages of alternative designs for highway systems, merging expressway with arterial highway planning, expressway interchange design, and joint planning of highways with rapid transit.

The diffusion of simulation effort that began with the shift from Detroit to Chicago continued in direct and indirect fashion during and following the completion of the Chicago study from 1955 to 1961. During the Chicago study period studies were commenced in Washington, D.C. (1955), Pittsburgh (1958), Minneapolis – St. Paul (1960), and Toronto (1959) [34]. All these later studies drew ideas and often moved personnel with experience from the Chicago study. Then in 1961 and 1962 a major migration of Chicago personnel occurred. A project was launched in 1961 to develop plans for six cities in upstate New York, beginning with the Niagara area comprising the cities of Buffalo, Niagara Falls, Tonawanda, and Lackawana [4: 319]. Roger Creighton became Director and John Hamburg, Associate Director of the Niagara Frontier Transportation Study 1961–1964. Subsequently they guided similar studies carried on for Rochester, Syracuse, Utica, Schenectady and Albany. In the process new persons were added to their staffs, some of whom like Eliot Rowe later migrated to still other locations to pursue highway-related simulations. Then, in 1966, Creighton and Hamburg jointly founded a firm to provide urban transportation consulting. Creighton opened an office in Delmar, New York, and Hamburg one in Bethesda, Maryland.

Meanwhile, in 1962, J. D. Carroll had left Chicago to assume leadership of the Tri-State Planning Commission located in New York City. Wilson Campbell, after serving as Director of the Chicago study from 1962 to 1968, moved to Albany to become Assistant Commissioner of Planning for the State of New York. Other moves included L. E. Keefer to Pennsylvania and Morton Schneider to San Francisco where in each case they continued to work on transportation planning problems.

The dominant mode of geographic diffusion of computer simulation as an intellectual technology, as represented by the growth and development of efforts beginning with the Detroit Metropolitan Area Traffic Study in 1953, was one of direct movement of people. Some transfer took place through the exchange of papers and in conferences. But the great bulk of the geographic transfer of applications occurred because individuals relocated.

Relocation took place for two major reasons. One was pull – the attraction of new major projects. The other was push. As projects neared completion not only did excitement drop but funds were less available as the job was probably considered done in the eyes of at least some sponsors. Other factors, by inference, were the desire for leadership of a project and the desire to be an independent consultant.

But despite various individual movements key members of the Detroit-Chicago staffs kept in close touch, serving as resources to one another and providing leads on projects. Their informal network relations thus remained active.

Diffusion of Concepts Across Substantive Topics

Computer simulation efforts within both the U.C.L.A. group from 1950 to 1954 and the Detroit group from 1953 to 1955 involved the diffusion of simulation across substantive topics. In each group the original topics were different. At

U.C.L.A. one of the subgroups that eventually became involved in traffic simulation was originally investigating electrical filter networks. This group was composed of four doctoral students headed by Professor Deforest Trautman. Trautman's office was next door to the office of Professor J. H. Mathewson, head of the U.C.L.A. Institute of Transportation and Traffic Engineering. The network group regarded themselves as applied mathematicians. They were intellectually and socially compatible and talked to each other ceaselessly on every subject under the sun. They attempted, among other things, to use mathematical analysis to investigate the properties of electrical filter networks but ran into mathematical intractabilities. So they built an electrostatic device to simulate transformations of the relevant mathematical expressions. That is, they used electrical fields to simulate, not filter networks, but the LaPlace transformations that were needed for inferring the networks' properties on the basis of mathematical models. An approach that reflected considerable ingenuity, since the simulator was a physical reality which was a second-order abstraction of another physical reality.

Another of the personnel associated with the Institute of Transportation and Traffic Engineering at U.C.L.A. was Daniel L. Gerlough. He had moved to U.C.L.A. in 1948 in order to work on the team creating SWAC, one of the earliest general-purpose digital machines. However, the SWAC project had not gotten underway by the time Gerlough arrived, so he took a temporary research job with the Transportation Institute. Temporary became less so after the expected job on SWAC was cancelled. Gerlough began modeling traffic behavior, and he thought about potential applications for simulators several times over the next few years. More than once, he wrote memoranda proposing that traffic flows could be simulated by means of electrical impedance networks.

The Network Group had been drawn together by a research contract from the Office of Naval Research. However, the contract funding was suddenly reduced in February 1953. Wishing to keep the Group together, the doctoral students began looking around for quarter-time jobs. Three of the students – Harold Davis, Jack Heilfron, and Arnold Rosenbloom – were hired by the Transportation Institute. Heilfron, who had worked on simulations of faults in electric power networks before he joined the Network Group, was asked to pursue traffic simulator studies, and three months later he proposed constructing a simulator. By September, the entire Group plus Gerlough were talking about traffic simulators.

One important factor in what happened was the stimulation of L. M. K. Boelter, Dean of Engineering. To quote four members of the Network Group:

"Boelter was a dean who was 20 years ahead of his time. I went to U.C.L.A. because of his interdisciplinary emphasis. He believed in mixtures across engineering fields and mixtures of engineering with the social sciences."

"Boelter was a kooky guy with kooky ideas. One was that we should solve some of our sociological problems with computers. . . . Our positions required that we respect Boelter's ideas. . . . Boelter was very perceptive of the consequences of the freeway system in southern California. He said, 'If you build a freeway, people will build houses along side it'."

"Boelter kept asking us when we'd be able to handle the traffic in some part of Los Angeles."

"The Dean needled Heilfron and Company. He tried to get people to do things. . . . One time the Dean had a meeting where he castigated people for having insufficient vision. I got really mad and asked him how we could come up with solutions before we had analyzed the problems, and he said, 'You can synthesize things without analyzing them'."

Meanwhile in 1953, in Detroit, Carroll's group was becoming deeply involved in conceptual problems posed by their metropolitan area traffic study. Even while immersed in travel survey data collection they were looking ahead to data analysis issues they would later confront. Simulation was not a concept they originally entertained to deal with such problems. Carroll met with Harry Goode at Willow Run in 1954 but later said that he did not fully grasp the import of Goode's ideas at the time. They also interacted with Ladislas Segoe, however, who did implant in their heads essential simulation ideas transferred from the field of hydraulics. Segoe pointed out that highways and pressurized water systems might have common properties. Both were subjected to varying demand conditions which caused important dynamic consequences for the behavior of the system. Both highways and water lines were characterized by intersections and, at times, by two-way flow conditions. Might it be possible to take advantage of these analogous properties? The Detroit group came to think so and began to incorporate these concepts into their thinking. The consequences were not immediate, partly because of their necessary preoccupation with other pressing project delivery deadlines and partly because they perceived no practical way to cope with the problems of representing traffic behavior either in a mathematical model or in a physical model. As the work progressed, however, first in Detroit and later in Chicago, and as they began to depend increasingly on computing equipment for data-handling purposes the real possibilities of machine simulation became more apparent to them. And they acted. In 1957 they simulated highway travel on a computer using as input survey-data on individually generated trips and fitting output to actual observed travel patterns.

Diffusion of Applications Across Substantive Topics

Our findings on diffusion of applications across substantive topics depend on inferences about work initiated by a group located in Pittsburgh and which carried out a transportation study for that city. The data are drawn from the history of computer simulation of human behavior by Starbuck and Dutton [6]. Inspection of the simulation history study turned up 26 Pittsburg-group publications involving thirteen different individuals as authors that dealt with eight

Table 2. Trace of diffusion of applications across substantive topics (table entries are numbers of publications)

Topics	1961	1962	1963	1964	1965	1966	1967	1968	1969	Totals
1. Taxation	1									1
2. Highways		2	1	2	1					6
3. Industrial siting		1				1				2
4. Urban renewal					1	1		1		3
5. Land use			1			1	1	2	1	6
6. Regional transportation						1	1		1	3
7. Budgeting							1	1	1	3
8. Public transportation								1	1	2
Totals	1	3	2	2	2	4	3	5	4	26

distinct applied topics. (See Appendix A.) These topics include municipal taxation, municipal budgeting, urban highway planning, urban public transportation, urban land use, industrial site location, urban renewal, and regional transportation. A trace of diffusion across these topics is shown in Table 2.

The history study classified each of the total 1921 simulations in two ways: (1) into one of five categories (A individuals, B interaction, C aggregation, D aggregation and interaction and G general), and (2) into a type which described its scientific level (types 1 to 8 are in decreasing order of scientific validity; type M is methodological). The Pittsburgh site publications fell across study categories and types as shown in Table 3. Compared to all highway-related studies the Pittsburgh studies were distributed over more categories but contained fewer highly rigorous scientific studies and more methodological and general discussion pieces. Thus one result of diffusion across substantive topics may be the generation of lower quality studies, at least initially.

Clearly the entrepreneurship of the Pittsburgh group aimed at the transfer of simulation as a technology to a wide variety of topics. This transfer extended not only to a range of topics within the content of Pittsburgh city administration but also to geographic spread to different projects in new locations such as the

Table 3. Classification of Pittsburgh site simulations by category and type

	Study Type			
Study Category	1-3	4-7	8-M	Totals
A				
B	3			3
C	1	3		4
D		5	10	15
G			4	4
Totals	4	8	14	26

Northeast Corridor transportation studies. There is evidence that this transfer was stimulated by needs for funding to support consulting activities by key principals and staff of the group.

Diffusion Across Computing Machines and Peripheral Devices

Diffusion across computing machines and peripheral devices occurred in all highway-related simulation sites previously discussed, including: U.C.L.A., Harmondsworth, Detroit, Chicago, and Pittsburgh. Sites that emphasized concepts tended to be more involved in exploring the relative advantages of analogue versus digital machines. And at U.C.L.A. and Harmondsworth different devices were actually constructed to test their performance. Chicago retained an outside consultant, Armour Research Foundation, to work with Peter Caswell from their own staff in developing analogue methods for traffic simulation. Although working analogue devices were assembled, and at U.C.L.A. one such device was awarded a patent, none of these devices withstood the competition from rapidly-evolving, general-purpose digital computers. At Harmondsworth, the slow working rate of their homemade Automatic Delay Computer, a digital device made from telephone relays, soon led them to transfer their operations first to the Automatic Computing Engine at the National Physical Laboratory and then to the British Pegasus II computer.

Peripheral output devices also received attention, especially at Chicago. There the group sought a more rapid way to examine the effects of alternative highway locations through the development of cathode-ray tube visual display output devices. Such a device, called the Cartographatron was developed, again with consulting aid from Armour [4: 34 and private communication from J. D. Carroll]. This device worked and was used to judge relative effects of different location alternatives. But the group relied on numerical printed and punched output for final analysis and documentation.

With the passage of time, simulation at these various sites became wholly dependent on the availability of larger, faster digital computers. The work increased both in complexity and in input-output requirements. Simulation at each of the sites moved steadily across successive generations of computers. Our data on computing budgets are sparse. Our impression is that total data-processing budgets rose but not nearly in the ratio of total data processing accomplished. Not only did unit data processing costs drop but programs and simulation methods became more efficient.

CONCLUSIONS

Beyond what the previous discussion already states we draw conclusions about diffusion of simulation as an intellectual technology on eight influence factors:

(1) availability of fund support, (2) visibility of consequences, (3) social support, (4) work site characteristics, (5) task group composition, (6) promotional agents, (7) communications media, and (8) computing hardware and software.

(1) *Fund support* Funds are important, both direct and indirect funds. Many of the projects discussed did not initially receive support for simulation itself. Funds came from budgets available at academic and field sites either for general purposes at the discretion of administrators or for specific applied results where simulation was one among a number of competing intellectual approaches to the problem. Later, as simulation became more respected, final applications included simulation as an explicit method. Thus this *avant guarde* technology was largely brought into being through internal group stimulation rather than because of externally imposed budget pressure or social recognition.

Levels of budgets in every instance where simulation flourished were less than handsome but more than meagre. Budget pressures were real and simulation had to justify itself in terms of results not promises. Thus a state that we term "minimal affluence" was present in successful work sites. Workers had enough but not too much budget to play with, and time competition both for applied results and for theoretical findings was real. These workers were conscious that other people were working at these same problems, and that sponsors had their necks stretched out – at least to a degree. These places were not like ivory towers.

Funds had both push and pull effects. At U.C.L.A. a reduction in O.N.R. funding pushed the network group out of their original line of work and transportation funds pulled them into highway simulation. Not without an intervening judgement about their competence, however, probably by both Mathewson and Boelter. For while the Transportation Institute had funds, it undoubtedly had available other personnel and projects from which to select. The network group was a wild bunch. But imaginative academic research administrators like to play hunches from time to time; they also feel responsibilities toward advanced doctoral candidates and towards intellectual innovation. J. D. Carroll harbored similar views, even in an applied setting. He said about one staff member, "He was in many ways a social misfit, but he had a penetrating mind and could produce huge bursts of creative energy." Carroll also said, "I was able to obtain some funds I could donate to efforts that were not limited to immediate delivery of project results. We were able to try things."

Clearly in Detroit, and later in Chicago, the pull of new funds from new locations, and in Pittsburgh the pull of new funds for new topics were major causes of diffusion of simulation. We further infer that the push of decreasing funds was present to some extend in these applied sites, as was true at U.C.L.A. In summary then the influence of funding depends on a push-pull process and upon a state of minimal affluence.

(2) *Visibility of consequences* A point emphasized by Ben-David [1] and by Gordon and Marquis [17] is that productive research sites are those where visibility of consequences is high. Gordon and Marquis found that visible sites are likely to be marginal ones, by which they mean close to the point of practice rather than close to places of purely academic inquiry. That finding also holds for our data on computer simulation development. Productive sites, in terms of quality as well as quantity of published results, were those close to or aimed at applications. Of course we do not know about unpublished results, but our assumption was that people who had something to say found a way to say it.

Although academic sites were not dominant, in every productive site academicians were either on the scene or close to it. Simulation efforts had academically trained leadership and drew on this training to recruit talent, to draw on the literature, and to retain consultants. Moreover, in many instances, these sites provided intensive field experience for advanced graduate students, often leading directly to the identification of earned degree research topics.

(3) *Social support* Social support was especially important at sites where concept diffusion took place. At U.C.L.A. Mathewson and Boelter both provided such support, but each gave a different type. Mathewson simply encouraged Gerlough, while Boelter at times openly needled the network group to get going on a realistic traffic problem and also induced Davis to offer a course on "The flow of discrete entities". As a result of teaching this course Davis began talking to Gerlough about how to simulate traffic on a digital computer. Gerlough proposed that a traffic simulation should be his own doctoral dissertation and enthusiastically went to work. Gerlough got further intellectual reinforcement from two non-U.C.L.A. sources. He went to Michigan to visit Harry Goode, and it is likely that the support was mutual. In 1951 Goode had cited, as one of simulation's potential uses, the estimation of the proportion of drivers who exceed a specified speed limit. However, it is not clear that at the time Goode intended to implement such a model or envisioned its potential. Gerlough then heard that a man named Davies, who had worked on traffic simulation, was visiting U.C.L.A. He went to see Davies and discovered Davies had been a key participant in a British traffic simulation that strongly resembled Gerlough's own. In fact it was the Harmondsworth Road Research Laboratory initiated effort. [20]

Strong social support from other group members, a culture that encouraged ideation, and freedom from unwanted distraction characterized much of the work at various simulation sites. Members of Carroll's group commented on the idea broth climate present in Detroit and in Chicago. Carroll's group received visits from members of the staff of the Bureau of Public Roads. The BPR also encouraged the discussion and diffusion of simulation project results through its periodic conferences and published proceedings. As Carroll said, "The BPR served as a 'typhoid Mary' in spreading ideas about highway and traffic simulations".

Clearly the Harmondsworth group was also prone to travel. Not only did Davies visit the U.S. but Hillier and other Road Lab staff visited Commonwealth locations like Australia. Australian highway conference proceedings carried their contributions, and later highway simulation studies began to appear in Australian journals, again studies that were most often generated in applied sites.

(4) and (5) *Work sites* were characterized by their applied locations in or near major urban areas, egalitarian relationships, leaders who pressed for results, high levels of staff member education and field practice, high energy levels, frequent in-group communication, frequent communication with other sites, frequent visits from behind-the-scenes sponsors and gadflys, and high attendance at professional meetings. *Task group composition* was closely related to work site characteristics. For instance, site location in major urban areas probably influenced self-selection of staff members as well as sponsor selection choices. Such locations are desired by some persons and not by others, and urban areas tend to be locations for major educational institutions.

(6) *Promotional agents* In highway simulation an operating government agency, the Bureau of Public Roads, played a major role. Its representatives made regular field trips. Comments of urban highway study staff members confirmed this, and also revealed that the very launching of urban highway studies might not have occurred without BPR's long-standing efforts to anticipate highway development needs. The BPR had long supported highway research and was one of the major architects behind federal legislation that established the Highway Trust Fund, one of whose provisions was to require the states to invest in planning as a condition for receiving highway construction money.

Another type of promotional agent was also discernible from the record, however, and that was the "gadfly academic". One thread that ran through the literature was the role of persons who wrote, "think pieces", who traveled to meetings, who quickly adopted new fashions in intellectual methodology, who applied for fund grants, who tried to interest doctoral students in *avant guarde* topics, and who held conferences on newly developing lines of inquiry. At first, in our studies of computer simulation we gave these efforts short shrift, thinking that these individuals were not major contributors to the advance of this new intellectual technology. Now we are less certain about this conclusion. These contributions are still not our favorites for prize awards. But this type of behavior may play a significant role in the development and diffusion of new intellectual technology. We have observed elsewhere [27] that work in all simulation categories began with methodological studies, and that there was a group of people who published long series of solely general and methodological pieces. We also observed a steady substitution over time of high-quality for low-quality scientific studies, with the proportion of methodological and general pieces becoming constant with time. It was as if these nonscientific pieces were acting as a filter membrane to drawn an increasing fraction of pieces into scientific space. Instances of such membranes are common in chemistry and in biology where ionization causes a pressure differential between two sides of a porous partition, and a one-directional filtrate flow results. Gadfly's in new intellectual technology may comprise part of a similar social membrane.

(7) *Communication media* that diffused highway-related simulations were in order of importance: (a) face-to-face meetings at applied sites, (b) conferences among highway researchers and field engineers, (c) working papers and correspondence, (d) conference proceedings, and (e) academic journals. The needs during the early period when highway simulation was emerging were for a common language and for social support. Face-to-face visits were more productive in serving each of these needs. Then working papers became a rapid and precise way for simulation workers to exchange ideas and developments. Conference proceedings also served a specific audience and were published rapidly. Publications in academiç journals involved not only long time-lags but also segmented academic readers working on similar problems into different audiences. Note that once highway traffic simulation proved feasible it diffused in only a few years between 1954 and 1958 across numerous applied geographic sites. It was some time thereafter, approximately three years later, that academic journals reflected the facts of this diffusion (see Figure 1).

(8) *Computing hardware and software* were not only a means for but also a stimulant of highway-related simulations. Gerlough was an applied mathematician who came to U.C.L.A. to work on the creation of the SWAC computer, and then developed an interest in traffic simulation. Not the other way round. A more common effect, however, was that large scale digital computer development in the 1950's made possible a level of data processing capability that people trying to do simulation already saw that they needed and thus quickly acquired.

Our findings about the diffusion of intellectual technology agree in many ways with aggregate level findings of others regarding the socio-economic processes involved in diffusion of innovations. When these processes are disaggregated, however, our data suggest that many important primary effects and secondary interaction effects are present that have not been explained, at least for the case of diffusion of intellectual technology.

Only a few persons such as Usher [31; 32] and Flueckiger [9] have examined disaggregate effects on a factual, case study basis, and these studies have focused on production and product technology more than on intellectual technology. We do not now have a strong theory to offer. But our conclusion is that one is

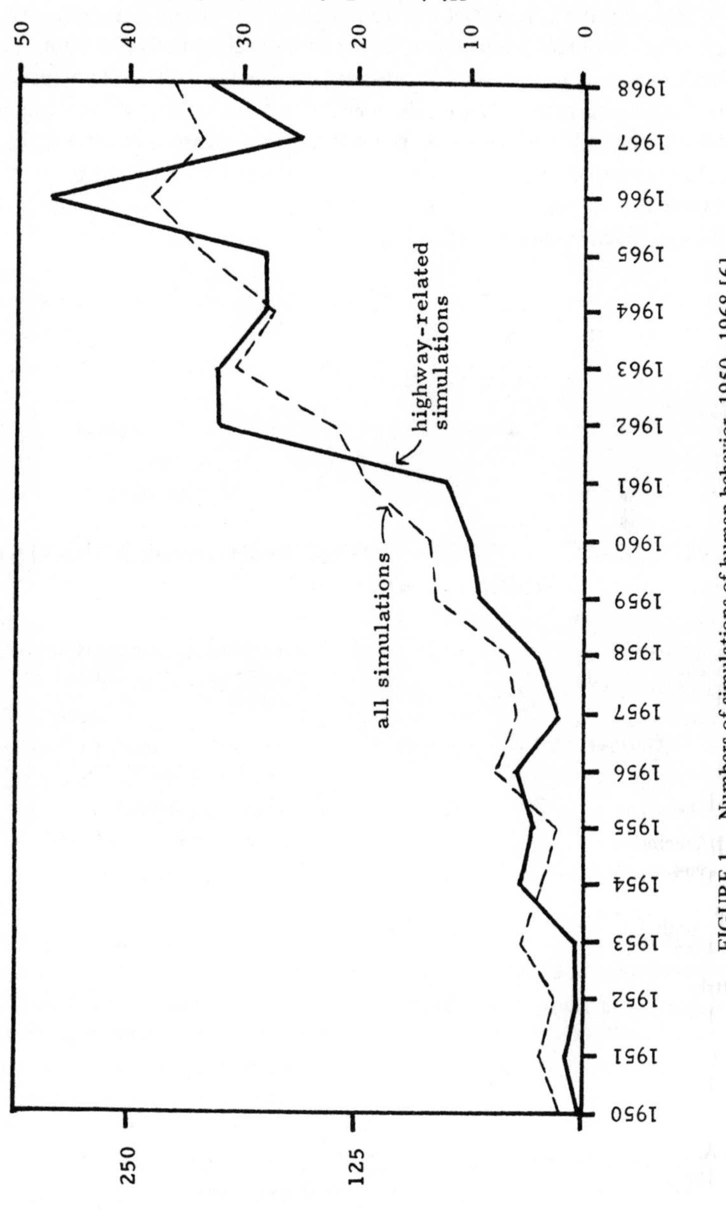

FIGURE 1 Numbers of simulations of human behavior, 1950–1968 [6].

needed. Present theories are based heavily on contradictory notions. One group of theories says that inventions are outputs of an evolutionary process. But a transistor is not a fruit fly; it has no potential for changing its own physical structure. Obviously humans are the critical variable. Individuals interact with tools developed, not only by themselves, but also by others. Humans change as a result, and one result is that each generation of users not only relearns but often adds to the total stock of tools. But present models are silent about this process. Another group of models introduces randomness into the innovation and diffusion process. But by definition random events have no cause. That premise doesn't hold out much hope for explaining effects.

ENDNOTE

[1] Quoted material is based on field interviews.

APPENDIX A. CITATIONS FOR PITTSBURGH GROUP SIMULATIONS FROM STARBUCK AND DUTTON, 1971

W. A. Steger 1961	8D (municipal taxation) Simulation and tax analysis, a research proposal.
Pittsburgh Department of City Planning 1962	MG (city highways) Data processing and simulation techniques.
I. S. Lowry 1963	8D (city highways) Location parameters in the Pittsburgh model.
S. H. Putman 1963	5D (industrial siting) Industrial location model.
J. P. Crecine 1964	5D (urban land use planning) TOMM (time oriented metropolitan model).
Pittsburgh Area Transportation Study 1963	2C (city highways) Final report, Volume 2: forecasts and plans.
N. J. Douglas, Jr. and W. A. Steger 1964	5D (city highways) Choices (s.d.s.e.) in a large-scale modeling effort: the Pittsburgh simulation model.
W. A. Steger and N. J. Douglas, Jr. 1964	8D (city highways) Simulation model.
W. A. Steger 1965	MG (city highways) Review of analytic techniques for the CRP.

W. A. Steger 1965	8D (urban renewal) The Pittsburgh urban renewal simulation model.
W. W. Bruck, S. H. Putman and W. A. Steger 1966	MD (Northeast Corridor transport) Evaluation of alternative simulation proposals: the Northeast Corridor.
S. H. Putman 1966	6C (industrial siting) Intraurban industrial location model design and implementation.
W. A. Steger 1966	5C (urban renewal) Analytic techniques to determine the needs and resources for urban renewal action.
W. A. Steger 1966	MD (urban land use planning) The realities of simulation exercises for city planning.
J. P. Crecine 1967	1B (municipal budgeting) A computer simulation of municipal budgeting.
D. D. Lamb 1967	MD (urban land use planning) Research of existing land use models.
S. H. Putman 1967	8D (Northeast Corridor transport) Modeling and evaluating the indirect impacts of the Northeast Corridor Systems.
J. P. Crecine 1968	1B (municipal budgeting) A simulation of municipal budgeting: the impact of problem environment.
J. P. Crecine 1968	7D (urban land use planning) A dynamic model of urban structure.
J. P. Crecine 1968	MG (urban land use planning) Computer simulation in urban research.
K. W. Heathington and G. J. Rath 1968	MG (public transportation) Computer simulation for transportation problems.
I. S. Lowry 1968	MD (urban renewal) Seven models of urban development: a structural comparison.
J. P. Crecine 1969	1B (municipal budgeting) Governmental problem solving: a computer simulation of municipal budgeting.
J. P. Crecine 1969	5D (urban land use planning) Spatial location decisions and urban structure: a time-oriented model.

J. S. Drake and L. A. Hoel 1969	MD (regional transportation) Some modeling considerations in statewide transportation planning.
K. W. Heathington and J. M. Bruggeman 1969	7C (public transportation) The use of computer simulation to analyze a demand-scheduled bus system.

REFERENCES

[1] Ben-David, J. and Zloczower, A., "Universities and academic systems in modern societies", *European Journal of Sociology,* **3:** 45–84, 1962.

[2] Campbell, E. W., "A mechanical method for assigning traffic to expressways", *Highway Research Board Bulletin,* **130:** 27–46, 1956; also *Traffic Engineering* **28,** No. 5: 9–14, 35, 1958.

[3] Cole, J. R. and Cole, S., "The Ortega hypothesis", *Science,* **178:** 368–375; 1972.

[4] Creighton, R. L., *Urban Transportation Planning.* Urbana: University of Illinois Press, 1970.

[5] Detroit Metropolitan Area Traffic Study, Report on the Detroit Metropolitan Area Traffic Study, Part II -Future Traffic and a Long Range Expressway Plan, March, 1956. Detroit: Detroit Metropolitan Area Traffic Study, 1956.

[6] Dutton, J. M. and Starbuck, W. H., "Computer simulation of human behavior: a history of an intellectual technology", *IEEE Transactions on Systems, Man, and Cybernetics, SMC–1,* **2:** 128–171, 1971.

[7] Dutton, J. M. and Starbuck, W. H., *Computer Simulation of Human Behavior,* New York: Wiley, 1971.

[8] Federal Highway Administration, *Stewardship Report on Administration of the Federal-Aid Highway Program 1956–1970,* Washington: U.S. Department of Transportation, 1970.

[9] Flueckiger, G. E., "Observation and measurement of technological change", *Explorations in Economic History,* **9,** No. 2: 145–177, Winter 1971–1972.

[10] Gerlough, D. L., Analogs and simulators for the study of traffic problems, in *Proceedings of the Sixth California Street and Highway Conference:* 82–83, Los Angeles: University of California Institute of Transportation and Traffic Engineering, 1954.

[11] Gerlough, D. L., Simulation of Freeway Traffic on a General-Purpose Discrete Variable Computer, Ph.D. dissertation, Los Angeles: University of California at Los Angeles, 1955.

[12] Gerlough, D. L., "Simulation of freeway traffic by an electronic computer", *Highway Research Board, Proceedings,* **35:** 543–547, 1956.

[13] Gerlough, D. L. and Mathewson, J. H., "Approaches to operational problems in street and highway traffic – a review", *Operations Research,* **4,** No. 1: 32–41, 1956.

[14] Glanville, W. H., "Road safety and traffic research in Great Britain", *Operations Research,* **3,** No. 3: 283–299, 1955.

[15] Goode, H. H., "Simulation – its place in system design", *Proceedings of the IEEE* (IRE), **39:** 1501–1506, 1951.

[16] Goode, H. H., Pollmar, C. H. and Wright, J. B., "The use of a digital computer to model a signalized intersection", *Highway Research Board, Proceedings,* **35:** 548–557, 1956.

[17] Gordon, G. and Marquis, S., "Freedom, visibility of consequences, and scientific innovation", *American Journal of Sociology,* **72,** No. 2: 195–202, 1967.

[18] Gruber, W. H. and Marquis, D. G. (Eds.) *Factors in the Transfer of Technology,* Cambridge: M.I.T. Press, 1969.

[19] Hughes, J. R. T., *Industrialization and Economic History: Theses and Conjectures.* New York: McGraw-Hill, 1970.

[20] Hillier, J. A., Whiting, P. D. and Wardrop, J. G., The Automatic Delay Computer, Harmondsworth, Middlesex: Road Research Laboratory, Research Note 2291, 1954.

[21] Mathewson, J. H., Trautman, D. L. and Gerlough, D. L., "Study of traffic flow by simulation", *Highway Research Board Proceedings,* **34:** 522–530, 1955.

[22] Mansfield, E., *The Economics of Technological Change,* New York: Norton, 1968.

[23] Price, D. J. de S., "The structures of publication and technology", in Gruber, W. H. and Marquis, D. G. (Eds.), *Factors in the Transfer of Technology:* 91–104. Cambridge: M.I.T. Press, 1969.

[24] Rogers, E. M. and Shoemaker, F., *Communication of Innovations,* 2nd ed., New York: Free Press, 1971.

[25] Rosenberg, N. (Ed.), *The Economics of Technological Change,* Harmondsworth, Middlesex: Penguin, 1971.

[26] Schon, D. A., *Technology and Change,* New York: Delta, 1967.

[27] Starbuck, W. H. and Dutton, J. M., "The history of simulation models", in Dutton, J. M. and Starbuck, W. H. (Eds.), *Computer Simulation of Human Behavior:* 9–102, New York: Wiley, 1971.

[28] Stark, M. C., Computer Simulation of Street Traffic, Washington: U.S. Department of Commerce, National Bureau of Standards Technical Note 119, 1961.

[29] Trautman, D. L., Davis, H., Heilfron, J., Ho, E. C., Mathewson, J. M. and Rosenbloom, A., *Analysis and Simulation of Vehicular Traffic Flow,* Los Angeles: University of California Institute of Transportation and Traffic Engineering, Research Report No. 20, 1954.

[30] U.S. Department of Transportation, *1972 Highway Statistics,* Washington: U.S. Department of Transportation, 1972.

[31] Usher, A. P., *A History of Mechanical Inventions,* New York: McGraw-Hill, 1954.

[32] Usher, A. P., "Technical change and capital formation", in National Bureau of Economic Research, *Capital Formation and Economic Growth:* 523–550, 1955.

[33] Utterback, J., "Innovation in industry and the diffusion of technology", *Science,* **183:** 620–626, 1974.

[34] Zettel, R. M. and Carll, R. R., *Summary Review of Major Metropolitan Area Transportation Studies in the United States,* Berkeley: University of California Institute of Transportation and Traffic Engineering, Special Report, 1962.

THE DUTIFUL DREAMER: REPRESENTATIONS IN MACHINES AND MORTALS

CHRISTOPHER R. LONGYEAR
University of Washington, Seattle

This paper deals with *representations* in artificial and natural cybernetic systems. Its relevance to communication and (self-) control in social processes may be considered from several levels. On an elementary level, even the most rudimentary devices must have a representation of their worlds of interaction for the most trivial communication and control. The manifestation of representations in various complex devices capable of richer interactions require correspondingly richer representations. On this first level, we can ask questions concerning the nature and limitations of representations.

At a second level, we find that rich enough systems must be able to represent recursively their own first-level representations (and those of others); furthermore, they must be able to represent recursively themselves (and others) representing themselves. (We note that von Foerster's address, "The Cybernetics of Cybernetics", [in this volume] finds here a representation of a representation.) We can ask questions at this second level concerning the nature and limitations of such recursively self- and other-representing representations.

At a third level, we find that second-level representations affect the communication and control processes of particular groups and of society in general. We can ask questions at this third level concerning the nature and limitations of second-level representations with respect to particular informational groups and human society in general.

First of all, what *is* a representation? Let us consider the once-humble thermostat. In his discussion of "the standard illustration" of negative feedback, Donald MacKay makes the following comment:

> The point of special relevance here is that insofar as an adaptive response is successful, the automatic movements of the goal-directed control-centre will in quite a precise sense mirror the current state of affairs, or rather those features of the state of affairs to which it is responding adaptively. The state of the control-centre is in effect a representation of those features of the state of affairs, just as much as a code-pattern of binary symbols could be. In each case we need a principle of interpretation, provided in one case by knowledge of the

evaluatory criteria in the other by the code book. In each case, given these, we can infer certain things about the state of affairs represented [28: 63].

What I want to emphasize at the moment is not merely that the state of the bimetallic switch is a representation of the thermostat's world of relevance. I hope we can take that for granted. I want to underscore that that representation is the *only* view of its world that the thermostat can have. To put it another way, the bimetallic switch has two states, say: open or closed. The thermostat's representation of its world, then, consists of two alternatives; either the world it represents is too cold or it is hot enough. (We are attributing labels that are meaningful to us here, but feel free to substitute any labels you like. For examples of keeping such distinctions very clear, see Meredith [33] or Maturana [30].) As long as the thermostat has no more representational ability than that, it cannot represent anything more elaborate to itself. It is consequently notoriously easy to fool. All we need to do is set a candle under the mechanism, for example, or to misconnect the wires. I want to emphasize the notion that the states of the thermostat are, in a sense, its total world. Thus, for example, the thermal (and only) world of a thermostat is represented as too cold – and the thermostat continues to believe it is too cold – when someone has stuck a matchstick between the contacts of the bimetallic switch. The thermostat will continue to believe it is too cold even if the house burns down, destroying thermostat and all. We must accept MacKay's claim that any adaptive mechanism must have some means of representing its world to itself.

At a different level, we may notice that people fiddle with thermostats. Twiddling may be due to an attempt to compensate for the device's inability to represent a world more complex than the either-hot-enough-or-too-cold world. Such a slightly more complex world might be one that recognizes the time of day (i.e. lower the heat at night) or the days of the week (i.e. stay cool on weekends). What such modifications amount to is a more complex interaction of systems. Thus, the clock-calendar system people use at home or at work may interact with the thermal world of the simple thermostat as a very simple mechanically coupled clock, or the thermostat might be adjusted by a vastly more complex human being, who is himself interacting with any number of complex systems, such as the global ecological system or the national economy. In the case of the clock-coupled thermostat, we might be tempted to think of the clock mechanism simply exercizing a hierarchically superior control over the thermostat, but we should remember that sometimes the thermostat overrides the clock, as it may when the temperature falls below some nocturnal minimum. However we think of the elaborated thermostat, we surely agree that a device which can represent and therefore interact with time as well as temperature is a more capable device than one which responds only to temperature. Anything able to change purposefully the thermostat's setting requires a system that can at least represent the alternatives as well as the concept of alteration. Such systems are at what we might call the second level of representations, for they must be able to represent the representations of their world of interaction.

Still another level must be invoked to notice that what people decide to do to their thermostats may have interaction with other systems (as with the ecology, with the economy, or with society in general). Anything able to represent the activities of second-level representations needs to represent the changing of representations of representations of alternative representations. This we might call the third level of representations. We have now traced superficially a sequence starting with simple representations of a cybernetic system, ending with recursively self- and other-representing representations capable of representing their own and others' modifications. We now return once again to the first level of representations as we turn our attention away from the fuel crisis to consider a more elaborate automaton. I shall next briefly describe a class of automata that is built up of elements which have the following properties. (See Figure 1.)

Each element is a little automaton in its own right. Each of these components consists of a set of strategies, procedures, processes, or programs (p) and a set of confirmations, conditions, contexts, or circumstances (c). Each element employs a similar strategy: each tries to satisfy its (c), using whatever (p) it happens to have stopped at the last time. The element keeps on trying one p after another, recycling through its entire repertoire of procedures until some set of its conditions are met. There are also inputs and outputs, some of which could be sensors

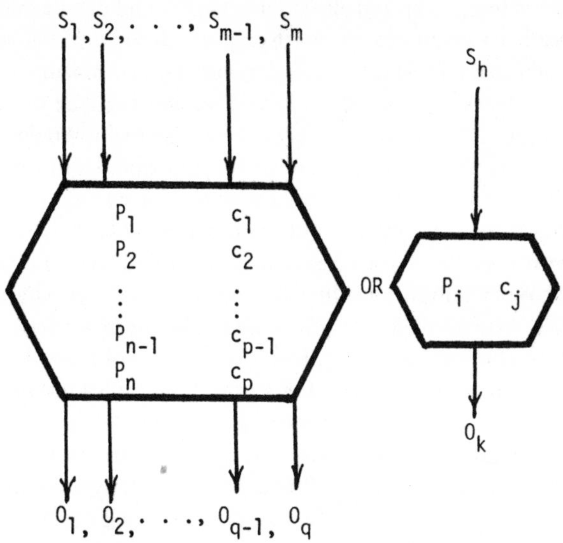

FIGURE 1 Dutiful Dreamer Component

or effectors. Notice, however, that none of these elements has an "off" switch; the only way an element can stop is to satisfy some set of its conditions.

If we now construct an increasingly complex assembly of such elements, we notice that any subassembly can in turn also be thought of as an element, with an admittedly more elaborate set of p's and c's. Suppose we go on assembling a more and more richly structured device which includes some sensors and some effectors. It begins to look like a robot that would, under the right conditions, become more efficient as its components stopped at the individual p's and c's which worked best.

Now, alas, there are also conditions that are *not* the right conditions. Let us look at some of these. For any element, an input that triggers it on will cause that element to go out of service until its particular conditions are satisfied. Its only effect until that time is to keep sending more and more signals out, hoping to find a situation in which its conditions can be satisfied. Furthermore, whenever the rate of input triggering signals s_i equals or exceeds the rate at which the element can satisfy its c_i, the element not only goes out of service, but it becomes more stupid, for it loses its previously learned most effective (p_i, c_i). If it is lucky, it may eventually regain that (p_i, c_i), but it is more likely to stop at some random (p_i, c_i) that happens to coincide with the satisfaction of that c. (Remember that we are here referring to a situation in which the triggering signals are *not* appropriate.) What is true for individual elements is also true for larger assemblies.

In spite of von Foerster's polemics against anthropomorphism, let me sin anthropomorphically for just a moment. Because of those polemics, let us agree that our lapse is temporary and superficial, and should not be taken too seriously [42]. To create a communicative context in which some of the automaton's functions might make immediate sense, let me ask you to think of some fairly elaborate set of usually automatic bodily movements related to the physical world. The act of sitting down can serve as our paradigm. We are able to sit down on an incredible variety of surfaces, in an amazing number of different circumstances, with an impressive repertory of postures. (Compare, for example, the automatic adjustments we make to sitting down on an unfamiliar upholstered airplane seat with those we make when using the more familiar hinged annual ring of Thomas Crapper's invention.) In short, we are awfully good at sitting down. Yet, compared with all the other things we do, how often do we sit down? Even for those of us who lead sedentary lives, the act of sitting down uses up a truly infinitesimal fraction of our days. Our robot, should it have learned to sit down, would soon forget how, for an inappropriate trigger to its sitting-down component would more often than not result either in that component cycling endlessly in its effort to satisfy its built-in conditions, or it would, by accident, think it had satisfied its conditions because of further random signals, in which case it would probably stop at some less appropriate p_i.

We would like our robot to learn to do things like sitting down, but we do not want it to become increasingly stupid as the rich variety of the world (or of its own internal interactions) sets components going in situations where individual c's cannot be satisfied. Can we arrange things suitably? An obvious way of reducing the inappropriate triggers is to reduce all the signals from the sensors. By reducing s_i's, we obviously reduce the ratio of the rate of s_i to the rate at which c_i can be satisfied, and therefore we lessen the situation that leads to growing stupidity. But consider a possible consequence: if we really shut off s_i inputs, some of which may be necessary to satisfy particular c_i's, there will be some elements that will never be satisfied. Thus, even when receptor signals are discontinued, the robot may not recover fully. To return to our anthropomorphic terms again, we may note that our robot may feel better when it goes to sleep, but it will remain stupid in some ways.

Suppose that we now deliberately produce a set of false signals whose purpose is to restore the otherwise jammed or mis-set elements to their effective p_i's. We could put it this way: by dreaming appropriately, our robot will stop being stupid. These "dreams" are in their own way also representations. They represent those skills preserved in the face of probabilistic evidence to the contrary. Economists may prefer to call deliberately false data "lies", but let me call them "dreams", for these "dreams" need not always be bad, though we might sometimes consider such dreaming a malfunction if the system were unable to tell the difference between false and genuine signals. It should be clear that he added capacity for "dreaming" achieves nothing by itself. But if control is exercized in the *selection* of dreams, then clearly a very important result in control is exerted both on the development of the robot and on its general activities. Such dream-control is clearly a second-level phenomenon. Just for the purposes of suggesting a third-level issue, we might note that the social consequences of such control, were it to involve members of a society, could be staggering. Would a society accept genetically transmitted "dreams", but not control of transmission exercized by an elite? Would a society tolerate the imposition of such "dreams" on uncooperative members?

Before leaving our dutiful dreamer, whose duty it is to dream to prevent becoming ever more stupid, and whose unwilling but inescapable duty it might become if imposed externally, let us look briefly at biological systems for suggestive parallels.

Michel Jouvet at Lyon [17; 18; see also bibliography in 16], has shown that sleeping cats dream. By surgically destroying the system that inhibits the bodily movements of dreaming cats, Jouvet has demonstrated that very young cats dream more than old cats; he has shown also that what they dream about is instinctive behavior that is otherwise not yet manifested. Thus, for example, a very young cat will dream about lapping before it does it; it will mentally practice fighting off a large assailant before it ever meets one. I will not speculate here on the enormous importance these findings have for psychology; I will only emphasize that such dreams are clearly representations. For the cat, it is biologically useful to practice its actions before the real world, with its compound hazards, demands them.

The only realistic source for such dreams I can think of must be genetic. We can imagine the consequences of modifying instinctive behavior and perhaps thereby the fundamental nature of cats, could we but control the transmission of dreams. To do so would immediately involve us in third-level considerations, for feline social structures would be inevitably affected by such manipulation. My point in raising these matters here is that we have a natural manifestation of what we found necessary in machines. The added observation that these representations occur in pre-verbal mammalia will do our claim for their necessity in verbal systems no harm.

Let us attend to human brains for a moment. One of the artifacts of that brain, we sometimes forget, is natural language. A product of no one knows how many brains interacting over uncounted centuries, human language may well turn out to be at once perhaps the largest and most complex system ever devised, perhaps the most crucial element in large social systems, and almost certainly the most essential part of knowledge structures in individuals and in human societies. Its nearest rival in complexity may be the human nervous system itself, impressive both for the huge number of elements out of which it is composed and for the enormous variety of complex structures it has. It is my intent here only to remind you that the human brain works by means of *memories*, that is to say, necessarily by means of *representations*. It should be obvious that the representations have to be richly structured enough to store the memories they do, and have to be adaptable enough to keep pace with a varied experience.

Let us now turn to natural-language question-answering computer systems. A set of assumptions and interests are represented by the discipline of computational linguistics as it borders on artificial intelligence. Those faced with the practical problems of getting a computer to answer questions in an actual language like English tend to approach the problem of implementing a grammar on a machine with a somewhat less doctrinaire view than that which characterizes much linguistics. Much current formal rationalistic linguistics, like that of the older structuralist school, is just not interested in the relation of language to belief systems or to the real world, nor is it concerned with matters of social interaction, nor does it choose to account for the appropriateness of a verbal event in a particular context or situation. [See 5; 9; 10; 11; 19; but cf. 8; 14; 15; 22]. In the first place, getting actual answers to real questions immediately plunges one into the murky area of particularity versus generality. We are in the heart of performance as well as in competence; or, to revert to an older set of terms, we are in *parole* as much as we are in *langue*.

One of the consequences of having to make a system able to store information of an *un*foreseen nature and to process language of an *un*predictable kind has been the development of language-processing systems which rely heavily on their data bases. [For examples, see 12; 25; 34; 38; 39; 40; 43].

The more interesting of such systems *create* their own data structures in memory on the basis of verbal conversation with human users as well as *exploit* those memories to understand verbal conversation. No single system will be here described, but it seems worth noting that the data bases of such systems must be *emergent* if the system is to adapt to unforeseen circumstances. Furthermore, the structures in which the information is to be retained must be rich enough to store appropriately the information of whatever complexity happens to occur. Since among other things one may want to use pronouns like "I" and "you" and "she", the system must be able to represent itself as well as others. One may want to refer to the activity of communication, so that the system must be able to represent itself and others communicating. Both abilities will have to be recursive in a system that hopes to keep up with any believable actual conversation. An example of a second-level activity is the representing of representations, as when we try to estimate the understanding of some subject matter by another. In order for me to write these sentences about how communication occurs, I must represent to myself a reader's representations of human conversation, his beliefs about computers, and any number of other particulars I assume are to be found in the reader's data base, if I may balance an anthropomorphic sin commited earlier by a momentary mechanomorphic lapse here. If my estimate was reasonably accurate (and I am considerably aided by the context-limiting setting of this Society and its particular scholarly objectives), communication will occur. If my guess was way off, communication will be seriously impeded. To talk about such things (let alone to hope to reach agreement about them!) requires us to represent the representing in various permutations.

But if the universe of discourse now becomes the interaction of other systems with the second-level attention to communication we just saw, then the system must be able to conjure up the third level of representations which represent those interactions with its own second-level interactions. It is to such third-level interactions that we now turn when we consider informational communities. Let me emphasize the distinction between a *linguistic* community and an *informational* community. Briefly, a *linguistic* community is one whose members largely share the same *language* (or linguistic mechanism). An *informational* community is one whose members largely share the same *view* of their world of discourse (or data base or context). Thus all members of a linguistic community might know how to type English, for example; but only those members who share the same data base of a given question-answering system can communicate effectively with the machine or with each other.

Humans are extremely adept at changing context. Just to touch on one aspect of our agility in manipulating a multiplicity of often very rapidly changing contexts, we need but remember the growing literature in sociolinguistics [e.g. 4;

14; 21]. According to Basil Bernstein, a major sociological difficulty in communication has less to do with the capacity to converse or the ability to shift among immediate contexts than it has to do with the capacity to shift from private, immediate, and local contexts to more widely available, public, and general contexts [4]. This distinction may be analogous to our relative success in creating data base systems for particular contexts, compared with our very rudimentary understanding of how we might choose one *appropriate* data base out of an array of available alternatives. (I might add a third-level paranthesis that I find the sociological implications of Bernstein's account much less attractive than those of the Steeg-Schulman paper presented at this conference. In misleading brevity, I summarize Bernstein's work as seeming to suggest that social class structures along with types of intellectual resources are transmitted from generation to generation by means of the language used at home, thereby virtually guaranteeing a lack of social mobility; for Steeg-Schulman, on the other hand, intelligent behavior is learned whenever a modest enough purchasing power is adequate to permit cybernetic processes to occur. In short, Bernard Shaw may have been right, as many of us suspected he was, when he commented in the "Preface" to *Major Barbara*, "The greatest of evils and the worst of crimes is proverty.")

Any given informational community is a "living" system by any objective standards: it adapts, it evolves, it reproduces copies of itself that are not necessarily exactly identical, and it dies. If an informational community becomes large, inefficiencies result, and the context fractionates. A staff that deals with shared problems can be an extremely productive and effective informational community as long as each member of the staff shares the world of common discourse. In a large group, this commonality cannot easily be maintained, for local interests and concerns will distort the commonality, or the size of the staff will interfere with the process whereby the commonality is achieved. Some part of the community will thus share a data base that is less and less general, until it is more accurate to speak of several informational communities than to characterize the group as a single informational community. The fractionalization inherent in the composition of a group such as the American Society for Cybernetics is a case in point. See, for example, the characterization of this society's informational interests in Klaus Krippendorff's "Preamble" to the "Call for Papers" which brought us here. (On this occasion of our tenth birthday, it is perhaps unkind to remind us that Margaret Mead suggested we might consider societal suicide ten years after founding this society. It may, however, be time to face up to her question, "When are we likely to need either death or transfiguration"? [31: 11].)

We were, however, reminded by von Foerster following the society's banquet that a too-adequate model, such as the Ptolemian cosmography, should arouse our suspicions. To this we might add Lewis's criticism of the extraordinarily long-lived model of physiology credited to Galen, which is not being submitted as a good simulation, it is being submitted as a bad model – not bad becuase it appears to be archaic or naive but bad because it lasted virtually intact for over 1400 years. This is perhaps one of the most damning statements one can make about a model. I propose that a good measure of a model's utility is how rapidly it leads to new hypotheses and new models, in other words, how rapidly it leads to its own rejection or at least to its own alteration [23: 110].

The shared world view of any given informational community will change rapidly if the community is informationally productive. Rapid change occurs in *small* informational communities. Therefore, efficient communication occurs only in *small* informational communities.

We induce new members into small informational communities; thus, we create new contexts when needed. We evoke in ourselves the appropriate contexts when we continue a conversation interrupted some time ago; thus, we re-create (or resurrect) contexts when needed. We learn something as a group when we are informationally productive; thus, we modify together the shared contexts of an informational community. But we do these things efficiently only when the informational community is relatively *small*. To put it another way, communication is effective only when much information is already shared.

Yet there are some orators or writers who apparently sometimes communicate to vast audiences. That ability suggests we should look for the data bases which seem to be shared by large groups. Whether a large group can be induced to *modify* its world of discourse as effectively as can a small group, is problematical. The "colonial lag" [29], that characterizes the conservatism of dialects peripheral to the cultural and commercial centers compared with the relatively faster changing prestige dialects of those centers suggests that in general, we would *not* expect very large groups to be as malleable informationally as small groups. To express the shared attitudes of multitudes is a very different thing from causing a change in their conceptual data structures.

We have seen that all interactive or adaptive devices require *representations.* I have suggested that certain kinds of question-answering systems have been perforce concerned with the nature of those representations related to natural language meanings, which are notoriously context-dependent. I should like now to suggest that the grammars that have been used with such systems may be of particular interest. I call "representational grammars" those grammars that make considerable use of natural language to structure their memories and make considerable use of their internal data bases to understand actual sentences. I believe that work in natural-language representational grammars suggests that we must attend to two types of capabilities, static and dynamic.

Statically, the data structures need to be rich enough to structure appropriately any unpredictable information. For example, information that is partly negated cannot be stored if the data structures are incapable of being so qualified [24].

Dynamically, the data base as a whole must adapt to unpredictable changes not only in informational detail but also in information community world view. Let us return for a moment to our simplest cybernetic device. If a thermostat does not respond in a time scale suitable to that of its thermal environment, its ability to control that environment is seriously impaired. We have already seen how the dutiful dreamer gets into difficulties when things it notices in its world get too much for it. It is similarly the case that a dynamic data base is essential for the continued well-being of interactive question-answering systems that structure their own data bases. If a system is not able to keep its data bases current in terms of the raw facts it deals with and also in terms of the changing world view of its information community's universe of discourse, then the system can not long be useful [39].

Consider next a time-sharing, multiple-access, multiple-user, interactive system such as REL, which must perforce make some attempt to identify the *user* so that appropriate data structures and their interpretive rules, vocabularies, and the like can be provided. This is frequently accomplished by explicit identification by name or by identifying number [12]. Such a technique suggests that one may wish to explore more widely the context that is appropriate for determining which of many data bases one will select. This is clearly at least a second-level representational matter. For parallels in other fields of cybernetics, we should recall that Gordon Pask briefly comments on general mechanical and biological attention-directing mechanicisms [36], and less formally, Arnheim discusses biological, especially human, sensory selection [3]. We might also note Kenneth Burke's comments on "Terministic Screens" as a means of invoking one context rather than another [6: 44–62]. Clearly, we may want to look further afield into *anything* in the external situation that might be indicative.

A very different kind of situational check is purely internal. If an interpretation of a string to be analyzed by a parser makes sense in the conversation where it occurs, we usually assume that the context is the right one. It is only when we no longer can make sense out of an analysis that we begin to have qualms about the appropriateness of our data base choice. As any real data base is bound to be messy, any one rule of grammar may have to decide, for example, whether a string should be interpreted metaphorically, literally, or some other way [26]. In any case, we notice that the traditionally impregnable walls between syntax, semantics, and pragmatics must be broached. I suggest that formal representations in data base structures such as those found in interactive question-answering (and information-storing) systems dealing with real events (even if they may be "toy events" to paraphrase Amarel) offer a framework to nail down some of the slippery notions we need, for such notions are completely context-dependent, but the context is completely specified by a given data base representation. But we should not forget that we might be wrong about even perfect internal consistency. Remember Ptolemy and Galen?

At any rate, it does seem as if there are some formal tools derived from computational linguistics and artificial intelligence, and especially from work with interactive question-answering and information-storing computer systems. These tools may assist us in digging out the various levels of representation inherent in any human informational situation. At the first level, we can determine whether any given formal structure is powerful enough to store appropriately some specified set of meanings. At the second level, we can investigate whether a given representation can successfully cope with representing its own representations and those of others; we can also determine whether the activity of representing can be adequately represented. At the third level, we can investigate the consequences to systems beyond the immediate informational community.

Clearly, an informational community must be adaptable to new situations that cannot in general be predicted. The study of what is *appropriate* or *relevant* to a given informational community seems to be one of the most important questions we all face, whether as biologists like Maturana we are trying to define an organism's ecological and informational niche [30], or as computer scientists like Amarel we are trying to define general procedures for finding appropriate contexts for problem-solving [2].

At the very first American Society for Cybernetics annual symposium in 1967, Saul Amarel made this plea, which I see as directed primarily to the first and second level questions:

> There is a close similarity between basic design problems in "robotics" and in question-answering systems. These are the problems of accepting information, up-dating it, structuring it, and using it efficiently in the generation of actions or answers, as the case may be.
>
> There is no doubt that implementation problems in "robotics" are important. By working on them we will learn how to combine into working wholes large and varied pieces of hardware and software. This is necessary for progress in intelligent systems, but it is by no means sufficient. The basic problems of representation, modeling, and search that are common to the design of any nontrivial robot or question-answering system are central to real progress in our area and, in my opinion, they require much more attention.
>
> In general, I think that in the area of artificial intelligence more effort should be directed to conceptual clarifications, and to critical studies (experimental and theoretical) of some of the basic problems that we face in the design of complex problem-solving information systems [1: 108].

The issues are still very much with us and still central to cybernetic concerns, though perhaps we are creeping gradually from level to level. The very title of this session, "Knowledge Structures and Society", suggests a multiplicity of levels of representations. Thus, cybernetic scholars may appropriately challenge the accuracy of the *raw evidence* of a paper presented here: does the representation in fact represent what it seems to? At the second level, we may appropriately question the *process* of representation: does a graph or a table, for example, adequately represent what it seems to, or, as a sequence of quotations might imply an unwarranted causal sequence of events, does it imply more (or less) than it should? And at the third level, we may appropriately challenge the *larger validity* of a process of representation in the context of wider historical or cultural systems, as when potentially racist implications could be drawn from a presentation.

Discussions may fail to come to grips with the individual issues partly because questions addressed to one level may be understood in reference to another level. Even if we could specify which level of representation we were challenging, we should still be prepared to shift up or down, for the levels are not independent. Inadequate first-level representations clearly can have consequences on the second level, as when inadequate data results in a distorted historiography. Second-order inadequacies can obviously have consequences on the third level, as when a distorted historiography interacts with anti-semitic feelings in the wider social system. Clearly, all three levels of representation are properly subject to criticism in *any* paper presented in a cybernetic context.

Though representations in artificial and natural systems may take varied forms, we have noticed at least three levels of representations in machines and mortals. Higher-level representations are definitely required for any nontrivial description of natural human language, and it may turn out that we need a recursive level-jumping ability. Even if we have such an ability, we may not always wish to exercize it.

It is a truism that you never get something for nothing. To reap the undeniable benefits of going to higher level systems, we must be prepared to pay a price. Fluent conceptual and pragmatic manipulation of a given level may be at the expense of ability to operate as well on some lower level. Such a price is one I think I should be willing to pay if I can, for I think the benefit usually outweighs the cost. If we blow each other up at one level, our attentions on any other level may be completely irrelevant, however. Whether we attended too low or too high is something I hope we never have to find out; but there is at least one situation where for a specific community of the most expert users of natural language, attention directed to the next-higher level of representation is clearly considered harmful. There may also be other circumstances in which we may have to pay more than we can afford in subtle coin. As X. J. Kennedy warns his fellow-poets in

Ars Poetica

The goose that laid the golden egg
Died looking up its crotch
To find out how its sphincter worked.
Would you lay well? Don't watch [20: 53].

ENDNOTE

[1] Acknowledgment for permission to quote is gratefully made to the MIT Press for the passage by MacKay, to Spartan Books for the passages by Amarel and by Lewis, and to Doubleday and Co. for the poem by Kennedy. The author also thankfully acknowledges an Award for Computer-Oriented Research in the Humanities from the American Council of Learned Societies, which supported some of the work reported here, a travel grant from the University of Washington to attend the ASC meetings in Philadelphia, and most helpful typing assistance from Sherry Laing and Shirley Hansen. Obligations to students and colleagues, who have helped me rectify and clarify some ideas, and to friends and teachers, among whom I am fortunate to number Albert H. Marckwardt, Frederick B. Thompson, and Warren S. McCulloch, are deeply appreciated.

REFERENCES

[1] Amarel, S., Problems of representation in artificial intelligence, *in* [**41**].

[2] Amarel, S., "Problem solving and decision making by computer: an overview", *in* [**13**: 279–329], 1970.

[3] Arnheim, R., *Visual Thinking,* Berkeley: University of California Press, 1969.

[4] Bernstein, B., "Social class, language and socialization", *in* [**7**: 102–110] and in his (Ed.) *Class, Codes, and Control,* vol. 1: 170–189, London: Routledge & Kegan Paul, 1971.

[5] Bloomfield, L., *Language,* New York: Henry Holt, 1933.

[6] Burke, K., *Language as Symbolic Action: Essays on Life, Literature, and Method,* Berkeley: University of California Press, 1968.

[7] Cashdan, A., and Grugeon, Elizabeth (Eds.), *Language in Education: A Source Book,* Prepared by the Language and Learning Course Team at The Open University, London: Routledge & Kegan Paul, 1972.

[8] Chafe, W. L., *Meaning and the Structure of Language,* Chicago: University of Chicago Press, 1970.

[9] Chomsky, N., *Aspects of the Theory of Syntax,* Cambridge, Mass.: MIT Press, 1965.

[10] Chomsky, N., *Studies on Semantics in Generative Grammar,* The Hague: Mouton, 1972.

[11] de Saussure, F., *Course in General Linguistics.* Bally, C., Sechehaye, A. and Reidlinger, A. (Eds.), Baskin, W. (Transl.), New York: Philosophical Library, 1959.

[12] Dostert, B. H., "REL – An information system for a dynamic environment", *REL Report,* No. 3, Pasadena: California Institute of Technology, 1971.

[13] Garvin, P. (Ed.), *Cognition: A Multiple View,* New York: Spartan Books, 1970.

[14] Halliday, M. A. K., "Language structure and language function", *in* [**27**: 140–165], 1970.

[15] Harrison, B., *Meaning and Structure: An Essay in the Philosophy of Language,* New York: Harper and Row, 1972.

[16] Hartmann, E. L., *The Functions of Sleep,* New Haven: Yale University Press, 1973.

[17] Jouvet, M., "The states of sleep", *Scientific American,* **216**: 62–72, 1967.

[18] Jouvet, M., "The role of monoamines and acetylcholine-containing neurons in regulating the sleep-waking cycle", *Ergebnisse der Physiologie Biologischen Chemie und experimentellen Pharmakologie,* **62**: 166–307, Berlin: Springer Verlag, 1972.

[19] Katz, J. J., *Semantic Theory,* New York: Harper and Row, 1972.

[20] Kennedy, X. J., *Nude Descending a Staircase,* Garden City, N.Y.: Doubleday, 1962.

[21] Labov, W., *The Study of Nonstandard English,* Urbana, Ill.: National Council of Teachers of English, 1970.
[22] Lamb, S., "Linguistic and cognitive networks", *in* [**13**: 195–222], 1970.
[23] Lewis, E. R., "Some biological modelers of the past", *in* [**37**: 109–126], 1966.
[24] Longyear, C. R., *The Semantic Rule,* Tempo 67TMP–55, Santa Barbara: General Electric Co., 1967.
[25] Longyear, C. R., "Computer simulation of natural language information processes", *Glossa,* **3**: 67–100, 1969.
[26] Longyear, C. R., Nature's Grammar: A Representational Theory of Language, (forthcoming).
[27] Lyons, J., (Ed.), *New Horizons in Linguistics,* Harmondsworth: Penguin, 1970.
[28] MacKay, D., *Information, Mechanism and Meaning,* Cambridge, Mass.: M.I.T. Press, 1969.
[29] Marckwardt, A. H., *American English,* New York: Oxford University Press, 1958.
[30] Maturana, H., "Neurophysiology of cognition", *in* [**13**: 3–23], 1970.
[31] Mead, M., "Cybernetics of cybernetics", *in* [**41**: 1–11], 1968.
[32] McCarthy, J. and Hayes, P. J., "Some philosophical problems from the standpoint of artificial intelligence", *in* [**35**: 463–502], 1969.
[33] Meredith, P., "Developmental models of cognition", *in* [**13**: 49–84], 1970.
[34] Minsky, M. L. (Ed.), *Semantic Information Processing,* Cambridge, Mass.: M.I.T. Press, 1968.
[35] Michie, D. and Meltzer, B. (Eds.), *Machine Intelligence,* 4, Edinburgh: Edinburgh University Press, 1969.
[36] Pask, G., "Cognitive systems", *in* [**13**: 349–405], 1970.
[37] Pattee, H. H., Edelsack, E. A., Fein, L. and Callahan, A. B. (Eds.), *Natural Automata and Useful Simulations: Proceedings of a Symposium on Fundamental Biological Models,* Washington, D.C.: Spartan Books, 1966.
[38] Rustin, R. (Ed.), *Courant Computer Symposium 8: December 20–21, 1971: Natural Language Processing,* New York: Algorithmics Press, 1973.
[39] Thompson, F. B., "Fractionalization of the military context", *Proceedings of the Spring Joint Computer Conference,* **25**: 219–221, 1964.
[40] Thompson, F. B., "English for the computer", *Proceedings of the Fall Joint Computer Conference,* **29**: 349–356, 1966.
[41] von Foerster, H., White, J. D., Peterson, L. J. and Russell, J. K. (Eds.), *Purposive Systems: Proceedings of the First Annual Symposium of the American Society for Cybernetics,* New York: Spartan Books, 1968.
[42] von Foerster, H., "Thoughts and notes on cognition", *in* [**13**: 25–48], 1970.
[43] Winograd, T., Procedures as a Representation for Data in a Computer Program for Understanding Natural Language. MAC TR–83, Project MAC. Cambridge, Mass.: Massachusetts Institute of Technology, 1971.

IMPLICATIONS FOR POLICY:

CONCLUSIONS OF THE CONFERENCE ON COMMUNICATION AND CONTROL IN SOCIAL PROCESSES, OCTOBER–NOVEMBER, 1974

Conference participants, on the assumption that a cybernetic orientation could contribute to the solution of contemporary problems of scientific and public policy, gathered at a concluding seminar out of which came the following recommendations:

I *Many emergent social problems arise from circular flows of information or are the product of unrestrained mutual causation. Their solution should be sought in the underlying control processes involved.*

Common to many current social problems is their sudden emergence. Population explosion, information overload, pollution, racial strife, the energy shortage, for example, are often perceived as "crises", not so much because they are entirely unpredictable but because of the failure of the Social Sciences to recognize their initially slow but usually exponential development before critical thresholds are reached. Research methods in the Social Sciences are largely geared to identify single or at best multiple causes of events but are relatively powerless when it comes to understanding the circular causal fabric of many social phenomena. Control engineering, on the other hand, has been preoccupied with morphostatic mechanisms seeking simple stabilities. Attempts at compensatory controls – as exemplified by responses to higher crime rates with more policemen – is likely to produce other unanticipated social problems not to mention the high costs involved and limited resources available for such efforts.

It is suggested that social problems emerging from the "vicious circles" of mutual causation, from the self reinforcing nature of social prejudices and from circular information flows can be anticipated by adequate techniques of analyzing positive feedback processes and that solutions to those problems might be found in the rearrangement of interdependent variables.

II *A better understanding of social change requires a greater focus on structural formulations of social processes.*

On the one hand, most social systems models assume continuous linear changes. Merely increasing the number of variables in those models – a recent trend that is exemplified by many global economic models or world models of resource uses, etc. – does not change this fundamental limitation. On the other hand, many social changes occur in steps and might be considered qualitative in nature;

revolutions, wars and catastrophies being extreme cases in point, more moderate examples are found in the social-structual adjustments to modern technology.

For example, the structual changes we are witnessing seem to be facilitated largely by the new technologies of communication and of information processing around which new industries, new institutional complexes, new organizational forms and new social practices have grown.

While linear models of social systems might prove appropriate under morphostatic conditions, they have proven incapable of understanding how quantitative changes evolve into structural changes; how morphogenesis takes place in society. To understand, predict and perhaps control ongoing, social processes, nonlinear models of structural change are needed and should be developed.

III *In policy decisions and in social research, attention should be directed to the social consequences of differential accessability and unequal distribution of information.*

While most decisional formulations assume knowledge to be equally distributed and sufficient for rational solutions to be forthcoming, differential access to modern information processing technologies and telecommunications has greatly enlarged the inequality in the distribution of information in society. And, by eroding the traditional socio-cultural controls on how knowledge is to be acquired and applied, this technology has favored the emergence of socio-technological controls that are hardly rationally comprehended or sufficiently understood. This unintended shift is manifest in numerous concerns such as the fear of the invasion of privacy, the fear of the increased institutional ability to confine the flow of public information, the fear of loss of democratic control.

Attention should be given to the unanticipated social changes in community and in social structure caused by unequal access to and distribution of information, to the political, social and scientific value of information, to the costs of selective information processing including the costs of violating privacy, and to the interests that the selective dissemination of information might serve.

IV *In planning policy, systems considerations should be given preference over sectional decision making.*

Most policy planning is oriented to the maximization of separate goals in separate sectors such as health, education, pollution, international relations, and the economy. These sectors are far from independent, however. Decisions in one sector are communicated and may alter the conditions for decision making in another. And, what serves one sector well may be detrimental to another. Sectional decision making, while politically expedient ignores the vital links that hold highly developed societies together and results in a sub-optimal quality of life.

Systems conceptions for policy decision that take account of these possibly complex interdependencies should be developed and cybernetics with its focus on feedback communication processes should be employed to achieve this end.

V *Contextual effects of technology transfer from industrialized to less industrialized countries should be examined cybernetically.*

Most decisions on the technology across national or cultural boundaries assume similar and desirable effects, whereas evidence indicates that despite well intended technical aid programs, differences in standard of living between affluent and poorer countries in most cases is increasing. Additionally, the political stability that such technology requires, tends in fact to be diminishing. To a degree better than chance, technology transfer has aided military dictatorships and the suppression of individual freedoms.

These discouraging facts are born out of a lack of understanding the culturally rooted knowledge base of societies other than our own, the communication networks that transform such knowledge into social action and that channel social action back into knowledge and the variables that are and that are not controllable in less industrialized countries. Technologies are the product of fertile socio-institutional environments with which they interact and through which they grow. Decisions on technology transfer should be preceded by an examination of how these "transplants" are taken by the "organism" the nature of the interactions they facilitate and the reorganizing processes they may set in motion. The development of adequate models of the social and political functions of technologies should be encouraged to make technology transfers more meaningful in the terms set forth by the aided countries themselves.

KRIPPENDORFF: Communication and control in Society